Energy Science and Applied Technology

Energy Science and Applied Technology

Edited by **Nora Ayling**

SYRAWOOD
PUBLISHING HOUSE
New York

Published by Syrawood Publishing House,
750 Third Avenue, 9th Floor,
New York, NY 10017, USA
www.syrawoodpublishinghouse.com

Energy Science and Applied Technology
Edited by Nora Ayling

International Standard Book Number: 978-1-68286-235-3 (Hardback)

Printed in China.

Contents

Preface

The study of energy science is aimed at promoting efficient usage of energy, research and development of cost effective and sustainable energy solutions. This book on energy science focuses on acquainting the readers with practical applications of the subject such as alternative sources of energy, environmental impacts, energy conservation, etc. This innovative and comprehensive book integrates the well-developed theory and practical applications of energy science. It will be well suited for both students and research scholars pursuing energy sciences or associated disciplines.

The information shared in this book is based on empirical researches made by veterans in this field of study. The elaborative information provided in this book will help the readers further their scope of knowledge leading to advancements in this field.

Finally, I would like to thank my fellow researchers who gave constructive feedback and my family members who supported me at every step of my research.

Editor

1

Use of furnace slag and welding slag as replacement for sand in concrete

Sreekrishnaperumal Thanga Ramesh*, Rajan Gandhimathi, Puthiya Veetil Nidheesh, Shanmugam Rajakumar and Subramani Prateepkumar

Abstract

In this project, a study was made to obtain low cost building materials using industrial wastes (welding and furnace slags). The objective of the study is to use these wastes in low-cost construction with adequate compressive strength. Different fine aggregate replacements have been studied by substituting 5%, 10%, and 15% of slag. The waste material was substituted for replacement of fine aggregates and for the preparation of concrete blocks. In this project, we have followed Indian standard methods and arrived at the mix design for M25 grade concrete. Experimental studies were conducted only on plain cement concrete. The preliminary studies were conducted by mixing the slag with the cement concrete cubes of standard sizes. The building material specimens were analyzed for compressive strength as per IS code. For the test and other specifications, it can be concluded that the welding and furnace slags can increase the strength of the concrete. The optimum compressive strength of concretes after 28 days has been found to be 41 N/mm^2 for 5% welding slag and 39.7 N/mm^2 for 10% furnace slag replacements. The results show that 5% of welding and 10% furnace slags replacement with sand is very effective for practical purpose.

Keywords: Slag, Sand replacement, Waste management, Low-cost building materials

Background

Since the large demand has been placed on building material industry especially in the last decade, owing to the increasing population which causes a chronic shortage of building materials, the civil engineers have been challenged to convert the industrial wastes to useful building and construction materials [1]. The challenge for the government is to reduce the waste's harmful impacts to both health and the environment [2]. Concrete industry is particularly important as it is not only responsible for consuming natural resources and energy but also for its capacity of absorbing other industrial wastes and by-products [3]. It is known that about 8 to 10 tons of fresh concrete waste can be produced every day from a concrete-batching plant with a daily output of 1,000 m^3 of concrete [4]. The choice of aggregates is important, and their quality plays a great role; they can not only limit the strength of concrete, but owing to their characteristics, they affect the durability and performance of concrete [5]. The worldwide consumption of sand as fine aggregate (FA) in concrete production is very high, and several developing countries have encountered some strain in the supply of natural sand in order to meet the increasing needs of infrastructural development in recent years [6]. So, there is large demand for alternative materials for fine aggregates in construction industry. To overcome the stress and demand for river sand, researchers have identified some alternatives for sand, namely scale and steel chips [7], waste iron [8], crushed granite fine [6], etc.

Environmental management in developing countries is a complex issue because environmental problems are linked with social and economic aspects, which must be considered in the development of any environmental program or regulation [9]. The problem of waste accumulation exists worldwide, specifically in the densely populated areas [10]. Scale, granulated slag, and steel chips are industrial wastes in the iron and steel industry and cause a nuisance both to the health and environment when not properly disposed [7]. Reuse of industrial solid waste as a partial replacement of aggregate in construction activities not only saves landfill space but also reduces the demand for extraction of natural raw materials [8]. Recent studies showed

* Correspondence: stramesh@nitt.edu
Department of Civil Engineering, National Institute of Technology, Tiruchirappalli, Tamilnadu Pin: 620015, India

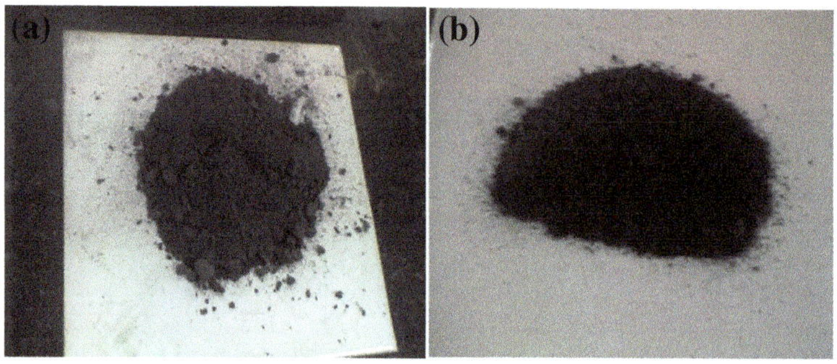

Figure 1 Samples of (a) WS (b) FS.

that slag can also replace with sand in concrete [11-13]. So, the present study focuses to replace sand by welding slag (WS) and furnace slag (FS) in the production of concrete. Properties of aggregates and slag were studied in detail. M25 concrete was designed for various percentages of slag, and compressive strength of concrete was found out.

Methods

Cement

Cement was obtained from single source and single batch from the reputed firm. 43 grade ordinary Portland cement with specific gravity of 3.1 is used in the casting of specimens.

Coarse aggregate (CA)

Crushed granite coarse aggregates of 20-mm maximum size were used.

Water

Potable water available in the NIT Campus conforming to the requirements of water for concreting and curing was used throughout the project.

Fine aggregate

Locally available river sand was used for the present study.

Welding slag and furnace slag

The WS and FS were obtained from local fabrication industries, and they are used to replace fine aggregate partially in the production of concrete. Figure 1 shows the nature and the physical structure of the WS and FS used for the investigation. The physical and chemical characteristics of the welding slag were determined in the laboratory as per standard methods.

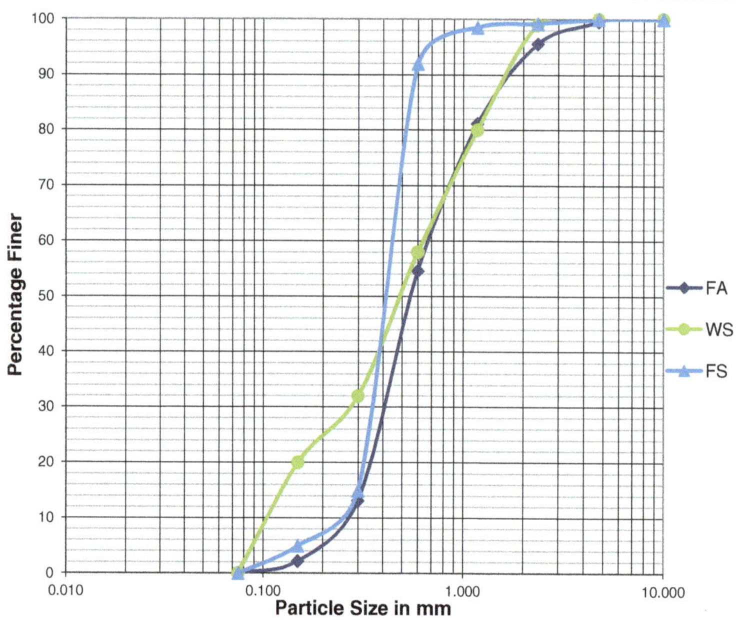

Figure 2 Sieve analysis of ingredients.

Table 1 Physical characteristics of ingredients

Sample number	Properties	FA	CA	WS	FS
1	Specific gravity	2.64	2.72	2.95	1.86
2	Fineness modulus	2.53	7.23	2.12	1.97

FA, fine aggregate; CA, coarse aggregate; WS, welding slag; FS, furnace slag.

Table 2 Chemical characteristics of WS

Metals	Concentration (mg/L)	
	Acid soluble	Water soluble
Fe	26.03	2.11
Mg	6.16	2.03
Zn	0.62	0.05
Al	63.9	5.49
Cu	0.13	0.02
Ca	50.82	22.28
Pb	0.12	0.04
Cr	0.28	0.04
K	2	0.68

Table 3 Chemical characteristics of furnace slag

Metals	Concentration (mg/L)	
	Acid soluble	Water soluble
Fe	4.79	0.04
Mg	3.06	0.54
Zn	0.12	0.01
Al	2.02	0.07
Cu	0.03	0.01
Ca	0.16	0.17
Pb	0.16	0.03
Cr	0.11	0.05
K	0.38	0.56

The X-ray diffraction (XRD) spectra were obtained by using a Rigaku Powder X-ray diffractometer (D-Max/Ultima III, Rigaku Corporation, Tokyo, Japan). XRD and scanning electron microscopic (SEM) analysis were carried out to identify the mineral composition and morphological features of welding slag. The prepared samples were exposed to X-ray with the 2θ angle varying between 10° and 80° with Cu Kα radiation. The applied voltage and current were 40 kV and 30 mA, respectively. SEM investigations of the welding slag samples were conducted in a Hitachi S-3000 H operated at 10 kV (Hitachi, Ltd., Tokyo, Japan). The SEM analysis was done at ×100 magnification to examine the morphological characteristics.

Preparation of test specimen

The experimental investigation has been carried out on the test specimens to study the strength properties as a result of replacing fine aggregate by WS and FS in various percentages namely 5%, 10%, and 15%. The test specimens were cast in steel molds. The inside of the mold was applied with oil to facilitate the easy removal of specimens. The raw materials were weighed accurately. The concrete was mixed thoroughly in dry condition. Initially, 75% of water was added to the dry mix, and they were mixed thoroughly. Then the remaining 25% of water was added to the mix. The mixing was continued until a uniform color was obtained. Fresh concrete was placed in the mold in three layers, and each layer was compacted using tamping rod. Then, the mold were placed in vibrator and compacted. The concrete specimen cast is a 150 × 150 × 150-mm cube. After 24 h from casting, the test specimens were taken out and placed in a curing tank, until the age of the specimens. The prepared specimens were analyzed for workability and compressive strength as per IS standards.

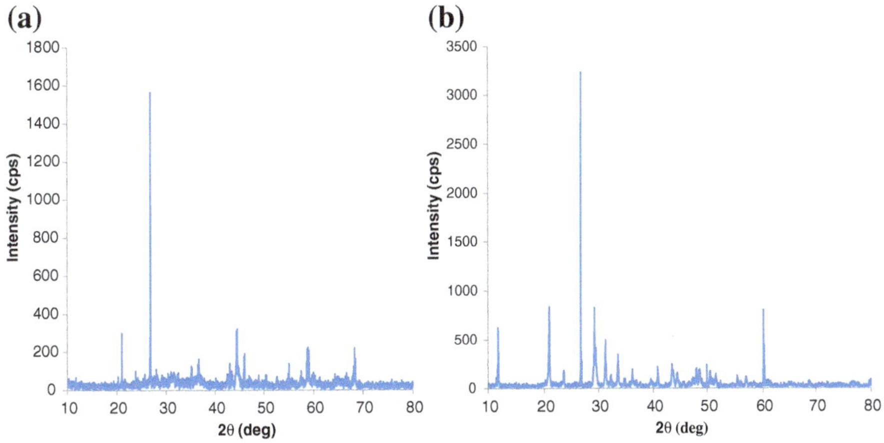

Figure 3 XRD analysis of (a) WS (b) FS.

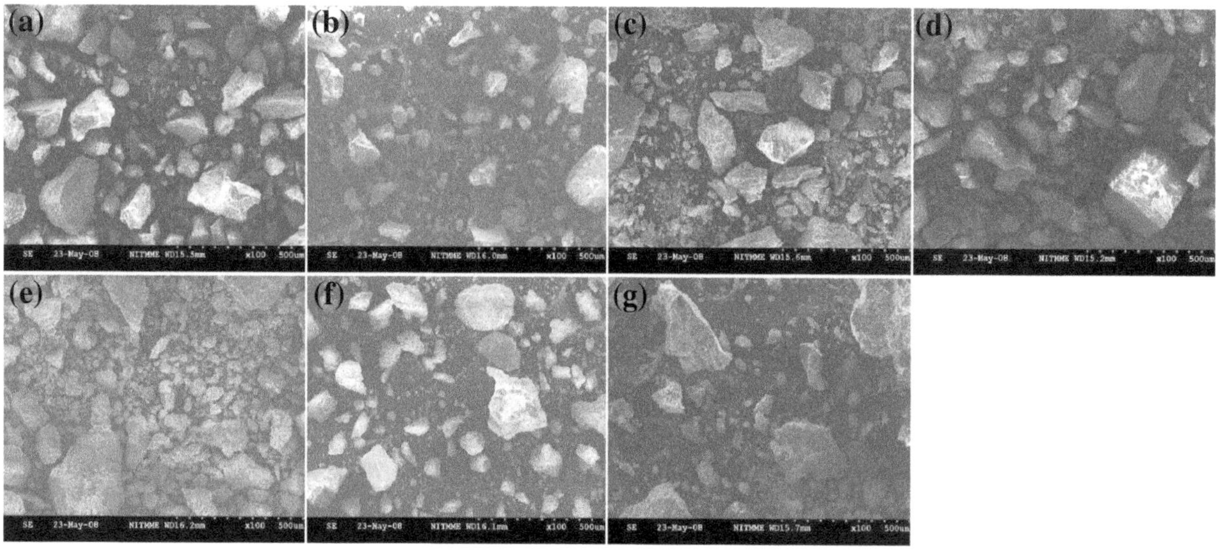

Figure 4 SEM analysis. (**a**) Controlled concrete, (**b**) 5% WS, (**c**) 10% WS, (**d**) 15% WS, (**e**) 5% FS, (**f**) 10% FS, and (**g**) 15% FS.

Results and discussion

Material properties

The size distributions of the materials used in the present study are shown in Figure 2. The size distribution of WS is following the same trend with that of FA. This shows that WS can be used as a replacement for FA in concrete, the FS shows a little bit deviation from sand.

The physical characteristics of ingredients used in the present study are given in Table 1. The specific gravity of WS is greater than that of CA and FA. But the fineness modulus of WS is less compared to CA and FA. The specific gravity and fineness modulus of FS is very less in all the cases. The chemical characteristics of WS and FS are given in Tables 2 and 3, respectively. The WS contains higher amount of Fe, Al, and Ca. These metals are highly soluble in acid than that of water, but the metal contents of FS are very less compared to WS. Since the water solubility of metal in WS and FS, is very less in nature, attack on steel due to this metal is negligible (except Ca and Al in the case of WS). The XRD of WS and FS are shown in Figure 3. The SEM images of controlled concrete and WS and FS added concrete are shown in Figure 4.

The mix was designed for M25 grade of concrete using IS method of mix design. The material properties used in the mix design are given in Table 4, and the mix proportion for making cubes for control concrete and various percentage of replacement of sand with WS and FS are given in Tables 5, 6, 7, and 8.

The variation of compressive strength for various percentage replacement of sand with welding slag for seventh and 28th day is shown in Figure 5. The seventh day compressive strength of concrete increases with the increase in slag concentration up to 10% of sand replacement. Seventh day compressive strength of concrete increases from 17 to 27.8 N/mm^2 for the WS concentration increment of 5% to 10%. Fifteen percent WS concentration also shows the seventh day compressive strength than that of controlled concrete. The seventh day compressive strength

Table 4 Material properties for concrete mix

Properties	Values
Specific gravity of cement	3.1
Specific gravity of water	1
Specific gravity of sand (FA)	2.64
Specific gravity of coarse aggregate (CA)	2.72
Dry rodded bulk density of coarse aggregate (kg/m^3)	1634.8
Fineness modulus of sand (FA)	2.53
Fineness modulus of coarse aggregate (CA)	7.23
W/C ratio for 43 grade cement	0.44
Fineness modulus of welding slag	2.12
Specific gravity of welding slag	2.95
Fineness modulus of furnace slag	1.97
Specific gravity of furnace slag	1.86

Table 5 Mix proportion for controlled concrete

Grade of concrete	Slump (mm)	Quantities per m^3 of concrete (kg)			
		Water (L)	Cement	Sand	Aggregate
M25	0 to 25	188	427.27	700.40	1,057.72
	25 to 50	194	440.91	672.95	1,057.72
	50 to 75	200	454.55	645.50	1,057.72
	75 to 100	206	468.18	618.04	1,057.72

Table 6 Mix proportion for 5% replacement of sand

Grade of concrete	Slump (mm)	Quantities per m³ of concrete (kg)					
		Water (L)	Cement	FS	WS	Sand	Aggregate
M25	0 to 25	188	427.27	24.67	39.13	665.38	1,057.72
	25 to 50	194	440.91	23.71	37.60	639.30	1,057.72
	50 to 75	200	454.55	22.74	36.06	613.22	1,057.72
	75 to 100	206	468.18	21.77	34.53	587.14	1,057.72

FS, furnace slag; WS, welding slag.

Table 7 Mix proportion for 10% replacement of sand

Grade of concrete	Slump (mm)	Quantities per m³ of concrete (kg)					
		Water (L)	Cement	FS	WS	Sand	Aggregate
M25	0 to 25	188	427.27	49.35	78.26	630.36	1,057.72
	25 to 50	194	440.91	47.41	75.20	605.65	1,057.72
	50 to 75	200	454.55	45.48	72.13	580.95	1,057.72
	75 to 100	206	468.18	43.54	69.06	556.24	1,057.72

FS, furnace slag; WS, welding slag.

Table 8 Mix proportion for 15% replacement of sand

Grade of concrete	Slump (mm)	Quantities per m³ of concrete (kg)					
		Water (L)	Cement	FS	WS	Sand	Aggregate
M25	0 to 25	188	427.27	74.02	117.40	595.34	1,057.72
	25 to 50	194	440.91	71.12	112.80	572.01	1,057.72
	50 to 75	200	454.55	68.22	108.19	548.67	1,057.72
	75 to 100	206	468.18	65.32	103.59	525.34	1,057.72

FS, furnace slag; WS, welding slag.

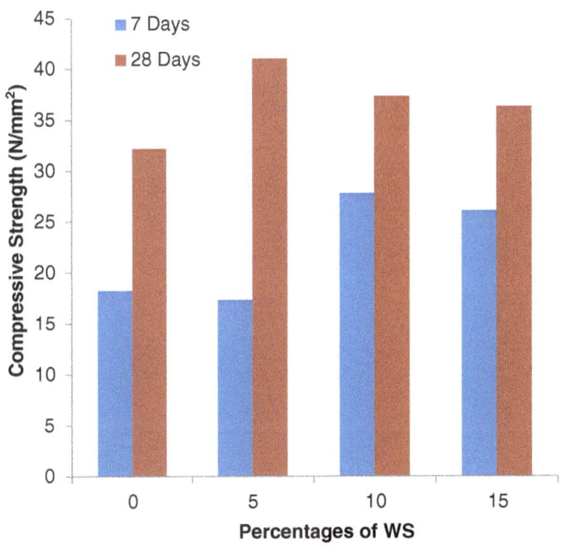

Figure 5 Relationship between WS percentage and compressive strength.

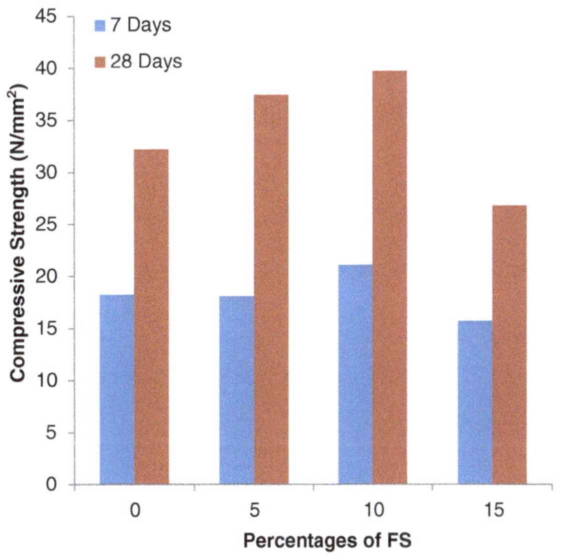

Figure 6 Relationship between FS percentage and compressive strength.

of controlled concrete is found to be 18.27 N/mm^2, but the 28th day compressive strength of concrete shows an optimum value at 5% WS concentration. The 28th day compressive strength of 5% sand replacement concrete has been found to be 41 N/mm^2. Then, the compressive strength of concrete is reduced to 36.3 N/mm^2 as expected earlier, while the 28th day compressive strength of controlled concrete is 32.2 N/mm^2. Also, 10% and 15% sand replacement shows more compressive strength than that of controlled concrete. This indicates that up to 15% sand replacement with WS will increase the compressive strength of concrete, but for practical purpose, 5% of WS is very effective.

Similarly, the variation of compressive strength for various percentages of FS for the seventh day and 28th day is shown in Figure 6. The seventh day compressive strength did not show any difference by the addition of 5% FS. But the compressive strength increases to 21.1 from 18.11 N/mm^2 for 10% FS addition, and after that it reduced to 15.7 N/mm^2 for 15% FS addition. The 28th day compressive strength of concrete increases with FS addition up to 10% and reduces. The 28th day compressive strength of 10% sand replacement concrete has been found to be 39.7 N/mm^2. This optimum value is less than that in the case of WS. Then, the compressive strength of concrete is reduced to 26.8 N/mm^2. Five percent sand replacement shows more compressive strength than that of controlled concrete. This indicates that up to 10% sand replacement with FS will increase the compressive strength of concrete. But for practical purpose, 10% of FS is very effective.

Conclusion

The furnace and welding slags have been utilized in the work by using it in the building materials as addition to concrete. WS and FS concretes showed better performance towards compressive strength. The compressive strength on seventh day of concrete cubes increases from 10% to 15% replacement of sand by WS than the reference materials. Similarly 10% of FS shows an optimum strength of 21.1 N/mm^2. The compressive strength on 28th day of concrete cubes increases from 5% to 15% of replacement of sand by WS than the reference materials. The optimum compressive strength of slag concretes has been found to be 41 N/mm^2 for 5% WS and 39.7 N/mm^2 for 10% FS. The results show that 5% of WS and 10% FS replacement with sand is very effective for practical purpose.

Competing interests
The authors declare that they have no competing interests.

Authors' contributions
STR gave guidance to SP, who conducted the experiments on furnace slag as a replacement material for sand. RG gave guidance to SR, who conducted the experiments on welding slag as a replacement material for sand. PVN analyzed the results and compiled them into a journal format. All authors read and approved the final manuscript.

Authors' information
STR is an associate professor in the Department of Civil Engineering at the National Institute of Technology, Tiruchirappalli. His research interests are solid waste management, water and wastewater treatment, and adsorption process. RG is an assistant professor in the Department of Civil Engineering at the National Institute of Technology, Tiruchirappalli. His research interests are water and wastewater treatment, solid waste management, and contaminant transport modeling. PVN is a Ph.D. scholar in the Department of Civil Engineering at the National Institute of Technology, Tiruchirappalli. SR and SP are M. Tech. graduates.

Acknowledgments
The authors are grateful to the Department of Civil Engineering, National Institute of Technology Tiruchirappalli, Tamilnadu for the project approval and support. The authors would like to express their thanks to the director of the National Institute of Technology Tiruchirappalli, Tamilnadu for approving the project.

References

1. Turgut, P, Algin, HM: Limestone dust and wood sawdust as brick material. Build Environ. **42**, 3399–3403 (2007)
2. Eddine, BT, Salah, MM: Solid waste as renewable source of energy: current and future possibility in Algeria. Int. J. Energy Environ. Eng. **3**, 17 (2012)
3. Almeida, N, Branco, F, Santos, JR: Recycling of stone slurry in industrial activities: application to concrete mixtures. Build Environ. **42**, 810–819 (2007)
4. Shi-cong, K, Bao-jian, Z, Chi-sun, P: Feasibility study of using recycled fresh concrete waste as coarse aggregates in concrete. Construct Build Mater. **28**, 549–556 (2012)
5. Neville, MA: Propriétés Des Bétons. Edition Eyrolles, Paris (2000)
6. Joel, M: Use of crushed granite fine as replacement to river sand in concrete production. Leonardo Electronic Journal of Practices and Technologies **17**, 85–96 (2010)
7. Alwaeli, M, Nadziakiewicz, J: Recycling of scale and steel chips waste as a partial replacement of sand in concrete. Construct Build Mater. **28**, 157–163 (2012)
8. Ismail, ZZ, AL-Hashmi, EA: Reuse of waste iron as a partial replacement of sand in concrete. Waste Manag. **28**, 2048–2053 (2008)
9. Silva, MAR, Mater, L, Souza-Sierra, MM, Corr^ea, AXR, Sperb, R, Radetski, CM: Small hazardous waste generators in developing countries: use of stabilization/solidification process as an economic tool for metal wastewater treatment and appropriate sludge disposal. J Hazard Mater. **147**, 986–990 (2007)
10. Batayneh, M, Marie, I, Asi, I: Use of selected waste materials in concrete mixes. Waste Manag. **27**, 1870–1876 (2007)
11. Zeghichi, L: The effect of replacement of naturals aggregates by slag products on the strength of concrete. Asian J. Civ. Eng. **7**, 2735 (2006)
12. Meenakshi Sudarvizhi, S, Ilangovan, R: Performance of Copper slag and ferrous slag as partial replacement of sand in Concrete. Int J Civ Struct Eng. **1**, 918927 (2011)
13. Alnuaimi, AS: Effects of copper slag as a replacement for fine aggregate on the behavior and ultimate strength of reinforced concrete slender columns. TJER **9**, 90102 (2012)

2

Life cycle assessment of producer gas derived from coconut shell and its comparison with coal gas: an Indian perspective

Chenicheri Chandroth Sreejith*, Chandrasekharan Muraleedharan and Palatel Arun

Abstract

Oil dependency and global warming have stimulated R&D activities on the utilization of secondary biofuels. This article investigates the suitability of coconut shell-derived producer gas (a secondary gaseous biofuel) as a substitute for coal gas from an environmental perspective using life cycle assessment (LCA) approach. Thermochemical gasification in an air-fluidized bed with steam injection is the gaseous fuel production process. LCA is carried out with respect to Indian conditions based on primary and modified Ecoinvent 2.0 data using IMPACT 2002+ environmental impact assessment methodology. The study indicates that coconut shell-derived producer gas life cycle is capable of saving 18.3% of emissions causing global warming potential, 64.1% of emissions causing ozone depletion potential, and 71.5% of nonrenewable energy consumption. The analysis of energy and exergy consumptions of the two life cycles reveal that the renewable fraction in the total energy demand is 62.9% for producer gas life cycle, while it is only 2.8% for coal gas life cycle. Allocation of the by-product, coconut palm residues, is the major responsible factor for this reduction in environmental burden. Based on the existing fertilization practice and the utilization of electricity from the Indian electric grid, substantial green house gas (GHG) emission savings cannot be achieved for producer gas life cycle over coal gas life cycle. However, the green electricity-aided catalytic gasification of coconut shell produced by organic farming can result in 43.4% reduction in GHG emissions, meeting European Union renewable energy directive.

Keywords: Life cycle assessment, Coconut shell, Gasification, Coal gas, Environmental impact

Background

Life cycle assessment (LCA) is a process of evaluating environmental burdens or benefits associated with the total life cycle of a product. This is conducted by identifying and quantifying the energy and materials used and waste products released into the environment [1]. It is performed in accordance with ISO 14040:2006 and ISO 14044:2006 standards. In simple terms, LCA is a holistic tool for evaluating the environmental impacts associated with a product, process, or activity. The ecological or environmental impacts include effects on the ecosystem, human health, and natural resources [2]. Generally, LCA is termed as a 'cradle to grave' approach since it accounts the environmental impacts throughout the product's life, right from raw material acquisition, processing, manufacturing, use, and finally its disposal [3]. The significance of LCA lies in the fact that it equips the policy makers and decision makers for adoption of suitable and sustainable energy supply systems. Increasing global concern due to air pollution and to limited oil reserves has generated much interest in environmental friendly alternatives to petroleum-based fuels [4]. This is more prominent in the context of the current substantial contribution of fossil fuels (over 85%) [5] to the global energy supply. Generally, biofuels produced from renewable energy sources (biomass) are considered to be an effective alternative to fossil fuels, and the concerned production technology is sustainable. To identify savings in energy and emissions from biofuel production and utilization, a thorough evaluation of the corresponding life cycle is to be carried out carefully [6]. LCA is an effective tool for this, which accounts for the relative environmental impacts of biofuel life cycle with respect to 'base case' such as fossil fuel-based life cycle. For example, for biodiesel, the

* Correspondence: ccsreejith79@gmail.com
Department of Mechanical Engineering, National Institute of Technology Calicut, NIT Campus (P.O.), Kozhikode 673601, India

base case may be petroleum diesel, and for producer gas derived from biomass, the base case may be coal gas. The adoption of LCA in the perspective of sustainable economy is more essential nowadays because of the boom in the installation of energy and heat production systems utilizing these sustainable energy techniques with the intention of meeting European Union (EU) targets of 20% renewable energy share by 2020.

The updated standard ISO 14044:2006 [7] specifies requirements and provides guidelines for life cycle assessment including definition of goal and scope of the LCA, life cycle inventory analysis (LCI) phase, life cycle impact assessment phase, life cycle interpretation phase, reporting critical review of the LCA, limitations of the LCA, relationship between the LCA phases, and conditions for use of value choices and optional elements.

LCA approach was adopted for assessing the environmental suitability of the utilization of various renewable energy resources such as solar PV, wind, and biomass by Manish et al. [8]. The base cases considered were coal-based thermal power generation and steam methane reforming for hydrogen production. The comparison of different options was made based on major indicators like life cycle cost, green house gas (GHG) emissions, and net energy ratio (NER). Secondary biofuels produced from unprocessed primary biomass-based materials are sustainable alternative to fossil fuels. Secondary biofuels may be of solid, liquid, or gaseous types which are produced by adopting various production routes. A state-of-the-art review of literature pertaining to liquid biofuels was reported by Nigam and Singh [9]. Producer gas is a secondary gaseous biofuel generated by the thermochemical gasification of biomass. NER and total energy ratio of the life cycle of rice straw-generated producer gas were evaluated by Shie et al. [10]. A comparison of biomass and coal in terms of several environmental indicators for integrated gasification combined cycles with [11] and without [12,13] CO_2 chemical absorption revealed the superiority of biomass. On further analysis of the potential of biomass to produce hydrogen, Koroneos et al. [2] reported the advantage of biomass gasification-syngas reformation-absorption route over biomass gasification-electricity generation-electrolysis route. Manish et al. [8] on reviewing the LCA studies on gasification with biomass and with biomass-coal blends for electricity generation reported an evident reduction in environmental impacts when biomass is used as energy source. However, the life cycle of biomass-based systems can have higher eutrophication potential [14] and acidifying emission potential. Consequences of using agricultural land for other purposes than food production [15] and the size and scale of the biomass-based combined heat and power (CHP) plant [16] are significant factors while utilizing biomass-based systems.

The choice of selected indicators for LCA can provide options for the improvement of the existing systems. Such a damage assessment based on the existing life cycle of natural gas combustion district heating system at a rural location in British Columbia was conducted by Pa et al. [17,18] recently. Similar region-specific LCA studies for woodchip-based 'green electricity generation' in Austria [1], post-consumer wood and forest residue-based gasification in a metropolitan area in Barcelona [19], and carbonization of several woody biomass (found plenty in Singapore) to charcoal [20] are worth mentioning. The comparatively lesser impacts on global warming, acidification, and eutrophication by the life cycle of biomass thermal technologies over that of biological routes were reported by Zaman [21] and Zhong et al. [22]. Iribarren et al. [23], on considering biofuel production systems from poplar biomass, stressed on the need for plantation and harvesting of the lignocellulosic biomass in terms of land occupation and fertilizer requirements. The utilization of hydrogen as fuels in vehicles and the progress in fuel cells have necessitated the need for LCA of the concerned hydrogen supply chains. Such analyses were reported for urban transportation with hydrogen vehicles [24] and for hydrogen refueling stations [25]. A comprehensive state-of-the-art review of environmental impacts of hybrid and electric vehicles was reported by Hawkins et al. [26].

Present study and its significance

India, one of the biggest agricultural countries in world, is in the process of setting up many large-and medium-scale biomass-based power plants for rural electrification, utilizing locally available feedstocks. Thermochemical gasification is a mature sustainable technology for gaseous fuel production from renewables [27,28]. The country has declared its commitments to the implementation of clean development mechanism, which is an arrangement under the Kyoto protocol. The abundantly available biomass materials in the subcontinent include eucalyptus wood, rice husk, paddy husk, coconut shell, cashew nut shell, and bamboo. The climatic conditions are not uniform in the country, making it suitable for these agricultural and forest biomass to grow in different parts of the country. For example, in the southern part of India, the most abundantly available biomass feedstock is coconut shell. Thus, utilization of coconut shells for bioenergy production for electrifying rural areas in southern India has significance. The environmental burdens associated with the utilization of these biomasses (and more) are largely different since irrigation schedule (depends on rain fall), fertilization schedule (depends on soil conditions), land requirement, yield of biomass per hectare, and heating value are different for these biomasses. Thus, it is vital to make assessments about various biomass feedstocks suitable for such power plants from an energy and environmental perspective. As a first step in the venture, coconut shell is taken for assessment in the present study, which will be followed by several other probable feedstocks mentioned already. Hence, the present work is a region-specific study for LCA of coconut shell-derived producer gas life cycle in the third largest coconut-producing country in the world. To the best of the authors' knowledge, no work has been reported on a comparative LCA of coconut shell-derived producer gas and coal-derived coal gas with respect to Indian conditions.

As the world is heading towards hydrogen energy economy, many attempts were reported for enhancing the hydrogen concentration in biomass-derived producer gas for the onward utilization in fuel cells and in hydrogen-fuelled vehicles. Steam reforming the producer gas generated by air-fluidized bed (AFB) gasification is one of the options for achieving this. Syngas (producer gas) can be made further rich in hydrogen by supplying more amount of steam to the AFB gasifier. However, the most common production method of steam utilizes fossil fuels which generate greenhouse gases and ozone-depleting gases, necessitating the optimization of steam utilization. Thus, the present study is significant as a 'well-to-tank' LCA approach to make comparison between two solid fuel-hydrogen-rich gaseous fuel life cycles, that is, the life cycles of fossil fuel-derived coal gas and its potential renewable alternative (coconut shell-derived syngas) are compared with respect to India, the world's largest low-ash-content coconut shell producer.

Location of the present study

Kerala, one of the southern states of India, is blessed with abundant rain fall in the range of 2,000 to 3,000 mm/year [29]. The fertile soil conditions in the region are well suited for coconut cultivation. Researches on a variety of coconut palm species have succeeded in developing new hybrids suiting varying conditions [30]. The specialty of coconut palm is that every part of it has valuable applications, as given in Table 1. The major use of coconut shell (main product) is for handicrafts and charcoal, the latter being one of the base materials for the manufacture of activated carbon. Coconut shell is also used as a primary biofuel for open combustion. The co-product (similar revenue to the main product) coconut oil can be used for edible and inedible (lubricant with nanoparticles addition) purposes. The by-product (lower revenue than that of the main product) palm residues find utilization as a fuel with energy content of 16 to 18 MJ/kg. The state is in the process of developing medium-scale gasification plants for rural electrification utilizing coconut shells as feedstocks for the past 2 to 3 years. The present study pertains to the conversion of coconut shell (primary biofuel) to producer gas (secondary biofuel), which has not yet been analyzed in this region on a life cycle basis.

Methods

Life cycle inventory analysis

Goal, scope, and functional unit

The major objective of the present analysis is to compare the environmental impacts of coconut shell-derived producer gas life cycle with that of the coal gas (generated by the gasification of coal) life cycle. Thus, the present LCA aims for a comparison of the relative environmental impacts of renewable and nonrenewable alternatives of energy production and quantifies the benefits from replacement of the latter with the former. In this study, contributions of various LCA phases to the overall environmental impacts are analyzed to identify the most significant phase in the life cycle. The reference functional unit for the study is 1 MJ energy content in the gaseous fuel (producer gas generated from coconut shell gasification or coal gas generated from coal gasification). This functional unit is selected since the energy content may be utilized for various purposes such as fueling transport system, production of electricity, and as a source of heat for systems like absorption refrigeration.

Producer gas life cycle

System boundaries. The coconut shell-derived producer gas life cycle is analyzed in the perspective of the agricultural scenario existing in the state of Kerala in India. The geographic boundary for the fossil fuel reference system is India except where resources are extracted and transported to India from other countries. A well-to-tank approach is adopted for the LCA because of the potential of the energy content of the generated gas for various

Table 1 Products, co-products, and by-products during coconut shell production

Category	Item	Uses
Product	Coconut shell	Charcoal, handicrafts, fuel for open combustion
Co-product	Coconut oil	Edible oil, lubricant
By-product (coconut palm residues)	Petioles and leaf	Fuel, hut thatching, organic manure
	Coconut seed cake	Cattle feed, organic manure
	Coconut husk	Fuel, organic manure, raw material for coir
	Stem	Piles, pillars, fuel

purposes, as mentioned in the section 'Goal, scope, and functional unit'.

The life cycle of the producer gas is presented schematically in Figure 1, including crushed coconut shell production and gasification. The various important subsystems include sand bed preparation, coconut cultivation, fertilizing, irrigation, coconut shell production, shell crushing, generation of gasification agents, and coconut shell gasification in a FB gasifier.

Coconut cultivation. Coconut palms are generally cultivated in rain-fed conditions and in irrigated conditions. Irrigation is mainly required in the summer season at predetermined intervals. Average rain fall in Kerala based on the data [29] available for the past 50 years is shown in Table 2. Based on the data, the duration of the summer season is taken as 3.5 months, and irrigation is required in this season. The common schedule [31] of irrigation is represented in Table 3.

Inventory data for coconut cultivation. The following data [31] for coconut cultivation are applied for the development of the sub-modules: sand bed preparation, coconut seedling, coconut plantation, irrigation, and fertilizing. Crude coconut oil and palm residues are the co-product and by-product, respectively, in the coconut shell production process.

- Coconut seedlings are grown in a nursery as per established specifications of planting.
- Seedling survival rate is assumed as 65%, and the percentage of healthy seedlings which can be replanted is taken as 75% as per data from local farmers.
- Selected seedlings are planted in pits of 1-m^3 size, and 5 kg of organic manure is used.
- The tree density is 175 trees/ha.

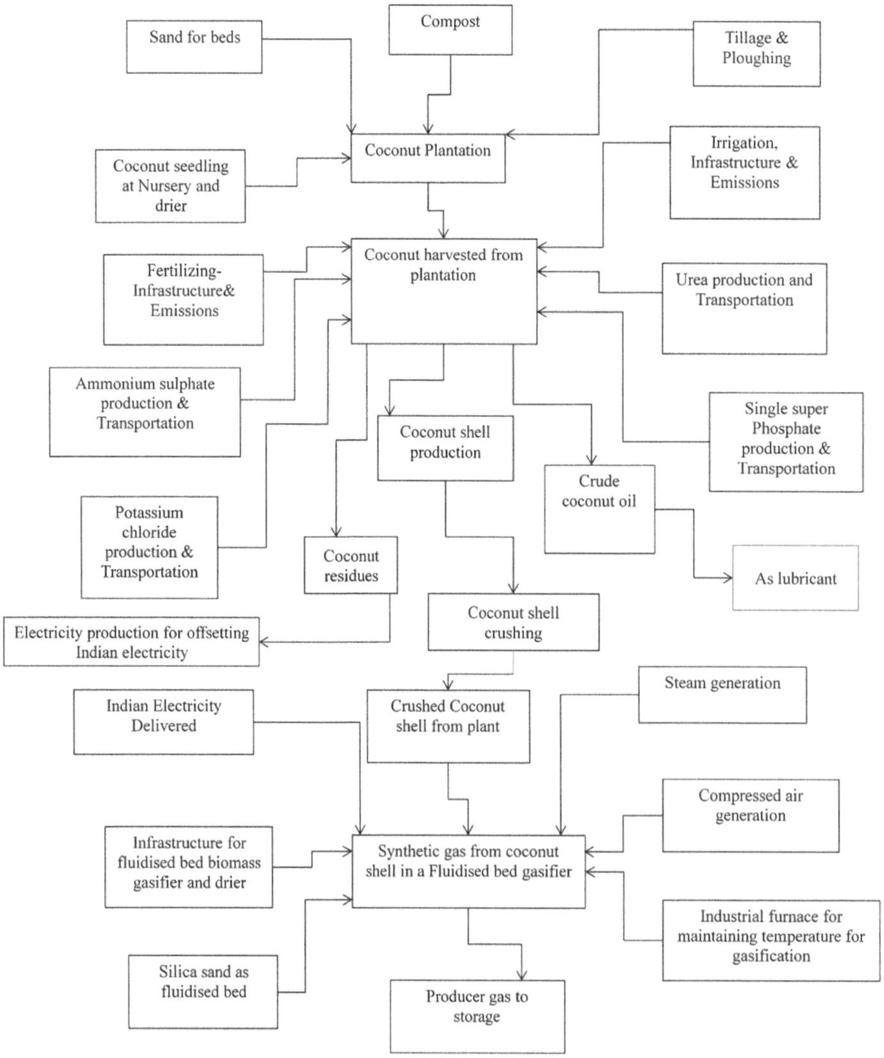

Figure 1 Schematic of producer gas life cycle.

Table 2 Average monthly rain fall in Kerala

Month	Rainfall (mm)
January	14.6
February	16.6
March	36.1
April	110.9
May	252.6
June	653.2
July	687.2
August	404.7
September	252.3
October	270.7
November	158.6
December	45.9

From [27].

- Irrigation and fertilization schedules are according to the principle of good management practices [31], as given in Tables 3 and 4, respectively.
- The coconut yield from a palm per year is 125 nuts, with an average weight of 1.6 kg for full nut and 0.5 kg for husked nut.
- The average quantity of dried copra from nuts is in the rate of 18 kg from 100 nuts (or from 50 kg of husked nut).
- The oil content of copra is about 70%, and the extraction efficiency is about 85%. Hence, expected oil recovery is about 60% of the weight of copra.
- Based on the above conditions, 4.7 kg of husked nuts (9.4 numbers equals approximately 10) is required to produce 1 kg of coconut oil.
- Also, 0.64 kg of shell is obtained from 1 kg of husked nut. It means that from a palm per year, 40 kg of nut shell can be obtained. Thus, 3.1 numbers of husked nuts are required to produce 1 kg of shell.
- Palm residues include the leaves and the dried leaf weighing 2 kg per tree on average and has a heating value of 16,500 kJ/kg.

In the coconut cultivation phase, chemical processes for the production of fertilizers, the utilization of these substances, physical infrastructure for coconut cultivation, and emissions from fertilizers and irrigation are included in the system boundary. Thus, almost all the sub-modules under the modules such as seedling, irrigation, composting, and fertilization in coconut cultivation are based on primary data. For fertilizer production, Ecoinvent 2.0 data were modified with respect to Indian electricity generation and transmission.

Coconut shell gasification. The technology for gaseous fuel production from solid feedstocks (coal or coconut shell) considered in this study is thermochemical gasification. Producer gas or coal gas with more hydrogen concentration is generated by steam injection into an AFB gasifier as mentioned in the section 'Present study and its significance'.

A gasification module is developed for the thermochemical treatment of crushed coconut shell. The producer gas production phase consists of sub-modules such as steam and compressed-air generation, sand preparation for its utilization as bed material, physical infrastructure for the gasification plant, and electric furnace for maintaining the reactor temperature. A carbon conversion efficiency of 100% for gasification is assumed, and the by-products of gasification (tar and ash) are neglected. The energy requirements (data modified with respect to Indian electricity mix) for all the relevant processes such as fertilizer production, coconut crushing, steam and compressed-air production, irrigation, and thermochemical gasification are included while developing the model.

Coal gas life cycle

Coal gas life cycle is used as the reference system for comparison with that of coconut shell-derived producer gas life cycle. The important sub-modules involved are coal transport from foreign mines, coal mining in India from open and underground mines, transport of Indian and imported coal to regional storage, and gasification of this coal mix to generate coal gas. India is the third biggest hard coal producer (526 million tonne/year) after China (2,971 million tonne/year) and the USA (919 million tonne/year). Eighty-five percent of the total coal utilization in the country is from domestic production, and the remaining 15% is imported from countries like Australia, South Africa, and North America [32]. Out of the domestic production, 85% is from open mining, and the remaining is from underground mining. Major domestic coal mining companies are Coal India Limited, Neyveli Lignite Corporation, and Singareni Collieries Limited. Thus, domestic coal is considered to be mined by these companies at locations in Dhanbad, Neyveli, and Hyderabad, respectively. The study pertains to Calicut, a northern city in Kerala, which is the regional storage

Table 3 Irrigation schedule for coconut cultivation

Age of palm	Quantity of irrigation (Liters per palm)	Schedule (in the summer season, 3.5 months)
First 2 years	1,110	Once in 4 days
Third year onwards	600 to 800	Once in 7 days

From [29].

Table 4 Fertilization schedule for coconut plantations according to the principles of good management practices

Age of coconut palm	Nutrient dosage	Quantity of fertilizer to be added (g)			
		Ammonium sulfate	Urea	Single super phosphate	Muriate of potash
3 months	1/10 of full dose	250	110	180	200
First year	1/3 of full dose	800	360	590	670
Second year	2/3 of full dose	1,675	720	1,180	1,340
Third year onwards	Full dose	2,500	1,080	1,780	2,010

From [29].

location taken for modeling of the transport module for coal (imported and domestic). The mode of transport of coal to regional storage location and the corresponding distances are given in Table 5. For coal transport, the Ecoinvent database is modified with respect to Indian electricity consumption for diesel production for fuelling the transport system (rail, road). The transports of the imported coal in other countries are modeled using the respective data source available in the Ecoinvent 2.0 database. Figure 2 shows the coal gas life cycle map.

Supporting process

A schematic of Indian electricity generation is shown in Figure 3. India's specific electricity generation module was created based on the national average annual proportion of electricity generation by fuel type. LCA modules were developed with contributions of 12%, 70%, 2%, 1%, and 15% of the gross electricity generation [33] in the country by natural gas, coal, nuclear, renewable, and hydropower plants, respectively. Due to the unavailability of Indian data in the Ecoinvent 2.0 database, western European data were modified with respect to Indian electricity share for various energy sources.

Similarly, the fuel data (coal and natural gas) for the power plants in the Ecoinvent database were modified incorporating the proportion of domestic production and import. The transport data were also modified accordingly.

Allocation procedure

Allocation is necessary when a process produces more than one valuable product. In such cases, it is necessary to divide the environmental impacts from the process between the products. Crude coconut oil and coconut residues are the co-product and by-product, respectively, of coconut shell production process. The utilization of energy in the coconut palm residues (by-product of coconut cultivation stage) and production of coconut oil are two major cases in the present LCA study. The 'division' of the environmental impacts may be done either by allocation or by system expansion. According to Gnansounou et al. [34], environmental impacts in a life cycle may even vary with the allocation method (mass, energy, carbon content, or economy) adopted. In this study, system boundary expansion method is adopted for allocating the co-product and by-product.

Agricultural residues of coconut plantation play a significant role in the energy analysis because of its availability in large quantity and its energy content. As mentioned in 'Inventory data for coconut cultivation', palm residues include the leaves and the dried leaf weighing 2 kg/tree on average and have a heating value of 16,500 kJ/kg. The tree density is 175 palm/ha. The thermal energy that can be generated from the palm residues is calculated from its lower heating value (LHV), followed by the computation of its conversion into electrical energy with an overall conversion efficiency (from palm residues to electric energy) of 25%. This assumption does not take into account the technology used for this conversion. Based on the inventory data for coconut cultivation provided in 'Inventory data for coconut cultivation', the net electricity generated from coconut palm residues is computed for the reference functional unit (1 MJ) in SimaPro. This amount of electricity is assumed to offset the corresponding Indian electricity production from coal and natural gas. Thus, the system boundary (as explained in 'System boundaries') is expanded to include the system of processes which are involved in the production of Indian electricity from coal and natural gas.

The environmental burdens are allocated for coconut oil by assuming it replacing commercial paraffin oil lubricant [35]. As mentioned in 'Inventory data for

Table 5 Mode of coal transport and distance to central location

	Mode of transport	Distance (km)	Remarks
Imported coal	Sea	36,336	Total distance from the three countries to the central storage location
	Rail	1,212	
	Road	110	
Domestic coal	Rail	3,994	Total distance from three domestic mining locations in India to the central storage location
	Road	377	

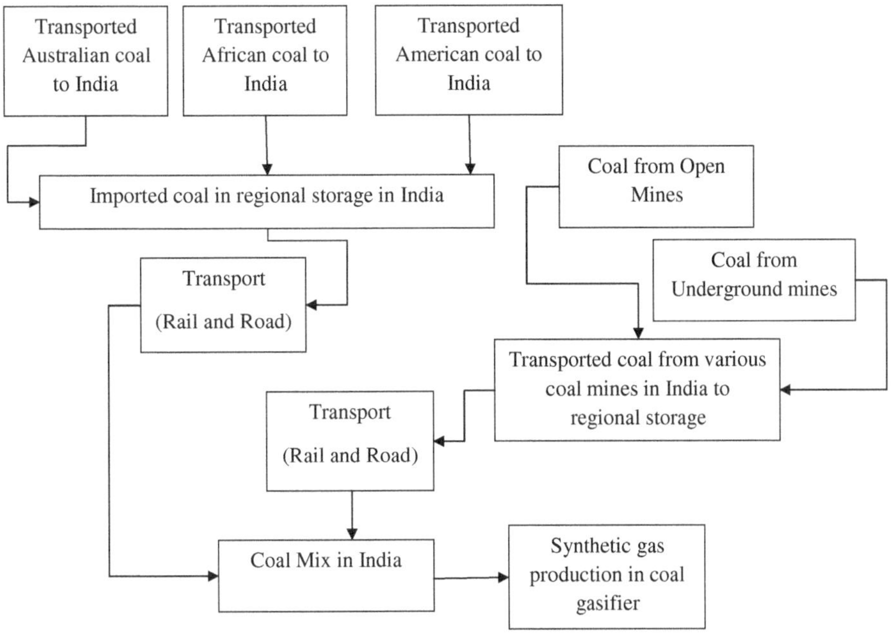

Figure 2 Schematic diagram of coal gas life cycle.

coconut cultivation', 4.7 kg of husked nuts (9.4 numbers equals approximately 10) is required to produce 1 kg (or 1.08 l) of coconut oil. Thus, corresponding to the generation of 1 MJ of syngas energy (reference functional unit), the net quantity of 'lubricant' (coconut oil) production can be computed and modeled in SimaPro. A separate module is created for paraffin oil lubricant production (modified Ecoinvent module), and the producer gas life cycle system boundary is expanded to include the system of processes in the commercial paraffin oil lubricant production. By proceeding like this, the net emissions associated with the production of producer gas with 1 MJ of energy are determined by subtracting the sum of the emissions (or impacts) associated with the saved fossil fuel-based electricity production and the saved commercial lubricant production from the emissions associated with the generation of producer gas.

Life cycle impact assessment

Following cutoff criteria are adopted in the study:

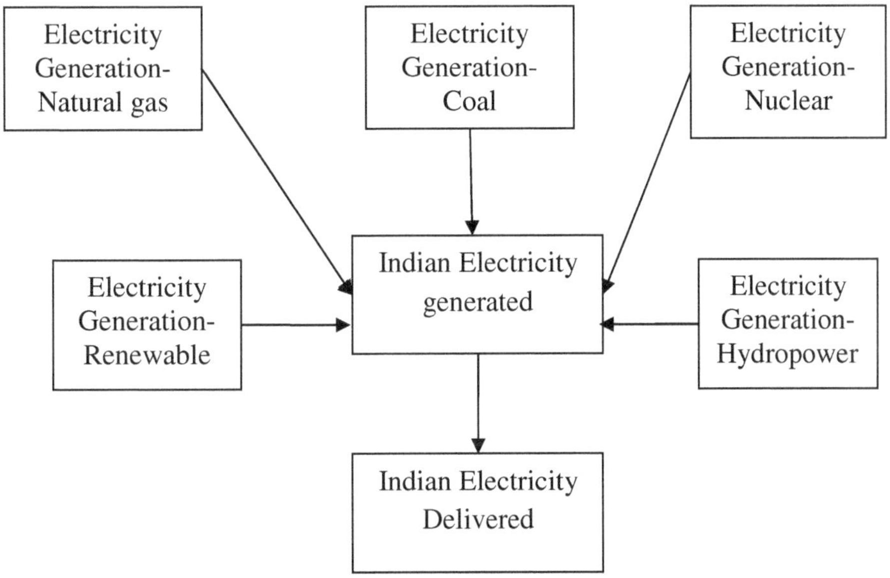

Figure 3 Modeling schematic for Indian electricity generation.

a. Since the maturity duration for coconut palm is 10 years, the study is conducted to analyze the impacts of producer gas life cycle corresponding to this duration.
b. Application of pesticides, insecticides, and herbicides for coconut cultivation is avoided due to lack of readily available local data.
c. CO_2 emissions from the utilization of producer gas are taken as nullified by the CO_2 absorption during the growth of coconut tree. This cutoff criterion is made in the context of unavailability of a well-established database to define the rate of sequestration of carbon by coconut palms.
d. Potential land use changes are not evaluated assuming the plantations are long-term existing coconut plantations.
e. Transport of personnel to carry out the different stages in the life cycle is not included.

Environmental impact assessment of the two life cycles is conducted by computational implementation of the inventories collected, using the software SimaPro 7.3.3® developed by Product Ecology Consultants (Amersfoort, The Netherlands) [36]. This software is widely used [8,12,21] for the inventory analysis and LCA of chemical and energy system chains. All the modules of the life cycle are developed using the Ecoinvent 2.0 database available in SimaPro. Tables 6 and 7 represent the corresponding module descriptions in SimaPro for producer gas life cycle and coal gas life cycle, respectively. Table 8 provides details of the inventory data (primary and sourced from

Table 6 SimaPro module descriptions for coconut cultivation and shell gasification in a FB gasifier

Module	Purpose
Sand for beds	Ecoinvent 2.0 module for production and related emissions of sand to be used in the nursery for coconut seedlings
Compost	Custom module for production and related emissions of compost to be used in planting coconut seedlings
Tillage and plowing	Modified Ecoinvent 2.0 module for land preparation for plantation and related emissions. Diesel source modified to Indian conditions
Coconut seedling at nursery	Custom module to represent the requirements for coconut seedlings at nursery
Coconut plantation	Custom module built from three inputs. Tillage, seedlings, and compost
Fertilizing: infrastructure and emissions	Modified Ecoinvent 2.0 module representing impacts of fertilizing 1 ha of plantation using the required machinery
Irrigation, infrastructure, and emissions	Modified Ecoinvent 2.0 module representing impacts of irrigating 1 ha of plantation using the required machinery
Ammonium sulfate production and transportation	Ecoinvent 2.0 module representing impacts of fertilizer production for usage in 1 ha of plantation over the lifetime
Urea production and transportation	Ecoinvent 2.0 module representing impacts of fertilizer production for usage in 1 ha of plantation over the lifetime
Potassium chloride production and transportation	Ecoinvent 2.0 module representing impacts of fertilizer production for usage in 1 ha of plantation over the lifetime
Single super phosphate production and transportation	Ecoinvent 2.0 module representing impacts of fertilizer production for usage in 1 ha of plantation over the lifetime
Coconut harvested from plantation	Custom module representing impacts associated with managing and operating the plantation. Energy from the combustion of biomass residues is used to offset Indian electricity delivered.
Coconut shell production	A custom module to represent the impacts associated with coconut shell production from the harvested coconuts
Crude coconut oil	Impacts of crude coconut oil production and related emissions in an oil extraction plant. Modified Ecoinvent 2.0 module is used with coconut harvesting and electricity utilization corrected to suit Indian conditions.
	Crude coconut oil (by-product) is allocated as lubricant for machinery with the addition of CuO nanoparticles.
Crushing of the shell	Impacts of grinding the coconut shell in biomass crusher for utilization in FB gasifier. Modified Ecoinvent 2.0 module is used with electricity utilization and input material quantity modified to suit Indian conditions.
Coconut shell gasification	Modified Ecoinvent 2.0 module for assessing the impact associated with crushed coconut shell gasification in a FB gasifier with steam and air as gasification medium
Steam for gasification	Modified Ecoinvent 2.0 module for steam generation. Modified with respect to Indian electricity mix
Compressed air for gasification	Modified Ecoinvent module for the assessment of impacts of producing compressed air at 2 bar pressure for fluidizing the bed

Table 7 SimaPro module descriptions for coal mining and gasification

Module	Purpose
Silica sand in the gasifier plant	Ecoinvent 2.0 module for production and related emissions of sand (SiO_2) to be used as bed material including the energy for drying the sand
Indian electricity delivered	Custom module to represent the impact of production and delivery of electricity for coal gasifier plant operation. The proportion of electricity generation is as given in Figure 4.
Steam generation at the site	Modified Ecoinvent 2.0 module for steam generation and related emissions. Electricity production and delivery modified to Indian conditions
Compressed-air generation	Modified Ecoinvent 2.0 module for compressed-air generation supply. The module includes a screw-type compressor, lubricating oil, electricity delivery, and transport of compressor and lubricating oil to the installation site.
Infrastructure for coal gasifier	Ecoinvent 2.0 module representing impacts of developing infrastructure for a typical coal gasifier. The dataset includes land use, building and facilities, dryer, gasifier, and gas treatment facility.
Industrial furnace for maintaining the temperature of the gasifier	Ecoinvent 2.0 module representing impacts of an oil furnace of 1-MW capacity. The module includes the production of oil, the infrastructure for the oil boiler, and transport to site.
Domestic coal mining and transport	Modified Ecoinvent 2.0 module representing impacts of domestic mining (underground and open) in India and its transport to a regional storage
Imported coal	Modified Ecoinvent 2.0 module representing impacts of coal imports from three countries, South America, South Africa, and Australia, in the ratio 2:1:7 to the regional storage location in India. Built-in Ecoinvent module for coal mining and transport in the respective countries were taken into account.
Coal transport to the plant site	Modified Ecoinvent 2.0 module including the impacts of road and rail transport in India.
Synthetic gas production in coal gasifier	Modified Ecoinvent 2.0 module representing impacts of gas production in fluidized bed coal gasifier with steam and air as gasifying medium

Ecoinvent) collected for model development. The impact assessment method selected is IMPACT 2002+ (IMPact Assessment of Chemical Toxics) available in the SimaPro methods library. IMAPCT 2002+ methodology links the LCI results to four broad damage categories: human health, ecosystem quality, climate change, and resources. The schematic of the IMPACT 2002+ framework [36] is shown in Figure 4. The following three specific environmental impact categories are selected in this study. Brief descriptions on the mentioned impact categories are given in the Appendix.

1. Global warming potential (GWP): Net emission of Intergovernmental Panel on Climate Change (IPCC)-identified greenhouse gases are calculated and expressed in terms of kilograms of CO_2 equivalent for a 100-year lifetime.
2. Ozone depletion potential (ODP): The emission of gases, mainly CFC-10, halon 1301, and halon 1211, is calculated and expressed in kilograms of CFC 11 equivalent.
3. Fossil fuel depletion potential (FDP): The degree of fossil fuel consumption is determined and expressed in terms of megajoules of energy surplus.

Results and discussion

In the following paragraphs, impact assessment results of the two life cycles are discussed in terms of the corresponding normalized unit for each impact category. The comparison between the life cycles and contributions of various sub-processes to these impacts are also discussed.

Global warming potential

A 100-year GWP value in kilograms of CO_2 equivalent, normalized by the functional unit of the study, i.e.,1 MJ heating value, is presented in Table 9. Results show that a saving of 18.3% in GHG emissions causing GWP can be obtained for the producer gas life cycle over coal gas life cycle. Process contributions of GHG emissions causing GWP in the two life cycles are given in Tables 10 and 11. From these tables, it is evident for the producer gas life cycle that the major contributions are from (1) steam generation, (2) coconut oil production, (3) fertilization, and (4) irrigation. Utilization of fossil fuel-based Indian electricity for steam generation and irrigation purposes is responsible for the emissions of greenhouse gases such as CO_2. The major source of potential emissions of fertilizer application is nitrogen oxides, computed according to Equation 1 [37].

$$N_2O_{\text{direct}} = \left[\sum_i m_{F,i}.\text{NC}_{F,i}\left(1 - \text{Frac}_{\text{gasf}}\right)\right] \text{EF}.\text{mw}_{N_2O}.\text{GW}_{N_2O} \quad (1)$$

where N_2O_{direct} is the direct N_2O emissions (tonne CO_2 equivalent), $m_{F,i}$ is the mass of synthetic type i fertilizer applied (tonne), $NC_{F,i}$ is the nitrogen content of synthetic

Table 8 Details of inventory data

Process/module	Data type and representativity to local conditions
Electricity generation	Primary data
	Electricity production mix is as follows: coal 70%, natural gas 12%, hydropower 15%, nuclear 2%,renewable 1%
	For each plant type, the import and domestic production of the corresponding fuel in India is accounted.
	For example, for coal power plants, the import/domestic production ratio in India is 8.5:1.5, whereas for the Ecoinvent European database, the ratio is 0.03:9.97.
	Similar modification is applied to other energy sources for a better representation of Indian conditions.
	For hydropower plants, the production capacity (kWh) is modified as the ratio between run-off river plant and reservoir plant in the ratio 0.2:0.8
Transport of coal	Modified Ecoinvent data
	For transport of fuels to a central location, the modules in the Ecoinvent database are modified incorporating the average distances by rail, road, and sea, as given in Table 5. Ecoinvent data are utilized for the development of modules for coal transport in the three countries from where coal is imported.
Infrastructure for electricity generation and transmission loss	Modified Ecoinvent data
	For the sub-process, infrastructure for electricity generation, the Ecoinvent database is modified by changing the land area required for each type of power plant for the process.
	For the sub-process transmission loss, modification is made in terms of Indian transmission loss of 35%.
Steam and compressed-air generation for gasification	Modified Ecoinvent data
	Western European data in the Ecoinvent database are modified with respect to electricity production and delivery in India.
Infrastructure for biomass gasification plant	Modified Ecoinvent data
	The Swiss database is modified with Indian electricity supply and vehicle transport (rail, lorry) suitable to Indian conditions.
Coconut shell production and crushing	Primary data
	A primary module with all the sub-modules developed based on the data corresponding to southern Indian conditions. The sub-modules include coconut plantation, irrigation, fertilization, composting, sand mix, tillage and plowing, crushing, etc. Allocation is made for crude coconut oil and coconut palm residues.

type i fertilizer applied (gN/100 g fertilizer), $Frac_{gasf}$ is the fraction that volatilizes as NH_3 and NO_X for fertilizers (dimensionless), EF is the emission factor for emissions (tonne N_2O), mw_{N_2O} is the ratio of the molecular weights of N_2O and N (tonne N_2O/tonne N), and GW_{N_2O} is the global warming potential for N_2O (kg CO_2 equivalent/kg N_2O).

The fertilizer application data given in Table 4 and inventory data provided in 'Inventory data for coconut cultivation' are incorporated. As per the equation, the direct N_2O emissions from the utilization of two nitrogen-containing fertilizers (urea and ammonium sulfate) are estimated as 0.00062 kg CO_2 equivalent for 1 MJ of syngas energy production.

For coal gas life cycle, the major emissions are caused by steam generation for gasification of coal, mining, and oceanic transport of the imported coal. The major contributing greenhouse gas to GWP is CO_2, with contributions of 95.9% and 96.2% of the total for producer gas life cycle and coal gas life cycle, respectively, as shown in Table 12. Power requirement for infrastructure and operation of the gasification plant being the same for these two life cycles, it is inferred that the decisive processes are coal mine operation and coal transport for coal gas life cycle.

The most significant contributions for coconut life cycle are from offsetting Indian electricity generation by coconut palm residues for by-product allocation. This process is favorable (GWP is negative) since fossil fuel-based electricity production is offset by the energy content (16.5 kJ/kg) of the residues, as mentioned in 'Allocation procedure'. In addition to this, allocation of the co-product (coconut oil) is also favorable since it replaces the commercial paraffin oil lubricant.

Two options are suggested for reducing the GHG emissions: (1) increasing the domestic mining capacity of coal, which will reduce the oceanic transport power requirement for importing the coal from other countries, as mentioned in 'Coal gas life cycle'; and (2) increasing the share of renewable and hydroelectric energy source contribution to electricity generation. At present, they are contributing only 16% of the total production capacity, as mentioned in 'Supporting process'. Promotion of these two energy sources, especially renewable bioenergy and small hydropower, shall result in reduction in the emissions of greenhouse gases for both the life cycles.

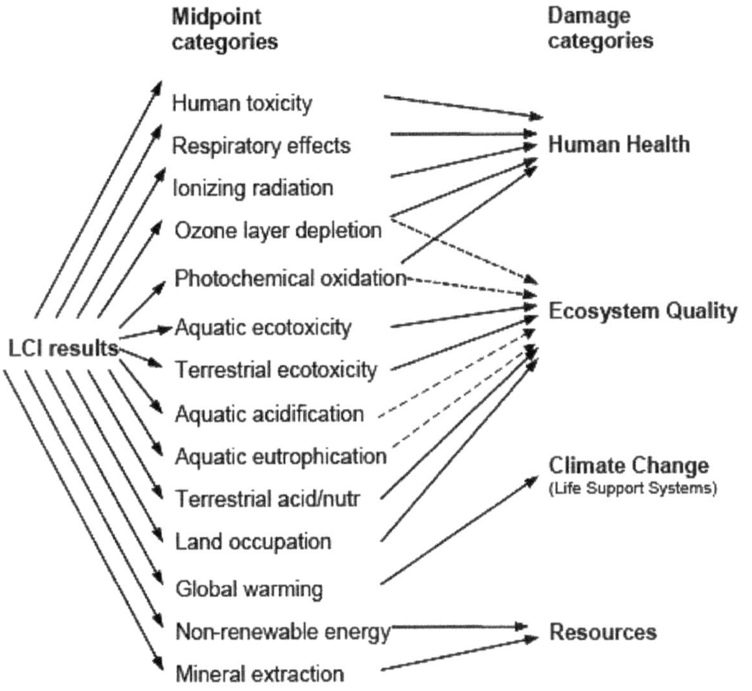

Figure 4 Overall scheme of IMPACT 2002+ framework.

Ozone layer depletion potential

Ozone layer depletion is a situation of serious concern as it will increase the amount of ultraviolet rays from the sun entering the atmosphere. The emissions of stratosphere ozone layer-depleting gases in the two life cycles are given in Table 13. The major gases responsible for the depletion of the ozone layer is CFC-10 (66.26% of ODP) for producer gas life cycle and halon 1301 (66.11% of ODP) for coal gas life cycle. The results provided in Table 9 show that ODP is reduced by 64.1% in the coconut shell-generated producer gas life cycle. The corresponding process contributions are provided in Tables 10 and 11. The by-product and co-product allocations are responsible for the reduction in ODP for producer gas life cycle, as evident from Table 10. Steam generation, fertilizer application, and irrigation are the significant contributors of ozone-depleting gas emissions for producer gas life cycle. Similarly for coal gas life cycle, in addition to steam production for gasifier operation, oceanic transport of coal (especially Australian coal for its large distance) and rail transport of domestic coal are decisive.

The major source of ozone-depleting gas emissions in both the cycles is the steam generation. The fossil fuel-based (natural gas and coal) electricity production for steam generation contributes 33% of the total ozone-depleting gas emissions. On analyzing the sub-processes, it is traced that 83% of the ODP of steam generation is attributed to the transport and storage of liquefied natural gas onshore and offshore for utilization in power plants. The liquefaction of natural gas is accomplished by cryogenic treatment, which is responsible for the emission of three gases, halon 1301, CFC 10, and CFC 114, to the atmosphere which deplete the ozone layer in the stratosphere. The remaining 17% is contributed by the utilization of diesel for the transport of hard coal to power plant locations. The diesel production stage (crude oil refining) involves many low-temperature extraction processes utilizing refrigerants, which emit ozone-depleting gases. It may also be noted that the other two major processes in producer gas life cycle are fertilizer production and irrigation, while transport of the coal through large distance (e.g., oceanic transport of

Table 9 Impact assessment of the two life cycles

Impact	Unit	Coal gas life cycle	Coconut shell-derived producer gas life cycle	Percentage of decrease for producer gas life cycle
GWP due to GHG emissions	kg CO_2 equivalent/MJ energy	0.133	0.108	18.3
Emissions causing ODP	kg CFC 11 equivalent/MJ energy	2.39E−8	8.6E−9	64.1
FDP	MJ energy surplus/MJ energy	4.19	1.4	71.5

Table 10 Process contributions for producer gas life cycle

Contributing life cycle process	GWP (kg CO_2equivalent/ MJ energy)	ODP (kg CFC 11 equivalent/ MJ energy)	FDP (MJ energy surplus/ MJ energy)
Gasification of coconut shell	1.21E−6	0	0
Electricity production and supply for gasification	6.2E−3	4.82E−10	0.0806
Steam production and supply for gasification	0.2068	8.3E−9	1.30
Compressed-air generation and supply for gasification	8.56E−3	6.64E−10	0.1116
Gasifier infrastructure establishment	8.06E−5	7.4E−12	1.15E−3
Crushing coconut in machine	4.16E−6	2.0E−12	6.58E−5
Electricity for crushing mill operation	2.45E−4	1.9E−11	3.18E−3
Crude coconut oil production (co-product allocation)	−0.098	−1.35E−9	−0.24
Fertilization machinery	2.53E−3	3.68E−10	0.04105
Irrigation	0.098	5.0E−9	0.98
Fertilizers	0.1009	1.19E−8	1.75
Coconut plantation (seedling, tillage, and compost)	2.89E−4	3.29E−11	3.6E−3
Electrical energy to main grid (by-product allocation)	−0.324	−9.46E09	−2.89
Furnace for gasifier operation	3.8E−6	2.46E−13	1.07E−4
Bed material, catalysts, etc.	1.86E−3	1.88E−10	0.034
Transport and maintenance of all the vehicle transport infrastructure	1.57E−3	2.6E−10	0.02701
Disposal of ash and sewage treatment	4.24E−5	1.006E−12	1.37E−6

the imported coal) is prominent for coal gas life cycle. The share of natural gas in electricity production and fueling (diesel) transport contributes ozone-harmful gases to these three sub-processes also.

Fossil fuel depletion potential or utilization of nonrenewable energy

Syngas produced from gasification of biomass feedstock is often cited as an effective alternative to fossil fuels, thus reducing the dependence on fossil fuel reserves. The utilization of nonrenewable energy or fossil fuel depletion potential (FDP) is expressed in megajoules of energy surplus. The FDP values for the two life cycles are shown in Table 9. The results show that a reduction of 71.5% is possible regarding fossil fuel consumption for coconut shell-produced gas life cycle, compared to coal gas life cycle. This significant reduction is attributed to two reasons: (1) utilization of renewable biomass (coconut shell) instead of fossil fuel (coal) and (2) allocation of coconut residues for offsetting fossil fuel-based power production capacity. The process contributions of both life cycles are given in Tables 10 and 11. Source-wise fossil fuel consumption is represented in Table 14 for both life cycles. The major fossil fuel is coal, with contributions of 84.96% and 77.62% for the life cycles, as shown in the table. For coal gas life cycle, the coal is utilized for electricity production (Indian electricity mix) and also as the feedstock in coal gasifier.

Contribution of life cycle stages to the impact emissions

Based on the impact values provided in Tables 10 and 11, the percentage share of major stages in the two life cycles to the impact emissions is shown in Tables 15 and 16. For producer gas life cycle, the major sources of life cycle emissions are generation of gasification agents (superheated steam and compressed air), production and utilization of synthetic fertilizers, and coconut cultivation and husbandry practices, as shown in Table 15. Utilization of fossil fuel-based electricity is used for the generation of steam and compressed air. This is responsible for the substantial share (38.97%, 32.58%, and 33.98% respectively to GWP, ODP, and FDP) of the stage (generation and supply of gasification agents) to the impacts. Nitrogen-containing fertilizer production and application is another life cycle stage which contributes 34.92%, 44.58%, and 40.52% respectively to GWP, ODP, and FDP. Considering these two factors, some improvement measures to the life cycle emissions are also discussed separately ('Sensitivity analysis').

For coal gas life cycle, the production of gasification agents and production and transport of domestic coal are the major contributors, as shown in Table 16. More than 85% of GWP and FDP is contributed by these two coal gas life cycle stages. Life cycle emissions from domestic coal production and transport dominate that of the corresponding imported coal since 85% of Indian coal is produced by domestic mining from open and underground mines, as mentioned in 'Coal gas life cycle'.

Table 11 Process contributions for coal gas life cycle

Contributing life cycle process	GWP (kg CO_2equivalent/ MJ energy)	ODP (kg CFC 11 equivalent/ MJ energy)	FDP (MJ energy surplus/ MJ energy)
Gasification of coal	1.09E−6	0	0
Electricity production and supply for gasification	5.58E−3	4.338E−10	0.07254
Steam production and supply for gasification	0.09612	7.47E−09	1.251
Compressed-air generation and supply for gasification	7.70E−3	5.976E−10	0.1004
Gasifier infrastructure establishment	7.25E−5	6.66E−12	1.04E−3
Imported coal			
Road transport	9.54E−5	1.53E−10	1.701E−3
Rail transport	1.13E−3	9.18E−10	0.021474
Electricity	5.02E−5	2.23E−11	1.152E−3
AU coal (production and ocean transport)	3.6E−3	4.482E−09	0.3092
SAf. coal (production and ocean transport)	3.29E−4	3.33E−10	0.04338
SAm. coal (production and ocean transport)	5.45E−4	6.89E−10	0.0779
Domestic coal (open mine and underground mine)			
Coal transport (road)	1.08E−3	1.735E−9	0.019278
Coal transport (rail)	6.82E−3	5.63E−9	0.130
Electricity	6.43E−4	5.04E−11	8.33E−3
Coal stock and infrastructure for mining	5.67E−3	9.34E−10	2.0898
Furnace for gasifier operation	3.42E−6	2.214E−13	9.61E−5
Bed material, catalysts, etc.	1.67E−3	1.692E−10	0.0306
Transport and maintenance of all the vehicle transport infrastructure	1.42E03	2.34E−10	0.0243
Disposal of ash and sewage treatment	4.07E−5	9.05E−13	1.23E−6

Electricity supply is a supporting process for many processes such as steam generation, fertilizer production, compressed-air production, furnace operation, crushing of coconut shell, infrastructure establishment for coal mining, etc. in the two life cycles. Fossil fuel-based electricity production is the major contributing process to the impacts for the life cycle stages mentioned. Table 17 lists the contribution of electricity production and supply to the emissions of the life cycles stages mentioned in Tables 15 and 16. From Table 17, it is evident that fossil fuel-based Indian electricity production almost completely controls the impacts from the stages coconut shell crushing (C) and gasification agent generation (D) in the producer gas life cycle. Similar is the case with stage 'C1' for the coal gas life cycle. Substantial electricity-consuming equipments such as crushing mill, boiler, and air compressor are responsible for this. It may also be noted that the emissions associated with coal transport dominates in the production and transport stage of the imported (A1) and domestic (B1) coal.

In short, it can be estimated from Tables 15 and 17 that the contribution of Indian electricity production and utilization to GWP, ODP, and FDP are 68.8%, 51.4%, and 56.3%, respectively, for the life cycle of producer gas. This is due to the fact that 82% of Indian electricity generation is a combination of coal and natural gas, as mentioned in 'Supporting process'. This fact necessitates switching over from fossil fuel-based electricity to renewable-based green electricity (discussed in 'Sensitivity analysis').

Other impact categories

In addition to the discussed impact categories of GWP, ODP, and FDP, IMPACT 2002+ links all types of LCI

Table 12 Major GHG emissions causing global warming

Producer gas life cycle			Coal gas life cycle		
Gas	GWP (kg CO_2equivalent)	% of total GWP	Gas	GWP (kg CO_2equivalent)	% of total GWP
CO_2	0.10473	95.96	CO_2	0.128	96.20
CH_4	0.00233	2.15	CH_4	0.0391	2.98
Dinitrogen monoxide	0.00062	1.52			
CO	0.0000217	0.02			

Table 13 Major gas emissions causing ozone layer depletion

Producer gas life cycle			Coal gas life cycle		
Gas	ODP (kg CFC 11 equivalent)	% of total ODP	Gas	ODP (kg CFC 11 equivalent)	% of total ODP
CFC 10	5.87E−9	66.26	Halon 1301	1.58E−8	66.11
Halon 1211	1.59E−9	18.49	CFC 10	5.95E−9	24.89
CFC 114	1.27E−9	14.76	Halon 1211	1.69E−9	7.10
			CFC 114	9.85E−10	1.90

results to four damage categories: human health, ecosystem quality, climatic change, and resource depletion. Under these damage categories, another eleven impact categories (brief description is provided in the Appendix) are also identified, and the life cycles are compared in terms of these categories also, as shown in Figure 5. The comparison is related to coal gas life cycle which is assumed to be having 100% impact. As evident from the figure, all the impact categories calculated show an edge for producer gas life cycle over coal gas life cycle. Major sources of the emissions responsible for these impacts are represented in Table 18. The difference in the impacts caused by these emissions is due to the contribution of electricity to the grid by coconut palm residues. This is because many of these emissions are resulting from the stack gas from fossil fuel-based power plants. The processes responsible for the emissions of these gases and ions are mentioned briefly for each of these impacts, as given below.

Carcinogenic and non-carcinogenic emission potential

The major source is the stack gas emissions from coal and natural gas power plants. Utilization of zeolite for various purposes in these plants makes carcinogenic emissions of elements like arsenic ion and PAH and non-carcinogenic emissions such as selenium. Infrastructure requirement for the power plant operation also contributes (<15%) to these emissions. However, the major emissions for coal gas life cycle are attributed to the tailing of coal in landfill from coal gasification plant (72%).

Respiratory organics emissions potential

A percentage of 91.2 of this impact is contributed by coal power plant emissions of volatile organic compounds and xylene. Emissions from the diesel-fueled vehicles and natural gas power plant emissions are also responsible for less than 8%. For coal gas life cycle, in addition to the 77% contribution from coal power plants and coal gasification plants, the oceanic freighter for coal transport is a critical entity emitting large proportions of methane and pentane.

Ionizing potential

Emission of species such as radon 222 and carbon 14 by nuclear power plants amounts to 73% of this impact. In the remaining, less than 21% is contributed by the emissions of radon 222 and cesium from coal-fired plants. Uranium enrichment process for fueling the nuclear power plant is also responsible for the emissions of radon 222.

Aquatic ecotoxicity potential

The major sources of this impact for both life cycles are the utilization of zeolite powder at natural gas and coal-fired power plants (52%), contamination of metals like Al and Cu in hot water released into the sea by the coal power plant (41%), and utilization of water for various processes in the life cycles (4%).

Terrestrial ecotoxicity potential

For producer gas life cycle, 93% of terrestrial ecotoxic emissions and 94.5% of terrestrial acidic emissions are contributed by solid waste emissions to the soil by coal-fired power plants to landfill. In the case of coal gas life cycle, the transport fuel (diesel) emissions are also responsible as its utilization is much larger than that for producer gas life cycle.

Land occupation

For producer gas life cycle, the occupation of arable land by coal power plant, hydroelectric power plant, and gasification plant share this impact in the ratio of 76%:14%:11%. However, for coal gas life cycle, land for open mines and

Table 14 Major nonrenewable energy sources causing fossil fuel depletion

Producer gas life cycle			Coal gas life cycle		
Material	MJ energy surplus	% of total FDP	Material	MJ energy surplus	% of total FDP
Coal	1.0867	77.62	Coal	3.56	84.96
Natural gas	0.2113	15.09	Natural gas	0.2416	5.77
Oil	0.04485	3.20	Oil	0.234	5.58
Uranium	0.0533	3.81			

Table 15 Percentage share of producer gas life cycle stages to GWP, ODP, and FDP

Stage in life cycle	Percentage contribution to the environmental impacts		
	GWP	ODP	FDP
Coconut cultivation and husbandry practices	22.73	18.32	22.25
Fertilizer production and application	34.92	44.58	40.52
Coconut shell production and crushing	0.09	0.09	0.08
Generation and supply of gasification agents	38.97	32.58	33.98
Coconut shell gasification in gasifier including infrastructure development and waste disposal	3.29	4.43	3.17

land for rail and road (for coal transport) are prominent (total 69%), leaving land utilization for power plant contributing less than 20%.

Aquatic acidification potential

The presence of sulfur-containing compounds in the stack emissions from coal-fired power plants and the nitrous oxide emissions from fertilizers for coconut cultivation are the major contributors to impact caused by aquatic acidifying emissions.

Aquatic eutrophication potential

Utilization of single super phosphate as fertilizer for coconut cultivation emits P_2O_5. For coal gas life cycle, coal tailing, emissions from coal power plant, and crude oil production are the major sources of aquatic eutrophication. In coal gas life cycle, the transport of imported coal by oceanic tanker also is a contributor since the emissions from transport fuel is very high.

Mineral extraction

Infrastructure for electricity production for each of the power plant and emissions from the utilization of zeolite powder are responsible for the extraction of minerals like Al and Cu. In addition, iron ore from mines are also responsible for the impact caused by mineral extraction for coal gas life cycle.

Energy and exergy consumption

The energy and exergy consumptions of the two life cycles in terms of renewable and nonrenewable energy sources are given in Table 19. The reason for the negative values of cumulative energy demand (CED) and cumulative exergy demand (CExD) for renewable biomass source is accounted when allocating the by-product coconut palm residues for electrical energy contribution to the main grid. Exclusion of this by-product increases the CED for producer gas life cycle to 2.3 MJ equivalents. In such case, the renewable fraction in total energy demand is 62.9%, while it is only 2.8% for coal gas life cycle. Substantial improvement in renewable fraction can be achieved by increasing the share of renewable fraction in total power production capacity of the country. To assess this, different scenarios are generated for the share of electricity production, and the results are given in Table 20. As expected, substantial increase in the renewable fraction of the cumulative energy consumption can be achieved by increasing the share of renewable energy in the Indian electric grid.

Sensitivity analysis

Sensitivity analysis is a systematic evaluation process for describing the effect of input variations in a system on the output [6]. The life cycle emissions of the producer gas are broadly divided into five contributors (A to E), as given below.

A. Coconut cultivation and husbandry practices (seedling, bed preparation, tillage, plowing, and irrigation)
B. Fertilizer production and application
C. Coconut shell production and crushing
D. Generation and supply of gasification agents (air and steam)
E. Coconut shell gasification in a gasifier including infrastructure development and waste disposal

Table 16 Percentage share of coal gas life cycle stages to GWP, ODP, and FDP

Stage in life cycle	Percentage contribution to the environmental impacts		
	GWP	ODP	FDP
Imported coal production and transport	4.33	27.62	10.86
Domestic coal production and transport	10.67	34.94	53.63
Generation and supply of gasification agents	76.12	33.76	32.25
Coal gasification in gasifier including infrastructure development and waste disposal	8.88	3.68	3.26

Table 17 Percentage share of electricity production and supply to the emissions of the life cycle stages

	Producer gas life cycle stage					Coal gas life cycle stage			
	A	B	C	D	E	A1	B1	C1	D1
GWP	57.6	26.8	93.8	99.8	1.6	2.1	29.3	93.8	1.5
ODP	61.1	17.1	98.1	99.3	2.3	0.8	7.2	98.1	2.3
FDP	63.1	19.9	99.5	99.7	7.9	0.2	0.03	99.5	7.9

A, coconut cultivation and husbandry practices; B, fertilizer production and application; C, coconut shell production and crushing; D, generation and supply of gasification agents; E, coconut shell gasification in gasifier including infrastructure development and waste disposal; A1,imported coal production and transport; B1,domestic coal production and transport; C1,generation and supply of gasification agents; D1,coal gasification in gasifier including infrastructure development and waste disposal.

Summary of GWP, ODP, and FDP of the life cycle in terms of these broad contributors (A to E) are illustrated in Figure 6 (Table 15 shows the corresponding percentage values). The contributions from by-product and co-product allocation are not represented. The existing fertilization practice (type and schedule) for coconut cultivation in the region and the utilization of gasifying agents for FB gasification of coconut shell are the major contributors to the discussed environmental impacts for the producer gas life cycle. The major reasons attributed to the higher sensitivity of these processes towards GWP are the utilization of fossil fuel-based electricity (68.8%) and the presence of nitrogen content in synthetic fertilizers (25.6%). Out of the 34.92% share of synthetic fertilizer production and utilization (Table 15) to total GWP, 26.8% is contributed by electricity from the grid (Table 17).

Scenario generation approach, which is widely used [38-40] in LCA studies for suggesting improvement options to environmental impacts, is adopted here to address the GHG emission savings. In the background of the information provided by the data presented in Tables 15 and 17, three scenarios were analyzed to account for their sensitivity on the base case GWP: (1) green electricity, (2) organic farming, and (3) catalytic gasification.

Green electricity

Complete replacement of fossil fuel-based electricity by renewable energy-based (green) electricity is not feasible in India in the near future, on behalf of the negligible current share (1%) of green electricity. Thus, generation of gasifying agents (air and steam) for gasification is taken as accomplished by energy source contribution in the following ratio: coal 50%, biomass 20%, hydropower 16%, natural gas 12%, and nuclear 2%. This means that the 20% share of fossil fuel is replaced by biomass in the Indian electric grid. A reduction of 11.1% in GHG emissions from the base case (GWP = 108 g CO_2 equivalent) was estimated. The GHG emission savings over coal gas case is 27.7%.

Organic farming

Application of synthetic fertilizers containing nitrogen emits nitrous oxides. A case in which 50% weight of synthetic fertilizers (ammonium sulfate, urea) replaced by organic fertilizers for coconut cultivation is generated. This results in improvement in GHG emissions by 5.2% from the green electricity-applied base case. The combined GWP improvement over coal gas is 31.5%.

Catalytic gasification

The lower heating value of syngas generated from steam-assisted AFB gasification of biomass ranges from 6 to 9 MJ/Nm^3 [41]. LHV depends on several factors such as design of the gasifier, temperature of steam and air, and particle size of the crushed biomass. In the present study, LHV is taken as 7 MJ/Nm^3. It was reported by Sutton et al. [42] that application of catalysts helps to improve the yield of hydrogen by 20% to 40% (depending

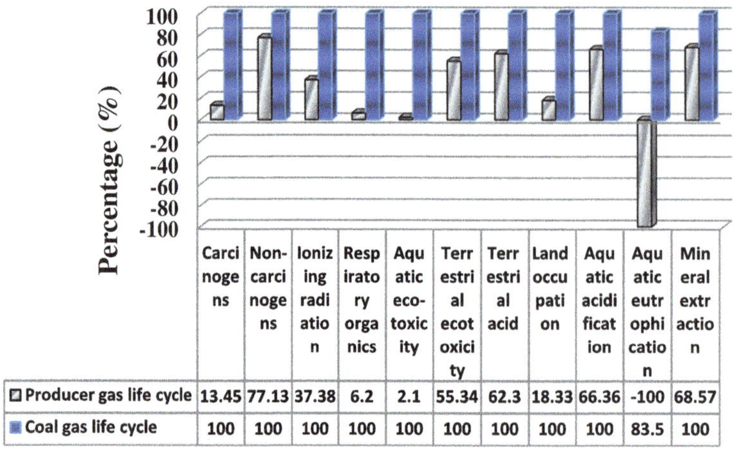

Figure 5 Percentage comparisons of the life cycles based on various emissions.

Table 18 Major emission/substances responsible for various impact categories and their contribution

Emissions	Producer gas life cycle		Coal gas life cycle	
	Emissions/substance	Percentage of total	Emissions/substance	Percentage of total
Carcinogens	Arsenic ion in water	41.46	Arsenic ion in water	62.87
	Tetrachlorodibenzene-dioxine	27.40	Aromatic hydrocarbons in air	21.41
	Aromatic hydrocarbons in air and water	25.01		
	Polyaromatic hydrocarbons	5.93		
Non-carcinogens (causing human diseases other than cancer)	Arsenic ion in water	73.16	Arsenic ion in water	63.59
	Tetrachlorodibenzene-dioxine	23.51	Selenium	13.06
			Barium	11.42
			Aluminum	5.75
			Tetrachlorodibenzene-dioxine	3.13
Ionizing radiation	Radon 222	66.96	Radon 222	76.51
	Carbon-14	26.86	Carbon-14	17.78
	Cesium −137	4.52	Cesium −137	4.97
Respiratory organics	Non-methane volatile organic compounds	26.41	Non-methane volatile organic compounds	85.50
	Xylene	46.08	Methane	3.78
	Pentane	9.47	Xylene	3.52
			Pentane	2.01
Aquatic ecotoxicity	Aluminum	97.03	Aluminum	95.03
	Copper ion	0.47	Copper ion	2.70
Terrestrial ecotoxicity	Aluminum	97.85	Aluminum	69.43
	Zinc	2.11	Copper	14.48
			Zinc	10.93
Terrestrial acid	Nitrogen oxides	76.54	Nitrogen oxides	79.61
	Sulfur dioxide	19.14	Sulfur dioxide	13.11
Land occupation	Occupation forest	40.43	Occupation forest	0.000237
	Occupation industrial area	51.43	Occupation industrial area	0.000133
Aquatic acidification	Sulfur dioxide	62.37	Sulfur dioxide	45.53
	Nitrogen oxides	31.7	Nitrogen oxides	35.24
Aquatic eutrophication	Phosphate	1.38	Phosphate	94.65
	Chemical oxygen demand	95.75	Chemical oxygen demand	4.19
Mineral extraction	Copper	79.94	Copper	79.22
	Aluminum	12.56	Aluminum	12.23

on the type of catalyst used), increasing the heating value correspondingly. Assuming a 30% improvement with catalyst application, the volume of syngas to be generated for 1 MJ of energy content (reference functional unit) could be reduced correspondingly. Incorporating this case (when all the three scenarios are implemented at the same time), 43.4% reduction in GHG emissions from the corresponding fossil coal gas case can be achieved. As reported by Singh et al. [39], this reduction in GWP ensures the sustainability of coconut shell-derived producer gas as a biofuel up to 2017, as defined by the EU renewable energy directive [43]. Figure 7 illustrates the GWP improvement accomplished under the scenarios discussed.

Comparison of the results with similar studies

Several studies reported on the comparative LCA of biofuels and their utilization systems with respect to fossil fuel-based systems. Some of the recent studies reported on LCA of biomass-derived syngas (or its utilization system) and are given in Table 21. Major relevant results drawn from these studies are also mentioned. According to Singh et al. [6], differences in approaches and assumptions taken under consideration in various LCA studies can lead to significant variations in the environmental impacts. It may be noted that direct comparison between the results is not recommended. This is because scaling of LCA balances to geographical scale may not be

Table 19 Energy and exergy consumption

Cumulative energy demand (MJ equivalent)			Cumulative exergy demand (MJ equivalent)		
Impact category	**Coal gas life cycle**	**Producer gas life cycle**	**Impact category**	**Coal gas life cycle**	**Producer gas life cycle**
Nonrenewable, fossil	4.235	1.37	Nonrenewable, fossil	4.08	1.36
Nonrenewable, nuclear	0.0632	0.0492	Nonrenewable, nuclear	0.127	0.0533
Renewable, kinetic	0.00322	0.00294	Renewable, biomass	0.00607	−1.98
Renewable, solar	0.00254	0.00248	Renewable wind, solar, geothermal	0.00595	0.0056
Renewable, potential	0.11	0.0978	Renewable, water	0.11	0.0978
Nonrenewable, primary	3.87E−7	9.12E−7			
Renewable, biomass	0.00638	−2.08			
Renewable, water	1.8	0.483			
Renewable, metals	0.00406	−0.00644			
Nonrenewable, minerals	0.00167	0.000704			

appropriate due to the variability in topography, soil and climatic conditions, and factors (economical and political) influencing land use [44,45].

Prospect of coconut shell in electrifying India

The purpose of any LCA study is to assess the system in the relevant and selected perspective, and hence, the suggestions and results hold good mostly in the selected system boundary only. The present study is applicable to India, which is the third largest coconut-producing country, behind the Philippines and Malaysia, having an area of about 1.78 million ha under the crop. Annual production is about 7,562 million nuts with an average of 5,295 nuts/ha. The state of Kerala, where the study pertains to, tops in production accounting 39% of the total production in the country. Coconut farming provides employment to nearly 10 million people and makes a contribution of nearly INR 70 billion to gross domestic product. The electricity sector in India had an installed capacity of 205.34 GW as of June 2012, the world's fifth largest [46]. In terms of fuel, coal-fired plants account for 70% (143 GW) of India's installed electricity capacity [47,48]. As per the data from the local agricultural university (as given in 'Life cycle inventory analysis'), 3.1 numbers of husked nuts are required to produce 1 kg of coconut shell. The average calorific value (LHV) of coconut shell ranges from 16.5 to 18 MJ/kg. Thus, based on the above mentioned annual production of nuts, by allocating 100% of the coconut shell for the production of power by converting it into syngas (producer gas), 40.25 to 43.9 GW of power could be produced. This is 19.6% to 21.4% of the total energy demand in the country, offsetting almost 35% to 38% of coal consumption. Careful planning and long-term policies on coconut cultivation encouragement, effective utilization of energy from coconut residues, and coconut shell gasification practice will definitely help to build an effective alternative fuel sector in India.

Conclusions

Two gaseous fuels (coconut shell-derived producer gas and coal gas) were compared in terms of their environmental impacts by conducting a LCA study with respect to Indian conditions. The environmental impacts are quantified as 18.3% lower GWP, 64.1% lower ODP, and 71.5% lower FDP, indicating better environmental performance of producer gas life cycle. In addition to this, comparisons based on several other impact categories also confirmed the edge of producer gas life cycle. Steam and compressed-air generation for gasifier operation and fertilizer production and application for coconut cultivation are the major contributors to the environmental impacts for producer gas life cycle. Substantial contribution

Table 20 Renewable fraction of CED for various electricity production scenarios

Share of total electricity production (%)					Renewable fraction of cumulative energy demand (%)	
Coal	**Natural gas**	**Hydroelectric**	**Nuclear**	**Renewable**	**Coal gas life cycle**	**Producer gas life cycle**
70	12	15	2	1	2.8	62.9
60	12	25	2	1	5.2	69.4
56	12	25	2	5	10.7	76.3
56	18	19	2	5	8.5	71.2
35	20	25	0	20	18.9	84.5

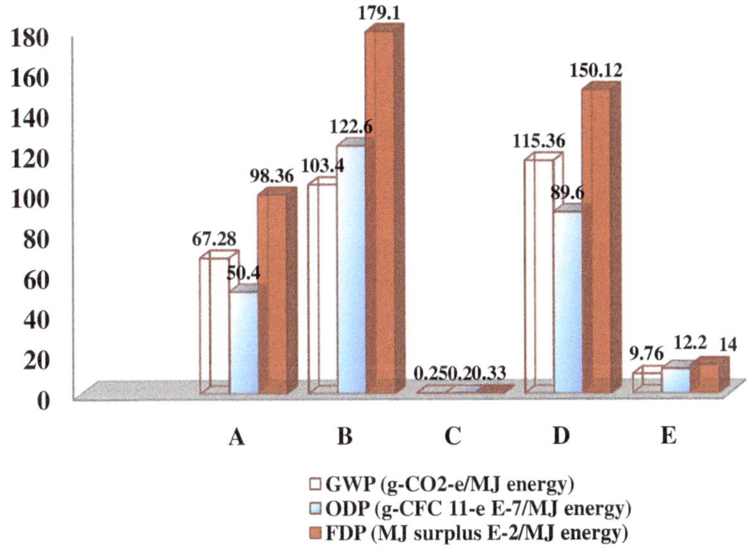

Figure 6 Major contributors to the life cycle impacts of producer gas. **A**, coconut cultivation and husbandry practices; **B**, fertilizer production and application; **C**, coconut shell production and crushing; **D**, generation and supply of gasification agents; **E**, coconut shell gasification in a gasifier including infrastructure development and waste disposal.

of the by-product allocation is the major factor responsible for lower emissions for the producer gas life cycle. Based on the existing fertilizing practice and the utilization of electricity from the Indian electric grid, substantial GHG emission savings cannot be achieved for producer gas life cycle. However, the green electricity-aided catalytic gasification of coconut shell produced by organic farming can result in 43.4% reduction in GHG emissions from fossil coal gas, meeting the EU renewable energy directive.

Appendix

Impact assessment methods and indicators

SimaPro contains several methods for the assessment of impacts of process and product stages. In these methods, multiple damage categories are taken into account, such as resource depletion, land use, climate change, and ecotoxicity. IMPACT 2002+ methodology proposes a feasible implementation of a combined midpoint/damage approach, linking all types of life cycle inventory results (elementary flows and other interventions) via 14

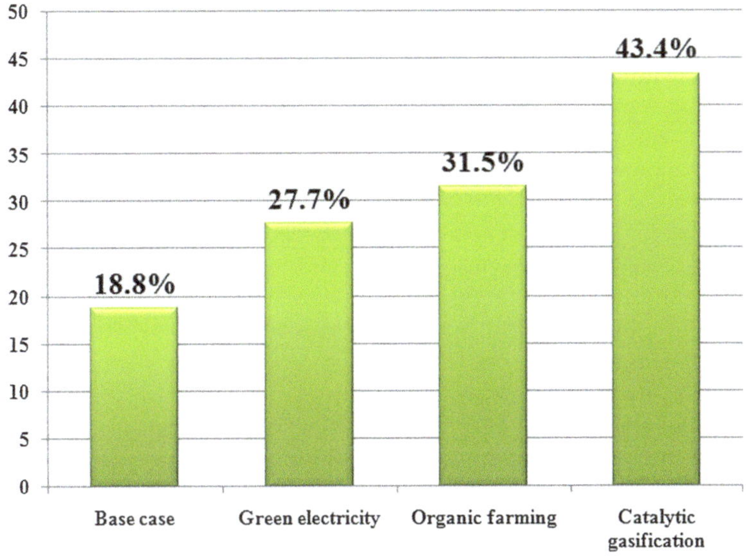

Figure 7 Percentage of GHG emission savings for producer gas life cycle under various scenarios. The scenarios are combined from left to right.

Table 21 Some recent studies on LCA of syngas generated by gasification of biomass

Criteria	Siegl et al. [1]	Roedl [14]	Guest [16]	Pa et al. [17]	Puy et al. [19]
Location	Austria	Norway	British Columbia	Barcelona metropolitan area	Sweden
Biomass material	Energy crops, manure, organic residues, wood chips	Forest residues	Wood wastes, wood pellets	Post-consumer wood, forest residues	Municipal solid waste
Reference functional unit	Kilowatt hour electricity generation in power plant which utilizes the syngas	Production of 1 MJ of thermal energy and 1 MJ of electrical energy in CHP plants	1 MJ of heat output	1,000 MJ of energy	1 ton of biomass
Impacts analyzed	GWP, ACD, TE, ET, ODP, ADP, PO, HT	GWP, PO, HT, AD	GWP, combined effect of the impact categories on human health	AD, GWP, ODP, HT, ACD, ET	AD, ACD, ET, GWP, ODP, HT, AE, TE, PO
Reference system	Anaerobic digestion power plant	Various capacity (micro, small, medium) gasification-CHP plants	Natural gas-based district heating system	Comparison between the two biomass life cycles	Sanitary land fill, incineration
Major relevant results[a]	64% to 83% reduction in GWP and 40% to 50% reduction in ODP	84% to 98% contribution of GWP from small- and medium-scale plants (for electricity)	82% and 83% reduction in GWP for wood wastes and wood pellets, respectively	34% reduction in GWP and 66% reduction in ODP for post-consumer wood biomass	3% reduction in GWP and 26% increase in ODP with respect to incineration
		59% to 72% contribution of ODP from small- and medium-scale plants (for electricity)			44% reduction in GWP and 46% reduction in ODP with respect to landfill

[a]Relevant in the context of the present study. AD, abiotic depletion potential; ACD, acidification potential; ET, eutrophication potential; GWP, global warming potential; ODP, ozone layer depletion potential; HT, human toxicity potential; AE, aquatic ecotoxicity potential; TE, terrestrial ecotoxicity potential; PO, photochemical oxidation potential; CHP, combined heat and power.

midpoint categories to four damage categories, as shown in Figure 4 [49].

For the assessment of the impact of GHG emissions, global warming potential of the greenhouse gases are expressed in terms of kilograms of CO_2 equivalent as per the IPCC fourth assessment report [50].

Ozone depletion potential values have been established mainly for hydrocarbons containing combined bromine, fluorine and chlorine, or CFCs. Here, one of the substances (CFC-11) has been adopted as the reference.

For assessment of the resource depletion using fossil fuel consumption, a term 'surplus energy' is employed in IMPACT 2002+. Mankind will always extract the best resources first, leaving the lower-quality resources for future extraction. The damage of resources will be experienced by future generations as they will have to use more effort to extract the remaining resources. This extra effort is expressed as surplus energy. Thus, fossil fuel depletion is expressed as surplus energy per extracted megajoule fossil fuel as a result of lower-quality resources.

Emission of carcinogenic substance causing human cancer is expressed as kilograms of C_2H_3Cl (vinyl chloride) equivalent. The normalization is based on the probability of the number of people from a group of 1 million who will develop cancer with the exposure to vinyl chloride. Impacts depend on the amount of the substance that will be transferred to water, air, or soil, each having different acceptable levels for the carcinogenic vinyl chloride.

Other midpoint impact categories in IMPACT 2002+ leading to the mentioned four damage categories are listed below [49] in terms of the equivalent units of known factors contributing to the corresponding impact:

1. Non-carcinogens: Human toxicity is stated in non-carcinogenic effects of equivalent mass of C_2H_3Cl (vinyl chloride).
2. Respiratory inorganics: Respiratory health effects are stated in equivalent units of inorganic particulate matter smaller than 2.5 microns.
3. Respiratory organics: Respiratory health effects are stated in equivalent mass of ethylene.
4. Ionizing radiation: Irradiating impacts are stated in equivalent units of becquerels of carbon-14.
5. Aquatic ecotoxicity: Freshwater ecotoxicity is stated in equivalent mass of triethylene glycol released into water (streams and lakes).
6. Terrestrial ecotoxicity: Land-based ecotoxicity is stated in equivalent mass of triethylene glycol released into the soil.
7. Terrestrial acidification/nitrification: Land-based acidification and nitrification are stated in equivalent mass of sulfur dioxide (SO_2) released into the air.

8. Land occupation: Units for land occupation are stated in square meters of arable land.
9. Aquatic acidification: Units are the same as for land-based acidification.
10. Aquatic eutrophication: Aquatic eutrophication is stated in equivalent mass of phosphate.
11. Mineral extraction: Mineral extraction impacts are stated in megajoules of surplus energy per kilogram of extracted substance.

Competing interests
The authors declare that they have no competing interests.

Authors' contributions
CCS conducted the life cycle assessment and drafted the manuscript. CM conducted the comparison between the life cycles, supervised the work and corrected the manuscript. AP performed the inventory collection, co-supervised the work and helped in drafting the manuscript. All authors read and approved the final manuscript.

Authors' information
CCS is a PhD research scholar in the Department of Mechanical Engineering of the National Institute of Technology Calicut, India. He obtained his B Tech and M Tech in mechanical engineering and refrigeration and cryogenic engineering, respectively, in 2000 and 2002. His research interests include biomass thermochemical treatments, life cycle assessment of renewable energy systems, and computational fluid dynamic studies on thermal and energy systems. He has published five papers in international journals and five papers in international conference proceedings. CM is a professor in the Department of Mechanical Engineering of the National Institute of Technology Calicut, India. He obtained his MSc Engg and PhD from the University of Calicut. His areas of interest include renewable energy systems, heat pipes, alternative refrigerants and fuels, and combustion. He co-authored two books and published more than 100 research papers in international journals and proceedings of conferences. AP is an assistant professor in the Department of Mechanical Engineering of the National Institute of Technology Calicut, India. He received his M Tech in energy management from the National Institute of Technology Calicut and his PhD in energy systems from the Indian Institute of Technology, Bombay. His research areas are mathematical modeling and optimal thermal design of various energy systems, energy management in thermal systems, and renewable energy utilization. He has published ten papers in peer-reviewed international journals and seven papers in conference proceedings.

References
1. Siegl, S, Laaber, M, Holubar, P: Green electricity from biomass, Part 1: environmental impacts of direct life cycle emissions. Waste Biomass Val **2**, 267–284 (2011)
2. Koroneos, C, Dompros, A, Roumbas, G: Hydrogen production via biomass gasification—a life cycle assessment approach. Chem. Eng. Proc **47**, 1261–1268 (2008)
3. Varun, Bhat, IK, Prakash, R: LCA of renewable energy for electricity generation systems—a review. Ren. Sust. Energy. Rev **13**, 1067–1073 (2009)
4. Merola, SS, Tornatore, C, Marchitto, L, Valenteno, G, Corcoine, FE: Experimental investigations of butanol-gasoline blends effects on the combustion process in a SI engine. Int. J. Energy Environ. Eng **3**, 6 (2012)
5. Self, SJ, Reddy, BV, Rosen, MA: Review of underground coal gasification technologies and carbon capture. Int. J. Energy Environ. Eng **3**, 16 (2012)
6. Singh, A, Pant, D, Korres, NE, Nizami, AS, Prasad, S, Murphy, JD: Key issues in life cycle assessment of ethanol production from lignocellulosic biomass: challenges and perspectives. Bioresour Technol **101**, 5003–5012 (2010)
7. ISO: International Standard, 1st edn. ISO, Switzerland (2006)
8. Manish, S, Pillai, IR, Banerjee, R: Sustainability analysis of renewables for climate change mitigation. Energy Sust. Dev **10**(4), 25–36 (2006)
9. Nigam, PS, Singh, A: Production of liquid biofuels from renewable resources. Prog. Energy Comb. Sc **37**, 52–68 (2011)
10. Shie, JL, Chang, CY, Chen, CS, Shaw, DG, Chen, YH, Kuan, WH, Ma, HK: Energy life cycle assessment of rice straw bio-energy derived from potential gasification technologies. Biores. Tech **102**, 6735–6741 (2011)
11. Corti, A, Lombardi, L: Biomass gasification integrated combined cycle with reduced CO_2 emissions: performance analysis and life cycle assessment (LCA). Energy **29**, 2109–2212 (2004)
12. Carpentieri, M, Corti, A, Lombardi, L: Life cycle assessment (LCA) of an integrated biomass gasification combined cycle (IBGCC) with CO_2 removal. Energy Conv. and Mngmnt **46**, 1790–1808 (2005)
13. Rafaschieri, A, Rapaccini, M, Manfrida, G: Life cycle assessment of electricity production from poplar energy crops compared with conventional fossil fuels. Energy Conv. & Mngmnt **40**, 1477–1493 (1999)
14. Roedl, A: Production and energetic utilization of wood from short rotation coppice—a life cycle assessment. Int. J. of Life Cycle Assess **15**, 567–578 (2010)
15. Kimming, M, Sundberg, C, Nordberg, A, Baky, A, Bernesson, S, Noren, O, Hansson, PA: Biomass from agriculture in small scale combine heat and power plants - a comparative life cycle assessment. Biomass Bioenergy **35**, 1572–1581 (2011)
16. Guest, G, Bright, RM, Cherubini, F, Michelesen, O, Stromman, AH: Life cycle assessment of biomass-based combined heat and power plants. J. of Ind. Ecology **15**, 908–921 (2011)
17. Pa, A, Bi, XT, Sokhansanj, S: A life cycle evaluation wood pellet gasification for district heating in British Columbia. Biores. Tech **102**, 6167–6177 (2006)
18. Pa, A, Jill, C, Bi, XT, Staffan, M, Sokhansanj, S: Environmental footprints of British Columbia wood pellets from a simplified life cycle analysis. Int. J. of Life Cycle Asses **17**, 220–231 (2012)
19. Puy, N, Rieradevall, J, Bartroli, J: Environmental assessment of post-consumer wood and forest residues gasification: the case study of Barcelona metropolitan area. Biomass Bioenergy **34**, 1457–1465 (2010)
20. Khoo, HH, Tan, RBH, Sagisaka, M: Utilization of woody biomass in Singapore: technological options for carbonization and economic comparison with incineration. Int. J. of Life Cycle Assess **13**, 312–318 (2008)
21. Zaman, AU: Comparative study of municipal solid waste treatment technologies using life cycle assessment method. Int. J. of Env. Sc. Tech **7**, 225–234 (2010)
22. Zhong, ZW, Song, B, Zaki, MBM: Life-cycle assessment of flash pyrolysis of wood waste. J. Clean. Prod **18**, 1177–1183 (2010)
23. Iribarren, D, Peters, JF, Dufour, J: Life cycle assessment of transportation fuels from biomass pyrolysis. Fuel **97**, 812–821 (2012)
24. Bartolozzi, I, Rizzi, F, Frey, M: Comparison between hydrogen and electric vehicles by life cycle assessment: a case study in Tuscany. Italy. App. Energy **101**, 103–111 (2012). doi:10.1016/j.apenergy.2012.03.021
25. Wulf, C, Kaltschmitt, M: Life cycle assessment of hydrogen supply chain with special attention on hydrogen refuelling stations. Int. J. of Hydrogen Energy **37**(21), 16711–16721 (2012). doi:10.1016/j.ijhydene.2012.03.028
26. Hawkins, TR, Gausen, OM, Strømman, AH: Environmental impacts of hybrid and electric vehicles—a review. Int. J. of Life Cycle Assess (2012). doi:10.1007/s11367-012-0440-9
27. Balat, M, Kirtay, E: Hydrogen from biomass–present scenario and future prospects. Int. J. of Hydrogen Energy **35**, 7416–7426 (2010)
28. Pollex, A, Ortwein, A, Kaltschmitt, M: Thermo-chemical conversion of solid biofuels. Biomass Conv. and Bioref. **1**, 21–39 (2011)
29. PRO Kerala: Kerala Climate. http://www.prokerala.com/kerala/climate.htm (2012). Accessed 4 March 2012
30. Coconut Development Board: http://www.coconutboard.nic.in/package1.htm (2012). Accessed 2 March 2012
31. Kerala Agricultural University: http://www.kau.edu (2012). Accessed 9 March 2012
32. Coal India Limited: Coal import export data. http://www.coalindia.in/default.aspx (2012). Accessed 17 March 2012
33. Central Power Research Institute: Govt. of India. www.cpri.in. Accessed 4 Oct 2012
34. Gnansounou, E, Dauriat, A, Villegas, J, Panichelli, L: Life cycle assessment of biofuels: energy and greenhouse balances. Bioresour Technol **100**(21), 4919–4930 (2009)
35. Thottackkad, MN, Perikinalil, RK: Kumarapillai, PN: Experimental evaluation on the tribological properties of coconut oil by the addition of CuO nanoparticles. Int. J. Prec. Eng. Manufact **13**(1), 111–116 (2012)
36. Product Ecology Consultants: Simapro Software. http://www.pre-sustainability.com/simapro-lca-software (2012). Accessed 24 March 2012

37. United Nations Framework Convention on Climatic Changes: Estimation of direct nitrous oxide emission from nitrogen fertilization. http://cdm.unfccc.int/methodologies/ARmethodologies/tools/ar-am-tool-07-v1.pdf (2007). Accessed 4 Oct 2012
38. Korres, NE, Singh, A, Nizami, AS, Murphy, JD: Is grass biomethane a sustainable transport biofuel? Biofuels Bioproducts Biorefin **4**, 310–325 (2010)
39. Singh, A, Nizami, AS, Korres, NE, Murphy, JD: The effect of reactor design on the sustainability of grass biomethane. Ren. Sust. Energy Rev **15**, 1567–1574 (2011)
40. Pandyaswargo, AH, Onoda, H, Nagata, K: Energy recovery potential and life cycle impact assessment of municipal solid waste technologies in Asian countries using ELP model. Int. J. Energy Environ. Eng **3**, 28 (2012)
41. Loha, C, Chatterjee, PK, Chattopadhyay, H: Performance of fluidized bed steam gasification of biomass – modeling and experiment. Energy Conv. Manag **52**, 1583–1588 (2011)
42. Sutton, D, Kelleher, B, Ross, JRH: Review of literature on catalysts for biomass gasification. Fuel Proc. Technol **73**, 155–173 (2001)
43. EU Renewable Energy Directive: Directive 2009/28/EC of the European Parliament and of the Council of 23 April 2009 on the promotion of the use of energy from renewable sources and amending and subsequently repealing Directives 2001/77/EC and 2003/30/EC. Official J. Euro. Un **16**, 16–62 (2009)
44. Schmer, MR, Vogel, KP, Mitchell, RB, Perrin, RK: Net energy of cellulosic ethanol from switchgrass. Proc. Natl. Acad. Sci. USA **105**, 464–469 (2008)
45. Gurgel, A, Reilly, JM, Paltsev, S: Potential land use implications of a global biofuel industry. J. Agric. Food Ind. Org **5**(2), 1–36 (2007)
46. Central Electricity Authority, Ministry of Power, Government of India: All India region-wise generating installed capacity of power. http://www.cea.nic.in/reports/monthly/executive_rep/jun12/8.pdf (2012). Accessed 6 Oct 2012
47. Ministry of Power, Government of India: Power sector at a glance "ALL INDIA". http://www.powermin.nic.in/indian_electricity_scenario/introduction.htm (2012). Accessed 6 Oct 2012
48. World Coal Institute: The coal resource, a comprehensive overview of coal. http://www.worldcoal.org (2011). 6 Oct 2012
49. Product Ecology Consultants, Washington: SimaPro Database Manual. www.pre-sustainability.com/simapro-lca-software (2012).Accessed 1 Oct 2012
50. IPCC: Fourth assessment report. http://www.ipcc.ch/pdf/assessment-report/ar4/wg1/ar4-wg1-chapter2.pdf(2007). Accessed 3 Oct 2012

3

Sustainable fermentation processing of two revalorized agro-industrial discards: carrot and brewer's yeast

Nora R Aimaretti[1,2*], Adriana L Clementz[1,2], Agustín Codevilla[2], María L Rojas[3] and Juan C Yori[1]

Abstract

An integrated and sustainable fermentation process was developed which enabled both the revalorization of two regional agro-industrial discards as well as by-product reuse. Carrot and brewer's yeast, which are commonly used for animal feed, were processed to obtain 77.5 L of ethanol, 450 kg of solid waste called bagasse, 970 L of liquid effluent called vinasse, and 39.8 kg CO_2 per each ton of discarded carrot. Results showed that the obtained bagasse was suitable for feeding 55 animals (calfs). The dilution of vinasse with fresh water (1:5) satisfied the requirements necessary to be used as beverage for the same number of animals, leaving a remnant which could be newly diluted (1:5) and used to irrigate a 0.025-ha carrot crop, the land dimension required to grow 1 ton of carrot.

Keywords: Discards, Carrot, Yeast, Fermentation, Ethanol, Sustainable

Background

First-generation biofuels, derived from agricultural products (corn, sugarcane, palm oil, rapeseed, or soybean), have produced an impact on grain prices, tightening the supply chain and the availability of land for food production [1]. The transformation of forests, savannahs, or pastureland into farms for biofuel production could lead to greater CO_2 emission than the emission reduction these biofuels could produce [2,3]. Consequently, serious damages to the ecosystems have been produced. Second-generation bioethanol is a biofuel made from nonedible raw materials [4]. Wastes or discards should no longer be seen as a burden but rather as another source of material such as energy fuel [1,5]. Thus, some organic waste may be incorporated as low-cost raw materials for second-generation bioethanol production, depending on the waste composition. Due to the wide variety of technological alternatives for the production of ethanol through fermentation, a global analysis of the process is required, leading to the development of a sustainable process, which cooperates with the final disposal of certain existing wastes in each region [6-9].

A particular case of a fermentable substrate is the one represented by the discards of carrot (*Daucus carota*) cultivation. The cultivated area in Santa Fe (Argentina) is approximately 1,500 ha, and its average yield is nearly 40 tons (t) ha^{-1}. Only 65%to 90% of carrot harvest meets the quality standards; consequently, during harvest time, 20 to 100 t of carrots with an optimal degree of freshness and maturity are discarded daily due to a sizing problem and then directed to animal feed [10]. Carrot is one of the most efficient crops in biomass accumulation [11], and it may be enzymatically hydrolyzed to increase yields [7]. In general, plants accumulate carbohydrates, e.g., starch. However, carrot is one of the few plants that accumulate free sugars into vacuoles (40% to 60% of total carbohydrates) as reserve carbohydrates. For free sugars, 95% is composed of sucrose, fructose, and glucose, which, together with terpenes, are the main determinants of carrot taste. Among the carbohydrates present in carrots, reducing sugars (RS) (fructose and glucose) are present in an equimolecular amount [12,13]. In addition, in the same geographical area, 1 t of *Saccharomyces cerevisiae* yeast cells is discarded daily by a beer manufacturing industry. So far, the elimination of these yeast

* Correspondence: naimaretti@fbcb.unl.edu.ar
[1]Instituto de Investigaciones en Catálisis y Petroquímica, Facultad de Ingeniería Química, Universidad Nacional del Litoral, Santa Fe 3000, Argentina
[2]Laboratorio de Investigaciones Aplicadas, Facultad de Química, Universidad del Centro Educativo Latinoamericano, Av Pellegrini 1332, Rosario 2000, Argentina
Full list of author information is available at the end of the article

cells has not been solved yet due to their elevated biochemical oxygen demand (BOD), and consequently, no process is applied to reuse them [10].

Taking into account the above considerations, the aims of this study are the following: (1) to develop a sustainable process for production of second-generation ethanol from two regional agro-industrial discards: carrot as substrate and brewer's yeast as biocatalyst, and (2) to qualitatively and quantitatively evaluate the effluents generated by the process in order to propose solutions to make it sustainable.

Methods

Raw material, handling, and storage

Discarded roots of carrot (DC) (*D. carota*) were collected in November to December from a packing shed in the Santa Fe area (31°25′S, 60°20′W), Argentina. These discards were by-products from carrot packing processes in which plant leaves were cut and whole roots were washed, dried, and selected before packing. DC were stored in shed under ambient conditions (20°C to 32°C, 50% to 60% humidity). During delivery of random representative samples obtained by sampling methods [10], carrots were packed in a polypropylene container and moved to the laboratory where they were used immediately.

Preparation and pretreatment of feedstock

Batches containing 0.5 kg of discarded carrot milled to a particle size less than 4 mm were dipped into water in a total volume of 1 L and mixed with the enzyme OptimaseCX255L (thermostable xylanase from *Trichoderma reesei*). The initial pH was then adjusted at 5.5, and the temperature was adjusted to 70°C, performing the hydrolysis during 2.5 h. After hydrolysis, they were compressed and filtered to obtain a liquid fraction and a solid residue. The pH of the liquid fraction was adjusted to 4.5 and was called carrot must (CM). The solid residue was called bagasse.

Microorganism

S. cerevisiae CCUB filtered and discarded by a local brewing industry after five operative cycles of use was obtained directly from filters. Whole yeast cells were kept in a sterile container, without nutrient addition, at 4°C and at saturation humidity for 4 days. For inoculum preparation, cells were suspended in an isotonic solution and counted by a differential direct-count method in order to calculate the volume required [14].

Fermentation conditions

Fresh must was used in every case, and its pH was adjusted to 4.5. The inoculums were adjusted to a value of 10^8 cell mL^{-1}. Batch fermentations were developed at 28°C in a 500-mL stirred tank bioreactor, equipped with controlled heating and stirring systems. The agitation speed was regulated at 100 rpm in all of the experiments. The fermentation progress was monitored following CO_2 production, which was collected in a gasometrical probe. Samples at different reaction times were taken and submitted to centrifugation. Supernatants were stored and conserved at −20°C. All experiments were performed in triplicate [7].

Recycle of the biocatalyst

For this experiment, a reactor was required and equipped with a filter which has a 3-μm pore size at the exit of products and a pump that allows the injection of nitrogen gas to facilitate the filtration process during the product output. When fermentation finished, the reactor content was filtered, leaving the cells retained in the filter. The reactor was then refilled with fresh must and then subjected to further agitation (200 rpm) for 10 min so that the yeast retained in the filter was dispersed again, and the must temperature reached 30°C to start a new fermentation cycle.

Analytical methods

Moisture

The moisture of carrots was determined using an approved method (44-15A) [10].

Sugar concentration

The concentration of RS was measured using the 3.5-dinitrosalicylic acid method. The concentration of total sugars (TS) was assayed by the same method after acid hydrolysis (1.2 mol L^{-1} HCl, at 65°C for 15 min), neutralization with 1 mol L^{-1} NaOH, and filtration [15].

Ethanol concentration

Ethanol concentration was determined by GC (Sigma 3B, dual flame ionization detector (FID) chromatograph, PerkinElmer Instruments, Branford, CT, USA). A FID and a packed column of Chromosorb 102 (2.0 m in length) were employed. The column oven was operated isothermally at 150°C, and the injection and detector ports were kept at 195°C and 220°C, respectively. Nitrogen was used as carrier gas with a flow rate of 30 mL min^{-1}, and the combustion gas was a mixture of hydrogen and air. Isopropanol (Anedra, Argentina, >99.9%) was used as internal standard [16]. In the chromatograph of every fermented sample, only two peaks appeared: one corresponding to the internal standard, and the other corresponding to ethanol.

Analysis of effluents

The analysis of effluents was performed according to the regulations applied in Argentina [17].

Parameters

Ethanol yield per carrot ($Y_{p/c}$) was considered as the ratio of total ethanol produced and used carrot (dry base) (g g^{-1}) [18]. Sodium absorption ratio (SAR) index represents the sodium adsorption and is calculated by the ratio of the sodium concentration to the square root of the halved sum of calcium and magnesium concentrations [19].

Statistical analyses

In order to examine the reproducibility of the experiment, three experimental runs of each batch operation were conducted, and determinations were performed in duplicate. The results were expressed as the average value and its standard diversion. Data were analyzed utilizing the one-way ANOVA procedure of the SPSS software. Differences between means were detected by Duncan's multiple range test. Data were considered significantly different when $P < 0.05$.

Results and discussion

Integration of the three steps of a fermentative process

In order to build a sustainable process of ethanol production from discarded carrots and yeasts, an efficient extraction of carrot sugars during the preparation of must is essential as well as an adequate fermentation process, which requires low energy demand and low equipment investment [9]. In the pursuance of these objectives, the options for the three main stages of the process are discussed.

Preparation of must

During CM preparation, the enzymatic hydrolysis of carrot tissue allowed us to obtain a must with 150.2 ± 0.6 g TS L^{-1}. This means that 151.7% of sugars were extracted, which could not have been achieved if the enzymatic hydrolysis of the tissues had not been carried out. The analysis of carbohydrate composition of the must showed that the sugars were sucrose, fructose, and glucose, all of them capable of being used by the yeast through alcoholic fermentation [7]. This stage left as discard 450 kg of bagasse from 1 t of processed DC as discard.

A viable alternative to favor hydrolysis during must preparation could be the optimization of a semicontinuous or continuous process with immobilized enzymes, which would decrease product inhibition occurring on the cellulases, thus decreasing their cost [20].

Fermentation

Fermentation of CM using free cells of *S. cerevisiae* CCUB discarded by a local beer industry allowed us to obtain an ethanol yield $Y_{p/c}$ = 0.403 g g^{-1} and 39.8 kg CO_2 which was equimolarly generated. In order to evaluate the possibility of increasing the lifetime of the biocatalyst, reducing the use of yeast cells, and avoiding the generation of a greater amount of residue, we studied the efficiency of the biocatalyst in consecutive fermentations, in which yeasts were recycled. Average concentrations of ethanol together with their standard deviations are shown in Figure 1 in which it can be observed that the ethanol yield remains constant during at least five recycles of cellular biomass, showing an increase in productivity (data not shown), since it reduces the lag phase time. Furthermore, it was observed that the carrot juice had no inhibitory effect on the strain of *S. cerevisiae* CCUB during this period. In addition to these results, further research is needed to assess the convenience of total, partial, or no replacement in order to ensure an appropriate inoculum, leaving the lysed cells as a nutrient of the must and increasing their nitrogen content.

Distillation

Finally, ethanol may be separated by simple distillation of the fermented must, leaving 970 L of a liquid effluent called vinasse per each ton of processed carrot. Subsequently, according to the use for which ethanol is intended, it may be subjected to various processes of purification and rectification. Thus, from the optimal integration of each stage, a yield of 77.5 L of second-generation bioethanol is obtained.

Characterization and disposal of process discards

The quantity and quality of discards generated by any process depend on the process conditions. Alternatives of reutilization of the bagasse and of the vinasse, depending on their chemical composition and their quantity, are shown. The remnant of yeasts and CO_2 are not analyzed in this work because the former are proposed to be recycled and the latter could be liberated to the atmosphere to be consumed by carrot plants during their growth.

Bagasse

This solid discard is composed of disintegrated carrot tissue, with an average moisture content of 67%. In a study where the primary focus was alternative food for the months of drought, Castillo and Gallardo [21] reported that carrot was classified as a voluminous nontraditional food that, due to its chemical composition (dry matter 7% to 30%, crude fiber 5% to 11%, and protein 4% to 16%), is considered a suitable food for incorporation into the diet of cattle, up to 30% in dry matter, without impairing the production of milk in cows. Furthermore, it is low in calcium and phosphorus, and its digestibility could be enhanced if it is combined with a high concentration of soluble carbohydrates (50% to 75%) as these cause high rates of ruminal fermentation. If necessary, the carrot content of proteins can be balanced with different sources of

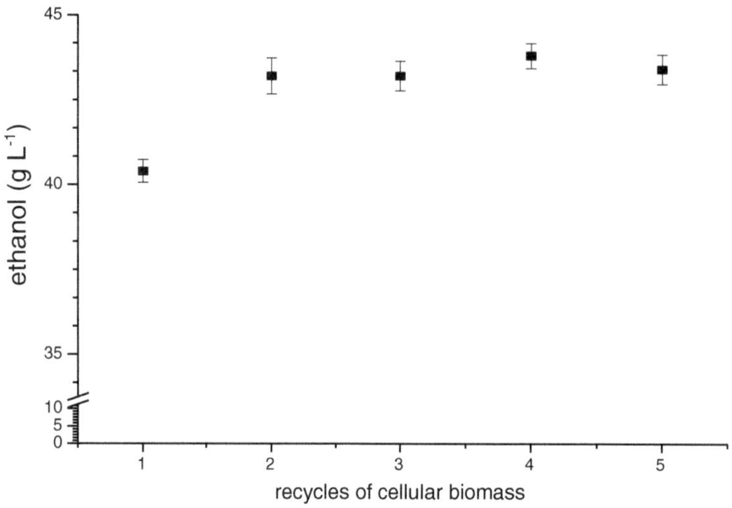

Figure 1 Ethanol concentration obtained through fermentation with recycles of cellular biomass.

nitrogen such as the yeasts used as biocatalysts in case they were discarded.

Taking into account the difficulties in bagasse transport, handling, and supply caused by its deterioration speed, this work proposes that bagasse be used to feed the animals in nearby fields, which are precisely the animals that are feeding on carrot discards today. Besides, the difficulties mentioned could be counteracted by keeping bagasse silage in combination with 15% to 25% of poultry farm waste [21]. Additionally, other alternatives for the reuse of the carrot bagasse could be its utilization as soil fertilizer or as substrate for composting.

Vinasse

Vinasse was characterized within the parameters established in law number 11220 of the Province of Santa Fe [17]. The results are shown in Table 1 (A) and are analyzed.

Taking into consideration the values of COD and BOD_5, it is clear that not only the organic load of the effluent is very high but also its biodegradability index (>0.5). Organic compounds could be reduced by applying a biological treatment such as the aerobic treatment of active mud or sewage if the final destination is a watercourse.

In this effluent, the remaining carbohydrates of the fermentation represent the most important source of organic load, and therefore, the control of this loss could reduce the environmental impact involving an increase in process efficiency and a significant reduction in the cost of final biological treatment. Furthermore, although the amount of fat detected in this effluent is very low due to the chemical composition of the substrate, its presence should be controlled so as not to influence on the effluent treatment.

Temperature control is very important because it influences on the rate of chemical and biochemical reactions, the solubility of gases and minerals, and the growth and respiration rates of aquatic organisms. The amber color, analyzed by sensory evaluation, is the consequence of the presence of beta-carotene; therefore, it is not expected to affect the passage of light.

In the same way, the sensory analysis of smell indicates that the slight yeasty odor is a result of the loss of biocatalyst in the last stage of filtering, while no harmful effects on the environment are produced. This should be a

Table 1 Characterization of liquid effluent according to law number 11220 of the Province of Santa Fe

Parameter		Value	Units
A	Chemical oxygen demand (COD)	44,000	mg O_2 L^{-1}
	Biochemical oxygen demand (BOD_5)	22,750	mg O_2 L^{-1}
	Total solid	22,600	mg L^{-1}
	Sedimentable solids (2 h)	<0.1	mL L^{-1}
	Fats	1.3	mg L^{-1}
	pH	4.4	
	Color	Amber	
	Smell	Slight	
	Temperature	80	°C
	Specific conductivity	9.68	dS m^{-1}
	Total hardness (Ca^{2+} + Mg^{2+})	700	mg CO_3Ca L^{-1}
B	Sodium	358.6	mg L^{-1}
	SAR index	18.9	SAR
	Potassium	2,270	mg L^{-1}
	Nitrogen	1,481	mg L^{-1}
	Phosphorus	150	mg L^{-1}
	Sulfate	810	mg L^{-1}

critical control point of the process so that the yeast content in the effluent can be minimized.

Of the inorganic compounds, calcium and magnesium cations, which are present in forming various types of salts, are the most important ones because they are responsible for the hardness of the waters. In addition, in relation to sodium, the SAR index was determined. The content of iron and remaining cations is generally not contemplated by the laws regulating effluent discharge. Nevertheless, the respective determinations shown in Table 1 (B) were performed in order to examine their nature and amount. The content of nitrogen and phosphorus is beneficial since they are known as the major nutrients or bio-stimulants of protists and plants so that, when they are not in the required proportions, they will be added to ensure proper biological treatment. Another important element is sulfur because it is required for protein synthesis. Toxic minerals like copper, lead, chromium, silver, arsenic, and boron were not analyzed since vinasse comes from the processing of carrot with water.

Therefore, in view of the composition of the liquid effluent of the process proposed (Table 1 (A)), it can be observed that in the event it is poured into a watercourse, it should be first cooled, neutralized, and biologically treated to reduce its organic load. However, its discharge would be subject to the dilution capacity of the receiving body and the final receiving body, as required by regional laws. Further research is needed to investigate whether the effluent liquid may be reused to be added to the enzymatic hydrolysis stage of must preparation. In this way, it may be possible to avoid the use of water to reduce the cost of correcting the pH, but the ionic strength must be taken into account because it could alter the activity of the biocatalyst. However, emphasizing the parameters in Table 1 (B) and considering that the fat content and odor are suitable, two alternative uses for the liquid effluent were evaluated: as irrigation water or as beverage for livestock. The characteristics of each of these alternatives are discussed.

Table 2 Yield of crop in ethanol production

Crop	Agricultural yield ($t\ ha^{-1}$)	Ethanol yield ($L\ t^{-1}$)	Ethanol yield ($L\ ha^{-1}$)
Sugarcane[a]	100	70	7,000
Corn[a]	8	370	2,960
Yucca[a]	20	180	3,600
Sweet sorghum[a]	35	86	3,010
Beet[a]	60	100	6,000
Carrot	40	53	2,120

[a]Adapted from Sanchez and Cardona [2].

Irrigation water Three important requirements of moisture of the carrot crops support the proposal of using vinasse as irrigation water. A progressively irrigated crop produces carrots of higher caliber because both the photosynthetic activity and the rate of sugar accumulation increase. The cultivation of carrot requires neutral or slightly alkaline grounds and is also very sensitive to salinity, to specific conductance values greater than 2 dS m^{-1} [22]. However, as disaggregated soils with not much organic matter are used, the salts present in irrigation decrease their importance as they do not accumulate in the soil.

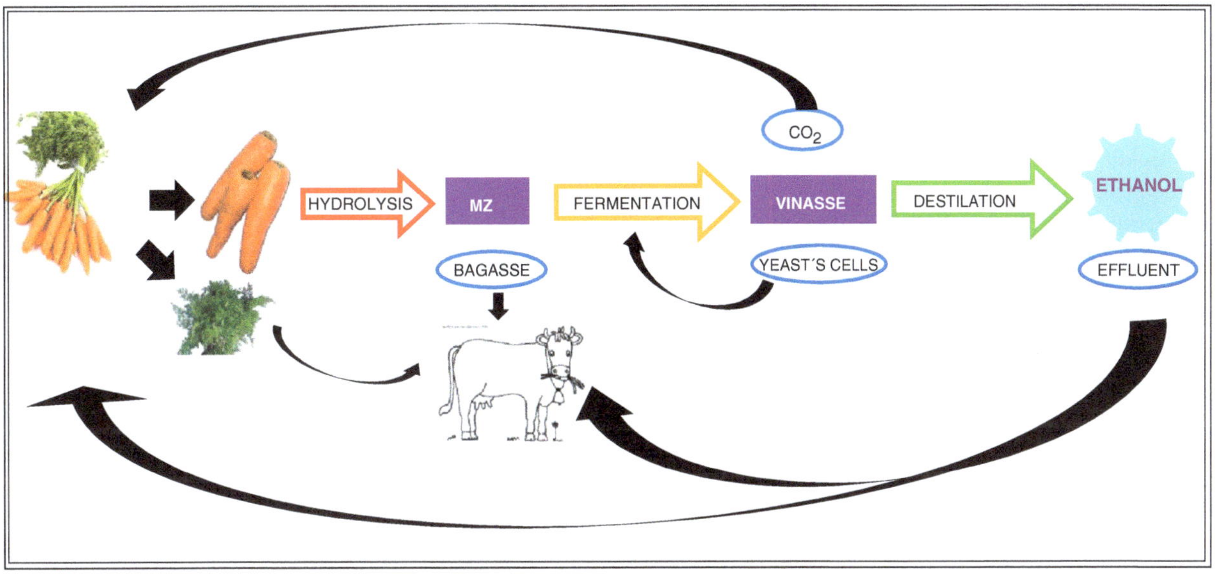

Figure 2 Integrated and sustainable process of ethanol production.

In this sense, the water quality indicator parameters developed by the consultative committee of the University of California and later extended by Ayers and Westcot [19] were analyzed. The combination of these parameters yields the degree of aptitude as irrigation water. According to these parameters, vinasse is defined as suitable for irrigation, with strict restrictions for its use due to its salinity, but there is no restriction if permeability is considered. In addition, it presents no toxic elements and has a good amount of bio-elements such as nitrogen, sulfur, and phosphorus, which are generally added as fertilizers to crops of vegetables like carrots. Since approximately 16 g L^{-1} of total solids are sugars, it is not convenient to use it directly for irrigation because it may stimulate the proliferation of some microorganisms and cause a mismatch in the soil microflora. A previous biological treatment is necessary after reducing its temperature.

As a consequence, the large amount of sodium and potassium ions could not affect the crop if the effluent is diluted to 1:10, which could be achieved by a combination with the rest of the water normally used for crop irrigation so that the pH would be consequently adjusted under these conditions. Therefore, water for irrigation would be obtained as enriched with nitrogen, phosphorus, potassium, calcium, and magnesium, which are the most recommended fertilizers for crops with good yields [22,23].

The volume obtained by diluting the liquid effluent produced by processing 1 t of carrot exceeds the requirement of water necessary to irrigate 0,025 ha of cultivation, which is the area required to harvest this amount (yield 40 t ha^{-1}), so it is necessary to explore some other application of the effluent in order to make the process sustainable.

Beverage for animals This alternative is based on the additional benefits that arise when using a beverage rich in sugar for animals that were fed with bagasse, improving digestibility while increasing the content of dietary phosphorus [21]. At this point, it is important to note that the liquid effluent can be used as a beverage for animals without submitting it to any prior biological treatment but requires a dilution of approximately 1:5 so that the content of sodium and potassium would not affect the health of livestock [24].

Design of an integrated and sustainable process

The integration of each stage allowed us to revalue one of the quantitatively most important agro-industrial discards of our region. Thus, it is possible to return to the soil and the ecosystem the majority of the elements extracted by carrots during their growth, making it a sustainable promising process for obtaining second-generation biofuel. As it can be observed in Figure 2, after harvesting 1 t of carrots, leaves were removed and used, together with the bagasse produced during CM preparation, to feed livestock. CM is then fermented in a batch process with biomass recycling from which fermented must and CO_2 are obtained, with the latter returning to the atmosphere. The yeast-free cells are reused in the same process or are employed for cattle feed if in excess. By distilling the fermented must, ethanol is separated and vinasse can be first diluted to 1:5 with water to reduce its salt content and correct its temperature and pH (which must be greater than 5.5). This gives a volume of about 4,850 L, of which 50% is required as a beverage for the same number of calfs, while the rest can be diluted again to 1:5 and used as irrigation of 0.025 ha of carrot crop.

To carry out this process, we propose the construction of a mobile module in which carrot processing operations take place: hydrolysis, fermentation, and distillation. This mobile module could be taken to the different packing shelves where carrot discards are generated. In turn, both the ethanol produced and the by-products obtained could be reused in the same temporary workplace.

Taking into consideration that the work area is characterized by its high yield per hectare of carrot, we conclude that the ethanol yield per hectare of cultivation is 3,100 L, making it comparable to the commonly employed energy crops for the production of ethanol presented in Table 2.

Conclusions

An integrated and sustainable fermentation process was designed to produce second-generation ethanol, revaluing two regional agro-industrial discards as well as by-product reuse. Discarded carrots and brewer's yeast, which are commonly used for animal feed, were processed to obtain 77.5 L of bioethanol, 450 kg of bagasse, and 970 L of vinasse per ton of discarded carrot. Bagasse was suitable for feeding 55 animals (calfs). Vinasse, after dilution with fresh water (1:5), satisfied the requirements necessary for beverage of the same number of animals, leaving a remnant which could be newly diluted (1:5) and used to irrigate a 0.025-ha carrot crop, the dimension of land required to grow 1 ton of carrot.

Abbreviations

CM: Carrot must; DC: Discarded carrot; RS: Reducing sugars; TS: Total sugars.

Competing interests

The authors declare that they have no competing interests.

Authors' contributions

NRA conceived the study, participated in its design, coordinated the tasks, helped in carrying out fermentations, and drafted the manuscript. ALC carried out the chemical analysis and interpretation of data and performed the statistical analysis. AC carried out the fermentations and interpretation of results and performed the statistical analysis. MLR participated in the design of the study and revised the results and manuscript critically for important intellectual content. JCY conceived the study, participated in its design, and helped draft the manuscript. All authors read and approved the final manuscript.

Authors' information

NRA is a biochemist. She received her Degree in Biotechnology and MSc in Food Science and Technology from UNL. She is a university professor at UCEL and received her Doctorate in Chemical Science from UNED. Her research interests are in the areas of biofuels and innovative food. ALC is an engineer in food technology and a professor and researcher in UCEL. AC is a food technologist and researcher in UCEL. MLR received her Degree in Chemical Sciences from CU and Doctorate in Chemical Science from UNED. She is a professor and researcher in UNED. JCY is a chemical engineer from UNL. He received his Doctorate in Chemical Science from UNED and is a professor and researcher in UNL.

Acknowledgments

The authors wish to thank Carolina Ybalo, Agustín Codevilla, and Mercedes Escorcia for their valuable cooperation. Thanks are also given to Agencia Santafesina de Seguridad Alimentaria for the use of their equipment. This work was carried out with the financial support of Project ALI121 from the Latin American Educational Centre University (UCEL).

Author details

[1]Instituto de Investigaciones en Catálisis y Petroquímica, Facultad de Ingeniería Química, Universidad Nacional del Litoral, Santa Fe 3000, Argentina. [2]Laboratorio de Investigaciones Aplicadas, Facultad de Química, Universidad del Centro Educativo Latinoamericano, Av Pellegrini 1332, Rosario 2000, Argentina. [3]Departamento de Química Inorgánica y Química Técnica, Facultad de Ciencias, Universidad Nacional de Educación a Distancia, Madrid 28040, Spain.

References

1. Pandyaswargo, A, Onoda, H, Nagata, K: Energy recovery potential and life cycle impact assessment of municipal solid waste management technologies in Asian countries using ELP model. Int. J. EnergEnviron. Eng. **3**, 28 (2012)
2. Sánchez, O, Cardona, C: Trends in biotechnological production of fuel ethanol from different feedstocks. Bioresour. Technol. **99**(13), 5270–5295 (2008)
3. Solbrig, O, Adámoli, J: Agro y Ambiente: una agenda compartida para el desarrollo sustentable. Foro de la cadena agroindustrial Argentina, Buenos Aires (2008)
4. Tan, KT, Lee, KT, Mohamed, AR: Role of energy policy in renewable energy accomplishment: the case of second-generation bioethanol. Energ. Pol. **36**, 3360–3365 (2008)
5. Eddine, B, Salah, M: Solid waste as renewable source of energy: current and future possibility in Algeria. Int. J. EnergEnviron. Eng. **3**, 17 (2013)
6. Pippo, W, Luengo, C: Sugarcane energy use: accounting of feedstock energy considering current agro-industrial trends and their feasibility. Int. J. Energ. Environ. Eng. **4**, 10 (2013)
7. Aimaretti, NR, Ybalo, CV, Rojas, ML, Plou, FJ, Yori, JC: Production of bioethanol from carrot discards. Bioresour. Technol. **123**, 727–732 (2012)
8. Prasad, S, Singh, A, Joshi, HC: Ethanol as an alternative fuel from agricultural, industrial and urban residues. Review. Resour. Conserv. Recycl. **50**, 1–39 (2007)
9. Santek, B, Gwehenberger, G, Ivancic Santek, M, Narodoslawskyb, M, Horvata, P: Evaluation of energy demand and the sustainability of different bioethanol production processes from sugar beet. Resour. Conserv. Recycl. **54**, 872–877 (2010)
10. Aimaretti, N: Desarrollo de un proceso sostenible para la producción de bioetanol a partir de desechos agroindustriales. Doctoral thesis, Universidad Nacional de Educación a Distancia, Madrid (2011)
11. Diamantopoulou, LK, Karaoglanoglou, LS, Koukios, EG: Biomass cost index: mapping biomass-to-biohydrogen feedstock costs by a new approach. Bioresour. Technol. **102**(3), 2641–2650 (2011)
12. Simon, P: Domestication, historical development, and modern breeding of carrot. Plant Breed. **19**, 157–190 (2000)
13. Suojala, T: Variation in sugar content and composition of carrot storage roots at harvest and during storage. Scientia Horticulture **85**, 1–19 (2000)
14. Aimaretti, N, Ybalo, C: Valorization of carrot and yeast discards for the obtention of ethanol. Biomass Bioenergy **42**, 18–23 (2012)
15. Yu, J, Zhang, X, Tan, T: Optimization of media conditions for the production of ethanol from sweet sorghum juice by immobilized *Saccharomyces cerevisiae*. Biomass Bioenergy **33**, 521–526 (2009)
16. Ratnam, B, Narasimha Rao, M, Damodara Rao, M, Subba Rao, S, Ayyanna, C: Optimization of fermentation conditions for the production of ethanol from sago starch using response methodology. World J. Microbiol. Biotechnol. **19**, 523–526 (2003)
17. ENRESS: Transformación del sector público de agua potable, desagües cloacales y saneamiento. Ley de la Provincia de Santa Fe No 11220. Gobierno de la Provincia de Santa Fe, Ente regulador de servicios sanitarios, Santa Fe (1996)
18. Colin, R, Bjorn, K: Basic Biotechnology, 2nd edn. Cambridge University Press, Cambridge (2002)
19. Ayers, R, Westcot, D: Water quality for agriculture: irrigation and drainage paper 29 (Rev. 1). Food and Agriculture Organization (FAO) of the United Nations, Rome (1985)
20. Olsson, L, Soerensen, H, Dam, B, Christensen, H, Krogh, K, Meyer, A: Separate and simultaneous enzymatic hydrolysis and fermentation of wheat hemicellulose with recombinant xylose utilizing *Saccharomyces cerevisiae*. Appl. Biochem. Biotechnol. **130**, 117–129 (2006)
21. Castillo, A, Gallardo, M: Alimentos no tradicionales en ganado lechero: consideraciones prácticas para su utilización. Información para extensión E.E. A INTA Rafaela **88**, 25–32 (1989)
22. Simon, P, Freeman, R, Vieira, J, Boiteux, L, Briard, M, Nothnagel, T, Michalik, B, Kwon, Y: Carrot. In: Prohens, J, Nuez, F (eds.) Vegetables II Handbook of Plant Breeding, Fabaceae, Liliaceae, Solanaceae, and Umbelliferae, vol 2, part 4, pp. 327–357. Springer, New York (2008)
23. García, M, Rocha, B, Gallo, A, Galván, G, Borges, A, Záccari, F: Comportamiento agronómico y calidad de poblaciones locales de zanahoria en el sur del Uruguay. Agrociencia Uruguay **16**(1), 86–96 (2012)
24. Faust, S, Aly, O: Chemistry of Water Treatment. Ann Arbor Press, New York (1998)

Fuel economy of a vehicle as a function of airspeed: the concept of parallel corridors

Charles H Heide and Pirooz Mohazzabi*

Abstract

In this investigation, it is shown that when a highway vehicle is driven at a constant speed on a level road, what actually determines its fuel economy is the airspeed of the vehicle rather than its ground speed or its speedometer reading. A theoretical model is then developed for the first time that describes the fuel economy of a highway vehicle as a function of its airspeed. The functional form of this model turns out to be in good correlation with the experimental data that have been collected by the authors over the last 2 years. Finally, the concept of parallel corridors is introduced and a design for highway construction is suggested that can significantly improve fuel economy of vehicles where traffic is heavy, by establishing a controlled driving environment. The results of this investigation apply to any vehicle using any source of energy.

Keywords: Fuel economy; Airspeed; Highway vehicles; Parallel corridors

Background

Although the invention of internal combustion engines (ICE) dates back to before 19th century, their widespread applications were hindered until mid 1850s when commercial drilling and production of petroleum began [1]. Since then, fossil fuel has been the main source of energy in transportation and travel. However, as the world became more aware of the limited supplies of fossil fuel and resource depletion as well as economical and ecological impacts of its usage, concerns began to grow. As a result, for many decades, at least as far back as the time of Great Depression in the USA, the nation has been concerned with fuel efficiency and conserving energy resources. However, the issue is not quite as simple and straightforward as it seems and there are challenges. For instance, improving fuel efficiency triggers an economic reaction causing additional travel that partially offsets the original energy saving. In other words, as the energy becomes cheaper, it provides an incentive to increase its use, a phenomenon known as the 'rebound effect' [2].

In general energy usage has direct impacts on the economy [3]. Furthermore, almost in every country in the world, transportation takes up a large portion of the fossil fuel consumption. For example, in the USA today, 70% of the total oil consumed is directed to fuels used in transportation - gasoline, diesel, and jet fuel [4]. In Nigeria, 96% of the total gasoline consumed and 40% of the diesel fuel consumed go to road transportation [5]. In addition to economical consequences, fossil fuel consumption results in the emission of toxic gases (CO, CO_2, NO_x, SO_2, HC, and PM) [6] that affects local and regional air quality, specially in urban areas, which can pose serious health hazards and cause other ecological issues. Today the emission of CO_2, which is directly related to the consumption of carbon-based fuel, is regarded as one of the most serious threats to the environment through greenhouse effect [7], and transportation alone is responsible for about 21% of the global CO_2 emission [8].

Shortages of gasoline and other petroleum fuels from time to time, as well as ecological concerns, have renewed interest in identifying factors that affect fuel efficiency of vehicles. To this end, various proposals for reducing fuel consumption have been made. These include simple modification of vehicle components such as replacement of bias-ply tires with radial tires and development of alternative fuel sources and new engine designs to optimize combustion. In addition, a noticeable trend toward smaller and more efficient vehicles has appeared in the last few decades. This is done through reducing vehicle weight by using new materials and by improving vehicles' exterior design to achieve lower aerodynamic drag [9]. Thus,

*Correspondence: pirooz.mohazzabi@uwp.edu
Department of Mathematics and Physics, University of Wisconsin-Parkside, Kenosha, WI 53141, USA

fuel efficiency has become a selling point in the automotive industry. Although most of the proposals for conserving fuel resources have been focused around vehicle design changes, some external factors have been considered as well. The national reduction of highway speed to 55 mph (88.5 km/h), mandated in the USA in the 1970s, is one example. Other examples include maintaining a constant highway speed or maintaining a constant throttle setting [10].

Oil dependency and global warming have stimulated research and development activities toward utilization of secondary biofuels and renewable energy sources including solid waste and vegetable oil [11-13]. In addition, alternative fuel sources and engines have become popular in recent years, and the endeavor toward production of the so called 'greener' vehicles continues. These include electric, hybrid, and fuel cell vehicles. Today, hybrid passenger cars are widely in use and quite popular. Electric cars have zero emission but suffer from the disadvantage of having to be recharged frequently with a long recharging time and having a relatively short range. The fuel cell vehicles, on the other hand, combine the advantages of an electric vehicle, i.e., almost zero emission, high efficiency, and silent operation, with the advantages of conventional internal combustion engine vehicles, i.e., long range and short refueling time. In addition, they have about twice the efficiency of an internal combustion engine during a typical driving cycle [14]. Although hydrogen is the most common fuel, other fuels such as natural gas and methanol have occasionally been used in fuel cells [15]. Hydrogen fuel cell vehicles, however, are very expensive due to the cost of the platinum used in them as catalyst, which has hindered their large-scale production. While fuel cell vehicles may someday replace those with internal combustion engines, various disadvantages, specially the cost of the fuel cells, mean that the market is unlikely to see their large-scale production for decades [16].

Returning to the internal combustion engine vehicles, other factors that influence fuel consumption and exhaust emission in transportation are those related to driving style and independent driving pattern [6,7,17,18]. It has been shown that nine driving pattern factors have important effect on fuel consumption and emission. These factors include acceleration and deceleration, gear changing, as well as the speed of the vehicle [18].

As a result of the ever-increasing cost of fossil fuel as well as the air pollution concerns, vehicle fuel consumption and emission have become two critical aspects in the transportation planning process of highway facilities. To this end, several mathematical models have been developed that predict vehicle fuel consumption and emissions using instantaneous speed and acceleration [19-21]. It is a well-known fact that fuel consumption (and emission) increases substantially as the speed of a vehicle increases, an issue that has been extensively discussed in the literature [22,23]. Some references have referred to this increase as being 'exponential' [24] although technically, this statement is not correct.

Although the effect of speed on fuel consumption of a vehicle has been extensively investigated, the effect of its air speed has not been studied. This is because all of the investigations in the past have focused on vehicles moving in still air, in which case the ground speed becomes the same as the airspeed. However, if the vehicle moves upwind or downwind, the results would be quite different. In fact, any experienced truck driver would testify that fuel consumption of the vehicle varies substantially between driving upwind or downwind all day and, to some extent, driving through crosswind. In this article we show that what really determines the fuel consumption of a vehicle under normal driving conditions is the airspeed of the vehicle rather than its ground speed. However, as mentioned above, if the vehicle travels in still air, its airspeed and ground speed are obviously the same.

The objective of this paper is to develop a theoretical model for the rate of fuel consumption of a vehicle (especially of a highway vehicle) as a function of its airspeed, taking into account the fuel consumption due to the moving parts of the vehicle as well. To the best of our knowledge, the effect of these factors on fuel consumption of a vehicle has not been investigated or addressed in the literature, not at least from a quantitative point of view. We then compare this model to the actual experimental data that we have collected and show that they are in good correlation. We also show that this model is consistent with those previously developed as well as the existing experimental data for vehicles moving in still air. Finally, based on our results, we suggest a new design for highway construction which could substantially reduce the rate of fuel consumption of vehicles on busy highways.

Methods

Theoretical model

The fuel consumed by a vehicle is responsible for four different types of work: (a) vehicle acceleration, which increases the kinetic energy of the vehicle, (b) vehicle climbing a hill, which increases the gravitational potential energy of the vehicle, (c) work done against all internal frictional forces as well as the rolling friction of the wheels, and (d) work done against air resistance. Here we are not interested in parts (a) and (b). We only consider cases where the vehicle travels on a straight level road with a constant speed.

Consider a vehicle traveling on a straight level road, which we take to be the x direction, at a constant speed. Assuming no air resistance and no internal friction between the moving parts of the vehicle, according to the Newton's first law of motion, the net force on the vehicle would be 0 and the vehicle would consume no fuel. But in reality, there is internal friction between the moving parts as well as the rolling friction of the tires. However, these frictional forces are all proportional to the normal forces acting between the moving parts, and the proportionality constants are all, to a good approximation, independent of the speed of the moving parts [25-27]. Therefore, for a vehicle traveling on a straight level road, to a good approximation, the contribution from these frictional forces is a constant k_0, independent of the speed of the vehicle. The value of k_0 depends on the details of the design of the vehicle.

We now turn our attention to air resistance. For a solid object moving in a fluid, the functional form of the force of air resistance is determined by the dimensionless Reynolds numbers [28,29],

$$Re = \frac{L\rho v}{\eta} \tag{1}$$

which is the ratio of inertia to viscous forces on the object. In this equation, L is a characteristic length of the object, ρ is the fluid density, and η is its dynamic viscosity. In general, when $Re << 1$, the viscous forces are much greater than the inertia forces and the latter can be neglected. In this case the magnitude of the drag force is proportional to the first power of the speed of the object relative to the fluid,

$$F_d = k_1\, v \tag{2}$$

which is known as the Stokes regime. On the other hand, when $Re >> 1$, the inertia forces dominate and the viscous forces can be neglected. In this case the drag force is in the Newton regime and its magnitude is given by a quadratic equation,

$$F_d = k_2\, v^2. \tag{3}$$

For intermediate values of the Reynolds number, both the linear and the quadratic terms are present.

For a typical vehicle, the characteristic length is of the order a few meters. The density of air at sea level and normal temperatures is about 1.3 kg/m^3, and its dynamic viscosity is about 1.8×10^{-5} Pa s. Therefore, for a typical vehicle even at speeds as low as a few kilometers per hour, say 5 km/h which is about 1.4 m/s, the Reynolds number is of the order of 10^5 to 10^6, which is much greater than unity, and the force of air resistance on the vehicle is in the Newton regime.

Therefore, for a vehicle traveling on a straight level road (the x direction) with velocity v, the total retarding force is given by

$$F = -k_0 - k_2 v^2 \tag{4}$$

where the first term is the contribution from all internal frictions in the vehicle as well as the rolling friction of the wheels, and the second term is the contribution from air resistance. Once again we stress that in this equation v is the airspeed of the vehicle, not its ground speed.

The work done by a force $\vec{F}$ on an object when the object undergoes a differential displacement $d\vec{r}$ is given by [30]

$$dW = \vec{F} \cdot d\vec{r} \tag{5}$$

Because in the case of our vehicle,

$$\vec{F} = -(k_0 + k_2 v^2)\,\hat{i} \text{ and } d\vec{r} = dx\,\hat{i} \tag{6}$$

where $\hat{i}$ is a unit vector in the direction of motion (the x direction), Equation 5 reduces to

$$\frac{dW}{dx} = -(k_0 + k_2 v^2). \tag{7}$$

Since dW/dx is the energy consumed by the vehicle per unit distance, the quantity $-(dx/dW)$ is proportional to the fuel economy Φ of the vehicle, defined as the ratio of the distance traveled to the volume of gasoline burned [31,32]. It is worth pointing out here that this quantity has occasionally been incorrectly referred to as 'fuel consumption rate' in the literature [33-35]. In terms of Φ, therefore, Equation 7 reduces to

$$\Phi = \frac{1}{c_0 + c_2 v^2}. \tag{8}$$

Note that in this equation we have used new constants c_0 and c_2, instead of k_0 and k_2, because Φ is only proportional to $-(dx/dW)$, not equal to it. This proportionality depends on a number of factors, including the efficiency of the engine and the limitations set by the second law of thermodynamics. This is because in an internal combustion engine, the chemical energy of the fuel is first converted into heat before it is converted into mechanical energy.

Equation 8 relates the fuel economy of a vehicle traveling on a straight level road to its *airspeed*. The constants c_0 and c_2 are specific to the details of the construction of the vehicle and vary from vehicle to vehicle. Before further investigation of this equation, for the sake of simplicity, we rewrite it in a slightly different form,

$$\Phi = \frac{a}{b + v^2}, \tag{9}$$

where the new constants a and b are related to c_0 and c_2 by $a = 1/c_2$ and $b = c_0/c_2$. The graph of this equation has the correct functional form for the experimental data that have been collected for various vehicles [24]. Equation 9 also conforms to the functional form of the regression model of Rakha et al. for fuel consumption rate as a function of speed [36] as well as the Oak Ridge National Laboratory experimental data [37], both for vehicles moving in still air.

Results and discussion

During several trips from Wisconsin to Arizona and back in the 1990s, one of us (CHH) noticed a marked difference in fuel consumption between the westward trips out and the eastward return trips. Based on this observation, a 2003 Chevy Suburban was equipped with a Pitot tube [38] and an airspeed indicator (Wag-Aero 0-120 mph). The Pitot tube was mounted in the front of the vehicle, flush with the front bumper and center of the radiator. The device was calibrated by driving the vehicle in still air and comparing the speedometer readings with the airspeed indicator. The agreement was remarkably good, within 2 mph (3.2 km/h). The vehicle was also equipped with an electronic device which was plugged into the OBD2 engine analyzer to read out the instantaneous fuel economy.

After several pre-runs and confirming that downwind driving results in a better fuel economy than upwind driving, from 2010 to 2013, the vehicle was driven from Wisconsin to Arizona and back. The driving mode was a steady ground speed of 70 mph (speedometer reading) or about 113 km/h, looking for flat sections of road to eliminate error due to the change of altitude. A total of 383 measurements were taken on the fuel economy of the vehicle at various airspeeds. The results are shown in Figure 1.

As can be seen from Figure 1, there is a considerable amount of scatter in the experimental data, which will be explained later. Nevertheless, the trend is quite obvious; the fuel economy decreases as the airspeed increases. A nonlinear least-squares fit of Equation 9 to the experimental data gives the following values for the constants a and b:

$$\begin{cases} a = (1.804 \pm 0.044) \times 10^5 \left(\frac{\text{km}}{\text{l}}\right)\left(\frac{\text{km}}{\text{h}}\right)^2 \\ b = (1.421 \pm 0.058) \times 10^4 \left(\frac{\text{km}}{\text{h}}\right)^2. \end{cases} \quad (10)$$

The graph of this function is also plotted in Figure 1 as a solid line. To see the goodness of the fit more clearly, we grouped the experimental data in airspeed intervals of 5 km/h. The result together with the theoretical curve with parameters given by Equation 10 is shown in Figure 2, which clearly shows that the theoretical function fits the experimental data quite well. The error bars in this figure correspond to the standard deviations of the means in averaging the airspeeds over 5-km/h intervals [39].

To see the correlation between ground speed and fuel consumption, we plotted the ground speed (speedometer reading) versus the fuel economy of our vehicle in Figure 3. Since all our data were collected at a ground speed of 70 mph (113 km/h), Figure 3 shows that there can be a large variation of fuel economy for the same ground speed,simply because the airspeed can be very different. In our case, with a ground speed of 113 km/h, the fuel economy varied from about 4.6 to 11.0 km/l. Therefore, if one collects only one data point at each ground speed, there would be essentially no correlation between the fuel economy and the ground speed. However, if we take several readings of the fuel economy at each ground speed and consider their average, we can see some correlation between the two, although this correlation is not as strong and well defined as that in Figure 1. This is due to the fact

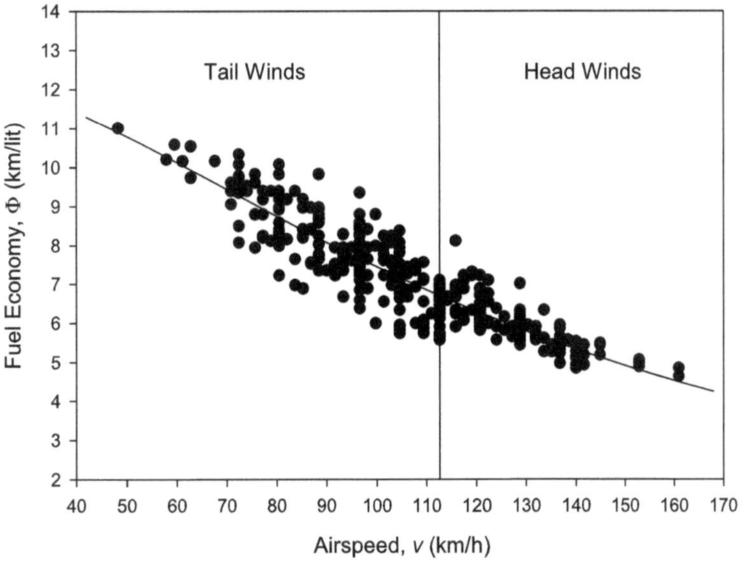

Figure 1 Fuel economy as a function of airspeed for a 2003 Chevy Suburban. Fuel economy as a function of airspeed for a 2003 Chevy Suburban driven at a constant ground speed of 113 km/h (70 mph). The markers are the experimental data and the solid curve is the nonlinear least-squares fit of Eq. (9) to the experimental data. The vertical line shows the ground speed of 113 km/h.

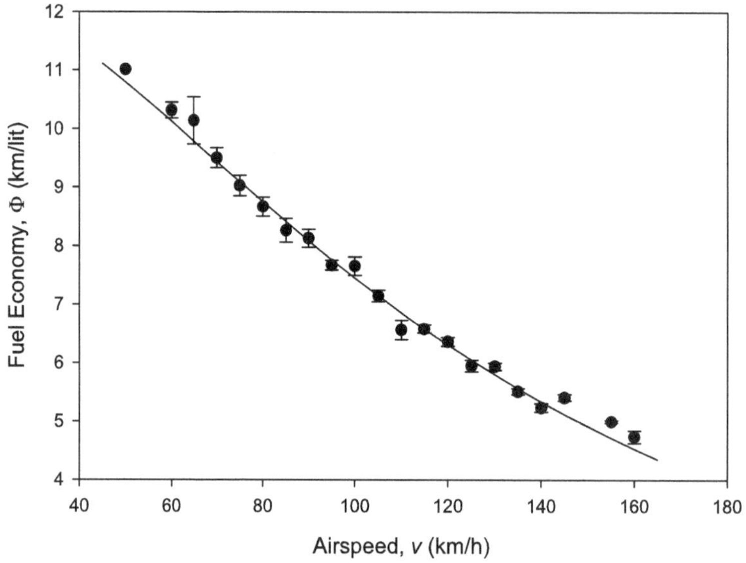

Figure 2 Fuel economy as a function of airspeed with data collected at 5-km/h intervals. Same as in Figure 1, except that the experimental data are collected in airspeed intervals of 5 km/h. The error bars indicate the standard deviations of the means as a result of averaging over 5-km/h intervals. The solid curve is the same as that in Figure 1.

that through the averaging process, ground speeds simply become the same as airspeeds.

Finally, to confirm the above results, we equipped a 2005 Volkswagen Passat for the test and collected over 200 measurements at various airspeeds, but at a ground speed of 60 mph (about 97 km/h). These measurements confirmed all the results obtained for the Chevy Suburban, but of course, with different parameters for Equation 9. The graphs for the Volkswagen Passat are qualitatively nearly identical to those for the Chevy Suburban, therefore we do not duplicate them here.

Conclusions

The results of this investigation show that our theoretical model correctly describes the experimental data on fuel economy of a vehicle as a function of airspeed. The theoretical model as well as the experimental data clearly show that the fuel economy of a vehicle does indeed depend

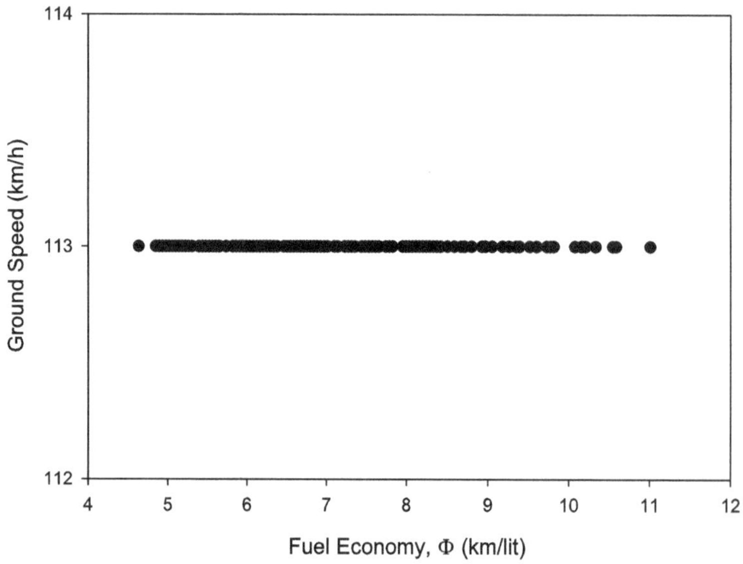

Figure 3 Ground speed as a function of fuel economy for a 2003 Chevy Suburban.

on its airspeed. The ground speed of the vehicle, on the other hand, plays little role in its fuel consumption unless the vehicle travels in still air, in which case the ground speed becomes the same as the airspeed. Furthermore, in addition to the air resistance, there is always a contribution to fuel consumption from the moving parts of the vehicle resulting in frictional forces between them as well as the rolling resistance of the wheels. This contribution, however, is a constant, independent of the speed for a given vehicle. The fuel consumption due to this contribution is equal to a/b, which is obtained by setting $v = 0$ in Equation 9. For the vehicle tested (Chevy Suburban), we have

$$\frac{a}{b} = \frac{(1.804 \pm 0.044) \times 10^5}{(1.421 \pm 0.058) \times 10^4} = 12.7 \pm 0.8 \frac{\text{km}}{\text{l}} \tag{11}$$

which would be the fuel economy of the vehicle if it traveled with an airspeed of 0.

Unlike in an aircraft, in an automobile, the Pitot tube is parallel to the airstream only when driving exactly upwind or downwind. Consequently, the automobile Pitot tube responds only to the parallel component of the airstream. Thus, in a 90° crosswind, the airspeed indicator in an automobile would read the same as the speedometer, regardless of the wind speed. This phenomenon contributes to data error, because a strong crosswind requires front wheels 'crabbing' into the wind to hold the road lane, contributing to more fuel consumption, while the airspeed indicator shows the same as the speedometer. This is one reason for the scatter in our data as shown in Figure 1. Another is the difficulty of finding a level stretch of road long enough to get stable airspeed and readings of the fuel economy.

In view of the above results, the question is how can the fuel economy of vehicles be improved? Of course, the first factor is to streamline the vehicle. This has been done by all manufacturers to the point that all new automobile profiles resemble each other. Streamlining the vehicle simply reduces the drag coefficient. Thus, the value of k_2 in Equations 3 and 4 are lowered as much as possible, which maximizes a and minimizes b in Equation 9 accordingly. The second factor is to control the driving environment in order to expose vehicles to lower airspeeds. This is where we introduce the concept of *parallel corridors*. To this end, an enormous opportunity exists on heavily traveled roads to reduce fuel consumption by significant amounts without changing the vehicle or driving habits. In addition, parallel corridors could eliminate crosswind effects as well.

Recently, on a trip west, one of us (CHH) drove on the middle lane of seven lanes of traffic leading east out of Phoenix, Arizona, on Highway 60. While driving in the center lane between closely spaced vehicles, all traveling at 70 mph (113 km/h), the airspeed indicator registered 30 to 40 mph (48 to 64 km/h), illustrating the fact that vehicles carry air along with them.

Several years of driving experience with an airspeed indicator has revealed that natural conditions sometimes resemble the parallel corridor concept. In the mountainous states, for example, lanes are sometimes separated and flanked by cuts or berms, where a substantial reduction in airspeed is noted if traffic is heavy. This unique situation was first observed on Interstate 17, between Phoenix and Flagstaff.

The first phase might be to construct a 10-km test strip on a busy lane to confirm the benefits. Design guidelines could be established for exits, snow accumulation, and barrier heights, etc. A cost-benefit ratio could be obtained to determine payback time. Various barrier types might also be evaluated. In some areas, a single center barrier might suffice.

Finally, for anyone wishing to explore the parallel corridor concept personally, we would suggest mounting the indicator in such place that both airspeed and speedometer could be read at the same time. It is customary in mounting the Pitot tube (any tube will do) that the first bend behind the opening to the air be inclined upward to prevent water (from rain) to accumulate and block pressure. In our case, a small diameter silicone rubber tube connected the Pitot tube to the airspeed indicator.

The results of this investigation apply to any vehicle regardless of its size, shape, engine design, or type of fuel consumed as well as electric vehicles.

Abbreviations

km/h: kilometers per hour; km/l: kilometers per liter of fuel; mph: miles per hour.

Competing interests

Both authors declare that they have no competing interests.

Authors' contributions

Both authors contributed equally in this investigation and writing the paper. Both authors read and approved the final manuscript.

Authors' information

CHH received his degree in Industrial Engineering in 1947, from Iowa State College. As a child during the Great Depression of 1929, he became motivated and impressed with the need for resource conservation as he grew up. He was actively involved in the Southeastern Wisconsin industries in various areas, including making fuel pumps for oil burners and making silicon rubber parts for the medical industry. He retired in 2007. As an amateur pilot, CHH became an expert in aviation and flight dynamics. He has always been interested in and an advocate of energy conservation.

PM is a professor of physics at the University of Wisconsin-Parkside. He received his PhD in Materials Science and Engineering from the University of California, Berkeley, in 1975. He has published extensively in a wide variety of areas, ranging from bicycle stability to cancer research. His current research interest is thermodynamics and statistical mechanics of small systems.

Acknowledgements

We would like to thank Adam J. Reck, Krista J. Reck, and Emma H. Jung for recording the data. We would also like to thank Kathryn H. Thompson for her help during the preparation of this manuscript.

References

1. Wikimedia Foundation Inc: History of the internal combustion engine. http://en.wikipedia.org/wiki/History_of_the_internal_combustion_engine. (2013) Accessed 7 April 2013.
2. Small, KA, Van Dender, K: Fuel efficiency and motor vehicle travel: the declining rebound effect. Energy J. **28**, 25–51 (2007)
3. Hall, CAS, Klitgaard, KA: Energy and the Wealth of Nations: Understanding the Biophysical Economy Springer Science + Business Media, LLC, New York (2012)
4. Institute for Energy Research. Petroleum: Percent of US transportation sector. http://www.instituteforenergyresearch.org/energy-overview/petroleum-oil/. (2013) Accessed 7 April 2013
5. Badmus, I, Osunleke, AS, Fagbenle, RO, Oyewola M O: Energy and exergy analyses of the Nigerian transportation sector from 1980 to 2010. Int. J. Energy Environ. Eng. **3**, 23 (2012)
6. Van Mierlo1, J, Maggetto, G, van de Burgwal, E, Gense, R: Driving style and traffic measures—influence on vehicle emissions and fuel consumption. in Proc. Instn Mech. Engrs, Part D: J. Automobile Eng. **218**, 43–50 (2004)
7. Ericsson, E, Larsson, H, Brundell-Freij, K: Optimizing route choice for lowest fuel consumption – potential effects of a new driver support tool. Transportation Res. Part C. **14**, 369–383 (2006)
8. Gorham, R: Air Pollution from Ground Transportation: An Assessment of Causes, Strategies and Tactics, and Proposed Actions for the International Community. The Global Initiative on Transport Emissions. A Partnership of the United Nations and the World Bank, New York, USA (2002)
9. Song, G, Yu, L, Wang, Z: Aggregate fuel consumption model of light-duty vehicles for evaluating effectiveness of traffic management strategies on fuels. J. Tranp. Eng. **135**, 611-618 (2009)
10. Schwarzkopf, AB, Leipnik, RB: Control of highway vehicles for minimum fuel consumption over varying terrain. Transportation Res. **11**, 279–286 (1977)
11. Sreejith, CC, Muraleedharan, C, Arun, P: Life cycle assessment of producer gas derived from coconut shell and its comparison with coal gas: an Indian perspective. Int. J. Energy Environ. Eng. **4**, 8 (2013)
12. Eddine, BT, Salah, MM: Solid waste as renewable source of energy: current and future possibility in Algeria. Int. J. Energy Environ. Eng. **3**, 17 (2012)
13. Raj, MT, Kandasamy, MKK: Tamanu oil - an alternative fuel for variable compression ratio engine. Int. J. Energy Environ. Eng. **3**, 18 (2012)
14. Ekdunge, P, Råberg, M: The fuel cell vehicle analysis of energy use, emissions and cost. Int. J. Hydrogen Energy. **23**, 381–385 (1998)
15. Wikimedia Foundation, Inc: Fuel cell (2013). http://en.wikipedia.org/wiki/Fuel_cell. Accessed 17 April 2013
16. Manzie, C, Watson, H, Halgamuge, S: Fuel economy improvements for urban driving: hybrid vs. intelligent vehicles. Transportation Res. Part C. **15**, 1–16 (2007)
17. van der, Voort, M, Dougherty, MS, van Maarseveen, M: A prototype fuel-efficiency support tool. Transportation Res. Part C. **9**, 279–296 (2001)
18. Ericsson, E: Independent driving pattern factors and their influence on fuel-use and exhaust emission factors. Transportation Res. Part D. **6**, 325–345 (2001)
19. Biggs, DC, ARFCOM: Models for Estimating Light to Heavy Vehicle Fuel Consumption. Australian Road Research Board, Victoria (1988)
20. Ahn, K, Rakha, H, ASCE M, Trani, A, ASCE M, Van Aerde, M: Estimating vehicle fuel consumption and emissions based on instantaneous speed and acceleration levels. J. Transp. Eng. **128**, 182–190 (2002)
21. Akçelik, R, Smit, R, Besley, M: Calibrating fuel consumption and emission models for modern vehicles. Paper presented at IPENZ transportation group conference, Rotorua, March 2012
22. De Vlieger, I: On-board emission and fuel consumption measurement campaign on petrol-driven passenger cars. Atmos. Environ. **31**, 3753–3761 (1997)
23. Tong, HY, Hung, WT, Cheung, CS: On-road motor vehicle emissions and fuel consumption in urban driving conditions. J. Air & Waste Manage. Assoc. **50**, 543–554 (2000)
24. MPG for speed. Speed kills MPG. http://www.mpgforspeed.com/. (2013) Accessed 16 March 2013
25. Halliday, D, Resnick, R, Walker, J: Fundamentals of Physics. 7th edition, p. 119. Wiley, New York, (2005)
26. Barger, V, Olsson, M: Classical Mechanics: A Modern Perspective. 2nd edition, p. 5. McGraw-Hill, New York, (1995)
27. Wikimedia Foundation Inc: Rolling resistance. http://en.wikipedia.org/wiki/Rolling_resistance. (2013) Accessed 16 March 2013
28. Dodge, RA, Thompson, MJ: Fluid Mechanics, pp. 426–427. McGraw-Hill, New York, (1937)
29. Mohazzabi, P: Falling and rising in a fluid with both linear and quadratic drag. Can J. Phys. **88**, 623–626 (2010)
30. Marion, JB, Thornton, ST: Classical Mechanics of Particles and Systems. 4th edition, p. 78. Saunders, New York, (1995)
31. Ross, M: Automobile fuel consumption and emissions: effects of vehicle and driving characteristics. Annu. Rev. Energy Environ. **19**, 75–112 (1994)
32. Tom Murphy: MPG for electric cars?. http://physics.ucsd.edu/do-the-math/2011/08/mpg-for-electric-cars/. (2011) Accessed 6 April 2013
33. Farlex Inc: The free dictionary by Farlex. Fuel consumption rate. http://www.thefreedictionary.com/fuel+consumption+rate. (2013) Accessed 28 March 2013
34. Dictionary.com: Fuel consumption rate. http://dictionary.reference.com/browse/fuel+consumption+rate. Accessed 28 March 2013
35. Definitions.net. STANDS4 LLC, Fuel consumption rate. http://www.definitions.net/definition/fuelconsumptionrate. (2013) Accessed 28 March 2013
36. Rakha, H, Van Aerde, M, Ahn, K, Trani, AA: Requirements for evaluating traffic signal control impacts on energy and emissions based on instantaneous speed and acceleration measurements. Transportation Research Board 79th Annual Meeting, January 9–13, 2000, Washington, D.C
37. West, B, McGill, R, Hodgson, J, Sluder, S, Smith, D: Development of data-based light-duty modal emissions and fuel consumption models. Soc. Automotive Eng. Paper No. 972910 (1997)
38. Encyclopedia Britannica Inc. Pitot tube. http://www.britannica.com/EBchecked/topic/462113/pitot-tube. (2013) Accessed 16 March 2013
39. Gould, H, Tobochnik, J: An Introduction to Computer Simulation Methods. 2nd edition, p. 355. Addison-Wesley, New York, (1996)

5

Optimization of two-step transesterification production of biodiesel from neem (*Azadirachta indica*) oil

Olugbenga Olufemi Awolu[1*] and Stephen Kolawole Layokun[2]

Abstract

This work investigated the optimization of biodiesel production from neem (*Azadirachta indica*) oil using a two-step transesterification process and determination of the qualities of the neem oil biodiesel. This was with a view to establish its production and viability potentials. Biodiesel production was carried using a two-step transesterification process. The first step was carried out using 0.60 *w/w* methanol-to-oil ratio in the presence of 1% *w/w* H_2SO_4 as an acid catalyst in 1 h at 50°C. The second step was base (NaOH) transesterification of the product from the first step using conditions specified in the optimization design. The central composite design optimization conditions for the second step were temperature (45°C to 65°C), catalyst amount (0.45% to 1.45% *w/w*), reaction time (45 to 65 min), and methanol/oil molar ratio (1.5 to 7.5). The physicochemical properties of the neem biodiesel were carried out using standard methods, while the fatty acid profile was determined using gas chromatography. Optimized biodiesel yield of 89.69% was produced at reaction time of 65 min, catalyst amount of 0.95 g, temperature of 55°C, and methanol/oil molar ratio of 4.5:1. The values for the physicochemical properties are 0.05% moisture content, 0.9 specific gravity at 25°C, 5.5 mm^2/S kinematic viscosity, 207 mg KOH/g, 70.7 g I_2/100 g iodine value, 55.31 cetane number, 39.85 MJ/Kg calorific value, 4 pour point, 8 cloud point, and 110 flash point. These values conform to international standards, in particular, American Society Testing Materials (ASTM).
It can be concluded that neem biodiesel showed a general compliance with known standards judging from its physicochemical properties and the engine test. These, coupled with its high yield, attested to the production viability and efficiency of neem biodiesel using two-step transesterification process.

Keywords: Optimization; Biodiesel; Transesterification; Cetane number; Gas chromatography

Background

Diesel fuels have an essential function in the industrial economy of developing and developed countries. The continued use of fossil fuel due to industrialization and domestic purposes especially in advanced nations resulted in environmental pollution which had contributed immensely to the present climate change, and its attendant effects had made it necessary to the development of alternative sources of energy [1]. This alternative fuel has been found to be technically feasible, economically competitive, environmentally acceptable, and readily available [1]. In addition, the fast depletion of fossil fuel and its high cost of procurement are factors that has made research into biodiesel, an alternative energy source that is clean, nontoxic, and a renewable necessity.

Biodiesel, either alone or in blends with fossil fuel, will help meet the world energy demands hitherto not yet sufficient. In addition, it is a renewable energy source which is unlike fossil fuels that is finite. Biodiesel obtained from energy crops also produces favorable effects on the environment, such as a decrease in acid rain and greenhouse effect caused by combustion [2]. Due to these factors and to its biodegradability, the production of biodiesel is considered an advantage than to that of fossil fuels [2]. Biodiesel is also nontoxic, eco-friendly, and useful in CO_2 recycling over short periods [3,4]. Biodiesel (fatty acid methyl esters (FAMEs)) derived from a variety of vegetable oils,

* Correspondence: oluawolu@futa.edu.ng
[1]Department of Food Science and Technology, Federal University of Technology, Akure, 23401, Nigeria
Full list of author information is available at the end of the article

animal fats, and waste oils may eventually be applicable as a petro-diesel and chemical substitute. Also, biodiesel evidences many salient properties, most notably cetane number, flash point, and volumetric heating value which have also been shown to be comparable to those of petro-diesel [5-8].

Many researchers have concluded that vegetable oils and their derivatives hold promise as alternative fuels for diesel engines with the only exception that the use of raw vegetable oils for diesel engines can cause numerous engine-related problems due to their high viscosity, low volatility, and high cetane number which can lead to severe engine deposits, injector coking, and piston ring sticking [9,10]. However, these effects can be reduced or eliminated through transesterification of the vegetable oil to form an alkyl ester [9,11].

The process of transesterification removes glycerin the triglycerides and replaces it with the alcohol used for the conversion process [12]. This process decreases the viscosity but maintains the cetane number and the heating value. It has also been shown that the use of biodiesel obtained from the mixture of oils for biodiesel production should be encouraged [13]. The result of the work showed that CO and HC emissions of the diesel engine when operated with the mixture as compared with high-speed diesel (HSD) were reduced by 10.97% to 21.16% and 38.76% to 47.6%, respectively. In addition, care has to be taken to ensure the storage stability of biodiesel when exposed to high temperature and air [14].

Biodiesel had been produced using homogeneous acid and base catalysts [15,16]. A two-step process which involves acid-catalyzed transesterification process and followed by base-catalyzed transesterification process has been used in converting vegetable oil with high free fatty acid (FFA) value. The first step was acid esterification and pretreatment to remove FFA in the oil, which is mainly a pretreatment process and could reduce the FFA [17,18].

The neem tree (*Azadirachta indica*) is a tropical evergreen with a wide adaptability, native to India and Burma, and been transplanted to Africa, the Middle East, South America, and Australia [19]. Neem (*A. indica*) tree is a native to tropical and semitropical regions with origin in Europe and later domesticated in Asia. It is extensively found in India and Indonesia [20]. It has been estimated that Indian neems bear about 3.5 million tons of kernels each year and that, in principle, about 700,000 tons can be recovered [4]. About 34 tons of neem seeds were exported from India in 1990 [21]. Neem oil has been extracted, and the yield is encouraging [20,22].

Response surface methodology (RSM) is a useful statistical technique which has been applied in research into complex variable processes. It employs multiple regression and correlation analyses as tools to assess the effects of two or more independent factors on the dependent variables [23]. It has been used in methanolysis optimization of sesame (*Sesamum indicum*) oil to biodiesel [24].

This work seeks to optimize neem biodiesel production from neem oil using response surface methodology. The optimization of the neem biodiesel production via a two-step transesterification process will be designed using a central composite design model of the response surface methodology. The effects of variables such as the neem oil/methanol ratio, reaction time, reaction temperature, and catalyst amount on the biodiesel yield will as well be determined in order to find out the optimum conditions required for the neem biodiesel production. The physicochemical properties and fatty acid profiles of the neem biodiesel produced will be determined using appropriate established procedures. This will help in establishing the conformity of the neem biodiesel to standards, hence its suitability as alternative fuel source.

Methods

Raw materials and reagents

The major raw material used was neem oil which was purchased from Zaria, Nigeria. The reagents are all of analytical grade.

Biodiesel production from neem oil

Experimental design

In order to optimize the central composite experimental design (CCD), a five-level-four-factor central composite design was employed for this study, which generated thirty experimental runs. The factors investigated in this study were reaction time (minutes), catalyst amount (g), temperature (°C), and oil/methanol ratio. The coded and uncoded factors (X_1, X_2, X_3, and X_4), and levels used are shown in Table 1.

Table 1 Factors and their levels for central composite design for neem biodiesel production

Variable	Symbol	Coded levels				
		−2	−1	0	1	2
Temperature (°C)	X_1	45	50	55	60	65
Catalyst amount (wt%)	X_2	0.45	0.70	0.95	1.2	1.45
Reaction time (min)	X_3	45	50	55	60	65
Oil/methanol ratio	X_4	1.5	3.0	4.5	6.0	7.5

Preheating of neem oil

The neem oil was heated fairly strongly at 80°C for 30 min using Gallenkamp magnetic stirrer thermostat hotplate (Weiss Technik, England). The aim was to reduce the viscosity of the neem oil.

Transesterification reaction

A two-step transesterification reaction was used. This was due to the fact that neem oil had a high free fatty acid value; hence, a modified method of Hanny and Shizuko [17] was used. This method included acid transesterification followed by base transesterification.

Acid transesterification The transesterification was carried out using 50 ml of methanol and 0.2 ml of concentrated H_2SO_4 mixed together inside a 250-ml conical flask. The conical flask was inserted into a water bath at 50°C. This mixture was later added to 200 ml of warmed (preheated) neem oil inside a 500-ml conical flask and placed on a magnetic stirrer with heater, continuously stirred, for 1 h for the acid transesterification to take place. The process was repeated for all the thirty samples.

Base transesterification The product of the reaction in acid transesterification was used for the base transesterification. The calculated amount of NaOH (catalyst) and methanol was added for each reaction for the temperature and reaction time specified, as shown in Table 2.

Separation of biodiesel from glycerin

After the base transesterification process, the biodiesel was allowed to settle for at least 24 h inside a separating funnel to allow clear separation of the biodiesel from glycerin. The layer on the top is the biodiesel, while the bottom layer is the glycerin. The biodiesel separation was carried out by decantation as the glycerin was drained off while the biodiesel remained.

Biodiesel washing and drying

Warm distilled water at 50°C was added to the separated biodiesel, and the mixture was shaken vigorously. The water was allowed to drain through the bottom of the separating funnel. This was carried out five times until a clear biodiesel was obtained. Anhydrous $CaCl_2$ was added to the biodiesel and heated gently at 50°C. The anhydrous $CaCl_2$ was later separated from the biodiesel to obtain a clean dry neem biodiesel. The volume of the biodiesel obtained from each sample was determined, while the percentage of biodiesel was calculated.

Statistical analysis of neem biodiesel

The data obtained in the experiments (Table 2) were analyzed using response surface methodology so as to fit the quadratic polynomial equation generated by the Design-Expert software version 8.0.3.1 (Stat-Ease Inc., Minneapolis, MN, USA). The quality of the fit of the model was evaluated using analysis of variance (ANOVA). The fitted quadratic response model is as described in Equation 1.

$$Y = b_o + \Sigma_{i=1}^{k} b_i X_i + \Sigma_{i=1}^{k} b_{iJ} X_i^2 + \Sigma_{i_1<J}^{k} \Sigma_j^k b_i X_i X_j + e, \tag{1}$$

where Y is response factor (% yield), and i and j denote linear and quadratic coefficients, respectively. b_o is the intercept, b_i is the first order model coefficient, k is the number of factors, and e is a random number.

Physicochemical analysis of neem oil biodiesel

The physicochemical analyses of the neem oil biodiesel were carried out using the Association of Official Analytical Chemists methods [25]. The analyses carried out include pour point determination (ASTM method), cloud point determination (ASTM), flash point determination (ASTM D93) using Pensky-Martens flash point tester, refractive index determination (using ATAGO Refractometer, RX-5000CX, Tokyo, Japan), specific gravity using specific gravity bottle according to AOAC method [25]), moisture content [25], viscosity (using Rion viscometer VTO4F, Netherlands), acid value [25], iodine value by Wijs method [25], peroxide value [25], saponification value [25], higher heating value (HHV) determination calculated model developed by Demirbas and Kara [26], cetane number (CN) determination using correlation reported by Patel [27] and Krisnangkura [28].

Characterization of neem oil and neem biodiesel

Fatty acid composition was determined by gas chromatography (HP 6890 powered with HP ChemStation Rev. A.09.01 [1206] software, Agilent Technologies, Inc., Santa Clara, USA). A 50 mg of the oil sample was hydrolysed for 5 min at 95°C with 3.4 ml of the 0.5 M KOH in dry methanol. The mixture was neutralized using titrimetric method by using 0.7 M HCI and then methylated using 3 ml of 14% boron trifluoride in methanol. The mixture was heated for 5 min at 90°C to achieve complete methylation process. The fatty acid methyl esters were extracted thrice from the mixture with redistilled n-hexane. The hexane extract was concentrated to 1 µl for gas chromatography analysis, and 1 µl was injected into the injection port of gas chromatography.

Table 2 Central composite design five-level-four-factor response surface study

Run	Factor 1 X_1 (°C)	Factor 2 X_2 (wt.%)	Factor 3 X_3 (min)	Factor 4 X_4	Response 1: % yield		Residual
					Actual	Predicted	
1	−1.000	−1.000	−1.000	−1.000	74.09	73.59	0.50
2	1.000	−1.000	−1.000	−1.000	73.50	73.28	0.22
3	−1.000	1.000	−1.000	−1.000	70.90	71.20	−0.30
4	1.000	1.000	−1.000	−1.000	68.00	68.53	−0.53
5	−1.000	−1.000	1.000	−1.000	84.88	85.32	−0.44
6	1.000	−1.000	1.000	−1.000	81.00	80.70	0.30
7	−1.000	1.000	1.000	−1.000	74.90	74.37	0.53
8	1.000	1.000	1.000	−1.000	68.00	67.40	0.60
9	−1.000	−1.000	−1.000	1.000	69.00	69.66	−0.66
10	1.000	−1.000	−1.000	1.000	64.50	65.14	−0.64
11	−1.000	1.000	−1.000	1.000	79.50	79.91	−0.41
12	1.000	1.000	−1.000	1.000	73.42	73.04	0.38
13	−1.000	−1.000	1.000	1.000	74.50	74.08	0.42
14	1.000	−1.000	1.000	1.000	65.50	65.26	0.24
15	−1.000	1.000	1.000	1.000	75.50	75.78	−0.28
16	1.000	1.000	1.000	1.000	64.00	64.61	−0.61
17	−2.000	0.000	0.000	0.000	75.50	75.26	0.24
18	2.000	0.000	0.000	0.000	63.71	63.77	−0.062
19	0.000	−2.000	0.000	0.000	63.26	63.31	−0.054
20	0.000	2.000	0.000	0.000	60.50	60.27	0.23
21	0.000	0.000	−2.000	0.000	87.50	86.86	0.64
22	0.000	0.000	2.000	0.000	89.69	90.15	−0.46
23	0.000	0.000	0.000	−2.000	73.50	74.02	−0.52
24	0.000	0.000	0.000	2.000	68.00	67.30	0.70
25	0.000	0.000	0.000	0.000	67.50	67.70	−0.20
26	0.000	0.000	0.000	0.000	68.00	67.70	0.30
27	0.000	0.000	0.000	0.000	67.50	67.70	−0.20
28	0.000	0.000	0.000	0.000	68.00	67.70	0.30
29	0.000	0.000	0.000	0.000	67.22	67.70	−0.48
30	0.000	0.000	0.000	0.000	68.0	67.70	0.30

Results and discussion

Optimization of neem oil biodiesel yield via response surface methodology

Table 2 showed the coded factors considered in this study, the experimental results, and the predicted values. The final equation in terms of coded factors for the central composite response surface quadratic model is shown in Equation 2.

The results of the ANOVA analysis for the response surface model (Equation 2) are shown in Table 3. The model's *F*-value of 292.39 implies that the model is significant. The *p* values <0.05 indicate that the model terms are significant. In this case, the linear coefficients X_1, X_2, X_3, and X_4; the cross-products, X_1X_2, X_1X_3, X_1X_4, X_2X_3, X_2X_4, and X_3X_4; and the quadratic coefficients, X_1^2, X_2^2, X_3^2, and X_4^2 are significant model terms (Table 3).

$$Y = 67.70 - 2.8X_1 - 0.76X_2 + 0.82X_3 - 1.68X_4 - 0.59X_1X_2 - 1.08X_1X_3 - 2.14X_2X_3 + 3.16X_2X_4 - 1.83X_3X_4 + 0.45X_1^2 - 1.48X_2^2 + 5.20X_3^2 + 0.74X_3^2 \qquad (2)$$

The effects of the factors such as reaction time, catalyst amount, temperature, and oil/methanol ratio on the percentage neem oil biodiesel yield were shown in Figures 1, 2, 3, 4, 5, and 6. Temperature had the greatest effect on the biodiesel yield, followed by the catalyst amount, reaction time, and oil/methanol ratio. The ANOVA analysis had also indicated the order of significance, giving temperature the factor that most significantly

Table 3 ANOVA for response surface quadratic model

Source	Sum of squares	*df*	Mean square	*F*-value	*p* value
Model	1,506.29	14	107.59	292.39	<0.0001
X_1	197.97	1	197.97	538.01	<0.0001
X_2	13.91	1	13.91	37.80	<0.0001
X_3	16.25	1	16.25	44.17	<0.0001
X_4	67.84	1	67.84	184.36	<0.0001
X_1X_2	5.53	1	5.53	15.04	0.0015
X_1X_3	18.51	1	18.51	50.31	<0.0001
X_1X_4	17.66	1	17.66	48.00	<0.0001
X_2X_3	73.15	1	73.15	198.78	<0.0001
X_2X_4	159.96	1	159.96	434.71	<0.0001
X_3X_4	53.33	1	53.33	144.92	<0.0001
X_1^2	5.63	1	5.63	15.31	0.0014
X_2^2	59.92	1	59.92	162.84	<0.0001
X_3^2	741.88	1	741.88	2,016.12	<0.0001
X_4^2	15.00	1	15.00	40.76	<0.0001
Lack of fit	4.94	10	0.49	4.26	0.0618
Pure error	0.58	5	12		
Corrected total	1,511.81	29			

df degree of freedom.

affected the oil yield, followed by catalyst amount, reaction time, and oil/methanol ratio.

This order of the effect of the factors on the percentage biodiesel yield agrees with that in the work of Jeong et al. [23], where their statistical response via a five-level-three-factor central composite ANOVA analysis showed that the temperature had the highest effect on the biodiesel yield with Prob > *F* value of 0.001, followed by catalyst amount with Prob > *F* value of 0.004, while neem oil/methanol oil ratio had Prob > *F* value of 0.072. In addition, the squares of the temperature, catalyst amount, and oil/methanol ratio were significant in the work by Jeong et al. [23], while only the squares of temperature and catalyst amount were significant in the present work. This could mean that those factors whose squares are significant have better effect on the biodiesel yield than those with only a linear factor that is significant. For example, reaction time, in this case, has a significant linear factor according to ANOVA, while its square factor is not significant. Also, all the cross-product terms were significant in this work and in the work of Jeong et al. [23]. In addition, a low lack of fit was noted according to the analysis of factors through ANOVA. This indicates that the model represents the actual relationship of all the parameters, which are well within the selected ranges (Table 2). In actual fact, the Prob > *F* value of 0.007 for the model is an indication of the significance of the model.

Figure 1 showed the effect of catalyst amount (X_2) and reaction time (X_3) on the biodiesel yield while keeping the other factors constant. The lowest yield was obtained around 56 min, irrespective of the catalyst amount. Highest yields were obtained at 50 and 60 min of base transesterification reaction. The catalyst amount of 0.70 and reaction time of 60 min gave the maximum yield of about 74%. At 60 min, the yield decreases with the increase of catalyst amount, while at 50 min, the yield increases slightly with the increase of catalyst amount. The effect of reaction time seems a not too important factor in biodiesel yield, as also observed by Zhang and Huang [29].

Figure 2 showed the effect of temperature (X_1) and reaction time (X_3) on the biodiesel yield. At 50 min, the yield was almost constant even with increasing temperature, but at 60 min, the yield decreases slightly with increase in

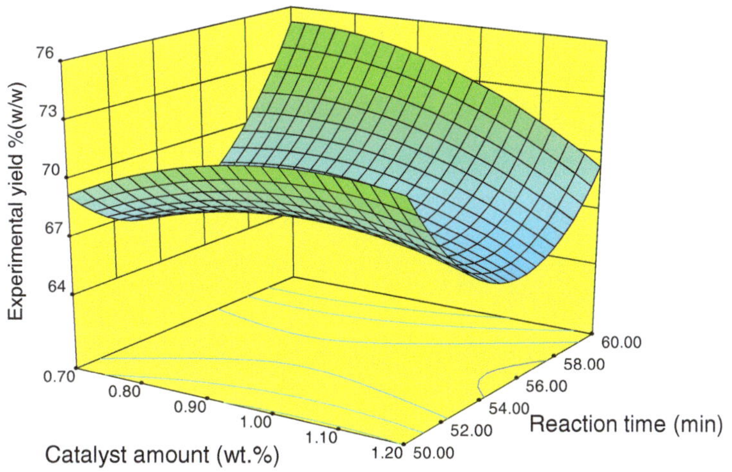

Figure 1 3D representation of effect of catalyst amount and reaction time on biodiesel yield.

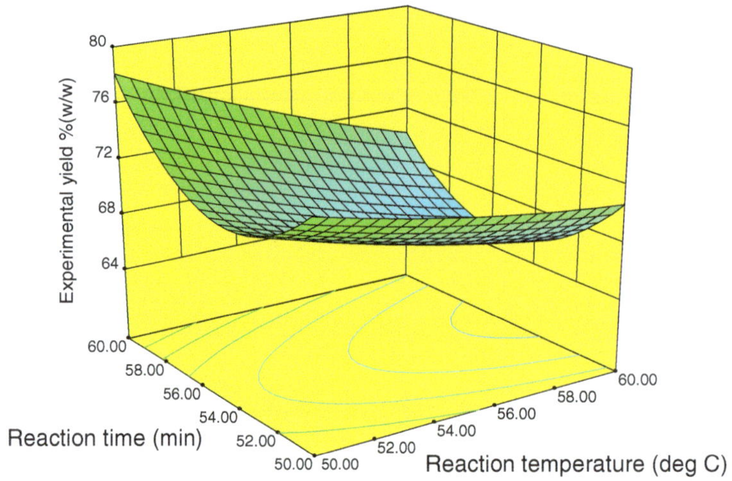

Figure 2 **3D representation of effect of temperature and reaction time on biodiesel yield.**

temperature (from about 76% to 70%). At all temperature ranges, the effect of time on the yield is the same: high at 50 and 60 min, while low at 55 min. It can be deduced that temperature had a slight effect on biodiesel yield, with the lowest temperature having the highest yield. This is actually advantageous because biodiesel production especially from neem requires low temperature, hence low energy cost.

The effect of temperature (X_1) and catalyst amount (X_2) is shown in Figure 3. The effect of temperature is stronger than that of the catalyst amount, which is also in agreement with the result of ANOVA. As previously observed, the highest yields were obtained with lowest catalyst amount in all cases. The effect of catalyst amount is slight, except in Figure 4 where its effect is more pronounced in the presence of methanol/oil molar ratio.

The effect of catalyst amount on the biodiesel yield is slightly pronounced, as shown in Figure 4. The highest yield was obtained both at the lowest catalyst amount (0.70) and lowest methanol/oil molar ratio (3.0). Conversely, the lowest yield was obtained at highest methanol/oil molar ratio (6.0) and lowest catalyst amount (0.70). This implies that at lowest catalyst amount, increasing the methanol/oil molar ratio decreases the biodiesel yield.

Figure 5 showed the effect of oil/methanol ratio (X_4) and reaction temperature (X_1). The lowest conditions in both cases produced the highest yield. As the oil/

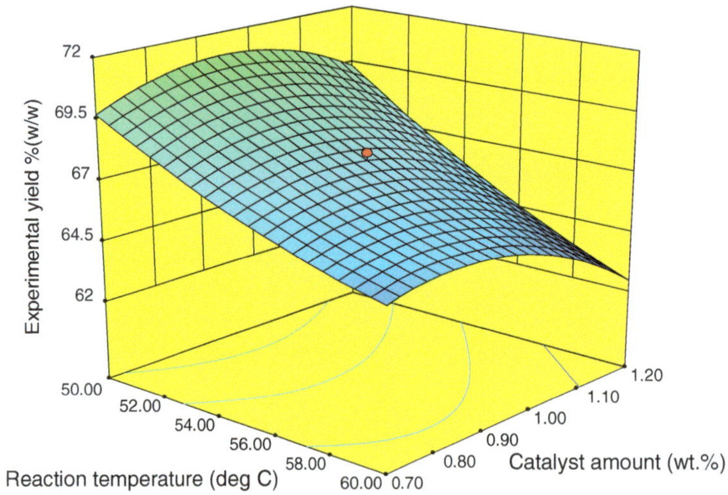

Figure 3 **3D representation of effect of temperature and catalyst amount on biodiesel yield.**

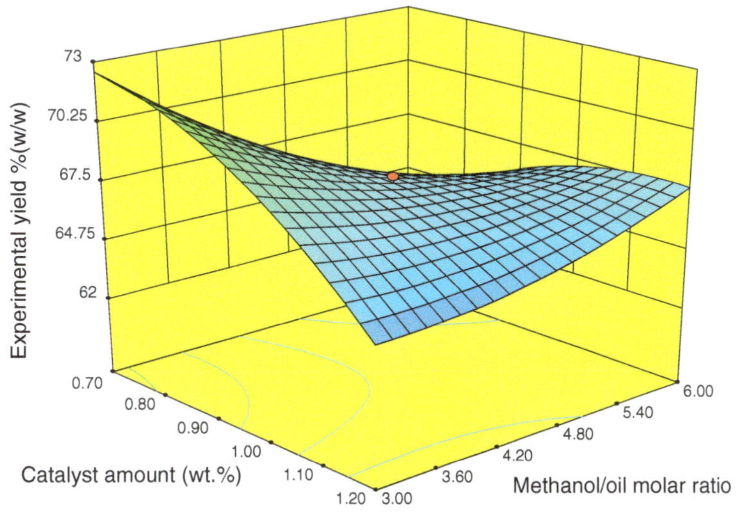

Figure 4 3D representation of effect of neemoil-methanol ratio and catalyst amount on biodiesel yield.

methanol ratio increases, oil yield decreases. At the lowest methanol/oil ratio of 3.00, the yield is almost constantly high, irrespective of the temperature.

The highest reaction time of 60 min and the lowest methanol/oil molar ratio gave the highest yield of about 78%, as shown in Figure 6. At 60 min of base transesterification, the biodiesel yield decreases slightly with increasing methanol/oil ratio, while at 50 min, the yield is almost constant.

Taking together all these results, the optimized biodiesel yield of 89.69% was produced at reaction time of 65 min, catalyst amount of 0.95 g, temperature of 55°C, and oil/methanol ratio of 4.5. The result of this work showed that the maxima yield were obtained at lowest factor values as seen in Figures 1, 2, 3, 4, 5, and 6. This will definitely have economic advantage on neem oil biodiesel production as low energy cost, low catalyst amount, low methanol/oil ratio, and low temperature are able to produce high biodiesel yield. This was in agreement with the works of Meher et al. [1], Jeong et al. [23], and Marchetti and Errazu [30].

Physicochemical analysis of neem oil biodiesel

The properties of the biodiesel obtained from this work were shown in Table 4. It is compared with the American standard (ASTMD6751) and European specification (EN14214). The density, flash point, iodine number, cetane number, pour point, moisture content, and the calorific value conform to the approved international standards.

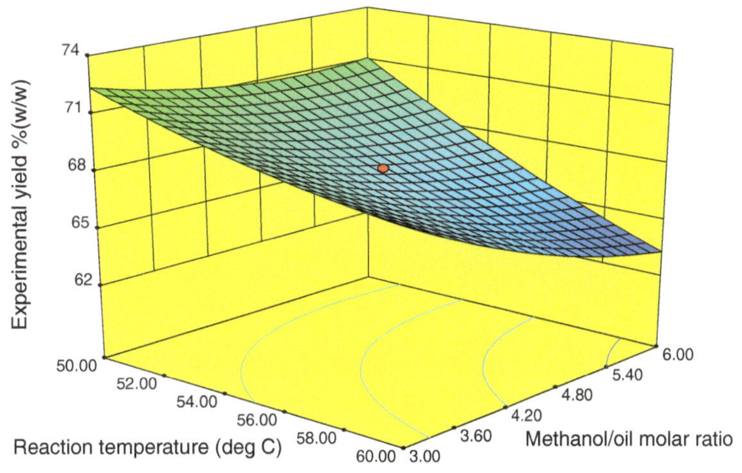

Figure 5 3D representation of effect of neemoil-methanol ratio and temperature on biodiesel yield.

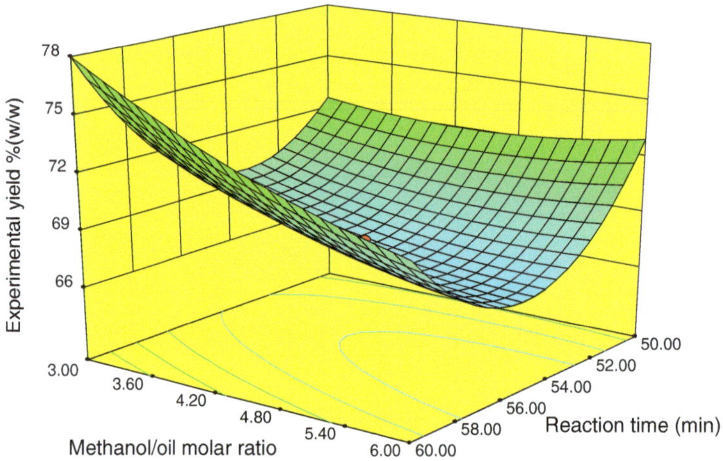

Figure 6 3D representation of effect of neemoil-methanol ratio and reaction time on biodiesel yield.

The viscosity also conformed to the American standard. The amount and type of the fatty acid content in the biodiesel are the major factors that determine the viscosity of biodiesel. In addition, the neem oil and neem oil biodiesel compared with vegetable oil from other sources.

Gas chromatography analysis of neem oil biodiesel

The result of the chromatography analysis of neem oil biodiesel was shown in Table 5. The percentage of unsaturated fatty acid is 80.02%, while the percentage of saturated fatty acid is 19.97%. It has been shown that biodiesel fuel with more unsaturated fatty acid composition has more density but has less viscosity, lower cetane number, and heating value. It also has lower thermal efficiency compared to high saturated fatty acid composition. It emits lower HC and CO, and less smoke compared to highly saturated biodiesel fuel.

Conclusion

The maximum neem oil methyl ester (NOME) conversion yield was validated as 85.13% (*w*/*w*) under the optimal reaction condition of 3:1 methanol/oil molar ratio, 50°C reaction temperature, 0.70% catalyst concentration, and 60-min reaction time. The fuel properties of NOME satisfied both the ASTM D 6751 and DIN EN 1424 standards. Thus, the present study demonstrates the usefulness of RSM for optimum conversion of *A. indica* seed oil to NOME. It also suggests that the seed oil could be used effectively as feedstock for NOME production.

Table 4 Physicochemical analysis of neem oil biodiesel compared with standards

Properties	Values	ASTM specification	EN 14214
Physical state at 25°C	Liquid	-	-
Percent moisture content	0.05	0.05 max	-
Specific gravity at 25°C	0.900	0.87 to 0.90	0.86 to 0.90
Kinematic viscosity (mm^2/s)	5.5	1.9 to 6.0	3.5 to 5.0
Saponification value (mg KOH/g)	207	-	-
Iodine value (g I_2/100 g)	70.5	-	120 max
Cetane number	55.31	40 min	51 min
Calorific value (MJ/kg)	39.89	-	-
Pour point (°C)	4	−15 to 10	-
Cloud point (°C)	8	−3 to 12	
Flash point (°C)	110	100 min	120 min

ASTM American society for testing materials, *EN* European standard for biodiesel, *max* maximum, min minimum.

Table 5 Fatty acid profile of neem oil biodiesel

Fatty acid	Percent composition
Unsaturated fractions	
Oleic acid (C18:1)	18.17
Linoleic acid (C18:2)	61.28
Linolenic acid (C18:3)	0.51
Erucic acid (C22:1)	0.00
Palmitoleic acid (C16:1)	0.06
Saturated fractions	
Palmitic acid (C16:0)	10.82
Stearic acid (C18:0)	9.15
Arachidic acid (C20:0)	0.00
Behenic acid (C22:0)	0.00
Lignoceric acid (C24:0)	0.00
Others	0.01
Unsaturated/saturated ratio	80.02:19.98
Total	100%

Competing interests
Both authors declare that they have no competing interests.

Authors' contributions
OOA designed the work, carried out the laboratory work, analyzes and carried out the statistical analysis of the result, and wrote the manuscript. SKL supervised the work, and corrected the write ups. Both authors read and approved the final manuscript.

Authors' information
OOA is a Lecturer II from the Department of Food Science and Technology, Federal University of Technology, Akure, Nigeria. He holds a Ph.D. degree in Chemical Engineering from Obafemi Awolowo University, Ile Ife, Nigeria. His research interests are in the area of food rheology, renewable energy from food wastes and underutilized crops, and optimization of food processes and extrusion processes. SKL is a Professor of Chemical Engineering, Obafemi Awolowo University, Ile Ife, Nigeria, a Fellow of Nigerian Society of Chemical Engineers (FNSChE), a Fellow of Nigerian Society of Engineers (FNSE), and a member of Biochemical Engineering Subject Group (BESG) of Ichem E, UK. He studied Chemical Engineering (Diplom-Ingenieur) in the Polytechnic Institute of Bucharest Romania (1964 to 1970). In 1971, he proceeded to the University College London from where he bagged M. Sc. D.U.C in Biochemical Engineering in 1972. He obtained Ph.D. DIC in Biochemical Engineering (1975) from the Imperial College London. He is generally interested in reactors and bioreactors, in particular. This captures microbial reaction kinetics and its application in design, modeling, and optimization. It also includes microbial production of metabolites (bioethanol, biobutanol, etc.), biodiesel through alcoholysis and enzymatic route, biopolymer production, and bioremediation. He is currently interested in the development of processes for lignocellulosic bioethanol production which also has to do with hydrolytic enzyme production.

Acknowledgements
Authors acknowledge Mr Igbe Festus of the Department of Biochemistry and Mr Oshosana of Department of Mechanical Engineering, Federal University of Technology, Akure, for assistance in carrying out some laboratory analyses.

Author details
[1]Department of Food Science and Technology, Federal University of Technology, Akure, 23401, Nigeria. [2]Department of Chemical Engineering, Obafemi Awolowo University, Ile Ife, Nigeria.

References
1. Meher, LC, Vidya Sagar, D, Naik, SN: Technical aspects of biodiesel production by transesterification–a review. Renew. Sustain. Energy Rev. **10**, 248–268 (2006)
2. Antolin, G, Tinaut, FV, Briceno, Y, Castano, V, Perez, C, Ramirez, AI: Optimisation of biodiesel production by sunflower oil transesterification. Bioresour. Technol. **83**(2), 111–114 (2002)
3. Graboski, MS, McCormick, RL: Combustion of fat and vegetable oil derived fuels in diesel engines. Prog. Energ. Combust. **24**, 125–164 (1998)
4. Jeong, GT, Park, DH, Kwang, CH, Lee, WT, Sunwoo, CS, Yoon, CH, Choi, BC, Kim, HS, Kim, SU, Lee, UT: Production of biodiesel fuel by transesterification of rapeseed oil. Biotechnol. Appl. Biochem. **114**, 747–758 (2004)
5. Ma, F, Hanna, MA: Biodiesel production: a review. Bioresour. Technol. **70**, 1–15 (1999)
6. Lang, X, Dalai, AK, Bakhshi, NN, Reaney, MJ, Hertz, PB: Preparation and characterization of bio-diesels from various bio-oils. Bioresour. Technol. **80**, 53–62 (2001)
7. Usta, N: Use of tobacco oil methyl ester in turbocharged indirect injection diesel engine. Biomass. Bioenerg. **28**, 77–86 (2005)
8. Jeong, GT: Park, DH: Response surface methodological approach for optimization of enzymatic synthesis of sorbitan ethacrylate. Enzyme Microb. Technol. **39**(3), 381–386 (2006)
9. Perkins, LA, Peterson, CL: Durability Testing of Transesterified Winter Rape Oil (*Brassica napus* L.) as Fuel in Small Bore, Multi-Cylinder, DI, CI Engines. Society of Automotive Engineers, Warrendale (1991)
10. Scholl, KW, Sorenson, SC: Combustion of Soybean Oil Methyl Ester in a Direct Injection Diesel Engine. Society of Automotive Engineers, Warrendale (1993)
11. Zhang, Q, Feldman, M, Peterson, C: Diesel Engine Durability when Fueled with Methyl Ester of Winter Rapeseed Oil. American Society of Agricultural Engineers, Washington, DC (1988)
12. Van Gerpen, JH, Hammond, EG, Yu, L, Monyem, A: Determining the Influence of Contaminants on Biodiesel Properties. Society of Automotive Engineers, Warrendale (1997)
13. Raheman, H, Jena, PC, Jadav, SS: Performance of a diesel engine with blends of biodiesel (from a mixture of oils) and high-speed diesel. Int. J. Energy Env. Eng. **4**, 6 (2013). doi: 10.1186/2251-6832-4-6
14. Mazumdar, P, Borugadda, VB, Goud, VV, Sahoo, L: Effect of storage parameters on stability of *Jatropha*-derived biodiesel. Int. J. Energy Env. Eng. **4**, 13 (2013). doi:10.1186/2251-6832-4-13
15. Noureddini, H, Zhu, D: Kinetics of transesterification of soybean oil. J. Am. Oil Chem. Soc. **74**(11), 1457–1463 (1997)
16. Narasimharao, K, Lee, A, Wilson, K: Catalyst in production of biodiesel: a review. J. Biobased Mater Bio. **1**, 19–20 (2007)
17. Hanny, JB, Shizuko, H: Biodiesel production from crude *Jatropha curcas* L. seed oil with a high content of free fatty acids. Bioresour. Technol. **99**, 1716–1721 (2008)
18. Awolu, OO, Odedeji, D: Production of fatty acid methyl ester (biodiesel) from waste cooking oil using transesterification methods. App. Trop. Agric. **16**(1), 23–29 (2011)
19. Ogbuewu, IP: Physiological responses of rabbits fed graded levels of neem (*Azadirachta indica*) leaf meal. M.Sc. thesis. Federal University of Technology, Owerri (2008)
20. Liauw, MY, Nathan, FA, Widiyanti, DI, Indraswati, N, Soetaredjo, FE: Extraction of neem oil (*Azadirachta indica A. Juss*) using n-hexane and ethanol: studies of oil quality, kinetic and thermodynamic. J. Eng. Appl. Sci. **3**(3), 49–54 (2008)
21. National Research Council (NRC): Neem: A Tree for Solving Global Problems. National Academic Press, Washington, DC (1992)
22. Awolu, OO, Obafaye, RO, Ayodele, BS: Optimization of solvent extraction of oil from neem (*Azadirachta indica*) and its characterizations. J. Sci. Res. Rep. **2**(1), 304–314 (2013)
23. Jeong, GT, Yang, HS, Park, DH: Optimization of transesterification of animal fat ester using response surface methodology. Bioresour. Technol. **100**, 25–30 (2009)
24. Betiku, E, Adepoju, TF: Methanolysis optimization of sesame (*Sesamum indicum*) oil to biodiesel and fuel quality characterization. Int. J. Energy Env. Eng. **4**, 9 (2013). doi: 10.1186/2251-6832-4-9
25. AOAC: Official Methods of Analysis, 17th edn. Association of Official Analytical Chemists, Inc, Maryland USA (2000)
26. Demirbas, A, Kara, H: New options for conversion of vegetable oils to alternative fuels. Energ. Sour. Part A **28**, 619–626 (2006)
27. Patel, V: Cetane Number of New Zealand Diesel Report. Office of Chief Gas Engineer, Energy Inspection Group, Ministry of Commerce Press, Wellington, New Zealand (1999)
28. Krisnangkura, K: A simple method for estimation of cetane index of vegetable oil methyl esters. J. Am. Oil Chem. Soc. **63**(4), 552–553 (1986)
29. Zhang, XW, Huang, W: Optimization of the transesterification reaction from cottonseed oil using a statistical approach. Energ. Source Part A **33**, 1107–1116 (2011)
30. Marchetti, JM, Errazu, AF: Esterification of free fatty acids using sulfuric acid as catalyst in the presence of triglycerides. Biomass. Bioenerg. **32**, 892–895 (2008)

Assessment of micro-wind turbines performance in the urban environments: an aided methodology through geographical information systems

Antonio Gagliano*, Francesco Nocera, Francesco Patania and Alfonso Capizzi

Abstract

Among renewable energy sources, the electrical generation from micro-wind turbines has not yet disclosed its huge potential especially in urban settings. Increasing the spread of micro-wind turbines not only promotes the decentralized generation of energy, but also helps tackle fuel poverty and to achieve reductions in the emission of greenhouse gases (GHGs).

This work proposes an innovative methodology to exploit wind flow fields, calculated by means of computational fluid dynamic (CFD) codes in urban environments, within the geographical information system (GIS) platform. In this way, the platform of users is amplified, even non-specialist users, that can utilize wind data to evaluate the potential production of electricity using micro-wind turbines. A pilot study was conducted for assessing the applicability of the approach in a Sicilian city. The results of this case study show the energy yield produced from a building-mounted wind turbine (BUWT). The developed methodology permits to enrich the information usually stored in the GIS platform allowing to supply useful information about suitable sites where micro-wind energy plants can be installed and to assess the production of renewable energy in the urban settings.

Keywords: Renewable energy; Urban environment; Wind turbine; GIS

Background

Power distributed generation is widely seen as a beneficial development in terms of both energy security and decarbonization for the electricity production.

In an urban environment, micro-generation systems [1] represent an opportunity in terms of development for renewable energy sources, research, technological innovation, and resource efficiency. It is current opinion [2,3] that governments' policies should promote the adoption of widespread micro-scale power generation in the urban environments.

Furthermore, the integration of energy and environmental models in urban planning tools is strongly recommended for the development of sustainable communities [4].

In this context, geographical information systems (GISs) are today commonly used in urban and regional scale planning.

The GIS software involves the use of some combination of digital maps and georeferenced data that describe different features of a given area at different scales. It is used for diverse applications: from urban to infrastructure planning, social and economic analysis, and recently also to increase the penetration of renewable energy technologies in urban areas. GIS can support the process of renewable energy source (RES) integration in many ways, such as finding potential areas for solar, biomass, and wind power plants; estimating the RES potential at macro- and micro-scales integrating physical modeling; and analyzing the technical, economic, and social impact of RES projects [5-7].

Therefore, it is interesting to exploit the specific features of the GIS software in order to calculate spatial energy intensity, identify potential sites or land constraints, as well as provide statistical analysis in a straightforward way [8-10].

* Correspondence: agagliano@dii.unict.it
Department of Industrial Engineering, University of Catania, Viale Andrea Doria, 6, 95125 Catania, Italy

The GIS is becoming a very important tool for supporting decision-making in intelligent cities on different sectors. In the era of smart cities, the digital maps and geospatial databases are integrated in workflows in land management, urban planning, and transportation [11].

Currently, micro-wind turbines (WTs) are an emerging technology energy generation in urban area, so there is relatively little empirical knowledge concerning their performance and the corresponding energy yield potential, particularly in the built environment [12,13]. Moreover, micro-WTs with roof-mounting options have recently become available, but their suitability for roof-mounting is still a questionable matter, particularly in urban environments, due to the complexity of the wind distribution [12,14]. One of the barriers for the diffusion of WTs in urban settlements is the difficulty to estimate the wind resource, which is highly site-specific and less predictable than the solar resource. As an example, available wind maps like the 'Wind atlas of Italy' [15] or the Wind Energy Resource Atlas of the United States [16] do not take into account local topographical features such as the presence of buildings, trees, and other obstacles, which cause the variation of wind velocity.

The aim of this paper is to present a GIS-based methodology able to support decision making in energy supply from RESs and energy policies at the urban level. GIS-based planning systems allow planners and citizens to quickly and efficiently create and test alternative energy supply scenarios.

The authors hope that this methodology will be able to support a higher penetration of renewable energy technologies in urban areas.

To achieve this objective, the following steps were implemented: measurement of wind velocity for the site characterization, computational fluid dynamic (CFD) simulation for the calculation of the wind flow within the built area, construction of the digital elevation model (DEM) of the urban area, transferring of wind flow data into the GIS software package.

As a result, local administrations may benefit from an urban wind map in the GIS environment that can be easily consulted, in order to set the wind data necessary for calculating the yearly energy production for a given WT. These issues will be shown in this paper and applied to a pilot study.

Methods

Wind energy

The power generated by a wind turbine can be expressed as [17]

$$P = \frac{1}{2} C_p \rho v^3 f(v), \tag{1}$$

where ρ is the air density, v is the wind speed, $f(v)$ is the wind speed distribution, C_p is the power coefficient of the rotor. The power coefficient is defined as the ratio between the energy extracted and the wind energy available, and its maximum value is 16/27 (Betz limit). The real C_p is much lower than the Betz limit: especially for micro-WTs, values in the range 0.2 ÷ 0.25 can be assumed. The value of C_p also includes mechanical and electrical losses; thus, the real value is always lower than the theoretical limit.

Wind speed is the most important parameter to be considered when addressing the design and the operation of wind turbines, since its probability density distribution greatly affects the performance. The typical two-parameter Weibull is a flexible distribution that is useful for describing unimodal frequency distributions of wind speeds at many sites [18].

The probability density function of Weibull can be expressed as follows:

$$f(v) = \frac{k}{A}\left(\frac{v}{A}\right)^{k-1} e^{-\left(\frac{v}{A}\right)^k}, \tag{2}$$

where $f(v)$ is the probability of observing a wind speed v, k is the Weibull distribution shape factor, and A is the Weibull distribution scale factor.

The Weibull distribution is important for the assessment of the wind energy potential and wind characteristics; therefore, it is necessary to determine its parameters properly. In this case, the Weibull distribution not only agrees better with wind speed data, but also represents the wind power potential much more accurately.

The two parameters A (m/s) and k (dimensionless) that administer the profile of the Weibull distribution curve are determined from the long term wind data for the candidate site. Higher values of A mean that the distribution is spread over a wider range. On the other hand, if the value of the shape parameter k is between 1 and 2, then the distribution is tilted towards lower wind speeds whereas higher values of k provide a distribution skewed towards higher wind speeds [19-21]. Several techniques can be used to determine the Weibull parameters to fit Weibull distribution to the measured data at a certain location [22,23].

Shape and scale factors can be determined by the mean wind speed standard deviation method using the following equations:

$$k = \left(\frac{\sigma}{V_{av}}\right)^{-1.086}, \tag{3}$$

$$A = \frac{V_{av}}{\Gamma(1 + {}^{1}/_{k})}, \tag{4}$$

where Γ is the gamma function, V_{av} is the mean wind speed and σ is the standard deviation calculated by

$$V_{av} = \frac{1}{N}\sum_{i=1}^{N} V_i, \tag{5}$$

$$\sigma = \left[\frac{1}{N-1}\sum_{i=1}^{N}(V_i - V_{av})^2\right], \tag{6}$$

where N is the number of data.

The annual energy can be calculated by multiplying the power output of the wind turbine *P(v)* for each wind speed for the total number of hours when the velocity v occurs.

$$E_v = 8,760\sum f(v)P(v) \tag{7}$$

Power curves *P(v)* depict the relationship between wind speed and power in the range of operation of the turbine. It should be noted that the above equation does not take into account the dynamics of the turbine in responding to rapid changes in wind speed; furthermore, it neglects the effects of rotor dynamics and the various mechanical devices used to avoid overspeeding and the damage of WTs for high wind speeds.

Micro-wind turbines

WTs can be classified into horizontal axis wind turbines (HAWTs) and vertical axis wind turbines (VAWTs). HAWTs are very sensitive to the changes in wind direction and to the turbulence phenomena which cause a negative effect on their energy performance. Rotor needs to be positioned in the wind direction by means of a tail or a yaw motor. Optimum locations for HAWTs are open areas with smooth air flow and few obstacles.

The variation of wind direction has less negative effects for VAWT because they do not need to be positioned into the wind direction. VAWT are usually designed for the deployment in the urban environment. Mertens et al. [24] conclude that certain VAWTs are preferable to HAWTs for roof-mounting upon (high) buildings. VAWTs do not suffer as much from reduced energy outputs, as a result of frequent wind direction changes, whereas HAWTs must yaw and track the wind to be able to extract energy economically.

Although long-term data about the performance of VAWTs are currently not widely available, VAWTs have (in theory) several advantages over HAWTs such as follows:

- Less maintenance as they can operate at lower rotational speed; there is no yawing mechanism to turn the blades towards the wind (fewer and slower moving parts); the generator is often placed under the rotor or on the ground, and so it is easy for repair and maintenance;
- Less aerodynamic noise as they operate at lower tip speeds;
- Turbulence and winds from all directions are handled more effectively, which can be a very important factor for WTs installed in residential area;

Integration of WTs in urban area must be considered from the aesthetic point of view, and all the installed energy systems should be adapted to the building architecture.

Usually, regulations do not permit WTs mounting on roof top of buildings in conservation areas or in world heritage sites [25]. Building-mounted wind turbines (BUWTs) must overcome some fundamental issues that are less binding for other micro-generation technologies, such as mounting method, vibration transmission to building, noise effect, flicker affecting neighbouring houses, shadows and reflections, maintenance access to the turbine, electromagnetic and/or electrical interferences, driver distraction, and so on. In this context, it is useful to report a Summary of Planning Guidelines suggested by the UK Government for the installation of micro-wind turbines [3].

Wind turbines on normal buildings are permitted if

- < 3-m above ridge (including the blade) and diameter of blades < 2 m;
- Internal noise < 30 dB;
- External noise < 40 dB;
- 'Garden' noise < 40 dB;
- Up to four turbines on buildings > 15 m (as with antennas);
- Vibration < 0.5 mm/s.

No roof top-mounted turbines will be allowed on buildings in conservation areas or world heritage sites. Concerning noise pollution, in Italy rules are very strict in the urban environment and set limits for the urban noise, both diurnal and nocturnal, through the equivalent continuous sound level (LAeq). Typically, cumulative turbine noise limits are daytime 55 dB(A) and night time 45 dB(A) [26].

BUWT installations can also perform a complementary function beside photovoltaic (PV) plants considering the seasonal anticorrelation in the time pattern of the wind and solar resources [27].

Computational fluid dynamic

CFD is applied in areas ranging from cooling nuclear reactors to characterization of environmental pollution. In this study, CFD will be used to investigate wind dynamics around urban areas and to identify sites that have suitable wind energy potential for electric power production. There is a wide range of researches based on numerical simulations using CFD to investigate the natural wind characteristics in urban settings [28-31].

Other studies tackle the influence of various calculation conditions (such as the size of computational domain, grid resolution, boundary conditions, choice of turbulence model, and so on) on the results of CFD simulation [32].

CFD codes solve complete the three-dimensional Navier–Stokes equation for describing the motion of a real fluid. The final expression of the Navier–Stokes equation is originated from the stress–strain relationship of Newtonian fluids:

$$\tau_{ij} = P\delta_{ij} - \mu\left(\frac{\partial v_j}{\partial x_i} + \frac{\partial v_i}{\partial x_j}\right), \tag{8}$$

where

- τ_{ij} is the tensor of the deformations with normal j and direction i;
- P is the pressure;
- δ_{ij} is the Kronecker delta which is zero if $i = j$ and 1 if $i \neq j$; and
- μ is the viscosity.

The term in brackets represents the rate of angular deformation of the fluid. The Navier–Stokes equation is obtained from the motion equation which is given below:

$$\rho\frac{dv_i}{dt} = \rho f_i - \frac{\partial \tau_{ij}}{\partial x_j} \tag{9}$$

where

- ρf_i is the mass forces (gravitational forces) and
- $\frac{\partial \tau_{ij}}{\partial x_j}$ is the term which considers the tensor changes in internal fluid volume.

The Navier–Stokes written in its classical formulation takes the form

$$\rho\frac{dv}{dt} = \rho f - \nabla P + \mu\nabla^2 v, \tag{10}$$

where

- ∇P = pressure forces
- t = time
- v = wind velocity

The Navier–Stokes equation is usually solved numerically because of their non-linearity by using the processing technique called 'at large vortices' (Large Eddy Simulation-LES).

In this study, the 3-D wind flow modeling and visualization software Virtual Wind [33] was used to calculate the wind flow field in the urban area.

The large eddy simulation is solved with the Smagorinski model (sub-grid scale) that uses a second-order scheme for the convective terms in the equations of moment and a second-order centered difference scheme for diffusive terms [34].

The corrector step, which requires the continuity equation, needs the solution of the Poisson equation for pressure. The virtual wind uses a direct FFT method to solve this equation, making this step not iterative. The resolution scheme uses the predictor-corrector method and as corrector uses a Poisson fast Fourier transform solver [35].

Wind flow in the built environment

A crucial factor in the development, siting, and operation of a BUWT is the ability to assess and characterize available wind resources.

Even small reductions of velocities cause significant loss of power, as Equation 1 indicates, so extreme care has to be taken in locating the optimal installation spot for turbines. Wind flow patterns change significantly in any direction in built environment: areas separated by very small distances can be characterized to very different wind speeds. Buildings cause flow separation, which can lead to zones of high turbulence intensity and velocity gradients, which could be problematic for micro-wind turbines. Wind acceleration near the buildings can be exploited to gain advantage for energy generation, e.g. orienting the buildings towards the most common wind directions and/or introducing a sloped side to increase the wind speed. The relative heights and orientations of buildings have also a profound effect on the wind flow. Ledo [36] pointed out that the reasons behind the limited installation of micro-WTs in urban areas are the low wind speeds, high levels of turbulence and relatively high aerodynamic noise levels generated by the turbines. Blackmore [37] and Rafailidis [38], asserted that one of the main factors affecting the performance of roof-mounted wind turbines is the building roof shape (roof effect). Other studies investigated the mounting location and analyzed the flow around pitched, pyramidal, flat, and gabled roofs with a view to developing small wind turbine sitting guidelines focusing on the mounting height [36,39,40].

The disturbed flow zones must be identified and avoided when choosing the site of installation of BUWTs. Local direction, speed, and wind turbulence depend on the shapes of buildings [41].

Wind potential in urban areas can essentially be determined in three different ways [42]:

- *In situ* measurements at least over a 1-year period directly at specific sites;
- Wind tunnel experimental tests;
- CFD calculations.

In situ measurements are probably the most accurate tool for assessing wind flow in a specific site, especially in the case of energy retrofitting of existing buildings for which the use of BUWTs is foreseen. However, this approach requires months or years of data acquisition, and the major drawback of this method is the high cost.

On the other hand, wind tunnel tests require the construction of an accurate physical model of the area around the buildings and the collection of distribution data of wind speed. This approach is slightly faster but also significantly more expensive. In addition, it is difficult to accurately model the effect of turbulence in a wind tunnel because it is limited by its size, and the results could be affected by significant errors.

On the contrary, the CFD approach is a practical and effective alternative to the other two approaches. CFD commercial packages have gained a good credit both in research and academic communities as well as in engineering consulting societies, thanks to their capability of providing an insight into the flow field around the buildings. Moreover, the first two approaches determine the wind velocities only in few points at the same time; instead, the CFD analysis allows to calculate the flow field within the whole investigated area. Thereby, the CFD simulations, in conjunction with a detailed model of the urban environment, allow to simulate the wind flow field around the buildings where the BUWTs could be installed. The CFD simulations do not provide information about wind intermittency, transients, etc.

Overall, it is very hard to predict wind speed in a specific site within the built-up areas. However, some general patterns can be identified to ensure the placement of the turbine into the undisturbed region, outside of the areas of turbulence:

- WTs should be positioned near the center of the roof where the wind is stronger;
- WTs should be positioned so as to meet the most frequent wind direction;
- The mast of the turbine should be approximately 50% taller than the surrounding objects;
- The lowest position of the rotor has to be at least 30% of the building height above the roof.

The wind flow field in the urban area is still extremely difficult to predict even for simple surrounding environments. Nevertheless, despite the widespread use of CDF simulations, the prediction accuracy and many factors that might affect simulation results are not yet thoroughly understood [43].

The GIS-based methodology

In this paper, a GIS-based methodology was developed as useful support to improve the diffusion of micro-wind turbines in urban environment. The proposed methodology needs the knowledge of several data, requires the use of different software as well as their reciprocal integration. The general pattern is depicted in the following:

Wind site characterization

The assessment of the production capacity of specific WTs, as well as the evaluation of the related technical and economic issues, requires data regarding wind speed and direction of a certain statistical significance acquired through systematic recordings over a period of 1 year or more. The analysis has to provide an estimation of the marginal distribution of wind intensity and direction and the Weibull statistical distribution.

Wind climatology characterization has to be developed starting from available anemometric data.

GIS mapping

The three-dimensional (3D) digital models, normally used for GIS applications, use square mesh (DEM or DTM) or triangular mesh. The triangular irregular network (TIN) is a vector-based representation (a digital data structure) used for the representation of a surface made up of irregularly distributed nodes and lines with three-dimensional coordinates (x, y, and z) that are arranged in a network of non-overlapping triangles. An advantage of using a TIN over a raster DEM in mapping and analysis is that the points of a TIN are distributed variably, based on an algorithm that determines which points are most necessary to an accurate representation of the terrain. ArcGIS uses Delaunay's triangulation method [44] to construct these triangles. On the other hand, TIN meshes are appropriate for contexts where morphology is gentle and continuous without sudden changes of height above the ground, but they do not supply a good representation of complex morphology that is characteristic of urban areas (presence of buildings). Therefore, a TIN-modified model suitable for urban environment was developed considering the terrain map and also the building elevation [8]. The ESRI software ArcGIS Desktop v.9.2 with Spatial Analyst and 3D Analyst extensions was used [45].

Interface between CFD and GIS

A crucial point of the procedure is the process of interface between the output files of the CFD simulations and the GIS environment: interaction between the CFD code and the GIS software is essential. The output files of CFD simulation are 'raster' files which contain the information on the values of wind velocity; wind velocities are represented with square-colored pixels in accordance with a default scale. The GIS software is not able to directly read these files, so it is necessary to modify them from RGB images (bands of color: red, green, blue) into gray scale images as shown in Figure 1.

After this preliminary operation, the following procedure has to be developed in the GIS environment:

- The CFD files have to be geo-referenced using the specific procedure that manages geographic data. Subsequently, to the process of geo-referencing, new raster files are generated; therefore, geometrical data are stored within a regular matrix (GRID) organized as a set of rows and columns. The size of the cells establishes the level of spatial resolution. Each cell of the GRID contains its own height above ground level (GRID-quote) and a set of 'information' that can be a particular geographic feature or other data, such as the wind speed. Then, the raster dataset can be displayed or rendered in many different ways.
- The 'stretched renderer' procedure displays the continuous raster cell values across a gradual ramp of colors. The stretched renderer works well when a large range of values is displayed such as in spectral imagery, aerial photographs, or elevation models. In this way, the gray scale, which represents the field of wind velocity (CFD output file), is modified, defining a rule to describe the variation of velocity into a value of velocity at each cell. Choosing a linear variation for the velocity implies that the black color is associated to the minimum value of velocity and the white color is associated to the maximum value of velocity. The results of these operations are raster files which provide for each cell the value of wind velocity calculated through CFD simulations.
- Additional raster files can be generated using the function 'Raster Calculator' of the plugin 'Spatial Analyst' of ArcGIS® [45], which permits to execute mathematical operations using the data contained in two or more raster files. Therefore, it is possible to create new raster files, which give for example the average value of the wind speeds calculated by means of the CFD simulations considering different wind directions.

In conclusion, the proposed procedure generates raster files which contain the wind speed at each point P (x, y, z) of the investigated urban area. Figure 2 shows the render of a raster file which provides the wind flow field within the built-up area.

CFD simulations

The Virtual Wind software was used to calculate the whole flow field in the investigated urban area. The choice to use a simple CFD code, like the Virtual Wind which is not intended for high-end CFD analysis, has the aim to allow, also for users who are not specialists, a quick prediction and comprehension of wind flows in complex environments.

The Virtual Wind needs to be supplied with predefined terrain geometry, which must obey a limited set of restrictions (i.e., the border of geometry to be recognized, as terrain cannot overlap itself). Virtual Wind works when connected as a plugin to Google Sketch up; thereby, it is quite simple to model the built up area.

Sáenz-Díez Muro et al. [41] used the Virtual Wind code to optimize the implementation of mini wind turbine in the urban environment.

Moreover, this CFD code was used as an internal analysis tool within its parent company, RWDI, for

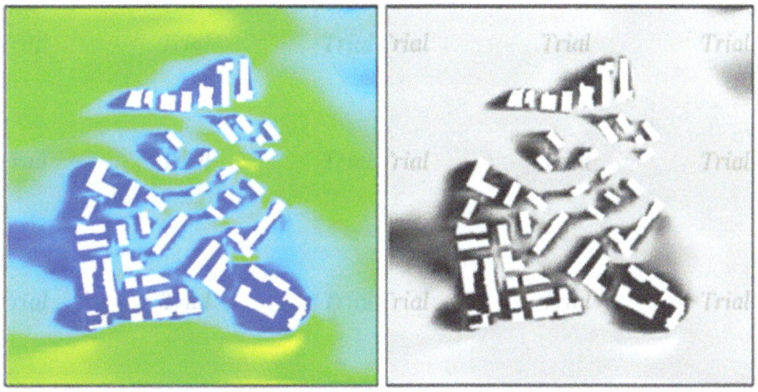

Figure 1 Example of transformation from RGB to grayscale image.

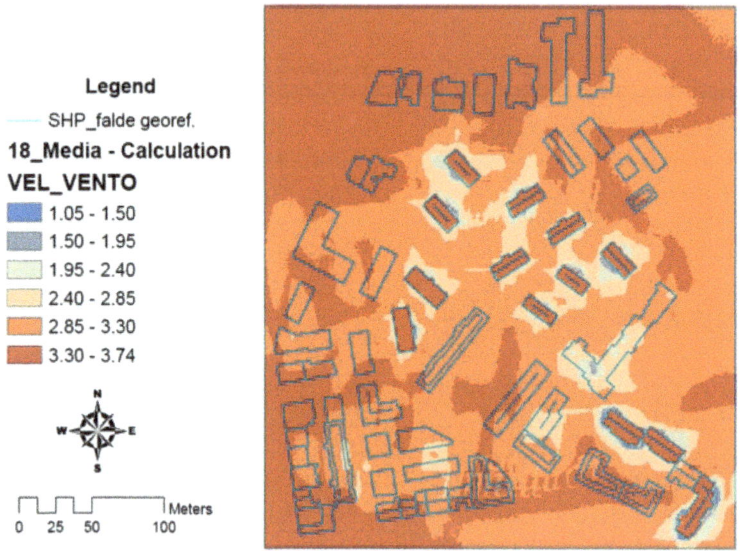

Figure 2 Wind velocity field.

approximately 5 years before it was released commercially. It has been run on hundreds of internal projects [46].

BUWTs energy production

The above procedure provides the average wind velocity in each point of the built-up area. Therefore, it is possible to calculate, once the parameters of the Weibull distribution are defined, the wind velocity distribution by Equation 2. Hence, the yearly energy production of a BUWT is calculated by Equation 7.

Comprehensively, a methodology has been developed that can provide a very detailed information about the wind velocity in urban areas.

A fundamental feature of the methodology is its high geometrical resolution; indeed, a spatial resolution of 1.0 m × 1.0 m is adopted, which is far more accurate than other wind atlases, based on air flow models that account for the effects of topography using an horizontal scale of 1 km^2 (e.g., NOABL 2012) [47].

Pilot study

The proposed methodology was tested and verified by means of a first pilot study. In the following, the authors will report the results obtained for an urban area of the municipality of San Cataldo, a city located in the centre of Sicily (latitude: 37°29′0″N, longitude: 13°59′0″E). The wind data were provided by two weather stations installed within an urban area of San Cataldo, for a period of about 3 years. These weather stations provide both speed and wind direction, measured by means of anemometers at 20.0 m above ground. The wind rose gives the following information:

- The wind velocity has not prevailing directions;
- The investigated urban area has a weak wind potential.

The above-mentioned wind measurements were used to calculate the average wind speed (V_{av}) along the four cardinal directions (Table 1).

The lowest mean wind speed (2.20 m/s) was measured in the North sector; in the East, West, and South sectors, the mean wind speeds are quite similar between them, and they are all below 3.0 m/s. Amongst the most important recorded anemological data, the frequency distribution of wind speed is of specific interest.

According to the measurements of wind speed, it is necessary to use estimation methods to derive the parameters k and A on the basis of available data of the mean wind velocity V_{av}. Considering the peculiarity of the built-up areas, three Weibull probability functions were compared, which used the same value of shape factor (k) = 1.4, and three different values of the distribution scale factor (A): $A_1 = 2.085$, $A_2 = 1.5\ V_{av}$, and $A_3 = 6.0$.

The above values were chosen in agreement with literature data [12]. Figure 3 shows the wind speed frequency

Table 1 Yearly average wind speed

Direction	North	West	South	East
Mean velocity V_{av} (m/s)	2.20	2.88	2.91	2.71

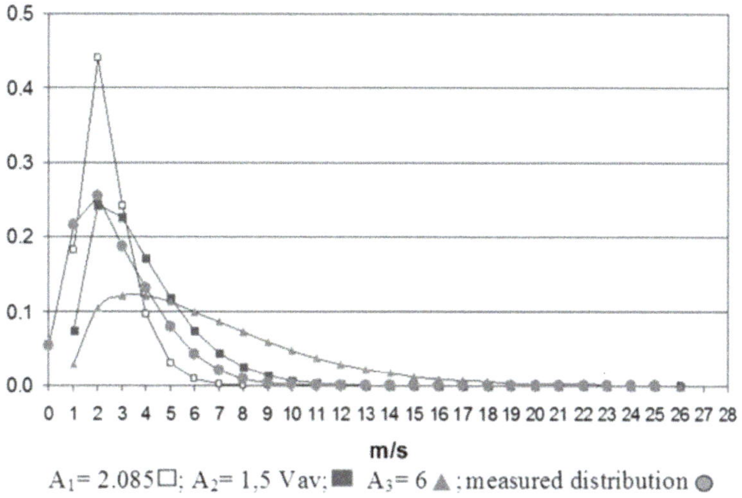

Figure 3 Measured and Weibull wind frequencies.

obtained both by measured data and by Weibull frequency distributions.

It can be noticed that

- The Weibull frequency distribution obtained, adopting the scale factor $A = 1.5\ V_{av}$, is the one which approaches better the measured wind frequencies; and
- Wind speed frequency is high around 2.0 ÷ 3.0 m/s, while low frequencies occur for speed above 7.0 m/s.

Results and discussion

CFD simulations require as input both wind data and the model of the geometry of the built-up area. The 3D models of the terrain, including the buildings, were created through Google Sketchup, a widespread CAD tool used for creating built-up area. Therefore, the 3D model of the terrain is easily imported in Virtual Wind as a plugin. The supply input parameters for the simulations were wind speed, simulation resolution, spatial extents, and monitored quantities. The size of the calculation domain is $x = 325$ m, $y = 360$ m, $z = 30$ m. Uniform computational grid was adopted across the whole calculation domain. Thermal stratification and density variations were not considered.

Figure 4 shows the 3D modelization of the investigated area.

To accurately simulate wind flow in the built environment, velocity profiles at the inlet boundary surface of the calculation domain must be accurately modelled.

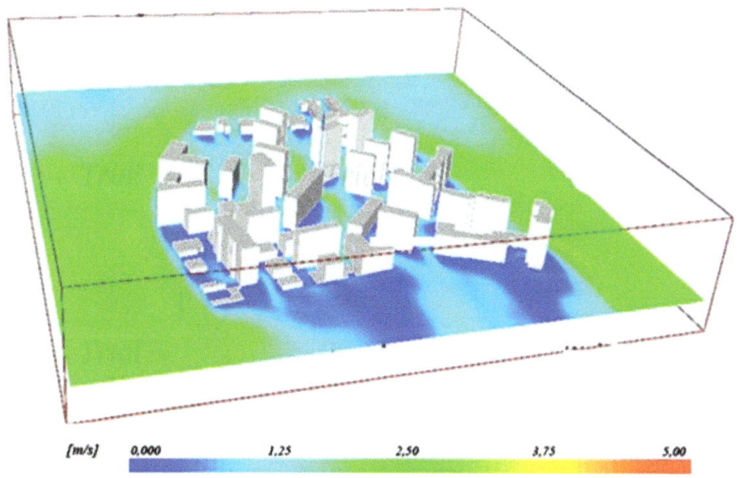

Figure 4 3D view of the calculation domain.

The vertical profile of wind velocity was calculated as a function of the roughness of the ground:

$$V_2 = V_1(h_2/h_1)^{\alpha}, \tag{11}$$

where V_2 is the wind speed at the height h_2, V_1 is wind speed at the original height h_1, and α is the surface roughness coefficient, assumed to be 0.28. It has been underlined that the power law profile with an exponent $\alpha = 0.28$ overestimates the velocity profile below the mean building height, while it underestimates the profile above the building [48]. CFD simulations provide the wind flow field, on the XY plane at whatever heights above ground level (AGL); thereby, it is possible to detect the most suitable locations for turbine mounting. Figure 5 shows the wind flow field in the case of wind blowing from West direction and $h = 6.0$ m. To investigate the effect of wind direction, simulations were run with different wind directions (North, East, South, and West).

The analysis of the wind flow field indicates that large streets with an aspect ratio $d/h > 1.5$ (d = street width, h = building height) are characterized by higher wind speeds, up to 3.45 m/s, and consequently, these zones could be the most suitable for the installation of micro-wind turbines integrated into the poles of public illumination.

Mounting small wind turbine systems to the corner or the side of a building may take advantage of the favorable flow between buildings.

On the contrary, narrow streets with an aspect ratio $d/h < 1.5$ are characterized by lower wind speeds. Figure 6 shows the wind flow field on a vertical plane.

It is possible to notice that the wind field is affected by the shape, height, and relationships that buildings have with each other and to visualize an effect of disturbance generated by the taller building and the extension of the disturbed area. Furthermore, it is possible to highlight the zones where there is a significant reduction of wind speed (blue color) and the zones where the wind speed is stronger (green or yellow color).

The analysis of the vertical profile of the wind velocity at the point 1 in Figure 7 was also performed, placed near a building 23.0-m tall.

It is possible to observe the abrupt reduction of the wind velocity in correspondence of the height of the adjacent building and the increase of the velocity above the height of 26.0 m, that is out of the disturbed zone.

With the aim to validate the numerical results of CFD simulations, a comparison with other researches and literature data were performed for similar geometries [49,50]. Such analysis indicates a good correlation and agreement between our results and bibliographic data. Moreover, the calculated values of wind velocity were compared with experimental measurements in two points (anemometric stations), observing differences in the order of 10 ÷ 15%. Obviously, a more extensive campaign of experimental measurements could further confirm the reliability of this study.

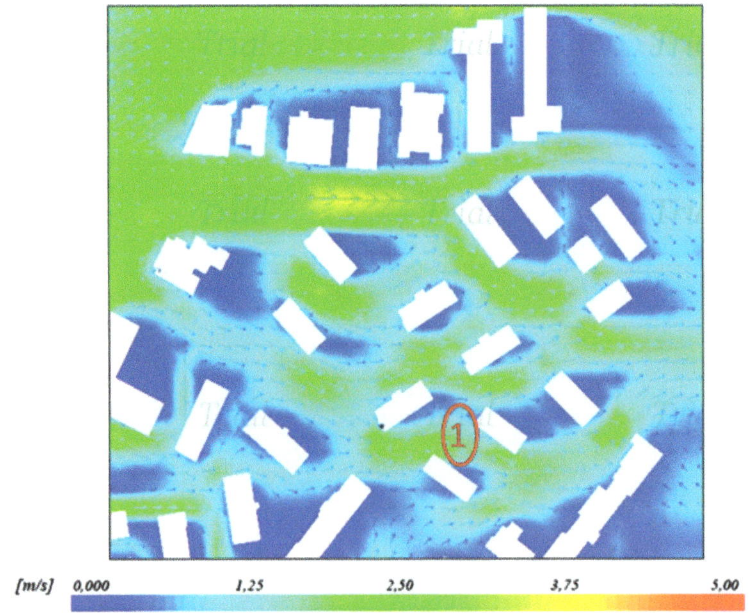

Figure 5 Wind flow field at a horizontal plane.

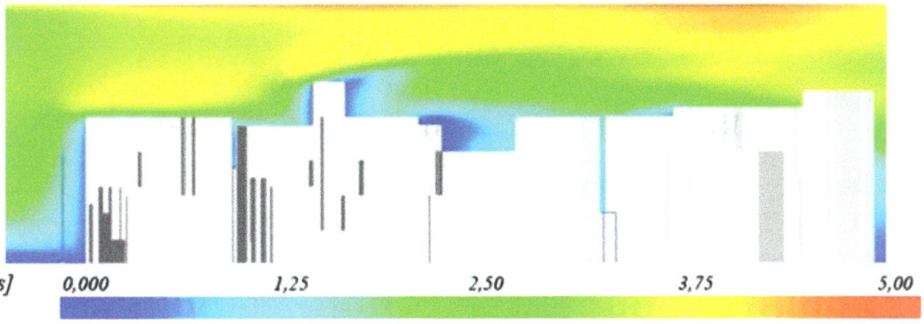

Figure 6 Wind flow field on vertical plane.

GIS mapping

The digital elevation model of the urban area was built using data carried out from large scale vector mapping (scale 1:2,000) that provides both the data of the morphology of the terrain and the data related to the anthropic elements (buildings, plants, vegetation, and so on). Otherwise, in the 'classical' approach, which represents each building with a single polygon which is the gutter surface of the building, the adopted model introduces a second polygon, placed very close to the ground surface. This second polygon acts as a 'break line' in the 3D modelling. It is 'shrinked' with an offset of 5.0 cm from the polygon which represents the building, and it allows to differentiate the buildings and the ground surface. This strategy has the advantage to create a very reliable and well representative TIN of the built-up area (Figure 8).

The accuracy of the raster data is strongly related to the GRID resolution where these data are recorded. In this study, the GRID is characterized by a size mesh of 1.0 m × 1.0 m in the *XY* plane. Therefore, each cell of the GRID contains the information concerning its own quote (GRID-quote).

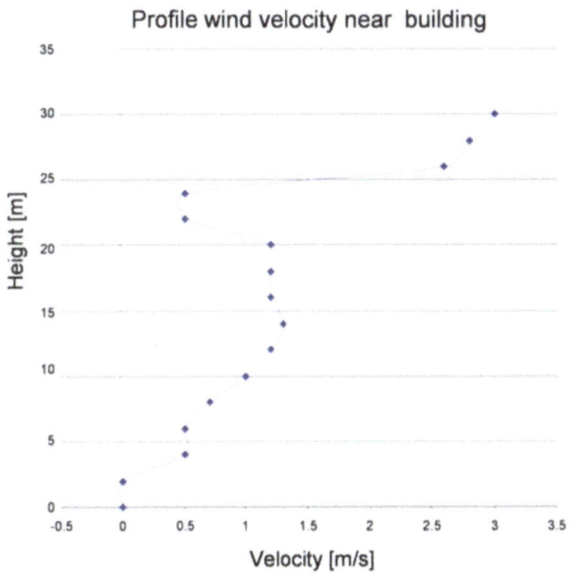

Figure 7 Vertical wind shear (point 1 in Figure 5).

Wind mapping

As previously discussed, a nodal point of the proposed methodology is the possibility to interface the output of CFD simulations with the GIS environment. Adopting the above illustrated procedure, it was possible to obtain at any spatial position of the built-up area, the mean wind velocity calculated by means of the CFD simulations. The overall result of the study is the possibility to generate the wind maps for any height AGL. Figure 9 shows the average wind velocity as stored into the GIS software.

Wind maps were elaborated for different heights AGL: 9.0, 18.0, and 24.0 m. The height of 9.0 m was considered as representative of BUWTs mounted on the poles for public lighting; instead, the heights of 18.0 and 24.0 m were considered as representative of BUWTs mounted on the roof of the buildings.

Wind maps, customized in the GIS software, supply the average wind speed and the needed parameters to define the wind velocity distribution (Weibull distribution), which are the most important data necessary to calculate the power output of BUWTs in a realistic way. Thereby, inhabitants of the district can use this tool to calculate the yearly energy obtainable by WTs in every point of the urban area.

The aesthetics of BUWTs has to be complementary rather than in contrast with the nearby buildings to ensure their acceptance in the urban context.

Figure 10 shows an example of aesthetic integration of BUWT pole mounted in the building environment of San Cataldo city.

In any case, both the visual and noise impacts of micro-wind turbines should be assessed by the Local Planning Authority.

Figure 8 Overlapping between DEM model and ortho-photos.

Annual energy production

The above procedure allows to import the wind maps for the investigated area in the GIS platform. Starting from that, a comprehensive analysis of the energy performances of three commercial models of BUWTs are developed. Power curves and main technical specifications of the selected WTs are reported in Table 2 and Figure 11.

As previously mentioned, within the investigated area, two alternatives for mounting the BUWTs were evaluated:

a) On the pole of public lighting ($h = 8.0$ m), where the average wind speed of 3.50 m/s was obtained consulting the GIS database; and

b) On the roof of buildings ($h = 24.0$ m), where the average wind speed of 4.80 m/s is available according to the GIS database.

The yearly energy production was calculated by Equation 3 as a function of the wind velocity distribution and the BUWTs manufacture power curves.

Table 3 shows the year-round performance of the three micro-wind turbine. It was assumed that positive energy output starts immediately when the wind meets the cut-in speed.

One can observe the following:

- A turbine mounted on the roof of the building takes advantage of the building height; therefore, the

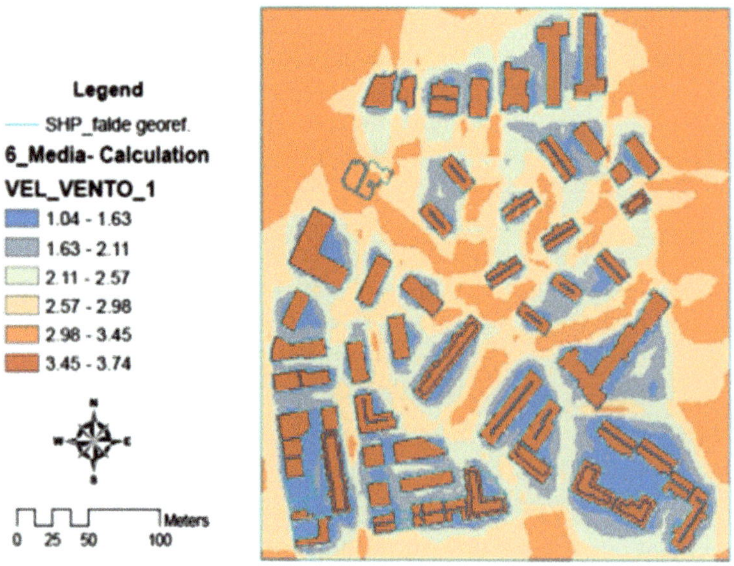

Figure 9 Average wind velocity at 6 m (AGL).

Figure 10 Photo-simulation of WT urban integration.

energy production is about 30% higher than for the pole mounting;

- UWT_1 and UWT_2 (VAWT) give more or less the same annual energy, with a difference of about 10%;
- UWT_3 that is a HAWT has the lowest annual energy production, and this is mainly due to its power curve that has poor performance for weak wind velocity. Furthermore, this WT has very large dimensions that are not always compatible for building mounting.

On the whole, the procedure allows to calculate the annual energy for each kind of WT. It is also possible to compare the performance of several WT and consequently to choose the most adequate for the installation in the specific site. The GIS platform also allows to provide for each building other information, i.e., the number of dwellers, the consumption of electric energy (daily, monthly, yearly), and so on. Such information can be useful to evaluate statistical match between demand and supply, which determines how much electricity will be exported, as well as for planning the installation of RES.

Economic and environmental benefits

An accurate economic analysis of micro-generation technologies is important as it allows prospective buyers to assess the financial outlay and the annual and total payback period of the turbine. The electricity generated by the BUWTs means direct savings for the owners through a lower electricity consumption bill, and if the BUWTs are eligible for feed-in tariff (FiT), it is also possible to earn an added export tariff.

In Italy, wind power plants with a nominal peak power up to 200 kW (micro-generation) can benefit from the all-inclusive tariff grant [51] and thereby receive the incentive of 30€ cents/kWh for 15 years. After the period of 15 years, the energy produced through BUWT will be traded to the national grid. The analysis of the economic viability was performed considering an energy production of 1,468.00 kWh, as reported in Table 3.

The following assumptions were fixed:

- The total cost of the system is 6,000.00€ (4,000.00€ WT + 2.000 for installation, electric connection) [52,53];
- The FiT is 0.30€ /kWh, within the first 15 years [51]; and
- The price is 0.09€/kWh for the following period [51].

The financial analysis indicates a payback time of 14 years which is lower than their estimated life-span of 20 years. The economic feasibility and the market potential for BUWTs is highly influenced by the reduction of WTs costs due to technical progress and by incentive policy. Electricity generated by BUWTs results also in a

Table 2 Technical specifications of micro-turbines

Turbine model	Rated power (kW)	Rated wind speed (m/s)	Cut-in speed (m/s)	Cut-out speed (m/s)	Tower blade height (m)	Typology of WT
WT_1	1.5	14.0	4.0	20.0	2.5	Darrieus
WT_2	1.5	15.0	3.0	21.0	4.0	Savonius
WT_3	2.3	22.0	4.5	22.5	9.0	Horizontal axis

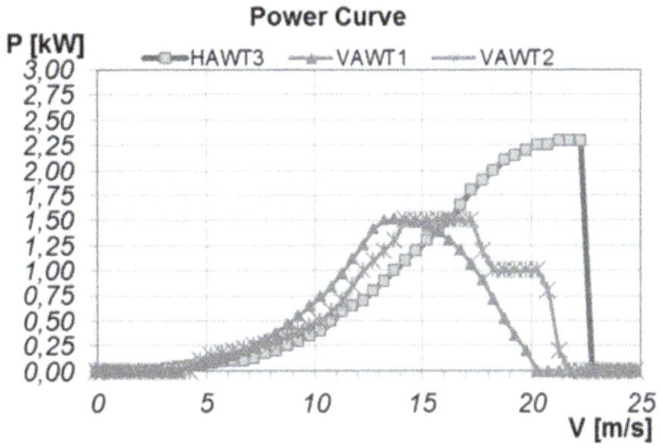

Figure 11 Power curves of selected WTs.

Table 3 Yearly energy production

	Yearly energy production (kWh/year)	
	Pole mounting	Building mounting
$BUWT_1$	628.0	1,468.0
$BUWT_2$	553.0	1,280.0
$BUWT_3$	393.0	910.0

reduction in the emission of greenhouse and acid gases. Carbon intensity of electricity varies greatly depending on fuel source; considering the carbon intensity of network electricity to be 0.527 kg CO_2/kWh [54], it was possible to calculate the ratio between cost investment and emission savings, which is 6.12€/kg CO_2.

Obviously, it is possible to obtain more seductive results with higher average wind velocity, or if more efficient wind turbines will be available on the market. The financial analysis indicates that micro-wind turbines are economically viable also in the built-up area and they can play a primary role in future urban energy supply, such as the PV plants which are widely diffuse in the building environment.

BUWTs should be capable of producing a significant proportion of the annual electricity demand; otherwise, they may become a primary aesthetic feature which would be unsatisfactory from both an environmental viewpoint and that of clients, planners, designers, occupants, and local inhabitants.

Considering that the average Italian domestic household consumes (excluding thermal uses) approximately 3,000÷4,000 kWh per year [55,56], the installation of just one BUWT can cover more or less 40% of the annual energy demand of a single domestic household. Thereby, to increase the capacity credit, that expresses how much 'conventional' power can be avoided or replaced by wind power [57,58], it is useful to install multiple WTs on multi-storey buildings.

Conclusions

The work provides a first step for assessing the feasibility of building integrated wind turbines, encouraging the incorporation of renewable energy sources in new and refurbished buildings, and promoting the adoption of micro-scale power generation in urban environments.

The developed methodology allows creating a useful database in the GIS platform that provides the wind distribution within an urban area by enabling fast and accurate spatial information for calculating the yearly energy production obtainable of micro-wind devices. In this way, owners/occupiers have the possibility to evaluate the effective potentiality of wind energy generation in urban areas.

An innovative aspect of this approach is mainly linked to its high productivity, which allows to automatically extend the evaluation of the potential production from a single site of specific interest to an entire urban area.

Moreover, the integration between GIS software and methodologies to assess potentiality of RES can be a useful tool to support energy conservation programs, planning of local RES energy resources, and implementation of energy strategies as combining micro-cogeneration and poly-generation.

Overall, the research can contribute to the diversification of energy sources and maximization of on-site RES penetration in urban areas in accordance with the objectives set by the EU.

Future developments of the research will be the comparison between the hourly household energy load and the hourly power provided by the BUWTs, with the aim to support the strategies for the dispatch and control of energy in a smart grid with stronger connection between energy demand and supply.

Competing interests

The authors declare that they have no competing interests.

Authors' contributions

AC collected the wind data and carried out initial analysis. The GIS elaboration and CFD calculations were performed by AG and AC. The 'Introduction' section was drafted by AG and FP. The 'Results and discussion' section was jointly drafted by AG and FN. All authors read, edited, and approved the manuscript to be submitted.

References

1. Bergman, N, Eyre, N: What role for microgeneration in a shift to a low carbon domestic energy sector in the UK? Energy Efficiency **4**, 335–353 (2011)
2. Bahaj, AS, Myers, L, James, PAB: Urban energy generation: influence of micro-wind turbine output on electricity consumption in buildings. Energy Buildings **39**, 154–165 (2007)
3. Peacock, AD, Jenkins, D, Ahadzi, M, Berry, A, Turan, S: Micro wind turbines in the UK domestic sector. Energy Buildings **40**, 1324–1333 (2008)
4. Medrano, M, Brouwer, J, Carreras-Sospedra, M, Rodriguez, MA, Dabdub Samuelsen, G: A methodology for developing distributed generation scenarios in urban areas using geographical information systems. Int. J. Energy Technol. Policy **6**, 413–434 (2008)
5. Voivontas, D, Assimacopoulos, D, Mourelatos, A, Corominas, J: Evaluation of renewable energy potential using a GIS decision support system. Renew. Energy **13**, 333–344 (1998)
6. Byrne, J, Zhou, A, Shen, B, Hughes, K: Evaluating the potential of small-scale renewable energy options to meet rural livelihoods needs: A GIS- and lifecycle cost-based assessment of Western China's options. Energy Policy **35**, 4391–4401 (2007)
7. Caiaffa, E, Marucci, A, Pollino, M: Study of sustainability of renewable energy sources through GIS analysis techniques. Comput. Sci. **7334**, 532–547 (2012)
8. Gagliano, A, Patania, F, Nocera, F, Capizzi, A, Galesi, A: A proposed methodology for estimating the performance of small wind turbines in urban areas. In: Smart Innovation, Systems and Technologies, 12th edn, pp. 539–548. Springer, Heildeberg (2012)
9. Gagliano, A, Patania, F, Nocera, F, Capizzi, A, Galesi, A: GIS-based decision support for solar photovoltaic planning in urban environment. In: Smart Innovation, Systems and Technologies, vol. 22, pp. 865–874. Springer, Heildeberg (2013)

10. Caputo, P, Costa, G, Ferrari, S: A supporting method for defining energy strategies in the building sector at urban scale. Energy Policy **55**, 261–270 (2013)
11. Verma, S: GIS-driven renewable energy framework for smart and sustainable cities. India Geospatial Digest (2013)
12. Allen, SR, Hammond, GP, McManus, MC: Prospects for and barriers to domestic micro-generation: a United Kingdom perspective. Energy **85**, 528–544 (2008)
13. Acosta, JL, Combe, J, Djoki´c, SZ, Hernando-Gil, I: Performance assessment of micro and small-scale wind turbines in urban areas. EEE Syst. J. **6**, 152–162 (2012)
14. Bussel, GJW, Mertens, SM: Small wind turbines for the built environment. In: Proceedings of the 4th European and Asian Wind Engineering Conference, Prague, 11–15 July 2005
15. RSE: Atlante Eolico Interattivo. http://atlanteeolico.rse-web.it/viewer.htm. Accessed 10 May 2013
16. NREL: Dynamic maps, GIS data, & analysis tools. http://www.nrel.gov/gis/wind.html. Accessed 10 May 2013
17. Burton, T, Sharpe, D, Jenkins, N, Bossanyi, E: Wind energy handbook. Wiley, New York (2001)
18. Carta, JA, Ramirez, P: Analysis of two-component mixture Weibull statistics for estimation of wind speed distributions. Renew. Energy **32**, 518–531 (2007)
19. Bhattacharya, P, Bhattacharjee, R: A study on Weibull distribution for estimating the parameters. J. Appl. Quant. Method **5**, 234–241 (2010)
20. Akdag, SA, Bagiorgas, HS, Mihalakakou, G: Use of two-component Weibull mixtures in the analysis of wind speed in the Eastern Mediterranean. Applied Energy **87**, 2566–2573 (2010)
21. Oyedepo, SO, Adaramola, MS, Paul, SS: Analysis of wind speed data and wind energy potential in three selected locations in south-east Nigeria. Int. J. Energy and Environ. Eng. **3**, 1–11 (2012)
22. Costa Rocha, PA, Coelho De Sousa, R, Freitas De Andrade, C, Vieira Da Silva, ME: Comparison of seven numerical methods for determining Weibull parameters for wind energy generation in the northeast of Brazil. Applied Energy **89**, 395–400 (2012)
23. Chang, TP: Performance comparison of six numerical methods in estimating Weibull parameters for wind energy application. Applied Energy **88**, 272–282 (2011)
24. Mertens, SM, van Kuik, van Bussel, G: Performance of an H-Darrieus in the skewed flow on a roof. J. Solar Energy Eng. **125**, 433–440 (2003)
25. Li, Z, Boyle, F, Reynolds, A: Domestic application of micro wind turbines in Ireland: investigation of their economic viability. Renew Energy **41**, 64–74 (2011)
26. Patania, F, Gagliano, A, Nocera, F, Galesi, A: The noise pollution in nature protected area of the Pergusa Lake. Int. J. Design Nature Ecodynamics **6**, 52–67 (2011)
27. Tina, G, Gagliano, S, Raiti, S: Hybrid solar/wind power system probabilistic modelling for long-term performance assessment. Solar Energy **80**, 578–588 (2006)
28. Lopes, AMG: Windstation a software for the simulation of atmospheric flows over complex topography. Environ. Model Software **18**, 81–96 (2003)
29. Araújo, AM, Arraes De Alencar Valença, D, Asibor, AI, Carvalho Rosas, PA: An approach to simulate wind fields around an urban environment for wind energy application. Environ. Fluid Mech. **13**, 33–50 (2013)
30. Patania, F, Gagliano, A, Nocera, F, Galesi, A, D'Amico, A: An applied research to forecast the trend of air pollution in particular geometries of streets: the real case of urban canyon. WIT Transactions Ecology Environ. **86**, 33–44 (2006)
31. Li, X-X, Liu, C-H, Leunga, DYC, Lam, KM: Recent progress in CFD modelling of wind field and pollutant transport in street canyons. Atmos. Environ. **40**, 5640–5658 (2006)
32. Yoshie, R, Mochida, A, Tominaga, Y, Kataoka, H, Yoshikawa, M: Cross comparisons of CFD prediction for wind environment at pedestrian level around buildings. Part 1: comparison of results for flow-field around a high-rise building located in surrounding city blocks. In: 6th Asia-Pacific Conference on Wind Engineering. Seoul (2005)
33. EERE: Virtualwind. http://apps1.eere.energy.gov/buildings/tools_directory/software.cfm/ID=579/pagename=alpha_list_sub?print. Accessed November 2013
34. Sagaut, P: Large Eddy Simulation for Incompressible Flows. Springer, Berlin (2006)
35. McLean, W: Poisson solvers. http://www.cs.northwestern.edu/~jet/Teach/2004_3spr_IBMR/poisson.pdf. Accessed July 2013
36. Ledo, L, Kosasih, PB, Cooper, P: Roof mounting site analysis for micro-wind turbines. Renew. Energy **36**(5), 1379–1391 (2011)
37. Blackmore, P: Siting Micro-Wind Turbines on House Roofs. IHS BRE Press, Watford (2008)
38. Rafailidis, S: Influence of building areal density and roof shape on the wind characteristics above a town. Bound. Lay. Meteorol. **85**(2), 255–271 (1997)
39. Abohela, I, Hamza, N, Dudek, S: Effect of roof shape on energy yield and positioning of roof mounted wind turbines. In: Proceedings of 12th Conference of International Building Performance Simulation Association, Sydney, 14–16 November 2011
40. Mertens, S: Wind energy in the built environment. Multi-Science Publishing Co. Ltd., Essex (2006)
41. Sáenz-Díez Muro, JC, Jiménez Macías, E, Blanco Barrero, JM, Pérez De La Parte, M, Blanco Fernández, J: Two-dimensional model of wind flow on buildings to optimize the implementation of mini wind turbines in urban spaces. In: ICREPQ'10, Granada, 23–25 March 2010
42. Paterson, DA, Apelt, CJ: Simulation of wind flow around three-dimensional buildings. Build. Environ. **24**, 39–50 (1989)
43. Yoshie, R, Mochida, A, Tominaga, Y, Kataoka, H, Harimoto, K, Nozu, T: Cooperative Project for CFD Prediction of Pedestrian Wind Environment in the Architectural Institute of Japan. J. Wind Eng. Ind. Aerod. **95**, 1551–1578 (2007)
44. Chew, PL: Constrained Delaunay triangulations. Algorithmica. **4**, 97–108 (1989)
45. ESRI: ArcGIS desktop tutorials. http://webhelp.esri.com/arcgisdesktop/9.3/index.cfm?TopicName=tutorials. Accessed 10 May 2013
46. EERE: Building Energy software tools directory. http://apps1.eere.energy.gov/buildings/tools_directory/software.cfm/ID=579/pagename_submenu=solar_climate_analysis/pagename_menu=other_applications/pagename=subjects. Accessed July 2013
47. NOABL: Wind speed database. http://www.decc.gov.uk/en/content/cms/meeting_energy/wind/windsp_databas/windsp_databas.aspx (accessed). Accessed July 2013
48. Agrawal, N, Gupta, VK, Gupta, A, Mittal, A: Comparison of codal values and experimental data pertaining to dynamic wind characteristics. J. Wind Eng. **9**(1), 33–53 (2012)
49. Plate, EJ: Methods of investigating urban wind fields-physical models. Atmos. Environ. **33**, 3981–3989 (1999)
50. Raciti, M, Castelli, A, Benini, E: Modeling, strategy and numerical validation of the turbulent flow over a two-dimensional flat roof. World Acad. Sci. Eng. Technol. **79**, 462–468 (2011)
51. President of Italian Republic: Legislative Decree 3 March 2011 N° 28. http://www.acs.enea.it/doc/dlgs_28-2011.pdf. Accessed 10 May 2013
52. Energy Saving Trust: Domestic small-scale wind information. http://www.est.org.uk/myhome/generating/types/wind/. Accessed 10 May 2013
53. Tutto green: Eolico domestico o minieolico: quali sono le offerte sul mercato? http://www.tuttogreen.it/eolico-domestico-o-minieolico-quali-sono-le-offerte-sul-mercato/. Accessed10 May 2013
54. IPCC: Carbon dioxide intensities of fuels and electricity for regions and countries. http://www.ipcc.ch/pdf/special-reports/sroc/Tables/t0305.pdf. Accessed 10 May 2013
55. Ruggieri, G: Consumi elettrici nel settore domestico. http://www.aspoitalia.it/attachments/220_Gianluca%20Ruggieri%20%20Consumi%20elettrici%20nel%20domestico.pdf. Accessed 10 May 2013
56. Autorità per l'Energia: Elettrica ed il Gas. I servizi di vendita. http://www.autorita.energia.it/it/consumatori/bollettatrasp_ele.htm. Accessed 10 May 2013
57. D'haeseleer, WD, Voorspools, KR: An analytical formula for the capacity credit of wind power. Renew. Energy **31**, 45–54 (2006)
58. Kelleher, J, Ringwood, JV: A computational tool for evaluating the economics of solar and wind microgeneration of electricity. Energy **4**, 401–409 (2009)

Development and performance evaluation of mixed-mode solar dryer with forced convection

Chandrakumar B Pardhi[1*] and Jiwanlal L Bhagoria[2]

Abstract

Based on preliminary investigations under controlled condition of drying experiments, a mixed-mode solar dryer with forced convection using smooth and rough plate solar collector was constructed. This paper describes the development of dryer considerations followed by the results of experiments to compare the performance of the smooth and the roughed plate collector. The thermal performance of solar collector was found to be poorer because of low convective heat transfer from the absorber plate to air. Artificial rib roughness on the underside of the absorber plate has been found to considerably enhance the heat transfer coefficient. The absorber plate of the dryer attained a temperature of 69.2°C when it was studied under no-load conditions. The maximum air temperature in the dryer, under this condition, was 64.1°C. The dryer was loaded with 3 kg of grapes having an initial moisture content of 81.4%, and the final desired moisture content of 18.6% was achieved within 4 days while it was 8 days for open sun drying. This prototype dryer was designed and constructed to have a maximum collector area of 1.03 m^2. This solar dryer been be used in experimental drying tests under various loading conditions.

Keywords: Mixed mode, Forced convection dryer, Solar collector, Grapes

Background

Drying may be required for several reasons. First and most often, water is removed from the fresh crop to extend its useful life. The dried product is later rehydrated prior to use in order to produce a food closely resembling the fresh crops, for example, in the use of dried vegetables. Second a crop may require drying so that it can be further processed. For example, many grains are dried so that they can be ground into flour. Third, fresh crops are sometimes dried so that a new product distinctly different from its original form can be produced.

Fundamentals of the drying process

Drying involves the removal of moisture and in thermal drying this is achieved through the application of heat to the product. The heat increases the vapor pressure of the moisture in the product above that of the surrounding air. Pressure and thermal gradient cause the moisture, both liquid and vapor, to move to the surface of the product. Evaporation takes place and water vapor is transferred to the surrounding air. This air may become saturated, but the process of drying continues if this surrounding moist air is replaced by less-saturated air.

Traditional sun drying

The traditional method of drying known as 'sun drying' involves simply laying the product in the sun. Major disadvantage of this method is contamination of the products by dust, destruction by insects and microorganism, and pecking by birds. Furthermore, some percentage will usually be lost or damaged during handling; it is labor-intensive, nutrients loss occurs, such as vitamin A, and is time-consuming. Lastly, the method totally depends on good weather conditions. The major advantage in the energy requirements for this open sun drying process is that the solar and wind energy is readily available freely in nature; hence, the capital requirement is marginal, making it the viable method of drying agricultural produce even in commercial scale especially in developing country. The safer alternative to open sun drying is drying in a solar dryer. This is a more efficient method of

* Correspondence: cpardhi@gmail.com
[1]Department of Mechanical Engineering, Sagar Institute of Science, Technology and Research (SISTec-R), Bhopal 462044, India
Full list of author information is available at the end of the article

drying which produces better quality products; but in this case, initial investments are required.

Solar dryers

A solar dryer is an enclosed unit to keep the food safe from damage from birds, insects, microorganism, pilferage, and unexpected rainfall. The produce is dried using solar thermal energy in a cleaner and healthier fashion. Basically, there are four types of solar dryers:

1. Direct solar dryers. In these dryers, the material to be dried is placed in a transparent enclosure of glass or transparent plastic. The sun heats the material to be dried, and heat also builds up within the enclosure due to the 'greenhouse effect.' The drier chamber is usually painted black to absorb the maximum amount of heat.
2. Indirect solar dryers. In these dryers, the sun does not act directly on the material to be dried thus making them useful in the preparation of those crops whose vitamin content can be destroyed by sunlight. The products are dried by hot air heated elsewhere by the sun.
3. Mixed-mode dryers. In these dryers, the combined action of the solar radiation incident on the material to be dried and the air preheated in solar collector provides the heat required for the drying operation.
4. Hybrid solar dryers. In these dryers, although the sun is used to dry products, other technologies are also used to cause air movement in the dryers. For example, fans powered by solar PV can be used in these types of dryers.

The drying process

The process of dehydration consists of removal of moisture from the produce by heat usually in the presence of a controlled flow of air (Figure 1). Initially, the produce to be dried is washed, peeled and prepared (if necessary), and placed on flat-bottomed trays that are placed into the dryer. The solar rays enter the cabinet through the cover material. Upon reaching the solar collector or the tray surface, they are converted into heat energy, raising the inside temperature. The heat energy is transferred to the produce to be dried. The heated produce gives out water vapor and dries up. Gradually the heated moist air goes up and leaves the drying chamber through the air outlet at the high end of the drier.

The efficiency of drying of the solar dryer is influenced by relative humidity in the air, the moisture content of the materials to be dried and their amount and thickness. The solar radiation intensity on the materials varies with seasons, time of the day, and length of exposure, ambient air temperature, and wind speed, which are important factors.

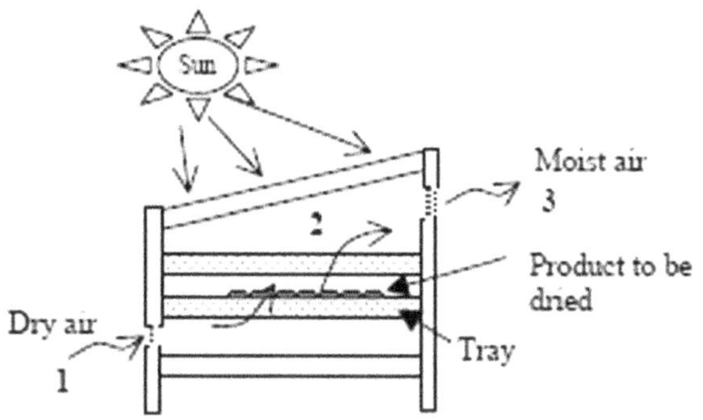

Figure 1 Drying process. (1) Cold dry air enters the drying chamber through the air inlet. (2) The solar rays enter the cabinet through the transparent cover material where they are converted into heat energy, thereby increasing the temperature inside. The heated food gives out water vapor and dries up. (3) Gradually, the heated moist air goes up and leaves the drying chamber through the air outlet at the high end of the dryer.

Literature review

The performance of the solar drying system is highly influenced by the performance of the collector. Therefore, several studies have been conducted in order to improve the performance of the solar dryer Belhamri [1] studied a simple efficient and inexpensive solar batch dryer for agriculture products. During periods of low sunshine a heater is used. Onion was chosen as the dried product because of its swift deterioration characteristic. The results showed that drying is affected by the surface of the collector, the air temperature, and the product characteristics. Muller et al. [2] designed and constructed a dryer with a collector area of 16.8 m^2 which is expected to dry 195.2 kg of fresh mango(100 kg of sliced mango) from 81.4% to 10% wet basis in 2 days under ambient conditions during harvesting period from April to June. Ismail et al. [3] designed and constructed a solar dryer based on preliminary investigations for mango slices drying under controlled conditions. The designed dryer with a collector area of 16.8 m^2 was expected to dry 195.2 kg of fresh mango (100 kg of sliced mango) from 81.4% moisture level to 10% on wet basis in 2 days under ambient conditions during harvesting period from April to June. Mujumdar et al. [4] studied briefly the emerging drying methods and selected recent developments applicable to postharvest processing. In their study, they included the heat pump-assisted drying with multimode and time-varying heat inputs, low and atmospheric pressure superheated steam drying, modified atmosphere drying, intermittent batch drying, osmotic pretreatments, microwave-vacuum drying etc. Bolaji et al. [5] developed a simple and inexpensive mixed-mode dryer from locally sourced materials. Bukola et al. [6] experimentally found out the performance evaluation of a mixed-mode solar dryer for food preservation. The temperature increase inside the drying cabinet was up to 74% for about 3 h immediately after 12noon. The drying rate and system efficiency were 0.62 kg/h and 57.5%, respectively. Sarsavadia [7] developed a solar-assisted forced convection dryer to study the effect of airflow rate (2.43, 5.25, 8.09 kg/min), air temperature (55°C, 65°C, 75°C), and fraction of air recycled (up to 90%) on the total energy requirement in drying of onion slices. Kumar et al. [8] used a natural convection mixed-mode solar dryer in performing the experiments on potato cylinders and slices of the same thickness of 0.01 m with respective length and diameter of 0.05 m to investigate the convective heat transfer coefficient. Sreekumar et al. [9] developed a new type of efficient solar dryer with an arrangement to absorb maximum solar radiation by the absorber plate. Abene et al. [10] studied experimentally to improve the efficiency-temperature rise couple of the flat plate solar collector by considering several types of obstacles disposed in rows in the dynamic air vein of the flat collector. Ramana Murthy [11] studied various aspects of solar driers applied to drying of food products at a small scale. Karim et al. [12] studied experimentally the effect of different operating variables on drying potential and drying time. Smitabhindu et al. [13] used a simulation and optimization model to minimize the drying cost per unit of dried banana.

Methods

Solar dryer design considerations

The following points were considered in the design of the natural convection solar dryer system:

1. The amount of moisture to be removed from the given quantity of grapes
2. Harvesting period during which drying is needed
3. The daily sunshine hours for the selection of the total drying time
4. The quantity of air needed for drying
5. Daily solar radiation to determine energy received by the dryer per day and
6. Wind speed for the calculation of air vent dimensions.

Construction of mixed-mode solar dryer

The materials used for the construction of mixed-mode solar dryer were inexpensive and easily obtainable in the local market. Figure 2 shows the essential features of the dryer consisting of the solar collector (air heater), drying cabinets, and drying trays.

Solar collector setup

The experimental setup is an open-flow loop that consists of a test duct with entrance and exit sections, a blower unit, control valve, orifice plate, various devices

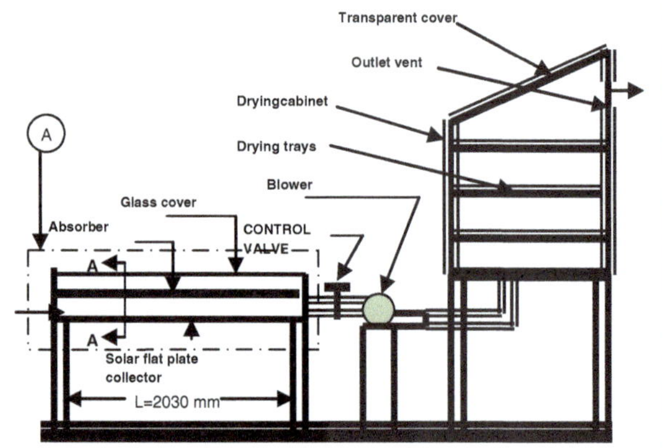

Figure 2 Schematic diagram of experimental setup.

for measurement of temperature, and fluid head. The flow system consists of an entry section, a test section, exit section, a flow meter, and a centrifugal blower. The setup consists of two identical wooden ducts: one is rough absorber duct and other one is smooth absorber duct. Each duct size is 2,030 mm × 200 mm × 25 mm (dimensions of inner cross section) and is constructed from wooden panels with 32 mm thickness. The test section length is 1,500 mm (33.75 D). The entry and exit lengths were 177 mm (2.3√WH) and 353 mm (5√WH), respectively, as shown in Figure 2.

The absorber plate material which is an aluminum sheet with 1,500 mm × 200 mm in size has a thickness of 3 mm which is painted black to increase the absorbing capacity of the plate. Another absorber plate is roughed with strips to provide obstacle to the path of air to obtain the maximum temperature. The solar collector assembly consisted of airflow channels enclosed by transparent cover (glazing).

The glazing is a single layer of 4-mm thick transparent glass sheet. It has a surface area of 200 mm by 1,500 mm. The outlet of duct is connected to the orifice meter with an inclined manometer to measure the mass flow rate of air. The outlet of the orifice meter is connected to the inlet of the blower. The outlet of blower is connected to the inlet of the cabinet dryer.

The drying cabinet

The drying cabinet, together with the structural frame of the dryer, was built from well-seasoned wood which could withstand termite and atmospheric attacks. An outlet vent was provided toward the upper end at the back of the cabinet to facilitate and control the convective flow of air through the dryer. The roof and the two opposite side walls of the cabinet are covered with transparent glass sheets of 4-mm thick, which provided

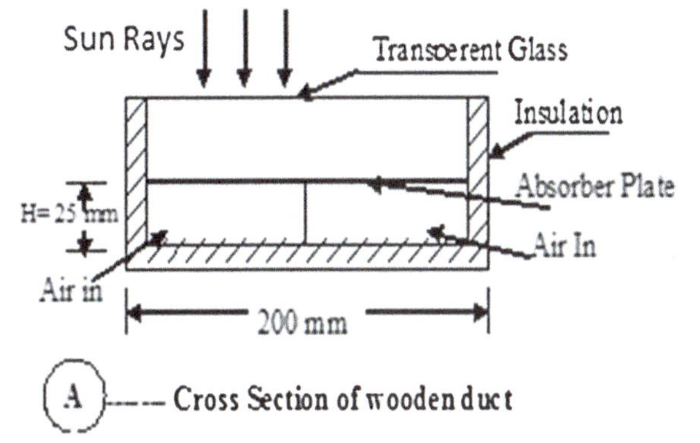

Figure 3 The drying cabinet.

additional heating due to greenhouse effect as shown in Figure 3.

Construction details

In the constructed solar drying cabinet, the materials used in the construction include 3/4 plywood, perspex glass, wooden bars, (for construction of the body) nails, wire mess, black paint, and hardwood (Figure 4). Perspex glass was used as glazing surface to cover the top and the sides. The top glazing measures 550 mm × 500 mm, and the each two side measures 640 × 520 mm. The plywood with 550 mm × 600 mm size was used in covering the base. The door of the dryer was made of the wood 520 mm × 500 mm, while the opposite side of the door was covered by 540 × 500 mm of the Perspex glass. The dryer skeleton was formed with wood raised 200 mm from the ground.

Drying trays

The drying trays are contained inside the drying chamber and are constructed from a double layer of fine chicken wire mesh with a fairly open structure to allow drying air to pass through the food items. The three trays were separated with a gap of 100 mm. The trays are arranged in a zigzag manner. The trays were made of wire mess and hardwood measuring 30 mm in depth and with dimension of 300 mm × 540 mm, as shown in Figure 5.

The orientation of solar collector

The flat plate solar collector is kept horizontally and oriented in such a way that it receives maximum solar radiation during the desired season of use. The best stationary orientation is south in the northern hemisphere and north in the southern hemisphere. Therefore, solar collector in this work is orientation facing south.

Experimental condition and assumption

We summarize the conditions during our experiment and our assumptions in Table 1.

Operation of the dryer

The dryer was a passive system in the sense that it had no moving parts. The sun rays entering through the collector glazing energizes it. The absorption of the rays is enhance by the inside surface of the collector that were painted black and the absorbed energy heats the air

Figure 4 Photograph of the experimental set.

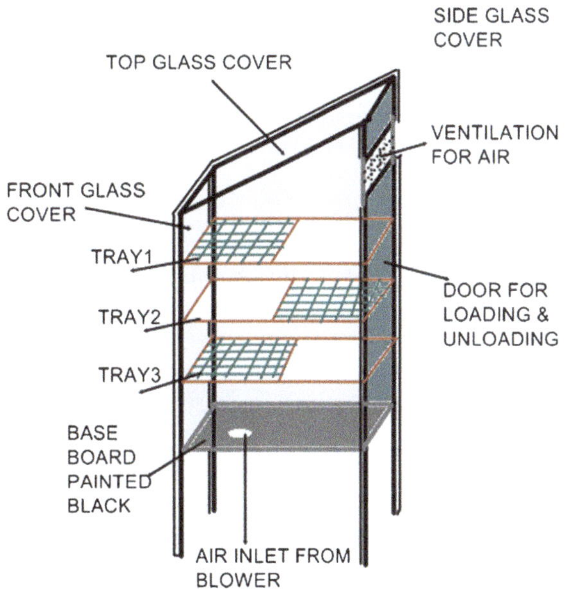

Figure 5 Three-dimensional view of the drying cabinet.

Table 1 Experimental condition and assumption

Item	Condition or assumption
Location	Bhopal
Crop	Grapes
Drying period	March to April 2009
Drying material quantity	3 kg
Loading rate (mp)	1 kg/tray
Initial moisture content, Mi	81.4% wb
Final moisture content, Mf	18.2% wb
Ambient air temperature, Tam	30°C (average for April)
Ambient relative humidity, RH_{am}	15% (average for April)
Max allowable temperature, T_{max}	65°C
Drying time (sunshine hours)	7 h (average for April)
Reynolds number, *Re*	3,000 to 12,000
Flat plate type	Smooth and rough
Test size	1500 mm length and 200 mm wide
Hydraulic diameter (D_x)	44.44 mm
Insulation, *I* (W/m^2)	700 to 900 W/m^2
Plate material	GI sheet

inside the collector. The greenhouse effect achieved within the collector drives the air current through the drying chamber. If the vents are open, the hot air rises and escapes through the upper vent in the drying chamber while cooler air at ambient temperature enters through the lower vent in the collector.

Therefore, an air current is maintained as cool air at a temperature T_a enters through the lower vents and hot air at a temperature *Te* leaves through the upper vent. When the dryer contained no items to be dried, the incoming air at a temperature T_a has a relative humidity H_a and the outgoing air at a temperature *Te* has a relative humidity H_e. Because $Te > T_a$ and the dryer contains no item, $H_a > H_e$. Thus, there is a tendency for the outgoing hot air to pick more moisture within the dryer as a result of the difference between $H_a > H_e$.

Dryer performance evaluation

The mixed-mode solar dryer shown in Figure 2 was tested during the months of March and April 2009 to evaluate its performance. During the testing period, the air temperatures at collector inlet, collector outlet, drying chamber, and ambient were measured by laboratory-type digital thermometer (accuracy ± 0.01°C) at regular interval of 1 h between the hours of 0900 and 1600 local time. The solar intensity was measured by means of a portable pyranometer placed horizontally and facing south. The dryer was loaded with grapes (1 kg in each tray) and the weight was measured at the start and at one-hour intervals thereafter. The initial weight and the final weight of grapes up to the stage when no further weight loss occurred were known. The dryer performance was evaluated and comparison was made with the result obtained with natural drying process and by forced convection with smooth and rough absorber plate.

Data reduction

Pressure drop calculation

Pressure drop was across the orifice plate measured by using the following relationship

$$\Delta Po = \Delta h \times 9.81 \times \rho m \times 1/5,$$

where ΔPo is the pressure difference, Δh is the difference of liquid head in manometer, and ρm is the density of mercury, i.e., 13.6×10^3

Mass flow measurement

Mass flow rate of air has been determined from the pressure drop measurement across the orifice plate by using the relation

$$m = \mathrm{Cd} \ \times \mathrm{Ao} \ \times \left[2\rho \ \Delta \mathrm{Po}/\left(1-\beta^4\right)\right]^{0.5},$$

where m is the mass flow rate of air (kg/s), Cd is the coefficient of discharge of orifice 0.62, Ao is the area of orifice plate (m^2), ρ is the density of air 1.157 kg/m^3, and β is the ratio of diameter (Do/Dp) = 26.5/53 = 0.5.

Velocity measurement

The velocity of the air flow V was measured by using the relation

$$V = m/\rho \mathrm{wh},$$

where m is the mass flow rate in kilograms per second, ρ is the density of air 1.157 kg/m^3, h is the height of duct in meters, and w is the width of duct in meters.

Reynolds number

Reynolds number is calculated as the ratio of viscous force to inertia force

$$Re = VD/\nu,$$

where V is the velocity of air (m/s), D is the hydraulic diameter (m), and ν is the kinematic viscosity, 16.70×10^{-6} m^2/s.

Heat gained by air

The heat transfer rate Q_a is given by

$$Q_a = m \ \mathrm{Cp}(\mathrm{To}-\mathrm{Ti}),$$

where m is the mass flow rate of air (kg/s), Cp is the specific heat of air (kJ/kg), To is the outlet temperature (°C), and Ti is the inlet temperature (°C).

The heat transfer coefficient for the heated test section has been calculated from

$$h = \mathrm{Qa}/A_{\mathrm{p}}\left(t_{\mathrm{pav}} - T_{\mathrm{fav}}\right).$$

where Tpav is the avearge temperature of plate and Tfav is the average temperature of fluid.

Thermal efficiency

The thermal efficiency is calculated by using the relation

$$\eta = \mathrm{Qa}/I\mathrm{Ap},$$

where Qa is heat gain by air (W); Ap is the area of plate of collector (m^2); and I is the solar insolation (W/m^2).

Removal of moisture

The amount of moisture in kilogram to be removed from the product was calculated using the following equation:

$$\mathrm{Mw} = \mathrm{Mp(Mi{-}Mf)/(100{-}Mf)},$$

where Mp is the initial mass of product to be dried (kg), Mi is the mass of sample before drying, and Mf is the mass of sample after drying.

Moisture content

The moisture content (MC) is given as

$$\mathrm{(MC)\%} = \mathrm{(Mi{-}Mf)/td} \times 100\%,$$

where Mi is the mass of sample before drying, Mf is the mass of sample after drying, and td is the drying time.

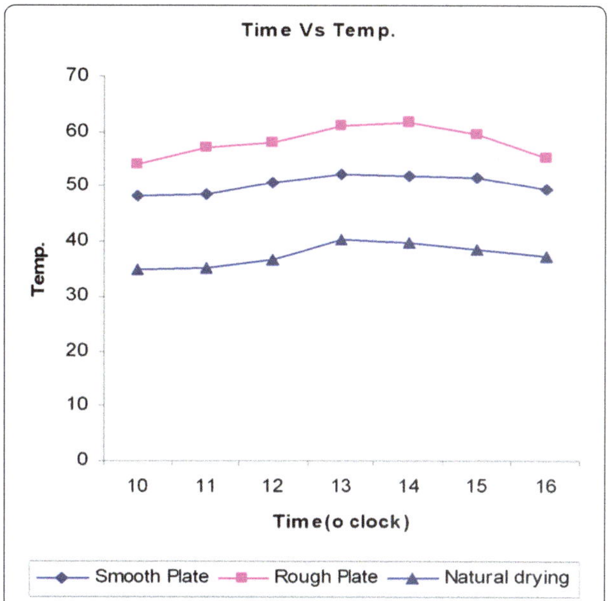

Figure 7 Variation of temperature with respect to time.

Results and discussion

The test is carried out without load to collect the data at various Reynolds number ranges from 3,000 to 15,000. The various parameters such as Nusselt number, Heat transfer coefficient, friction factor, and thermal efficiency are calculated at each set of Reynolds number.

Figure 6 shows the variation in thermal efficiency with Reynolds numbers of different values. As Reynolds number increases, the thermal efficiency also increases. The maximum efficiency is obtained at Reynolds number 12,000. Beyond Reynolds number 12,000, it is found that

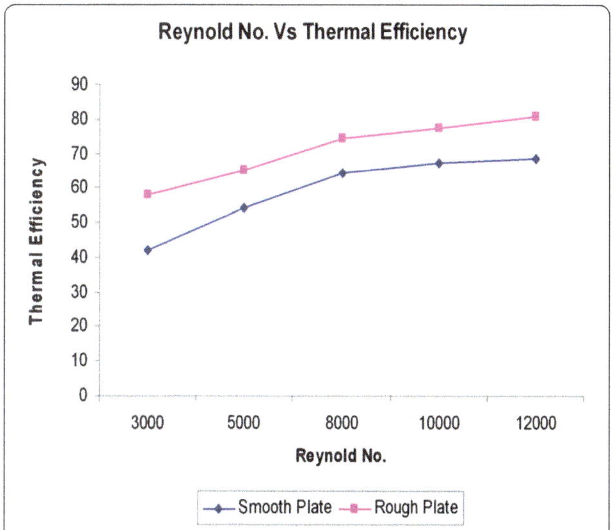

Figure 6 Variation of thermal efficiency versus Reynolds number.

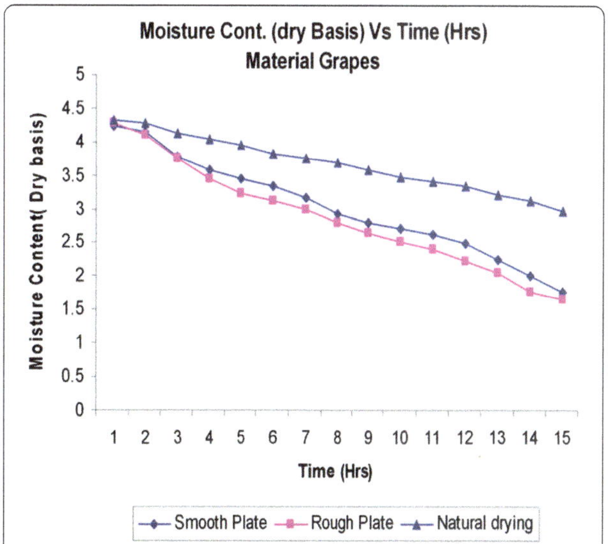

Figure 8 Variation of moisture content (dry basis) versus time.

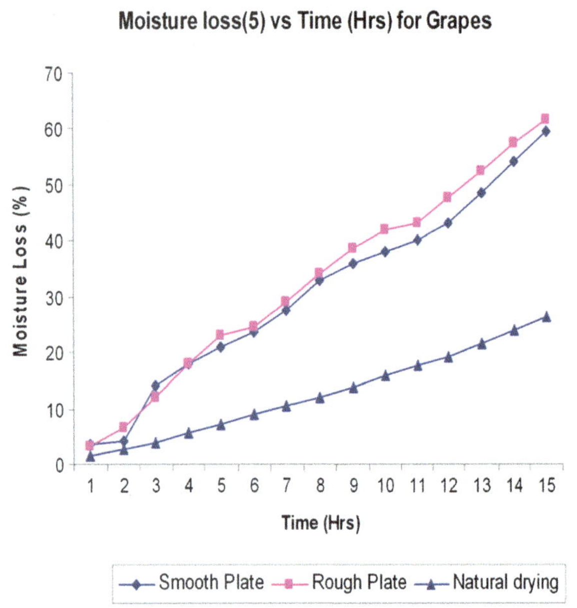

Figure 9 Variation of moisture loss (% basis) versus time (h).

the thermal efficiency decreases. Hence, the entire tests are carried out at fixed Reynolds number of 12,000.

Figure 7 shows a typical day results of the hourly variation of the temperatures in the solar collector and in the drying cabinet compared to the ambient temperature. The dryer is hottest about mid-day when the sun is usually overhead.

The plot shows temperatures versus the local time in hours for performance evaluation of grapes drying. The ambient temperature is quite low, varying from a minimum of 32°C to a maximum of 40°C. This is followed by the temperature of smooth plate which ranges from a minimum of 48°C to a maximum of 53°C. The temperature in the roughed plate is the highest, ranging from 54°C to 62°C. Thus, the hourly variation of the temperatures inside the cabinet and air heater is much higher than the ambient temperature during most hours of the daylight. The temperature increase inside the drying cabinet was up by 22°C for 72.0% of the time of exposure. This indicates a prospect for better performance than open-air sun drying.

Figure 8 shows the drying curve for grapes in the mixed-mode solar dryer. It was observed that the drying rate increased due to the increase in temperature. It was observed that the moisture removal rate is increased by using the roughed plate collector as compared to smooth plate and open sun drying. Moisture content (dry basis) of grapes reduces with time and is more in the roughed plate. Thus, it shows that the time period required for drying is greatly reduce by using roughed absorber plate.

Figure 9 shows the moisture loss percentage basis versus time period. The moisture loss for smooth plate and rough plate collectors is nearly almost the same with some small differences as compared to the natural (open sun) drying process. The moisture loss (% basis) is more in the roughed plate compared to the smooth plate. To remove the moisture (81.0%) from grapes by open sun drying, it takes around 7 to 8 days. By using mixed-mode solar dryer, this time is reduced to 3 to 4 days.

Conclusions

Based on the above results, the following conclusions can be drawn:

1. A simple and inexpensive mixed-mode solar dryer was designed and constructed using locally sourced materials.
2. The hourly variation of the temperatures inside the cabinet and air heater is much higher than the ambient temperature during the most hours of the day.
3. The temperature increase inside the drying cabinet was up to 22°C (64.5%) for most the hours in the noon time. The drying rate, collector efficiency, and percentage of moisture removed (dry basis) for drying grapes were 0.38 kg/h, 67.5%, and 85.4%, respectively.
4. The dryer exhibited sufficient ability to dry food items at a reasonable rapid time to a safe moisture level and simultaneously ensured a superior quality of the dried product
5. The drying of grapes in the open sun takes 7 to 8 days during clear sunny weather conditions. However, it only takes 4 to 5 days in the solar cabinet dryer under similar weather conditions. Also, the quality of dried grapes is remarkably better in cabinet dryer compared to open sun drying as the product is protected from dust and insects.
6. The drying rate of the dryer for grapes is 0.24 kg/day while that of open air is 0.1 g/day. This shows that the dryer performance is much better than open sun drying. The moisture to be removed from the grapes is 0.7 kg. The dryer efficiency is 31.0% per day.
7. The dryer is easy to build and required only semi-skilled laborer and limited facilities to fabricate. Thus, the dryer is suitable for use in urban as well as rural areas of the country.

Competing interests
The authors declare that they have no competing interests.

Authors' contributions
Both authors, CBP and JLB, contributed equally and significantly in writing the paper. Both authors read and approved the final manuscript.

Authors' information
CBP is an assistant professor in the Department of Mechanical Engineering in SISTec-R, Bhopal. JLB is an associate professor in the Department of Mechanical Engineering in MANIT, Bhopal.

Acknowledgment
The authors would like to thanks to ex-director of Regional Research Laboratory in Bhopal, for supporting in the revision and corrections as suggested by the reviewers. The authors would also like thank the reviewers for their valuable comments and suggestions.

Author details
[1]Department of Mechanical Engineering, Sagar Institute of Science, Technology and Research (SISTec-R), Bhopal 462044, India. [2]Department of Mechanical Engineering, Maulana Azad National Institute of Technology (MANIT), Bhopal 462051, India.

References

1. Bennamoun, L, Belhamri, A: Design and simulation of a solar dryer for agriculture products. J Food Eng. **59**, 259–266 (2003)
2. El-Shiatry, MA, Muller, J, Muhlbauer, W: Drying fruits and vegetables with solar energy in Egypt. AMA **4**, 61–64 (2002)
3. Akoy, E, Ismail, M, El-Fadil, AA: Design and construction of a solar dryer for mango slices. In: Proceedings of International Research on Food Security, Natural Resource Management and Rural Development-Tropentag. University of Bonn, Bonn, Germany (2006)
4. Mujumdar, AS, Law, CL: Drying technology: trends and applications in postharvest processing. Food and Bioprocess Technol. **3**(6), 843–852 (2010)
5. Bolaji, BO, Olalusi, AP: Performance evaluation of a mixed-mode solar dryer, AU Journal of Technology, a Publication of Assumption University, Bangkok. Thailand **11**(4), 225–231 (2008)
6. Bolaji, BO: Performance evaluation of a simple solar dryer for food preservation. In: Proceedings of the 6th Annual Engineering Conference of School of Engineering and Technology. Federal University of Technology, Minna, Nigeria (2005)
7. Sarsavadia, PN: Development of a solar-assisted dryer and evaluation of energy requirement for the drying of onion. Renew. Energy **32**, 2529–2547 (2007)
8. Tripathy, PP, Kumar, S: Modeling of heat transfer and energy analysis of potato slices and cylinders during solar drying. Applied Thermal Eng. **29**, 884–891 (2009)
9. Sreekumar, A, Manikantan, PE, Vijayakumar, KP: Performance of indirect solar cabinet dryer. Energy Conversion and Management **49**, 1388–1395 (2008)
10. Abene, A, Dubois, V, Le Ray, M, Ouagued, A: Study of a solar air flat plate collector: use of obstacles and application for the drying of grape. J. Food Eng. **65**, 15–22 (2004)
11. Ramana Murthy, MV: A review of new technologies, models and experimental investigations of solar driers. Renewable and Sustainable Energy Reviews. **13**(4), 835–844 (2009)
12. Karim, MA, Hawlader, MNA: Drying characteristics of banana: theoretical modelling and experimental validation. J. Food Eng. **70**, 35–45 (2005)
13. Smitabhindu, R, Janjai, S, Chankong, V: Optimization of a solar-assisted drying system for drying bananas. Renew. Energy **33**, 1523–1531 (2008)

8

Assessment of a practical model to estimate the cell temperature of a photovoltaic module

Rodolfo Araneo*, Umberto Grasselli and Salvatore Celozzi

Abstract

Instantaneous solar irradiance profiles or solar irradiation data collected with small time intervals (e.g., minutes) are usually required for the energy simulation of photovoltaic systems, especially as concerns the estimation of the cell temperature. However, meteorological stations and technical standards often provide just monthly average values of the horizontal daily solar irradiation; extensive climate databases which make available up to date hourly observation data, or satellite-derived data are seldom available. The goal of the present paper is to investigate the suitability and the accuracy of a methodology aimed at estimating the time profile of the cell temperature of a photovoltaic system on the basis of only the monthly mean values of the daily global irradiation on a horizontal surface. The methodology consists of a chain of well-established models which are applied one after another, in a step-by-step procedure, in order to derive the cell temperatures from the solar radiation data. In particular, we selected different models as possible candidates for each step of the methodology and compared their predictions with measured data to identify the most suitable ones. In addition, we tried several combinations of models in order to identify the most accurate combination. Comparisons with data measured in Rome confirm the suitability of the proposed approach and give information about its accuracy.

Keywords: Photovoltaic cell; Irradiation; Irradiance; Linear model; Transient cell temperature

Background

In the last century, it has become clear worldwide, the key role, that energy plays in human life [1,2]: it is at the base of our modern life and economy. Currently, the growing attention and increasing significance to environmental issues have stimulated countries to exploit renewable energy resources and to encourage their use [3-6]. Among them, solar photovoltaic (PV) energy is considered one of the leading potential sources of electricity for the 21st century [7-11]: it utilizes an abundant energy source (the sun), has no emissions, and can be easily integrated in buildings, and the cost of the installed kWp is decreasing and becoming more and more affordable with payback periods shorter and shorter. In fact, in the recent years, the generation of solar electricity from PV systems has penetrated the energy market in those countries where clear and stable policy for subsidies have been made.

The rapid growth of the solar industry has expanded the importance of PV system design and application for more reliable and efficient operation [12,13]. The design of PV systems in an economically optimal way [14-16] is usually done through detailed computer simulations [17,18]: transient analysis is useful when the different energy phenomena which take place in the heart of the production systems [19] should be accounted for, e.g., if they present some kind of solar tracking [20].

In performing such an analysis, the accurate evaluation of the operating temperature of the PV device (either a simple module or a PV/thermal collector of a building-integrated PV array) is of paramount importance [21] because several performance parameters of the PV system depend on it through the so-called temperature coefficients [22]. The derivatives can be determined for short-circuit current (I_{sc}), maximum-power current (I_{mp}), open-circuit voltage (V_{oc}), maximum-power voltage (V_{mp}), and maximum power (P_{mp}), as well as fill factor (FF) and, finally, efficiency (η). Particular attention is focused on the energy conversion factor of a PV system that is commonly described by the electrical efficiency η

*Correspondence: rodolfo.araneo@uniroma1.it
[1] D.I.A.E.E. – Electrical Engineering Division, "Sapienza" University of Rome, Via Eudossiana 18, Rome 00184, Italy

[23] defined as the ratio of the electricity generated to the global solar irradiation impinging on the collector's surface. Furthermore, the temperature coefficients for PV systems are directly related to the temperature coefficients for their individual cells, and thus, the so-called cell temperature T_c is the real key parameter to be identified because it affects, directly or indirectly, the energy conversion efficiency [24] of any PV system. In addition, it plays an important role in PV system design and sizing since often, the worst case-operating conditions dictate the array size.

More physical insights about the dependence of the electric energy conversion mechanisms on the cell temperature can be obtained looking at the solid state physics of the cell [25-27]: the energy bandgap of semiconductors and, consequently, the quantum conversion efficiency tends to decrease as the temperature increases [24,28] due to the fact that the interatomic spacing increases when the amplitude of the atomic vibrations increases as a consequence of the increased thermal energy. An increased interatomic spacing decreases the potential seen by the electrons in the material, which in turn reduces the size of the energy bandgap (a direct increase/decrease of the interatomic distance and, consequently, bandgap can be obtained also by applying high compressive/tensile stress [29,30]).

Despite its great importance, the cell temperature T_c depends on such a lot of parameters and weather variables that its correct evaluation is the critical point of any methodology aimed at properly sizing PV systems. A scan of the relevant literature [31,32] easily produces an impressive number of correlations expressing the cell temperature as a function of the pertinent weather variables, namely ambient temperature, local wind speed as well as the solar irradiance on the surface of PV systems, which are tilted towards the sun to maximize the amount of the incident solar radiation [25,26,33]. Radiation data, such as hourly direct and diffuse irradiance on the tilted surface, are usually required but smaller time steps could be necessary when load profiles vary with smaller time constants. However, meteorological stations and National or International Standards often provide only daily global irradiation data on horizontal surface (sometimes, they measure also the diffuse component), which are commonly summarized in monthly average values [34-36].

In this framework, the objective of the present paper is a task of great practical importance from an engineering point of view: we want to define a methodology for estimating the operating cell temperature T_c starting *only* from the monthly mean value of the daily global irradiation on a horizontal surface. The methodology consists of a chain of models that are well established in literature and that we apply in a step-by-step procedure to derive the cell temperature from the solar radiation data. In particular, we compare several models, focusing on how to couple them in order to obtain the best accuracy in the predicted data; finally, we investigate the suitability and the accuracy of the proposed methodology. In addition, we provide two new models for calculating the solar irradiance and the ambient temperature profiles. Comparisons with data measured in Rome confirm the applicability of the proposed approach and give information about the accuracy of the model.

The paper is organized as follows: in subsection 'Theory and models', the proposed procedure is illustrated; subsection 'Irradiance model' analyzes the radiation models used to predict the irradiance profiles on a tilted surface from the monthly average daily global irradiations on a horizontal surface; the models used to predict the PV cell temperature from the irradiance profiles are described in subsection 'Operating temperature of the photovoltaic cell'. Section 'Results and discussion' presents the relevant results; the measurement set-up is described in subsection 'Set-up', while in subsection 'Assessment of the models', a statistical analysis is performed on the results in order to assess the validity of the whole procedure, identifying and discussing the most accurate models. Finally, the main conclusions of the work are drawn.

Methods

Theory and models

The goal of the procedure described in the following subsections is to estimate the transient temperature of the photovoltaic cell during the day n_d of the year, placed on a surface S_t tilted γ_t with respect to the horizontal plane and rotated α_t with respect to the North-South direction, starting from the monthly average daily global irradiation value $\overline{H}$ on a horizontal surface (the configuration is illustrated in Figure 1). The procedure consists of five steps that are applied one after another in succession as shown in Figure 2. Each step is based on the application of well-established models which are compared and discussed in order to obtain an accurate and robust procedure.

Irradiance model

As is well known, the quantity of solar radiation reaching the Earth's surface during a day is governed by several factors: the solar elevation at noon, the duration of the day, the turbidity of the air, the total amount of water vapor in the air, and the type and amount of clouds [37]. When a direct measurement is not available, as often happens, H is taken as the monthly mean value $\overline{H}$ referred to the 'average day' of the month, i.e., the day when the solar declination δ is equal to the average value of the month [26,33]. The monthly average daily irradiation can be obtained from national standards or measurement campaigns or through an Angstrom-type regression equation which relates the ratio between monthly average daily global irradiation and monthly average daily extra-terrestrial irradiation at

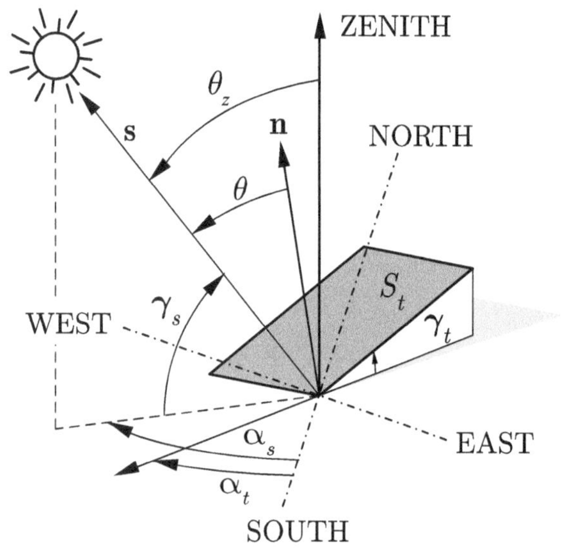

Figure 1 Geometrical configuration of PV cell placed on surface S_t. Tilted γ_t from horizontal plane and rotated α_t from North-South direction.

a particular location to the average fraction of possible sunshine hours [38].

From $\overline{H}$, the global hourly irradiation H_h on a horizontal surface during the day n_d of the month has to be somehow evaluated in order to compute the hourly global irradiation G_{Th} incident on S_T (for all the details see [33]), as

$$G_{Th} = B_{Th} + D_{Th} + R_{Th} \; , \tag{1}$$

where the direct B_{Th}, sky-diffuse D_{Th} and ground reflected (albedo) R_{Th} components are given by

$$B_{Th} = B_h r_b = (1 - k_d)\, r_b H_h \tag{2a}$$

$$D_{Th} = D_h r_d = k_d r_d H_h \tag{2b}$$

$$R_{Th} = \rho \frac{1 - \cos \gamma_t}{2} H_h \; . \tag{2c}$$

In Equation 2, B_h, D_h and $H_h = B_h + D_h$ are, respectively, the direct diffuse and global hourly irradiation on a horizontal surface, $r_b = \cos\theta / \cos\theta_z$ is the beam radiation factor, with θ the solar incidence angle on S_T and θ_z the zenith angle of the Sun ([33], chap. 1), k_d is the cloudiness index, $r_d = D_{Th}/D_h$ is the conversion factor for the hourly diffuse irradiation, and ρ is the ground reflectivity. It is assumed that the ground has very low reflectance, i.e., the reflected component is much lower than the sum of the direct and diffuse irradiation, so that an isotropic model may be used for the computation of the ground albedo. The relevant parameters in Equation 2, i.e., H_h, k_d, and r_d, are evaluated as described in the following subsections: 'Horizontal hourly global irradiation values', 'Hourly diffuse and direct irradiation', and 'Hourly diffuse irradiation on a tilted surface', respectively.

Once the hourly global irradiation G_{Th} is known, the irradiance profile $I_{gT}(t)$ on the tilted surface is finally computed as reported in subsection 'Solar irradiance profiles'.

Horizontal hourly global irradiation values

Several well-established methods have been proposed in literature to convert the average daily horizontal global irradiation H into horizontal hourly global values H_h. Most of them assume that the day is symmetric, i.e., with the same irradiation values for hours situated symmetrically with respect to the solar noon [39-41], while a few of them try to account for the asymmetries between morning and afternoon [42,43]. The characterization of a general distribution of hourly solar irradiation is made difficult by the non-stationary nature of the solar irradiation, which is affected by an unpredictable noise due to a variety of factors. Recently, sophisticated methods have been proposed, such as neural networks and autoregressive average models [44,45]. Though all these methods are reliable and show good accuracy, they make the formulation more

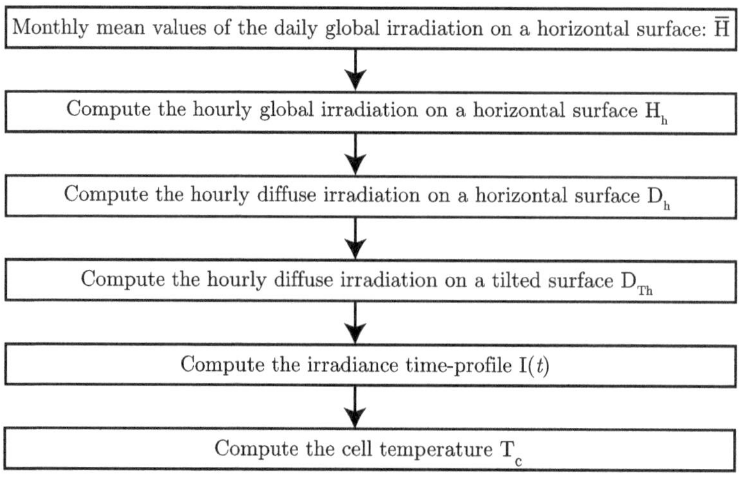

Figure 2 Steps of the proposed procedure.

difficult to be implemented, so that the simple but effective correlation model presented in [41] is used. The correlation has been confirmed to be compatible with results obtained for Canada, India, Israel, and lastly, Corsica [46] and to work best for clear days, when solar processes produce most of the output. According to this model, the hourly values H_h are computed as

$$H_h = r_t \overline{H}\ , \tag{3}$$

where the ratio r_t is given by

$$r_t = \frac{1}{2}\left(a + b\cos\frac{\omega_2+\omega_1}{2}\right) \cdot \frac{(\sin\omega_2 - \sin\omega_1) - \frac{\pi(\omega_2-\omega_1)}{180^\circ}\cos\omega_s}{\sin\omega_s - \frac{\pi\omega_s}{180^\circ}\cos\omega_s} \tag{4}$$

with

$$a = 0.4090 + 0.5016\sin(\omega_s - 60^\circ) \tag{5a}$$

$$b = 0.6609 + 0.4767\sin(\omega_s - 60^\circ)\ . \tag{5b}$$

In (4), ω_1 and ω_2 are the hour angles at the beginning and end, respectively, of the hour in question ($\omega = 15^\circ(12-h)$, with h as the solar time), and ω_s is the sunset hour angle for the day equal to $\cos^{-1}(-\tan\delta\tan\phi)$, with ϕ as the latitude of the location and δ as the declination for the day, which can be computed by means of the Cooper or Spencer formulas ([33], chap. 1).

In (3) the monthly mean value $\overline{H}$ of the global daily solar irradiation on a horizontal surface has been used and the coefficient r_t accounts for the specific day n_d of the month through the declination angle δ. The suitability and accuracy of (3) in the framework of the whole proposed methodology has been assessed in section 'Results and discussion'.

Hourly diffuse and direct irradiation

Since the pioneer work of Liu and Jordan [39] in the early 1960s, several models have been proposed in literature to evaluate the hourly average cloudiness index $k_d = D_h/H_h$, where D_h is the hourly sky-diffuse irradiation on a horizontal surface. A complete list of these models is beyond the scope of the present paper; comparative studies among the most frequently used correlations can be found in [47-49]. These correlations are usually expressed in terms of first- to fourth-degree polynomials relating the diffuse fraction k_d with the hourly clearness index $k_t = H_h/H_{0h}$, defined as the ratio of the hourly global solar irradiation H_h and the hourly extraterrestrial solar irradiation on a horizontal surface H_{0h} ([33], chap.1).

In the present work, we have compared the four models from Karatasou [50], Erbs [51], Miguel [52], and Reindl [53] against experimental irradiation values for the Italian location of Rome over a period of 2 years.

The Karatasou [50] model is a third-order polynomial correlation based on data from one location at Athens (Greece):

$$\begin{aligned} &k_d = 0.9995 - 0.05k_t - 2.415k_t^2 + 1.4926k_t^3 && k_t \le 0.78 \\ &k_d = 0.20 && k_t > 0.78\ . \end{aligned} \tag{6}$$

The Erbs [51] model is a fourth-order polynomial correlation based on data from four locations in USA:

$$\begin{aligned} &k_d = 0.9996 - 0.09k_t && k_t \le 0.22 \\ &k_d = 0.951 - 0.1604k_t + 4.388k_t^2 \\ &\quad -16.638k_t^3 + 12.336k_t^4 && 0.22 < k_t \le 0.80 \\ &k_d = 0.1652 && k_t > 0.80\ . \end{aligned} \tag{7}$$

The Miguel [52] model yields a third-order polynomial for k_d using a dataset from several sites in the North Mediterranean Belt (e.g., France, Greece, Italy, Portugal, and Spain):

$$\begin{aligned} &k_d = 0.9943 - 0.081k_t && k_t \le 0.21 \\ &k_d = 0.724 + 2.734k_t - 8.32k_t^2 + 4.967k_t^3 && 0.21 < k_t \le 0.76 \\ &k_d = 0.1766 && k_t > 0.76\ . \end{aligned} \tag{8}$$

Finally, Reindl [53] established a very simple correlation, studying the influence of climatic and geometric variables on the hourly diffuse fraction based on data measured at five European and US locations:

$$\begin{aligned} &k_d = 1.0234 - 0.248k_t && k_t \le 0.30 \\ &k_d = 1.45 - 1.67k_t && 0.3 < k_t \le 0.78 \\ &k_d = 0.1474 && k_t > 0.78\ . \end{aligned} \tag{9}$$

A comparative analysis of the predictions of the four models based on consolidated standard statistical parameters [54] reported in section 'Results and discussion' has showed that the physically based method proposed by Miguel correlates better with the data collected in Rome.

Once the hourly diffuse irradiation component D_h is known, the hourly direct irradiation B_h is then computed as $B_h = H_h - D_h$.

Hourly diffuse irradiation on a tilted surface

The conversion of the hourly horizontal diffuse irradiation D_h to hourly diffuse irradiation D_{Th} on the tilted surface S_T can be expressed as

$$D_{Th} = r_d D_h\ , \tag{10}$$

where r_d is the conversion factor.

The methods proposed in literature for the evaluation of r_d are classified into isotropic or anisotropic models. The isotopic models assume that the sky dome irradiates uniformly so that the diffuse radiation incident on the tilted surface depends on the fraction of the sky dome seen by it. The anisotropic models try to model the anisotropy of the diffuse radiation, decomposing it in different components, i.e., the circumsolar component

(Sun's aureole), the brightness of the horizon, and the isotropic component of the sky dome. Validation studies and testing of several well-established models to predict r_d have been performed in [55-57]. However, results are not always in agreement, and ranking these models according to their accuracy is not simple. In fact, their abilities to predict the diffuse radiation on a tilted surface are *a priori* function of the atmospheric conditions (clear, partially cloudy, or overcast sky), since they take assumptions about the isotropy or anisotropy of the sky dome. The way in which the anisotropy is accounted for makes them strongly dependent on the surface orientation (it has been observed that all models produce large errors for east-west-facing PV surfaces [55]). In addition, several models are based on empirical data obtained for some particular geographical locations, so that they show enhanced/poor accuracy when used in locations with similar/different irradiation conditions. Anyway, the use of well-known statistical indicators [54], and much more complex statistical analyses [56] performed on data collected from several locations in the world, have generally, but not always, indicated that the Ma-Iqbal [58], Reindl [59], Muneer [60], and Perez [61,62] models are those that give the most accurate predictions for small azimuth angle α_t (i.e., south-facing surfaces). In addition, in [56], it was found that the Ma-Iqbal model performs best under all sky conditions, clear and partially cloudily, whereas the Muneer model gives the best results for cloudy sky conditions.

In this framework, we have selected these four models and compared their predictions against experimental irradiation values, as reported in section 'Results and discussion'.

In the model proposed by Ma-Iqbal [58], the diffuse irradiance on an inclined plane is considered to be the addition of the circumsolar component coming from the direction near the solar disk and a diffuse component isotropically distributed from the rest of the sky. These two components are weighted according to an index of anisotropy which represents the transmittance through the atmosphere of direct irradiance. Unlike the original model of Hay [63] who defined his own sky clarity factor $F = B_h/H_{0h}$, Ma and Iqbal used the clearness index k_t as index of anisotropy. According to the Ma-Iqbal model, the conversion factor r_d can be computed as

$$r_d = k_t r_b + (1 - k_t)\frac{1 + \cos\gamma_t}{2}\,. \tag{11}$$

The Reindl model [59] assumes the linearity of the isotropic and circumsolar contributions to the diffuse radiation on a tilted plane according to the Hay model and adopts the same correction factor that takes into account the brightness of the sky near the horizon used in the Temps-Coulson model [64]. The conversion factor r_d is determined as

$$r_d = F r_b + (1 - F)\frac{1 + \cos\gamma_t}{2}\left[1 + \sqrt{\frac{B_h}{H_h}}\sin^3\frac{\gamma_t}{2}\right], \tag{12}$$

where $F = B_h/H_{0h}$ is Hay's sky clarity factor.

Muneer [60], partially following the model proposed by Gueymard [65], considers that the irradiance can be expressed as a linear combination of values between fully covered sky and cloudless sky which in turn is the addition of the circumsolar component and a hemispheric factor. The conversion factor r_d is expressed as

$$r_d = T_M(1 - F_M) + F_M r_b\,, \tag{13}$$

where T_M is Muneer's tilt factor defined as the ratio between the slope background diffuse radiation and the horizontal diffuse radiation given by

$$T_M = \frac{1 + \cos\gamma_t}{2} - \frac{2B}{3 + 2B}\left[\frac{\gamma_t\cos\gamma_t - \sin\gamma_t}{\pi} + \frac{1 - \cos\gamma_t}{2}\right], \tag{14}$$

and F_M is a composite anisotropic index, equal to Hay's sky clarity factor F for non-overcast conditions and 0 for overcast sky. In Equation 14, B is the radiation distribution index whose values depend on the particular sky and azimuthal conditions, and for the location for southern European locations, Muneer recommends the following correlation:

$$\frac{2B}{3 + 2B} = \pi\left(0.00263 - 0.712F - 0.688F^2\right)\,. \tag{15}$$

The Perez model is more computationally cumbersome than the others because it represents the isotropic diffuse, circumsolar, and horizon-brightening components with more details by using empirically derived coefficients. Yet, the model developed by Perez in [62] is considerably simpler, and more accurate, than the original model proposed in his first work [61]. In fact, the conversion factor r_d is computed as

$$r_d = (1 - F_1)\cos^2\left(\frac{\gamma_t}{2}\right) + r_b' F_1 + F_2\sin\gamma_t, \tag{16}$$

where F_1 and F_2 are, respectively, the circumsolar and horizon brightness coefficients

$$F_1 = \max\left[0, \left(f_{11} + f_{12}\Delta + \frac{\pi\theta_z}{180}f_{13}\right)\right] \tag{17a}$$

$$F_2 = f_{21} + f_{22}\Delta + \frac{\pi\theta_z}{180}f_{23}\,, \tag{17b}$$

where θ_z is the beam radiation factor, $\Delta = k_d k_t$ is the brightness parameter, and f_{ij} are tabulated statistically derived coefficients [62] depending on the clearness parameter

$$\varepsilon = \frac{1}{1 + 5.535\cdot 10^{-6}\theta_z^3}\left(\frac{D_h + \frac{B_h}{\cos\theta_z}}{D_h} + 5.535\cdot 10^{-6}\theta_z^3\right). \tag{18}$$

In Equation 16 r'_b is the modified beam radiation conversion factor defined as

$$r'_b = \frac{\max(0, \cos\theta)}{\max(\cos 85°, \cos\theta_z)} \,. \tag{19}$$

Solar irradiance profiles

The transient simulation of the temperature of photovoltaic modules requires the knowledge of continuous profiles versus time t of the global solar irradiance I_{gT} on the tilted surface during the day n_d. Starting from the hourly mean values of the global irradiation G_{Th}, various studies consider that the solar irradiance is distributed over the time with a constant repartition which, however, has been demonstrated to be an unrealistic hypothesis, since this assumption does not provide a precise idea of the different transient energy phenomena which take place in the heart of the solar system. In [17], it was proposed that a linear model which allows the determination of irradiance data, averaged on small time-step Δt (e.g., 1 min), from hourly irradiation values. Starting from this work, a second-order model is proposed here.

The solar irradiance $I_j(t)$, on the tilted surface, is assumed to vary in a quadratic manner between the beginning time h_{j-1} and the ending time h_j of the jth hour of the day (e.g., the first hour begins at time t equal to $h_0 = 00:00$ and ends at time $h_1 = 01:00$), i.e.

$$I_j = a_j t^2 + b_j t + c_j \,, \tag{20}$$

with $j = 1, 2 \ldots 24$ and $h_0 = 0, h_1 = 1, \ldots, h_{24} = 24$. It is evident that it is necessary to enforce three equations for every hour occurring between sunrise h_{SR} and sunset h_{SS} times since the number of unknowns for every hourly profile is equal to three. As shown in Figure 3a, if sunrise or sunset does not occur in the jth hour, the three conditions to be enforced are as follows:

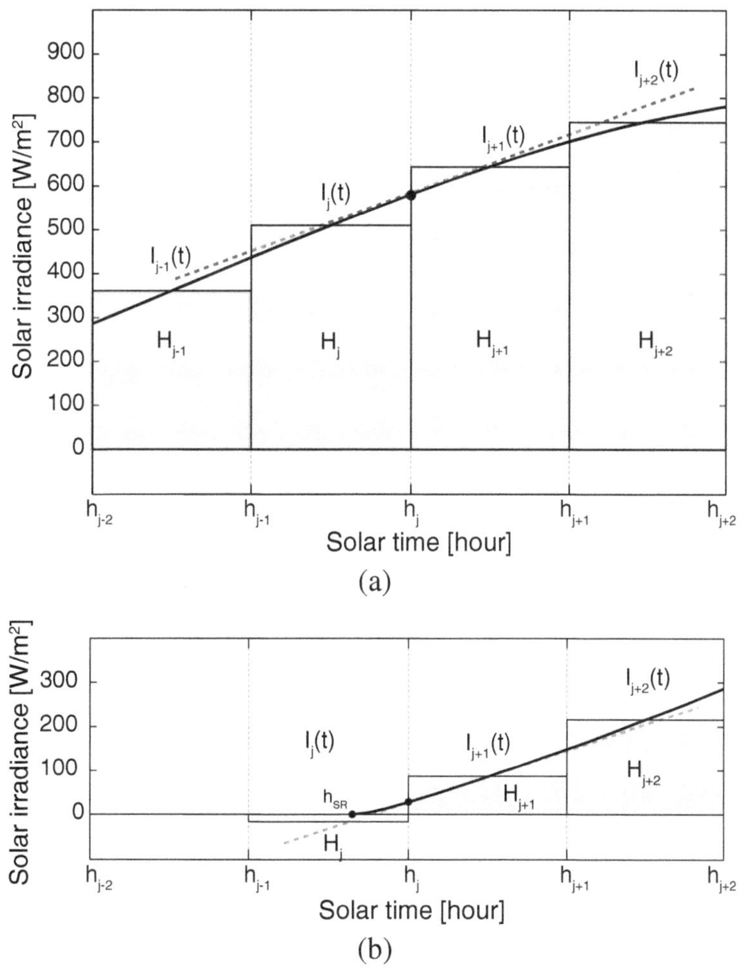

Figure 3 Illustration of quadratic model for approximating solar irradiance profile from hourly mean values of global irradiation. **(a)** Daylight hours. **(b)** Hour when sunrise takes place.

- the integration of the irradiance $I_j(t)$ over the hour must be equal to the hourly irradiation H_j (on tilted or horizontal surface),

$$\int_{h_{j-1}}^{h_j} I_j(t)\mathrm{d}t = H_j \; ; \tag{21}$$

- the profiles $I_j(t)$ and $I_{j+1}(t)$ and their first derivatives at every time h_j between two successive hours must be continuous

$$a_j h_j^2 + b_j h_j + c_j = a_{j+1} h_j^2 + b_{j+1} h_j + c_{j+1} \tag{22a}$$
$$2a_j h_j + b_j = 2a_{j+1} h_j + b_{j+1} \, . \tag{22b}$$

If sunrise or sunset occurs in the considered jth hour (Figure 3b), Equation 21 must be modified according to

$$\int_{h_{SR}}^{h_j} I_j(t)\,\mathrm{d}t = H_j \tag{23a}$$
$$\int_{h_{j-1}}^{h_{SS}} I_j(t)\,\mathrm{d}t = H_j \tag{23b}$$

because the model considers that the solar irradiance has been spread over the hour whereas it just took place from sunrise or until sunset.

Then, in the hour when sunrise (or sunset) takes place, the value of the irradiance profile at h_{SR} (or h_{SS}) must be set equal to zero, i.e., $I_j\,(t \le h_{SR}) = 0$ (or $I_j\,(t \ge h_{SS}) = 0$).

Operating temperature of the photovoltaic cell

The cell-operating temperature T_c is the proper temperature to use in order to predict the electrical performance of the PV module. Because of the internal processes which take place within the cells during their exposure to the sun, a large portion of the incident irradiance is degraded and released as heat. Standard heat transfer mechanisms must be accounted for to compute the appropriate energy balance on the cell/module leading to the prediction of T_c. At steady-state conditions, only convection and radiation mechanisms are usually considered, since they are prevalent on the conduction mechanism which merely transports heat toward the surfaces of the mounting frame (especially in the case of rack-mounting free standing arrays). A survey of the explicit and implicit correlations proposed in literature linking T_c with standard weather variables and material and system-dependent properties can be found in [32].

In the present work, we compare four different explicit equations against experimental data. Among the large number of correlations proposed in the literature, we have chosen the equations whose application appears to be the best and simplest. Hence, we have chosen equations that are explicit, depend on easily measurable parameters, and are of wide applicability. The four alternative correlations are [31,32,66,67]

$$T_c = T_a + \frac{I_{gT}}{I_{NOCT}}\left(T_{NOCT} - T_{a,NOCT}\right) \tag{24a}$$
$$T_c = T_a + \left(\frac{0.32}{8.91 + 2.0\,v_w}\right) I_{gT} \tag{24b}$$
$$T_c = 0.943\,T_a + 0.028\,I_{gT} - 1.528\,v_w + 4.3 \tag{24c}$$
$$T_c = T_a + 0.0138\,I_{gT}\,(1 + 0.031\,T_a)\,(1 - 0.042\,v_w) \tag{24d}$$

where T_{NOCT} is the so-called nominal operating cell temperature (the temperature T_{NOCT} is defined as the temperature of the cell at the conditions of the Nominal Terrestrial Environment (NTE): Solar irradiance $I_{NOCT} = 800\ \mathrm{W/m^2}$, ambient temperature $T_{a,\,NOCT} = 20°\mathrm{C}$, average wind speed is 1 m/s, nil electrical load, and free-standing mounting frame is oriented normal to solar noon), T_a is the ambient temperature and v_w is the wind speed (free stream wind speed in the windward side of the PV array).

Two remarks are necessary before applying the previous expressions. The cell temperature T_c may be higher than the back-side temperature T_b, which is the directly measured quantity, of a few degrees, being this difference dependent mainly on the module substrate material and the intensity of the solar irradiance $I_{gT}(t)$. The two temperatures are related through the simple linear expression [68]

$$T_c = T_b + \frac{I_{gT}}{I_{ref}}\Delta T_{ref} \, , \tag{25}$$

where I_{ref} is the reference solar irradiance of 1,000 $\mathrm{W/m^2}$ and ΔT_{ref} is the temperature difference under this reference solar irradiance. This temperature difference is typically ranging between 2°C and 3°C for flat-plate modules in an open-rack setting [68].

The evaluation of T_c requires an accurate estimation of the ambient temperature. Assuming it constant during the day and equal to the average value are unrealistic assumptions. Here, the following third-order approximation for a smooth transient from the minimum T_{min} to the maximum T_{max} temperatures registered during the day is assumed

$$T_a(t) = \frac{1}{(h_{SR} - h_{SS})^3}\left[(h_{SS} - t)^2(h_{SR} - 3h_{SS} + 2t)T_{max} + (3h_{SR} - h_{SS} - 2t)(h_{SS} - t)^2 T_{min}\right] . \tag{26}$$

It has been observed that the approximation (26) provides accurate agreement with the measured results during the hours from sunrise to sunset, when the PV system produces energy.

Results and discussion

The ability of the proposed procedure to predict the PV cell temperature is tested comparing the predicted results with experimental values measured in Rome during the 2 years 2010 and 2011. The four models selected for the computation of the cloudiness index k_d in subsection 'Hourly diffuse and direct irradiation', for the computation of the hourly diffuse irradiation on a tilted surface (i.e., the conversion factor r_d) in subsection 'Hourly diffuse irradiation on a tilted surface', and for the computation of the cell temperature T_c in subsection 'Operating temperature of the photovoltaic cell' are compared. Furthermore, we studied the best combination of these models in order to obtain the best accuracy of the predictions.

The accuracies have been assessed using well-known statistical indicators [54]: the mean bias error (MBE), the root mean square error (RMSE), and the coefficient of determination (R^2). They are defined as

$$\mathrm{MBE} = \frac{1}{N}\sum_{i=1}^{N} x_{p,i} - y_{M,i} \tag{27a}$$

$$\mathrm{RMSE} = \sqrt{\frac{1}{N}\sum_{i=1}^{N}\left(x_{p,i} - y_{M,i}\right)^2} \tag{27b}$$

$$R^2 = \frac{\mathrm{ESS}}{\mathrm{TSS}} = \frac{\sum_{i=1}^{N}\left(x_{p,i} - \bar{y}_M\right)^2}{\sum_{i=1}^{N}\left(y_{M,i} - \bar{y}_M\right)^2}, \tag{27c}$$

where N is the number of data points during each day, $x_{p,i}$ and $y_{M,i}$ are, respectively, the predicted and measured (observed) data points, $\bar{y}_M = \sum_{i=1}^{N} y_{M,i}$ is the mean value of the measured data, ESS is the explained sum of squares and TSS is the total sum of squares. The MBE determines whether the model underestimates or overestimates experimental data, the RMSE gives a measure of how well the model can predict the experimental data (it provides information on the short-term performance of the model by allowing a term-by-term comparison of the actual deviation between the estimated and the measured values), and R^2 provides a measure of how well the model explains observed phenomena (its value shows how well future outcomes can be predicted by the model).

Set-up

The irradiance and temperature data have been measured on the roof of the Department of Electrical Engineering, Faulty of Engineering, located in Rome, Italy (ϕ = 41°53′38″N, λ = 12°29′37″E) during the 2 years 2010 and 2011. As shown in Figure 4a, the set-up consists of 20 polycrystalline Si-modules mounted in free-standing manner, i.e., on racks placed above the roof. The modules are tilted $\gamma_t = 30°$ from the horizontal surface and are South-facing ($\alpha_t = 0°$). The modules are subdivided in two strings, each formed by the series connection of 10 modules (the upper and lower ones) which have been left open circuited during the temperature measurements. The manufacturer's specifications give T_{NOCT} = 48°C ± 3°C (fixed to 50°C , in the following) and ΔT_{ref} = 2°C (added to T_b which is the real measured quantity). Figure 4b shows the PT100 sensors (platinum resistance thermometers sensors) placed on the back of the PV module to measure the rear temperature T_b. Six sensors have been placed on three modules (two sensors per module) in order to investigate the uniformity of the temperature among different modules with the same exposition. An identical PT100 sensor has been used to measure the ambient temperature as shown in Figure 4c. The Yokogawa DC100 data collector unit has been used to collect and register the data on seven independent channels. The solar irradiance data were acquired using a class-1 pyranometer Delta-Ohm LP Pyra (Soesterberg, The Netherlands). The measurements have been performed with a real-time data acquisition step of 30 s (2,880 data points for each channel per day). According to the Guide JCGM [69], the expanded uncertainty $U(T_c)$ with a coverage factor $k = 2$ (corresponding to a confidence level $p = 95.45\%$) is ±0.81°C.

The monthly average daily irradiation values used in the following computations are {2.33, 3.28, 4.25, 5.29, 6.34, 6.74, 6.72, 5.87, 4.70, 3.61, 2.51, 2.12} kWh/(m² day), provided by national standards for the 12 months of the year, respectively. The albedo factor has been set equal to 0.2 (dark-colored, rough soil surfaces). No particular shading affected the PV system output.

Assessment of the models

As explained in the previous section, the correct modeling of the irradiance incident on the tilted surface of the PV module is required at the first step of the proposed procedure. Hence, it is mandatory to assess the accuracy of the selected models used for its prediction.

The diffuse and direct irradiation values on a horizontal surface predicted by the four models used for the computation of the cloudiness index k_d (subsection 'Hourly diffuse and direct irradiation') are reported in Figures 5 and 6, respectively, where they are plotted against observed values. The modeled results have been computed; considering that in the input of Equation 3, the aforementioned monthly average daily irradiation values $\overline{H}_h$ were provided by national standards in order to compute the hourly values of the global irradiation on a horizontal surface and then these global values were decomposed into diffuse and direct components. For each model, the figures also show the plot of the linear line of best fit (dotted line) as well as a line of the one-to-one correlation (solid line) for comparison purposes. Table 1 (Hourly horizontal diffuse irradiance and Hourly horizontal direct irradiance) presents the modeled statistics for the diffuse D_h and direct B_h irradiance estimates,

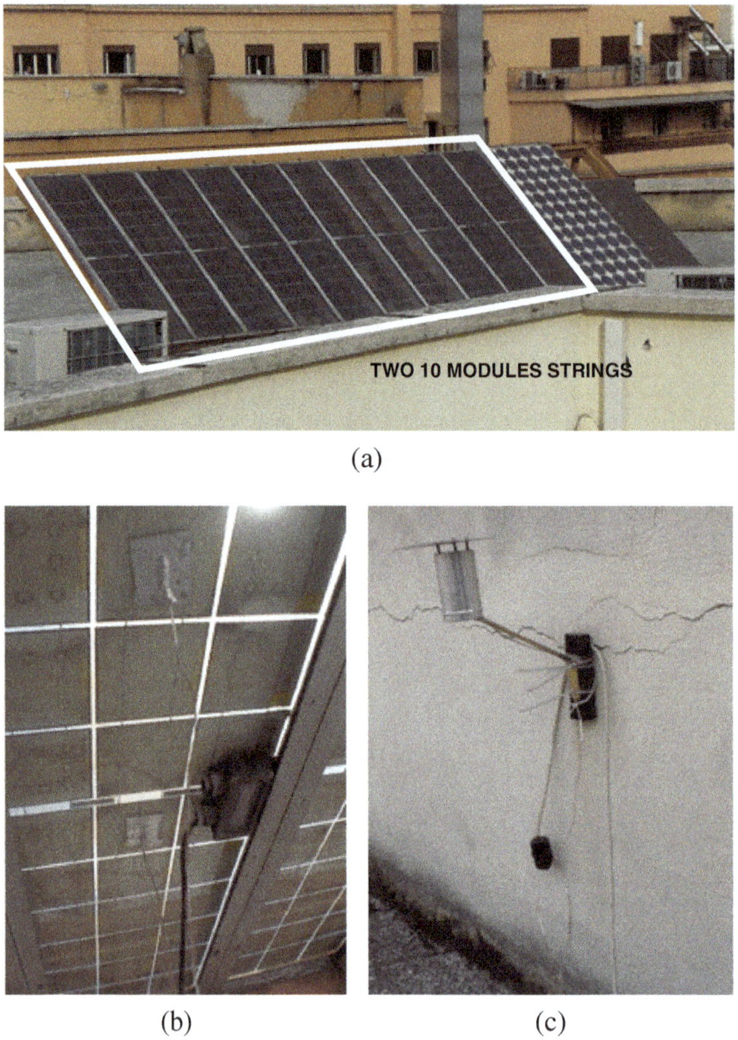

Figure 4 Test set-up. (a) Free-standing modules placed on the roof of the Electrical Engineering Department. **(b)** PT100 sensors used to measure the back temperature T_b of the modules. **(c)** PT100 sensor for the measurement of the ambient temperature T_a.

respectively, in comparison to the experimental data. The results indicate that all the models provide an acceptable agreement. The level of uncertainty for all the models, represented by the RMSE in W/m^2, varies between 25 and 50 W/m^2 for diffuse irradiance (33% to 61% of the average value) and between 80 and 125 W/m^2 for direct irradiance (27% to 43% of the average value). With regard to model bias, the results indicate that the level and the direction of the bias, represented by the MBE in W/m^2, vary among the four models. Results show that the Miguel model achieves the best match with the experimental data as indicated by the deviation of the line of best fit from the line of the one-to-one correlation and as confirmed by the higher R^2 values.

The models for converting the hourly horizontal diffuse irradiation D_h to the hourly diffuse irradiation D_{Th} on a tilted surface (subsection 'Hourly diffuse irradiation on a tilted surface') are compared in Figure 7 where the values predicted by the four selected models are reported along with the experimental data measured on the horizontal surface. The figure shows the predicted values of the global irradiance G_{Th} on the tilted plane of the PV panels plotted against measured values. In addition, Table 1 (Hourly global irradiance) presents the statistics for the global irradiance estimates. The results show that the Perez model gives the best estimate of the global irradiance on the tilted surface. The bias values (MBE) highlight that the four models exhibit a tendency to underestimate global irradiance values compared to the experimental data and that the four models achieve comparable values of R^2. The uncertainty (RMSE) ranges between 80 and 125 W/m^2.

The results of the four cell temperature models reported in Equation 24 are shown in Figure 8. The modeled results

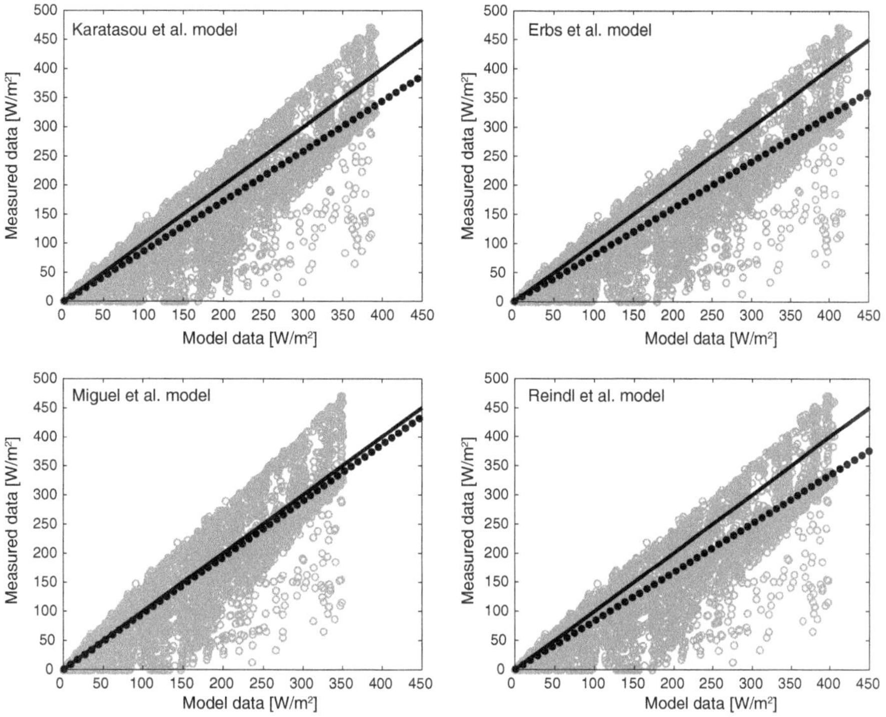

Figure 5 Experimental versus modeled hourly values of diffuse irradiation D_h on a horizontal surface for Rome. With a linear line of best fit (dotted line) and a line of one-to-one correlation (solid line).

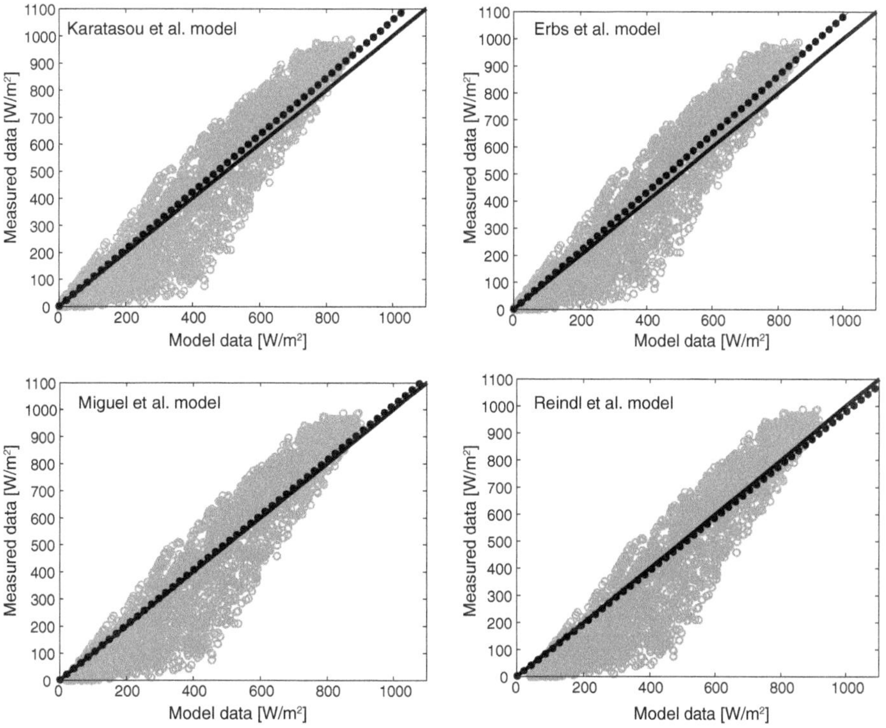

Figure 6 Experimental versus modeled hourly values of direct irradiation B_h on a horizontal surface for Rome. With a linear line of best fit (dotted line) and a line of one to one correlation (solid line).

Table 1 Irradiation and temperature statistics for model estimates against measured data

	Model	MBE	RMSE	R^2
Hourly horizontal diffuse irradiance D_h in W/m^2	Karatasou	+17.065	42.707	0.839
	Erbs	+20.332	47.642	0.799
	Miguel	+6.334	26.678	0.881
	Reindl	+18.443	44.678	0.823
Hourly horizontal direct irradiance B_h in W/m^2	Karatasou	−42.459	125.004	0.820
	Erbs	−21.398	80.770	0.818
	Miguel	−7.54	118.221	0.902
	Reindl	+35.431	96.254	0.828
Hourly global irradiance G_{Th} in W/m^2	Ma-Iqbal	−15.875	143.623	0.770
	Muneer	−9.169	142.357	0.784
	Perez	−0.320	137.428	0.815
	Reindl	−7.581	142.219	0.792
Cell temperature T_c in °C computed every 15 min	Equation 24a	−2.330	3.263	0.579
	Equation 24b	−0.978	7.349	0.645
	Equation 24c	−0.340	4.193	0.884
	Equation 24d	−0.530	8.005	0.820

have been computed every 15 min using the experimental global irradiation measured on the surface of the PV panels, using Equation 26 to estimate the ambient temperature and setting the wind speed v_w to the average value of the day (the weight of the wind speed in the used equations is not so high as to require a more accurate approximation). The results indicate that the third model provides the closer agreement with the measured results. The level of uncertainty for all the models varies between 4°C and 8°C (23% to 47% of the average value) while the bias level ranges between 0.8°C and 2.3°C. The bias direction is always negative, i.e., the models tend to underestimate the cell temperature. This is mainly due to several reasons: there are uncertainties in the T_{NOCT} calculation; difficulties arise on the correct computation of the instantaneous wind velocity v_w due to the fact that there are at least seven different definitions in use as explained in [31]; the operating temperature T_c itself shows some variations on the same module and among different modules. In fact, looking at Figure 9, which shows the temperature profiles on August 21st 2010, temperature differences of around 5°C can be observed depending on the thermocouple position. It should be underlined that the estimation of the T_c from the back temperature T_b with reasonable accuracy is here possible because the modules are standard crystalline ones. When modules are specially designed for some particular applications, extra care must be exercised.

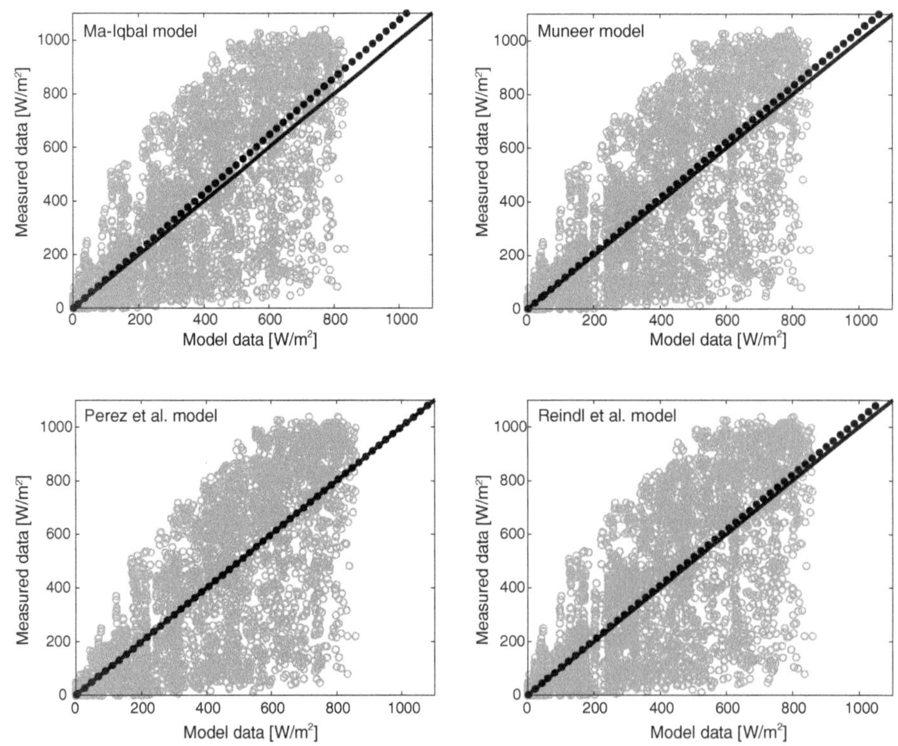

Figure 7 Experimental versus modeled hourly values of global irradiation G_{Th} on a tilted surface for Rome. With a linear line of best fit (dotted line) and a line of the one-to-one correlation (solid line).

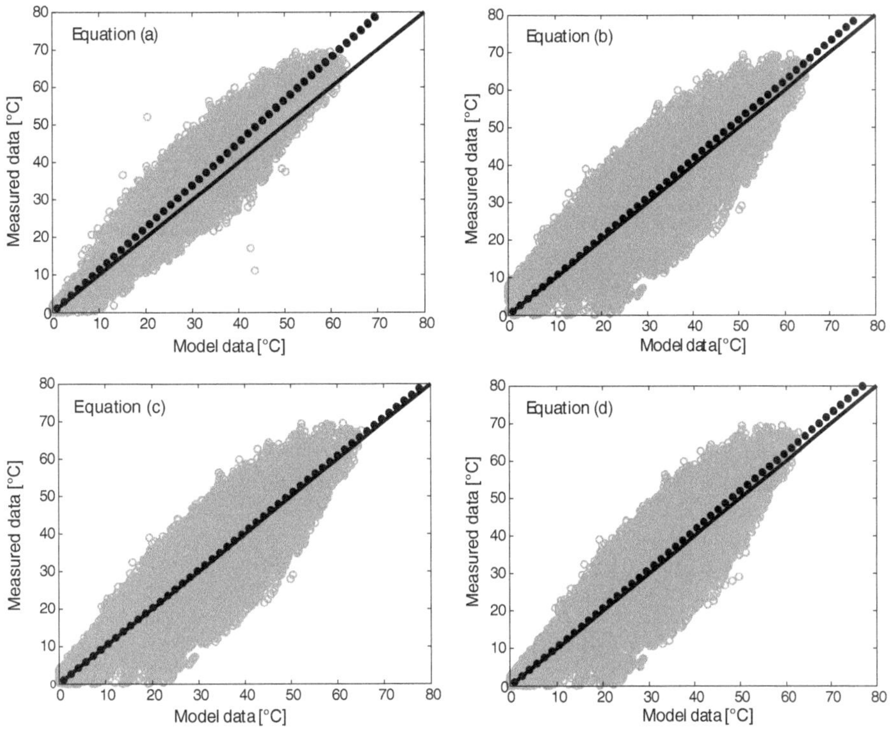

Figure 8 Experimental versus modeled hourly values of cell temperature T_c for Rome. With a linear line of best fit (dotted line) and a line of the one-to-one correlation (solid line).

It should be observed that a heating of the module is observed during the afternoon-evening period: this makes differences between T_c and T_a higher in the afternoon than in the morning for the same values of irradiance.

Figure 10 shows the comparison between the profiles of the ambient temperature T_a measured during two representative days and the profiles estimated through Equation 26 starting from average historical temperature data (http://www.ilmeteo.it). It can be observed that the accuracy is reasonable; when the day is particularly clear and the irradiation is high (e.g., June 21st), the measured profile rises a little faster than the predicted one. Yet, the differences remains always below 5°C.

Model combination results

From the previous results, it was found that the Miguel model, the Perez model, and Equation 24c achieve the best

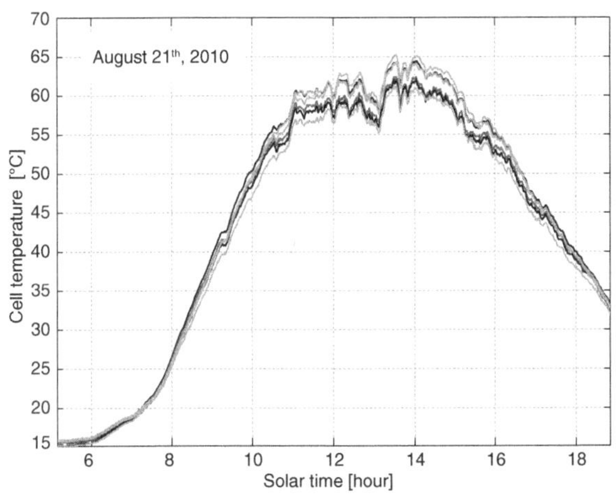

Figure 9 Differences in module temperatures on August 21st measured by six thermocouples.

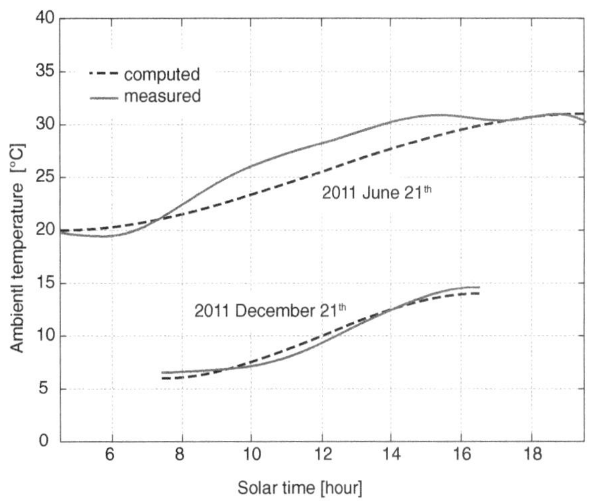

Figure 10 Comparison between estimated ambient temperature profile and measured data for two representative days.

Table 2 Temperature statistics for combination-modeled estimates against measured data

D_h model	D_{Th} model	T_c model	MBE	RMSE	R^2
Miguel	Perez	Equation 24c	−0.340	4.193	0.884
Karatasou	Muneer	Equation 24d	−0.384	4.524	0.792
Reindl	Reindl	Equation 24c	+0.415	5.142	0.715
Erbs	Ma-Iqbal	Equation 24a	−0.854	5.836	0.693

estimates of the diffuse irradiance on a horizontal surface, the diffuse irradiance on a tilted surface, and the cell temperature, respectively. However, it is possible to observe, especially for the global irradiance on a tilted surface and the cell temperature, that none of the models consistently outrank the other ones. Hence, we have tried all the possible combinations of the models used for the estimation of the hourly horizontal diffuse (direct) irradiance, the hourly global irradiance on the titled surface and the cell temperature, in order to identify the best combination. Table 2 presents the statistical results for the comparison between the most significant values predicted by some combinations and the measured values for the cell temperature T_c. The statistics confirm that the combination of the Miguel and Perez models and Equation 24c achieves the highest level of accuracy for the cell temperature.

Conclusions

The paper presents an exhaustive methodology for the estimation of the transient operating cell temperature of a PV system. The proposed procedure estimates the transient temperature of the photovoltaic cell during the day n_d of the year, placed on a surface S_T tilted γ_t from the horizontal plane and rotated α_t from the North-South direction, starting from the monthly average daily global irradiation value $\overline{H}$ on a horizontal surface. The study is considered of high interest since the values of horizontal solar irradiation are often the only data provided by meteorological stations.

The procedure consists of five steps that are applied one after another in succession:

- computation of the hourly global irradiation on a horizontal surface H_h;
- computation of the hourly diffuse irradiation on a horizontal surface D_h;
- computation of the hourly diffuse irradiation on a tilted surface D_{Th};
- computation of the irradiance time-profile $I(t)$;
- computation of the cell temperature T_c.

Several well-established models have been used at each step and the accuracy of different combinations of these models has been investigated in comparison with data measured in Rome. In addition, two new models for calculating the solar irradiance and the ambient temperature profiles have been also provided.

An error analysis, based on well-known statistical indicators, i.e., mean bias error (MBE), root mean square error (RMSE), and the coefficient of determination (R^2), reveals that the combination of the Collares-Pereira, Miguel, and Perez models, the proposed irradiation model, and Equation 24c achieves the highest level of model accuracy for the cell temperature. However, it has been observed that, concerning the global irradiance on a tilted surface and the cell temperature, none of the models consistently outrank the other ones.

Competing interests

The authors declare that they have no competing interest.

Authors' contributions

RA and SC developed the models; RA carried out the calculations; UG performed the measurements. All authors read and approved the final manuscript.

Acknowledgements

None.

References

1. Rosen, MA, Dincer, I: On exergy and environmental impact. Int. J. Energy Res. **21**(January 1996), 643–654 (1997)
2. Joshi, AS, Dincer, I, Reddy, BV: Performance analysis of photovoltaic systems: a review. Renew. Sustain. Energy Rev. **13**(8), 1884–1897 (2009)
3. Dresselhaus, MS, Thomas, IL: Alternative energy technologies. Nature. **414**(6861), 332–7 (2001)
4. Farret, FA, Simões, MG: Integration of Alternative Sources of Energy. Wiley-IEEE Press, New York (2006)
5. Eddine, B, Salah, M: Solid waste as renewable source of energy: current and future possibility in Algeria. Int. J. Energy Environ. Eng. **3**(17) (2012)
6. Lucchetti, E, Barbier, J, Araneo, R: Assessment of the technical usable potential of the TUM shaft hydro power plant on the Aurino River, Italy. Renewable Energy. **60**, 648–654 (2013)
7. Sharma, P, Harinarayana, T: Solar energy generation potential along national highways. Int. J. Energy Environ. Eng. **4**(16), 1–13 (2013)
8. Zahedi, A: Solar photovoltaic (PV) energy; latest developments in the building integrated and hybrid PV systems. Renewable Energy. **31**(5), 711–718 (2006)
9. Šúri, M, Huld T a, Dunlop, ED, Ossenbrink H a: Potential of solar electricity generation in the European Union member states and candidate countries. Solar Energy. **81**(10), 1295–1305 (2007)
10. Sharma, P, Harinarayana, T: Enhancement of energy generation from two layer solar panels. Int. J. Energy Environ. Eng. **3**(12), 1–9 (2012)
11. Jäger-Waldau, A, Szabó M, Scarlat, N, Monforti-Ferrario, F: Renewable electricity in Europe. Renew. Sustain. Energy Rev. **15**(8), 3703–3716 (2011)
12. Kolhe, M: Techno-economic optimum sizing of a stand-alone solar photovoltaic system. IEEE Transactions on Energy Conversion. **24**(2), 511–519 (2009)
13. Notton, G, Lazarov, V, Stoyanov, L: Optimal sizing of a grid-connected PV system for various PV module technologies and inclinations, inverter efficiency characteristics and locations. Renewable Energy. **35**(2), 541–554 (2010)
14. Grasselli, U: Probabilistic design of high quality power supply photovoltaic systems. In: Proceedings of the IEEE-IAS I&CPS May 2-6 1993, St.Petersburg, Florida, USA, (1993)
15. Grasselli, U: A probabilistic method for simulation and design optimization of stand-alone photovoltaic systems. In: Proceedings of the UPEC September 18-20, Iraklio, Crete, Greece, (1996)

16. Araneo, R, Lammens, S, Grossi, M, Bertone, S: EMC issues in high-power grid-connected photovoltaic plants. IEEE Trans. Electromag. Compat. **51**, 639–648 (2009)
17. Notton, G, Cristofari, C, Poggi, P, Muselli, M: Calculation of solar irradiance profiles from hourly data to simulate energy systems behaviour. Renewable Energy. **27**, 123–142 (2002)
18. Eltawil, Ma, Zhao, Z: Grid-connected photovoltaic power systems technical and potential problems – a review. Renew. Sustain. Energy Rev. **14**, 112–129 (2010)
19. Kim, SK, Jeon, JH, Cho, CH, Kim, ES, Ahn, JB: Modeling and simulation of a grid-connected PV generation system for electromagnetic transient analysis. Solar Energy. **83**(5), 664–678 (2009)
20. Femia, N, Lisi, G, Petrone, G, Spagnuolo, G, Vitelli, M: Distributed maximum power point tracking of photovoltaic arrays: novel approach and system analysis. IEEE Transactions on Industrial Electronics. **55**(7), 2610–2621 (2008)
21. Nordmann, T, Clavadetscher, L: Understanding temperature effects on PV, system performance. In: Proceedings of 3rd World Conference on Photovoltaic Energy Conversion, vol. 3, pp. 2243–2246, (2003)
22. King, DL, Kratochvil, JA, Boyson, WE: Temperature coefficients for PV modules and arrays : Measurement methods, difficulties, and results. In: 26th IEEE Photovoltaic Specialists Conference, 29 September - 3 October, pp. 1183–1186, Anaheim, CA, (1997)
23. Skoplaki, E, Palyvos, J: On the temperature dependence of photovoltaic module electrical performance: a review of efficiency/power correlations. Solar Energy. **83**(5), 614–624 (2009)
24. Wysocki, JJ: Effect of temperature on photovoltaic solar energy conversion. J. Appl. Phys. **31**(3), 571–578 (1960)
25. Lasnier, F, Gang-Ang, T: Photovoltaic Engineering Handbook, Adam Hilger, Bristol (1990)
26. Messenger, RA, Ventre, J: Photovoltaic Systems Engineering. 3rd edition. CRC Press, New York (2010). ISBN-10 1439802920 ISBN-13 978-1439802922
27. Solanki, CS: Solar Photovoltaics : Fundamentals, Technologies and Applications. PHI Learning Pvt. Ltd., New Delhi (2011). ISBN-13 9788120343863 ISBN-10 8120343867
28. Ferguson, LG, Fraasb, LM: Theoretical study of GaSb PV cell efficiency as a function of temperature. Solar Energy Mater. Solar Cells. **39**, 11–18 (1995)
29. Araneo, R, Falconi, C: Lateral bending of tapered piezo-semiconductive nanostructures for ultra-sensitive mechanical force to voltage conversion. Nanotechnology. **24**(26), 265707 (2013)
30. Araneo, R, Lovat, G, Burghignoli, P, Falconi, C: Piezo-semiconductive quasi-1D nanodevices with or without anti-symmetry. Adv. Mat. (Deerfield Beach, Fla.) **24**(34), 4719–24 (2012)
31. Skoplaki, E, Boudouvis, AG, Palyvos, JA: A simple correlation for the operating temperature of photovoltaic modules of arbitrary mounting. Solar Energy Mat. Solar Cells. **92**, 1393–1402 (2008)
32. Skoplaski, E, Palyvos, JA: Operating temperature of photovoltaic modules A survey of pertinent correlations. Renewable Energy. **34**, 23–29 (2009)
33. Duffie, JA, Beckman, WA: Solar engineering of thermal processes. 3rd edition. Wiley, New York (2006)
34. Li, H, Bu, X, Lian, Y, Zhao, L, Ma, W: Further investigation of empirically derived models with multiple predictors in estimating monthly average daily diffuse solar radiation over China. Renewable Energy. **44**, 469–473 (2012)
35. Mubiru, J, Banda, E: Monthly average daily global solar irradiation maps for Uganda: a location in the equatorial region. Renewable Energy. **41**, 412–415 (2012)
36. Muneer, T: Solar Radiation and Daylight Models. Elsevier Butterworth-Heinemann Ltd, Oxford, UK (2004)
37. Moharil, RM, Kulkarni, PS: Reliability analysis of solar photovoltaic system using hourly mean solar radiation data. Solar Energy. **84**, 691–702 (2010)
38. Page, JK: New sources of energy. In: Proceedings of the United Nations Conference on New Sources of Energy, Solar Energy, Wind Power and Geothermal Energy, Rome, 21 - 31 August 1961, vol. 4 p 378, (1961)
39. Liu, BYH, Jordan, RC: The interrelationship and characteristic distribution of direct, diffuse and total solar insolation. Solar Energy. **1**, 1–19 (1960)
40. Jain, PC: Comparison of techniques for the estimation of daily global irradiation and a new technique for the estimation of hourly global irradiation. Solar Wind Technol. **1**, 123–134 (1988)
41. Collares-Pereira, M, Rabl, A: The average distribution of solar radiation-correlations between diffuse and hemispherical and between daily and hourly insolations values. Sol Energy. **22**, 155–164 (1979)
42. Iqbal, M: A study of Canadian diffuse and total solar radiation data, II. Monthly average hourly horizontal radiation. Sol Energy. **22**, 87–90 (1979)
43. Jain, PC, Jain, S, Ratto, CF: A new model for obtaining horizontal instantaneous global and diffuse radiation from the daily values. Sol Energy. **41**, 397–404 (1988)
44. Wu, J, Chan, CK: Prediction of hourly solar radiation using a novel hybrid model of ARMA and TDNN. Sol Energy. **85**, 808–817 (2011)
45. Hocaoğlu, FO, Gerek, ON, Kurban, M: Hourly solar radiation forecasting using optimal coefficient 2-D linear filters and feed-forward neural networks. Sol Energy. **82**, 714–726 (2008)
46. Notton, G, Muselli, M, Louche, A: Two estimation methods for monthly mean hourly total irradiation on tilted surfaces from monthly mean daily horizontal irradiation from solar radiation data of Ajaccio, Corsica. Sol Energy. **57**, 141–153 (1996)
47. Jacovideis, CP, Tymvios, FS, Assimakopoulos, VD, Kaltsounides, NA: Comparative study of various correlations in estimating hourly diffuse fraction of global solar irradiation. Renewable Energy. **31**, 2492–2504 (2006)
48. Mondol, JD, Yohanis, YG, Norton, B: Solar radiation modelling for the simulation of photovoltaic systems. Renewable Energy. **33**, 1109–1120 (2008)
49. Jacovides, CP, Boland, J, Asimakopoulos, DN, Kaltsounides, NA: Comparing diffuse radiation models with one predictor for partitioning incident PAR radiation into its diffuse component in the eastern Mediterranean basis. Renewable Energy. **35**, 1820–1827 (2010)
50. Karatasou, S, Santamouris, M, Geros, V: Analysis of experimental data on diffuse solar radiation in Athens, Greece for building application. Int. J. Sustainable Energy. **23**, 1–11 (2003)
51. Erbs, DG, Klein, SA, Duffie, JA: Estimation of the diffuse radiation fraction for hourly, daily and monthly average global radiation. Sol Energy. **28**, 293–302 (1982)
52. Miguel, A, Bilbao, J, Aguiar, R, Kambezidis, H, Negro, E: Diffuse solar radiation model evaluation in the north Mediterranean belt area. Sol Energy. **70**, 143–153 (2001)
53. Reindl, DT, Beckman, WA, Duffie, JA: Diffuse fraction correlations. Sol Energy. **45**, 1–7 (1990)
54. Iqbal, M: An Introduction to Solar Radiation. Academic press, New York (1983)
55. Noorian, AM, Morandi, I, Kamali, GA: Evaluation of 12 models to estimate hourly diffuse irradiation on inclined surfaces. Renewable Energy. **33**, 1406–1412 (2008)
56. Evseev, EG, Kudish, AI: The assessment of different models to predict the global solar radiation on a surface tilted to the South. Sol Energy. **83**, 377–388 (2009)
57. Loutzenhiser, PG, Manz, H, Felsmann, C, Strachan, PA, Frank, T, Maxwell, G: Empirical validation of models to compute solar irradiance on inclined surfaces for building energy simulation. Sol Energy. **81**, 254–267 (2007)
58. Ma, CY, Iqbal, M: Statistical comparison of models for estimating solar radiation on inclined surfaces. Sol Energy. **31**, 313–317 (1983)
59. Reindl, DT, Beckman, WA, Duffie, JA: Evaluation of hourly tilted surface radiation models. Sol Energy. **45**, 9–17 (1990)
60. Muneer, T: Solar Radiation and Daylight Models for the Energy Efficient Design of Buildings. Architectural Press, Oxford (1997)
61. Perez, R, Stewart, R, Arbogast, R, Seals, J, Scott, J: An anisotropic hourly diffuse radiation model for surfaces: description, performance validation, site dependency evaluation. Sol Energy. **36**, 487–497 (1986)
62. Perez, R, Ineichen, P, Seals, R, Michalsky, JJ, Stewart, R: Modelling daylight availability and irradiance components from direct and global irradiance. Sol Energy. **44**, 271–289 (1990)
63. Hay, JE: Study of shortwave radiation on non-horizontal surfaces. Tech, Vol. 79-12. Atmospheric Environment Service, Downsview, Ontario (1979)
64. Temps, RC, Coulson, KL: Solar radiation incident upon slopes of different orientation. Solar Energy. **19**, 179–184 (1977)
65. Gueymard, C: An anisotropic solar irradiance model for tilted surfaces and its comparison with selected engineering algorithms. Sol Energy. **38**, 367–386 (1987)
66. Markvart, T: Solar electricity. Wiley, Chichester (2000)

67. Chenni, R, Makhlouf, M, Kerbache, T, Bouzid, A: A detailed modelling method for photovoltaic cells. Energy. **32**, 1724–1730 (2007)
68. King, D, Boyson, W, Kratochvil, J: Photovoltaic Array Performance Model. Tech. report, SAND2004-3535. Sandia National Laboratories, Albuquerque (2004). http://prod.sandia.gov/techlib/access-control.cgi/2004/043535.pdf
69. Joint Committee for Guides in Metrology: JCGM 100: Evaluation of measurement data – Guide to the expression of uncertainty in measurement. BIPM - Bureau International des Poids et Mesures Sevres (2009)

Load evaluation of horizontal-axis wind turbine rotor using coupled Beddoes near-wake model and free-wake method

Abdelfattah Bouatem[1*], Ahmed Almers[2] and Noureddine Boutammachte[2]

Abstract

Wind turbines operate mostly in yaw conditions that give rise to cyclic variations in aerodynamic forces applied on the blade. This induced load fluctuation is closely related to the upstream velocity field of the rotor and can be a significant source of fatigue and vibration. An accurate prediction of blade loading is considered the key in designing reliable and efficient wind turbines. The related calculation remains a complicated task to perform and requires enormous computing time. In this context, a numerical method is presented, aimed at evaluating the azimuthal fluctuation of the normal force. This method is obtained by coupling the Beddoes near-wake model and the free-wake method: the near-wake-induced velocities are calculated using Beddoes near-wake model with the far-wake contribution evaluated using the free-wake method. In addition, the unsteady effects on the aerodynamic coefficients are taken into account using the Beddoes-Leishman dynamic stall model. A computer code was developed, and numerical values were obtained in acceptable computational time. Results are compared with measurements performed in the NASA Ames wind tunnel.

Keywords: Wind turbine; Yaw; Skewed wake; Stall delay; Free wake; Near wake; Far wake

Background

To reach satisfactory levels of performance, the horizontal-axis wind turbine requires accurate predictions of the aerodynamic forces acting on the blades. However, in non-steady conditions, these aerodynamic load predictions remain a complicated task to perform because of the complex nature of the flow around the blades.

To model the wind turbine, a variety of mathematical models exist, such as the blade element method (BEM), computational fluid dynamics (CFD) method, and vortex method, each with different levels of accuracy and complexity.

The classical BEM method is obtained by coupling blade element theory with momentum theory [1]. This method assumes that the blade can be divided into several elements. The study is performed for each element by applying the momentum theory in the axial and tangential direction. The BEM method includes several approximations and limitations. Its validity may be extended using empirical corrections to take into account the finite number of blades, blade tip losses, and the cyclical variation of the axial induction factor in yaw conditions.

Currently, the use of the CFD techniques has experienced significant progress thanks to the improved performance of computers. Despite the accurate results obtained in most cases, CFD methods require huge computational resources and large memory.

The aim of the vortex theory is to model the wind turbine taking into account the geometry of the wake behind the rotor and its effect on the velocity field upstream [2]. The vortex theory principle is derived from the lifting line and lifting surface theories, which have been developed for airplane wings to determine wing loads and wake geometry. The flow is replaced by inviscid and incompressible fluid through an equivalent vortex system. The methodology is based on the Biot-Savart law to compute induced velocities and the vorticity transport theorem to shape the wake which is generally divided into near wake and far wake. The near wake consists of the trailing vortices issuing from the trailing

* Correspondence: Bouatem1@hotmail.com
[1]Department of Mechanical Engineering, Moulay Ismail University, ENSAM-MEKENS, P.O. BOX. 4024, Meknes 50003, Morocco
Full list of author information is available at the end of the article

edges. The far wake is reduced in two intense tip and root vortices.

The vortex method gives more accurate results than the BEM method and is relatively easy to implement compared with CFD methods. Consequently, the vortex theory can be a better choice for predicting the aerodynamic performance of a wind turbine. This methodology requires a perfect determination of the wake geometry. There are two main approaches to shape the wake. The first is known as 'prescribed wake.' According to this method, the wake geometry is known *a priori*. The second is the 'free-wake method,' which is based on the following steps: first, trailing vortices are detached from the blade and represented by Lagrangian markers; then, the wake-induced velocities are calculated using the Biot-Savart law at each Lagrangian marker to determine their new positions; finally, the Lagrangian markers are connected by a straight line to form the wake [3]. This method requires huge computational time, basically at the near-wake region. This can be explained by the large number of trailing vortices and straight lines which form the wake.

The main objective of this paper is to develop a numerical procedure that can predict aerodynamic loads without using great amounts of computer time which are generally required in the free-wake method. This procedure is obtained by combining the Beddoes near-wake model with the free-wake method. The Beddoes near-wake model is used to calculate the near-wake-induced velocities at the rotor plane. The free-wake method is used to shape the far wake in a simpler and faster way; the calculation begins with a generation of a rigid wake which is corrected to take into account the effect of wake-induced velocities.

The near-wake-induced velocities are added to the far-wake-induced velocities to calculate the circulation distribution along the blade. This iterative procedure continues until the rotor flow field takes a constant value. The wind turbine is assumed to be operating in yaw conditions. To take into account the unsteady aerodynamic effect on the blade loading, the Beddoes-Leishman dynamic stall model has been used. In this model, the deficiency in lift arising from the circulatory effect of shed vorticity is modeled using the indicial response function. To validate this numerical simulation, a comparison is made with measurements performed in the NASA Ames wind tunnel.

Methods

Beddoes near-wake model

The Beddoes near-wake model is designed to approximate the single blade near-wake contribution to the downwash using exponential functions. This model covers the first quadrant of the vortex life. After this first quarter of a revolution, the trailing vortices are supposed to be in the process of rolling-up [4]. The shed vorticity is not included in the Beddoes near-wake model but introduced in the calculation using the Beddoes-Leishman dynamic stall model.

As shown in Figure 1, the vortex element of length ds, strength Γ, and originating from the point B, induces the velocity dw at point A. Assuming that the geometry of the near-wake vortices can be approximated by a circular arc, the induced velocity dw can be expressed according to:

$$dw = \frac{\Gamma \cdot ds}{4 \cdot \pi \cdot r^2} \frac{1 - \left(1 - \frac{h}{r}\right) \cdot \cos(\Omega)}{\left[1 + \left(1 - \frac{h}{r}\right)^2 - 2 \cdot \left(1 - \frac{h}{r}\right) \cdot \cos(\Omega)\right]^{3/2}} \quad (1)$$

Beddoes suggested approximating the equation with a series of exponential functions [4]. The axial contribution at the *i*th time step to the downwash from one circular arc is given by

$$w\,(i) = \sum_{k=1}^{n} Z_K(i) \quad (2)$$

$$Z_K\,(i) = Z_k\,(i-1) \cdot e^{b_k \frac{\Delta\Omega}{\varphi}} + c_k \cdot D_w \cdot e^{b_k \frac{\Delta\Omega}{2 \cdot \varphi}}$$

$$\varphi = \frac{\pi}{4} \left| \left(1 + \frac{h}{2 \cdot r}\right) \ln\left(1 - \frac{h}{r}\right) \right|$$

$$D_w = \frac{\Gamma \left|\Delta\Omega \frac{r}{h}\right| \cos(\chi)}{4 \pi h \sqrt{1 + \left(\Delta\Omega \frac{r}{h}\right)^2}}$$

The Beddoes near-wake model is based on the assumption that the near-wake-trailed vortices remain in the rotor plane. Wang and Coton have developed an extension of the previous model to take into account a tilting angle χ [5]:

$$\tan(\chi) = \frac{V_\infty \cos\,(\Phi_y) + w_B}{\omega \cdot r + V_\infty \cdot \sin\,(\Phi_y) \cdot \sin\,(\Phi_r) + u_B} \quad (3)$$

where w_B and u_B are, respectively, the axial and tangential velocities at point B, ω the angular velocity, and V_∞ the free-stream velocity.

The near-wake tangential induced velocity (w) is calculated in the same way as the axial induced velocity:

$$w'(i) = \sum_{k=1}^{n} Z'_k(i) \quad (4)$$

$$Z'_K(i) = Z'_K(i-1) \cdot e^{b_k \frac{\Delta\Omega}{\varphi}} + c_k \cdot D_w \cdot e^{b_k \frac{\Delta\Omega}{2 \cdot \varphi}}$$

$$D'_w = \frac{\Gamma \left|\Delta\Omega \frac{r}{h}\right| \sin(\chi)}{4 \pi h \sqrt{1 + \left(\Delta\Omega \frac{r}{h}\right)^2}}$$

Each vortex is divided into several elements, and the total contribution of the vortex is calculated using the trapeze method at each time step. The circulation is

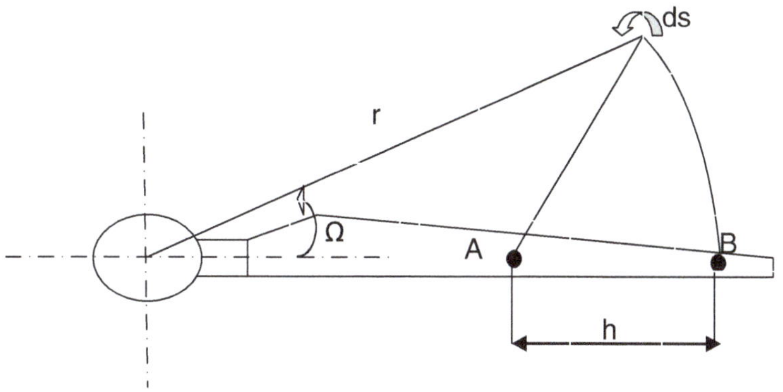

Figure 1 Beddoes near-wake geometry.

calculated at each blade element using the Kutta-Joukowski law [6]:

$$\Gamma = \frac{1}{2} V_{\mathrm{rel}} \cdot C \cdot C_{\mathrm{l}}(\alpha)$$

where $C_l(\alpha)$ is the lift coefficient, V_{rel} the relative velocity, C the cord, and Γ the circulation.

Far-wake modeling

This part of the wake consists of two vortices issuing from the blade tip and root. The trailing vortices shed from the leading edge sections are assumed to join together to form a strong tip and root vortex.

This rolling-up process takes place after the first-quarter revolution from their generation.

To model the far wake, the problem is investigated in two coordinate systems, as depicted in Figure 2, a blade-fixed Cartesian coordinate system (x,y,z) and a plane-fixed coordinate system (X,Y,Z) on the rotor disk.

Two different types of wake geometry are successively used during the calculation: a rigid wake and a free wake. The rigid wake is used to initialize the calculation and to give an initial geometry to the free wake. The free wake is based on the Prandtl lifting line theory. The vortices are represented by Lagrangian markers which are connected by straight line vortex segments, as represented in Figure 3.

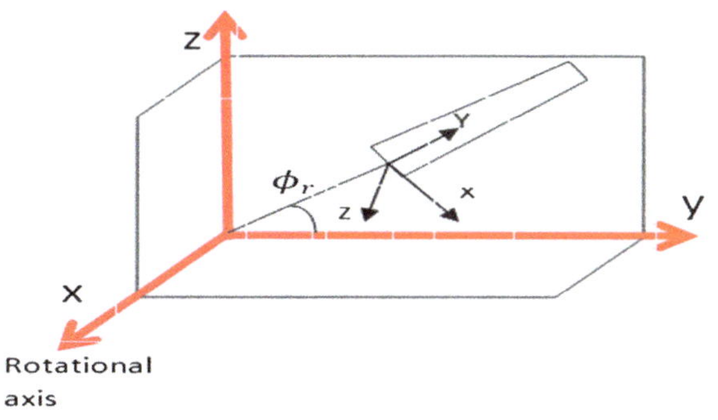

Figure 2 Coordinate systems for wind turbine rotor.

The positions of the markers are obtained by solving the transport equation using a second-order predictor/corrector scheme.

The rigid wake model assumes that the wake is convected by free-stream velocity. As a result, the flow behind the rotor consists of helical wake with constant pitch and diameter. The rigid wake is obtained from:

$$\vec{r}: \begin{cases} X_{\mathrm{V}} = V_{\infty} \dfrac{\Psi_{\mathrm{b}}}{\Omega} \\ Y_{\mathrm{V}} = R \cdot \cos(\Psi_{\mathrm{w}}) \\ Z_{\mathrm{V}} = R \cdot \sin(\Psi_{\mathrm{w}}) \end{cases} \tag{5}$$

where R is the blade radius and $\vec{r}$ is the position vector of collocation point (p) that can be represented using two variables Ψ_b and Ψ_w (being a measure of azimuth and a measure of wake age, respectively).

The initial helical geometry has to be distorted under the influence of induced velocities. The fundamental equation describing the transport of Lagrangian markers is

$$\frac{d\vec{r}}{dt} = \vec{W} \tag{6}$$

where $\vec{W}$ is the geometrical sum of local induced velocities generated by the different parts of the wake and the wind free-stream velocity.

For the rotating blade application, the governing Equation 6 can be written in the form of first-order partial differential equation:

$$\frac{\partial r}{\partial \psi_w} + \frac{\partial r}{\partial \psi_{\mathrm{b}}} = \frac{1}{\Omega} \cdot W \tag{7}$$

The induced velocity terms are highly nonlinear [7]. Therefore, the velocity transport Equation 7 cannot be solved analytically; numerical tools must be adopted. The solution is accomplished by discretizing the problem and transforming the governing Equation 7 into a finite

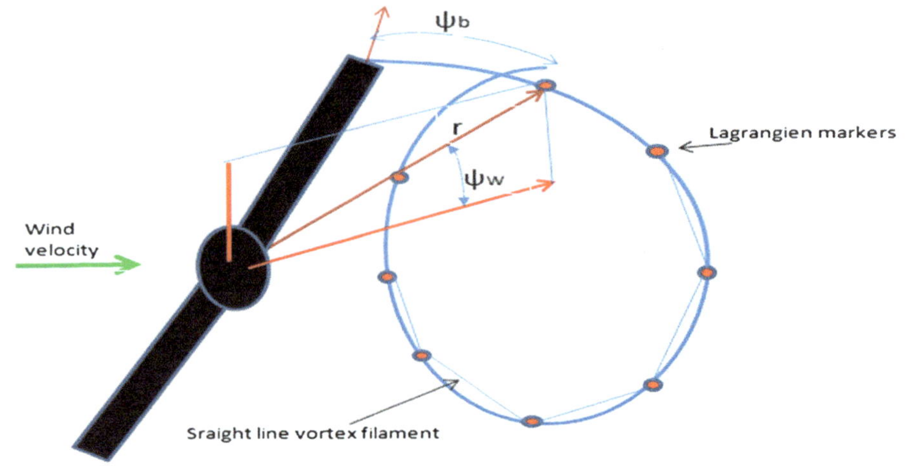

Figure 3 Representation of rotor wake.

difference equation. In the present work, a five-point central differencing scheme has been adopted. The discretized equation can be written as:

$$\overrightarrow{r_{i,j}} = \overrightarrow{r_{i-1,j-1}} + \left(\overrightarrow{r_{i,j-1}} - \overrightarrow{r_{i-1,J}}\right)\left(\frac{\Delta\Psi_w - \Delta\Psi_b}{\Delta\Psi_w + \Delta\Psi_b}\right) + \frac{2}{\Omega}\left(\frac{\Delta\Psi_w \times \Delta\Psi_b}{\Delta\Psi_w + \Delta\Psi_b}\right)\vec{W} \tag{8}$$

The Biot-Savart law is used here to calculate the induced velocities at Lagrangian markers to determine their new positions [8]. As illustrated in Figure 4, the induced velocity field generated by an infinitesimal element (dl) of the vortex line (MN) at any point is calculated using:

$$\overrightarrow{dV_{\text{induced}}} = \frac{\Gamma}{4\pi}\frac{\vec{dl} \wedge \vec{h}}{|h|^3} \tag{9}$$

The Biot-Savart law must be integrated to calculate the induced velocity generated by vortex line (MN). Referring to the Figure 4, Equation 9 can be written as

$$\overrightarrow{dV_{\text{induced}}} = \frac{\Gamma}{4\pi}\frac{\vec{r_1} \wedge \vec{r_2}}{\left|\vec{r_1} \wedge \vec{r_2}\right|^2}\left[\frac{\vec{r_0} \cdot \vec{r_1}}{\left|\vec{r_1}\right|} - \frac{\vec{r_0} \cdot \vec{r_2}}{\left|\vec{r_2}\right|}\right] \tag{10}$$

Note that at any iteration, the circulation Γ is assumed to be constant along the vortex. This parameter is equal to the maximum value along the blade. Furthermore, the representation of the tip and root vortex by straight segments reduces the accuracy of the calculation. Indeed, the Biot-Savart equation shows singular behaviors when point P is located very near or coincides with the vortex filament (Figure 4). These singularities can be removed by multiplying the Biot-Savart law with a viscous parameter K_V [9].

$$K_V = \frac{H^2}{\left(r_c{}^2 + H^4\right)^{1/2}} \tag{11}$$

Bhagwat and Leishman [10] suggested an empirical approach to calculate the growth of the viscous core radius:

$$r_c = \sqrt{\frac{4 \cdot \alpha \cdot \nu \cdot \delta \cdot \Psi_w}{\Omega}}$$

The eddy-viscosity coefficients δ is expressed as a function of the Reynolds number as:

$$\delta = 1 + a_1 \cdot R_e$$

where α is Lamb-Oseen constant and a_1 is an empirical parameter.

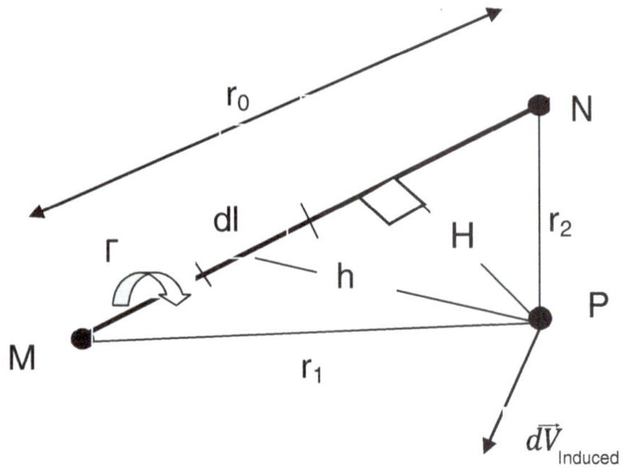

Figure 4 Velocity induced by a finite straight vortex.

Aerodynamic coefficients

The AERODAS model [11] is used to calculate the airfoil lift and drag coefficients, in both pre-stall and post-stall regimes, as functions of the angle of attack. The lift coefficient in the pre-stall regime is given by:

$$\text{if} \quad \alpha \geq A \qquad CL_1 = S1 \cdot (\alpha - A0) - RCL\ 1\left(\frac{\alpha - A0}{ACL_1 - A0}\right)^{N_1}$$

$$\text{if} \quad \alpha < A_0 \quad CL_1 = S1 \cdot (\alpha - A0) + RCL\ 1\left(\frac{-\alpha + A0}{ACL_1 - A0}\right)^{N_1}$$

where

$$RCL \quad 1 = S1 \cdot (ACL_1 - A0) - CL_{1\max}$$

$$N_1 = 1 + \frac{CL_{1\max}}{RCL1}$$

The drag coefficient in the pre-stall regime is obtained by following expressions:

$$\text{if} \quad (2 \cdot A_0 - ACD_1) \leq \alpha \leq ACD_1$$

$$CD_1 = CD_0 + (CD_{1\max} - CD_0) \cdot \left(\frac{\alpha - A0}{ACD_1 - A0}\right)^{M}$$

$$\text{if} \quad \alpha < (2 \cdot A_0 - ACD_1) \ \text{or} \ \alpha > ACD_1 \qquad CD_1 = 0$$

The lift coefficient in the post-stall regime is expressed as follows:

$$\text{if} \quad 0 < \ \alpha < ACL_1 \qquad CL_2 = 0$$

$$\text{if} \quad ACL_1 \leq \alpha \leq 92.0^\circ \quad CL_2 = -0.032 \cdot (\alpha - 92) - RCL\ 2 \cdot \left(\frac{92 - \alpha}{51}\right)^{N_2}$$

$$\text{if} \quad \alpha > 92.0^\circ \quad CL_2 = -0.032 \cdot (\alpha - 92) + RCL\ 2 \cdot \left(\frac{-92 + \alpha}{51}\right)^{N_2}$$

where

$$RCL2 = 1.632 - CL_{2\max}$$

$$N_2 = 1 + \frac{CL_{2\max}}{RCL2}$$

The drag coefficient in the post-stall regime is given by the following equations:

$$\text{if} \quad (2 \cdot A_0 - ACL_1) < \alpha < ACL_1 \qquad CD_2 = 0$$

$$\text{if} \quad \alpha \geq ACD_1 \qquad CD_2 = -CD_{1\max} + (CD_{2\max} - CD_{1\max})\ COS\left(\frac{90 - \alpha}{90 - ACD1} \cdot 90\right)$$

Figure 5 shows the calculated aerodynamic coefficient using AERODAS model.

Dynamic stall

Dynamic stall is one of the consequences of an unstable aerodynamic environment. When the angle of attack changes rapidly, the aerodynamic forces differ from their static values. Also, the cyclical variation of the angle of attack produces a significant hysteresis which affects the aerodynamic loads applied on the blade. Dynamic stall is of great importance in practice. In most cases, this is the first factor to consider in determining the performance of structures exposed to flows.

The Beddoes-Leishman models are semi-empirical equations based on airfoil indicial response. These indicial functions compute the non-steady aerodynamic coefficients as a function of angle of attack change in a sample time interval. This method is based on the superposition of three distinct parts [12]:

- Attached flow
- Separated flow
- Leading-edge separation.

Attached flow module In non-steady attached flow, the aerodynamic forces acting on an airfoil can be divided in two components, circulatory and impulsive. The circulatory component is linked to the existing vortices on the airfoil; it is given by [13]:

$$C^{c}_{N,n} = C_{N\alpha} \cdot \ \alpha_{eq,n} \tag{12}$$

$$\alpha_{eq,n} = \alpha_n \ - \ X_n \ - Y_n$$

$$X_n = X_{n-1} \cdot \exp(-b_1 \Delta S) + A_1 \cdot \Delta\alpha_n \exp(-b_1 \Delta S/2)$$

$$Y_n = Y_{n-1} \cdot \exp(-b_2 \Delta S) + A_2 \cdot \Delta\alpha_n \exp(-b_2 \Delta S/2)$$

$$\Delta S = \frac{2 \cdot U \cdot \Delta t}{C}$$

where $\Delta\alpha_n$ is the change in the angle of attack over a sample interval Δt, C is the cord, U is the free-stream velocity, and A_1, A_2, b_1, and b_2 are empirical constants.

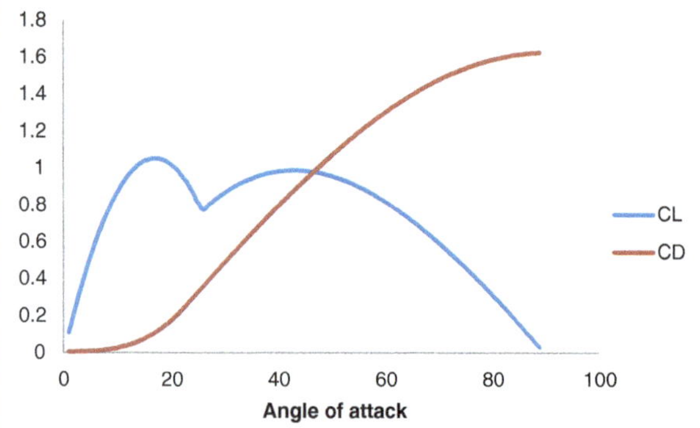

Figure 5 Calculated lift and drag coefficients for NREL S809.

The magnitude of the impulsive normal force coefficient can be obtained with:

$$C^{I}_{N,n} = \frac{3 \cdot C}{V_{eff}} \cdot \left(\frac{\Delta\alpha_n}{\Delta t} - D_n\right) \quad (13)$$

$$D_n = D_{n-1} \cdot \exp\left(-\frac{\Delta t}{K_\alpha T_1}\right) + \left(\frac{\Delta\alpha_n - \Delta\alpha_{n-1}}{\Delta t}\right) \exp\left(-\frac{\Delta t}{2K_\alpha T_1}\right)$$

$$T_1 = \frac{C}{A}$$

$$K_\alpha = \frac{0.75}{(1-M)} + \pi\ g^2 M^2 (A_1 b_1 + A_2 b_2)$$

$$g = (1-M^2)^{1/2}$$

where V_{eff} is the effective velocity, C represents the airfoil cord, A is the sound velocity, and M the Mach number.

The total normal force coefficient for attached flow is given by:

$$C^{P}_{N,n} = C^{I}_{N,n} + C^{C}_{N,n} \quad (14)$$

Trailing edge separation module Trailing edge separation is the gradual separation of the boundary layer from the surface of the aerofoil.

The relation between the flow separation point f and the normal force coefficient is given by the Kirchhoff relation [14]:

$$C_N = C_{N\alpha} \cdot \left(\frac{1+\sqrt{f}}{2}\right)^2 \alpha \quad (15)$$

where $C_{N\alpha}$ is the normal force coefficient slope near 0°. By inverting the Kirchhoff relation and using the airfoil's static characteristics, we obtain the static separation point as a function of angle of attack $f(\alpha)$.

In non-steady conditions, a delay must be introduced in the calculation of the normal force coefficient using:

$$C'_{N,n} = C_{N,n} - D_{p,n}$$

where the deficiency function is evaluated as follows:

$$D_{p,n} = D_{p,n-1} \cdot \exp\left(-\frac{\Delta S}{T_p}\right) + \left(C^{P}_{N,n} - C^{P}_{N,n-1}\right) \cdot \exp\left(-\frac{\Delta S}{2T_p}\right)$$

with

$$T_P = 9$$

It is now possible to evaluate the effective angle of attack which provides the same leading edge pressure in the steady case using the following expression:

$$\alpha_{f,n}(t) = \frac{C'_{N,n} - C_{N_0}}{C_{N_\alpha}}$$

where C_{N0} is the normal force coefficient obtained when the angle of attack is equal to 0°. Using the static separation point characteristic, presented previously, we compute the effective separation point:

$$f'_n = f(\alpha_{f,n})$$

The unsteady effect is taken into account in the calculation of the dynamic separation point according to:

$$f''_n = f'_n - D_{f,n}$$

The deficiency function is evaluated as:

$$D_{f,n} = D_{f,n-1} \cdot \exp\left(-\frac{\Delta S}{T_f}\right) + \left(f'_n - f'_{n-1}\right) \exp\left(-\frac{\Delta S}{2T_f}\right)$$

with

$$T_f = 5$$

Finally, using the Kirchhoff relation, we compute the contribution of the trailing edge separation in the calculation of the normal force coefficient as:

$$C^{f}_{N,n} = C_{N\alpha} \cdot \left(\frac{1+\sqrt{f''_n}}{2}\right)^2 \alpha_{eq,n}$$

Leading edge separation The leading edge separation takes place in the following condition [15]:

$$C'_{N,n} > C_{N,I}$$

where $C_{N,I}$ represents the critical value of the normal force, obtained with C_N corresponding to the static break in the pitching moment.

The contribution of the leading edge separation $C_{N,n}^{V}$ is calculated using:

$$C_{N,n}^{V} = \begin{cases} C_{N,n-1}^{V} \cdot \exp\left(\frac{-\Delta S}{T_v}\right) + (C_{V,n} - C_{V,n-1}) \cdot \exp\left(\frac{-\Delta S}{2T_v}\right) & \text{if } \alpha_n \frac{dC_V}{dt} \geq 0 \ \ and \ \ 0 < \tau_{Vn} < 2T_{vl} \\ C_{N,n-1}^{V} \cdot \exp\left(\frac{-\Delta S}{T_v}\right) & \end{cases} \tag{16}$$

$$C_{V,n} = C_{N,n}^{C}\ (1 - K_{N,n})$$

$$K_{N,n} = \left(\frac{1 + \sqrt{f_n''}}{2}\right)^2$$

The non-dimensional vortex time parameter $\tau_{v,n}$ is given by:

$$\tau_{v,n} = \begin{cases} \tau_{v,n-1} + \frac{dt}{c/2} \cdot V_{\text{eff}} \cdot 0,45 \quad if\ C_{N,n}' > C_{N,I} \\ 0 \quad if\ C_{N,n}' < C_{N,I}\ and\ \Delta\alpha_n > 0 \end{cases}$$

Finally, the aerodynamic coefficients are given as follows:

$$C_{N,n} = C_{N,n}^{f} + C_{N,n}^{V} + C_{N,n}^{I}$$

$$C_{C,n} = \eta \cdot C_{N\alpha} \cdot \sqrt{f_n''} \alpha_{\text{eq},n}^2$$

$$\eta \approx 0.95$$

$$C_L = C_{N,n} \cdot \cos(\alpha) + C_{C,n} \cdot \sin(\alpha)$$

$$C_D = C_{N,n} \cdot \sin(\alpha) - C_{C,n} \cdot \cos(\alpha)$$

where $C_{N,n}$ is the normal force coefficient, $C_{C,n}$ is the tangential force coefficient, C_L is the lift coefficient, and C_D is the drag coefficient.

To verify the dynamic stall model, Figures 6 and 7 show the comparison between numerical results and experimental data [16] for different values of reduced frequency $k = \omega c/2V$. The pitching motion is given by $\alpha = Am\sin(\omega pt) + \alpha m$. This comparison demonstrates the ability of the Beddoes-Leishman dynamic stall model to predict the non-steady aerodynamic coefficients.

Computational algorithm

The algorithm presented in Figure 8 describes the computational procedure. First, the calculation begins with the evaluation of the near-wake-induced velocities. Second, when the first-quarter revolution has passed, the vortices are assumed to be in the rolling-up process to form the far wake.

The far-wake geometry is initialized with a rigid wake; we assume at this stage that the wake behind a rotor consists of a helical wake with constant pitch and diameter, because the effects of induced velocities are neglected. Afterward, the induced velocities are calculated at different Lagrangian markers, and the wake is corrected by the implementation of a predictor/corrector scheme. Finally, this new geometry is used in the calculation of induced velocities at the rotor plane. This iterative process continues until obtaining a constant value of the induced flow and circulation of the rotor plane.

As illustrated in the algorithm, the circulation is used in the calculation of induced velocities. This parameter is evaluated in terms of the lift coefficient using the Kutta-Joukozsky law.

In the present work, the wake simulation is based on the assumption that the flow is inviscid. However, the calculation of the aerodynamic coefficients by the AERODAS model takes into account the viscous effect which permits the modeling of stall. Also, the aerodynamic coefficients are corrected by the implementation of Beddoes-Leishman dynamic stall model to take into consideration the non-steady effect.

Results and discussion

In this section, some illustrative examples are presented to demonstrate the capability of the proposed method to

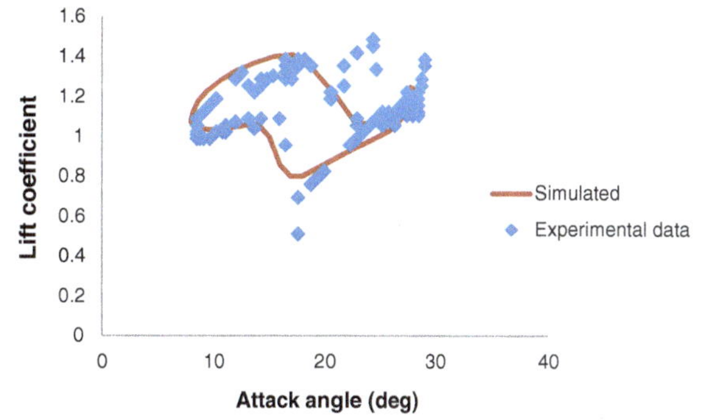

Figure 6 NREL S809 lift coefficient versus angle of attack; $\alpha = 20 + 10\sin(\omega t)$, $k = 0.025$, $M = 0.1$.

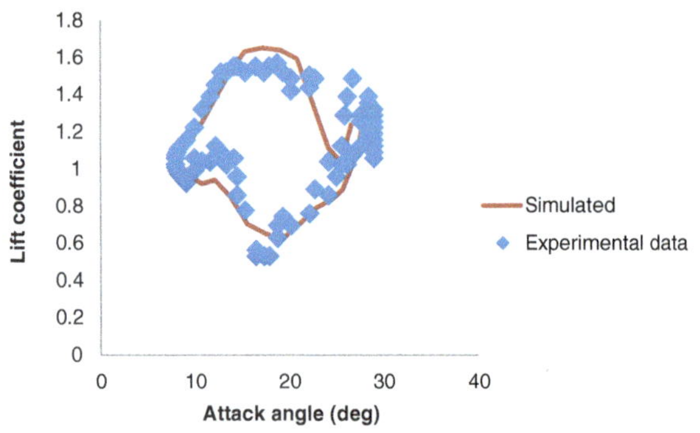

Figure 7 NREL S809 lift coefficient versus angle of attack; $\alpha = 20+10\sin(\omega t)$, $k = 0{,}051$, $M = 0.1$.

predict the variation of the normal force acting on the blade. Experimental results were obtained at the National Renewable Energy Laboratory in NASA Ames wind tunnel, where test conditions are rigorously controlled [17]. The NASA Ames wind tunnel possesses excellent test section flow quality. Flow speeds deviate by no more than 0.25% from the nominal value, whereas the flow velocity vector diverges no more than 0.5° from the test section axis. Turbulence intensity is typically no greater than 0.5%. The set speed value differed less than 0.1 m/s from the nominal value. The average pressure measurement system bias error was less than 0.22%.

Measurements were performed on a wind turbine operating in yaw conditions. This wind turbine was two bladed,

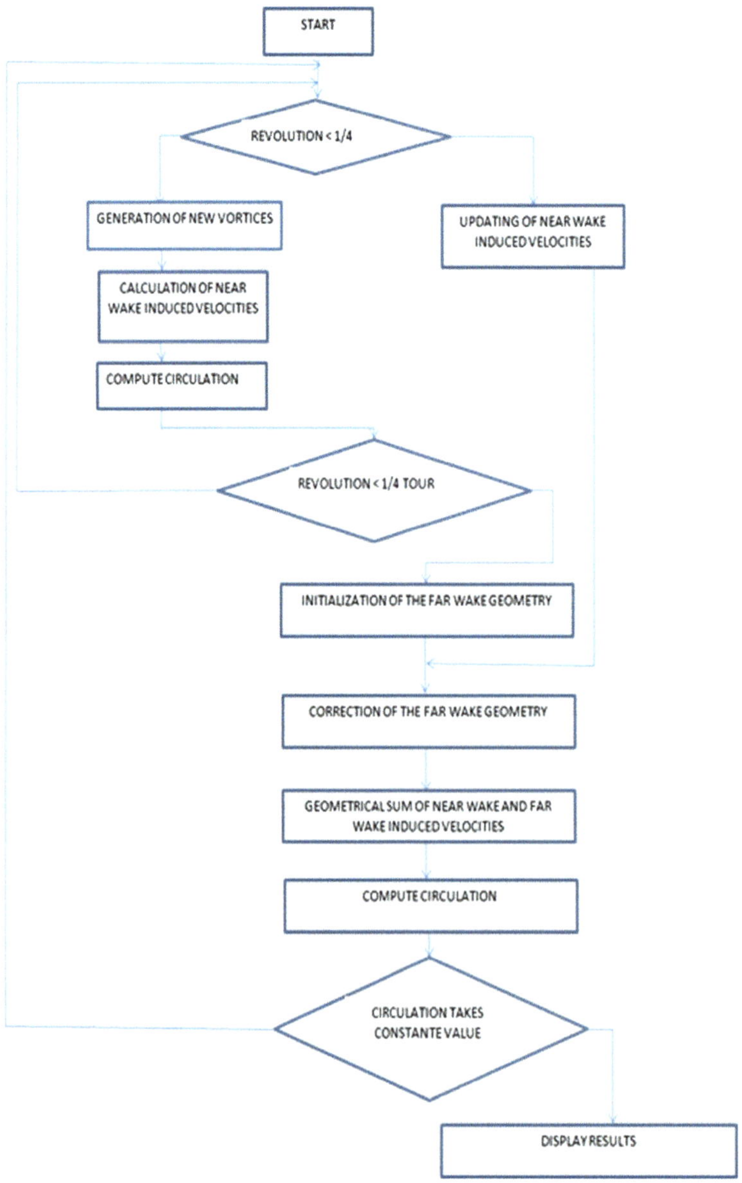

Figure 8 Computational algorithm.

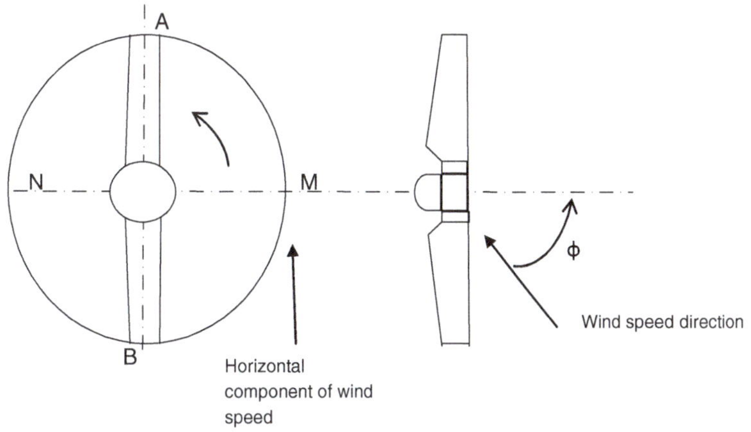

Figure 9 **Advancing and retreating effect.**

and it had a rotor diameter of 10 m. The yaw angle was equal to 30°, and the pitch angle was equal to 0°. The airfoil was of type S809 over the entire span [18].

Referring to Figure 9, the horizontal component of the wind velocity is not null in yaw conditions. This component must be taken into account in the calculation. Its projection in the blade-fixed Cartesian coordinate system can be expressed as $U \sin(\phi_y) \cos(\phi_r)$. Accordingly, the angle of attack is calculated with:

$$\alpha = a \tan\left(\frac{U \cos\left(\varphi_y\right) - U_{\text{induced}}}{\Omega\ r + W_{\text{induced}} + U \sin\left(\varphi_y\right)\ \cos\left(\varphi_r\right)}\right) - \beta$$

The relative wind velocity is calculated by using the following equation:

$$V_{\text{rel}} = \sqrt{\left(U \cos\left(\varphi_y\right) - U_{\text{induced}}\right)^2 + \left(\Omega\, r + W_{\text{induced}} \cdot U \sin\left(\varphi_y\right) \cos\left(\varphi_r\right)\right)^2}$$

where U_{induced} is the axial induced velocity and W_{induced} the tangential induced velocity.

In yaw conditions, we distinguish two main operating modes: the advancing and the retreating modes. The blade will be advancing in the half-plane (ANB) and retreating in the half-plane (BMA), as depicted in Figure 9. These mean that the horizontal component of the wind relative velocity is in opposition with the rotation vector in the half-plane ANB and in line with the rotation vector in the half-plane BMA. Consequently, the blade load, the relative velocity, and the angle of attack vary depending on the azimuthal position.

In the following, we represent the azimuthal variation of the normal force at 30% and 95% span of the blade to validate the algorithm and to better observe the influence of the skewed wake and the advancing and retreating behaviors.

Figures 10 and 11 show the distribution of the normal force at 30% and 95% span for an air speed of 5 m/s.

We observe that the calculated and measured curves are in perfect agreement. These are in phase and show the same tendencies.

The results presented in Figures 12 and 13 show the distribution of the normal force at 30% and 95% span for an air speed equal to 15 m/s. The agreement between

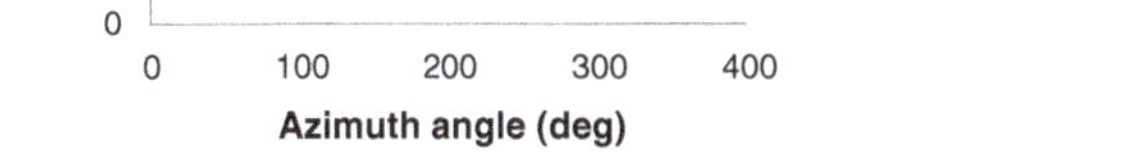

Figure 10 **Normal force distribution at 30% span, for wind speed of 5 m/s.**

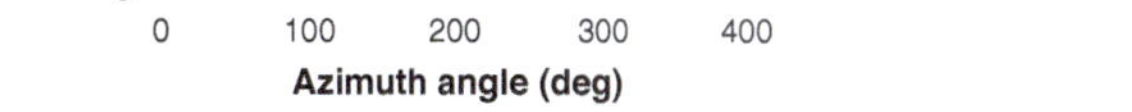

Figure 11 **Normal force distribution at 95% span, for wind speed of 5 m/s.**

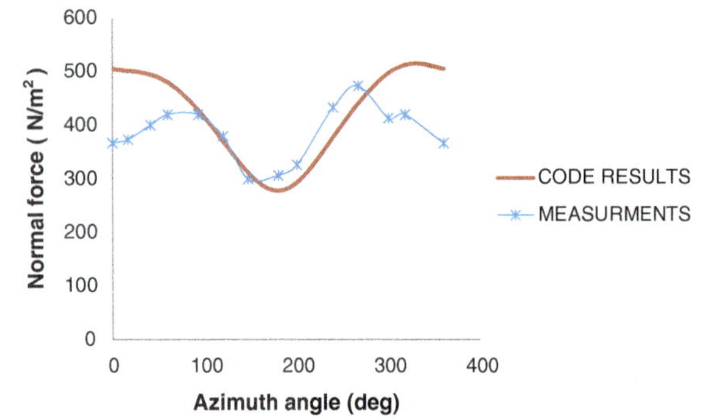

Figure 12 Normal force distribution at 30% span, for wind speed of 15 m/s.

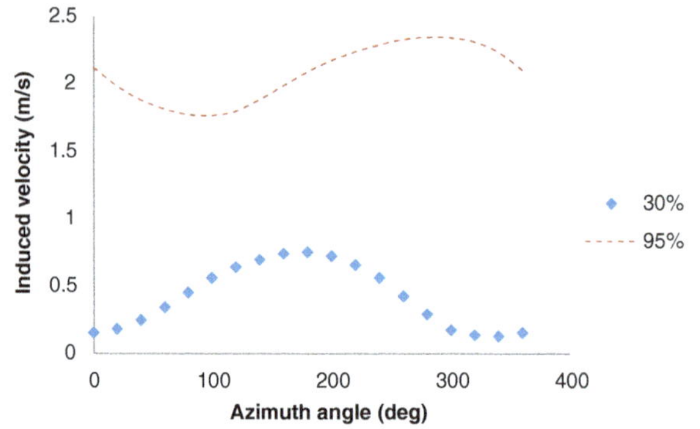

Figure 14 Azimuthal distribution of induced velocity for an air speed of 5 m.

measured and calculated curves is excellent in terms of shape despite the existence of an amplitude difference.

Prediction uncertainties, found at high wind speed, essentially originate from assumptions adopted in the modeling and the limits of models used to predict aerodynamic coefficient in the non-steady case. Furthermore, at high wind velocity, compressibility and turbulence become important. Consequently, the approach adopted in this work where the fluid is considered inviscid and incompressible can lead to significant discrepancies in results [17]. Also, the great number of empirical parameters used in the AERODAS model and the Beddoes-Leishman dynamic stall model could be a source of error. In this context, the study referenced in [19] confirmed that existing models show significant disparities at high wind velocity. This can be attributed to inconsistencies in empirical input parameters, which need to be derived from unsteady measurements for each and every airfoil, and over the appropriate ranges of the Reynolds and Mach numbers.

In contrast, the cyclical fluctuation of the normal force observed previously is mainly caused by the skewed wake effect and the advancing and retreating blade effect. To better illustrate the skewed wake effect, Figure 14 shows the variation of the induced velocities at 30% and 95% for an air speed of 5 m/s.

We note that there is a correlation between the distribution of the induced velocity presented in Figure 14 and the distribution of the normal force presented in Figures 10 and 11.

At the inboard section, the maximum value of the normal force is found between 180° and 360°. This behavior is consistent with the skewed wake effect because the minimum value of the induced velocity is found between 180° and 360°. For the outboard section, the maximum value of the normal force is found between 0° and 180°. This behavior is in line with the skewed wake effect presented in Figure 11 because the minimum of the induced velocity is found between 0° and 180°.

Generally speaking, for a wind speed of 5 m/s, the distribution of the normal force is mainly influenced by the skewed wake effect in spite of some disturbances caused by the contribution of the advancing and retreating blade effect.

Figure 15 shows the azimuthal variation of the relative velocity and horizontal component of the wind velocity for a wind speed of 15 m/s. We observe that the azimuthal variation of the horizontal component of the wind velocity

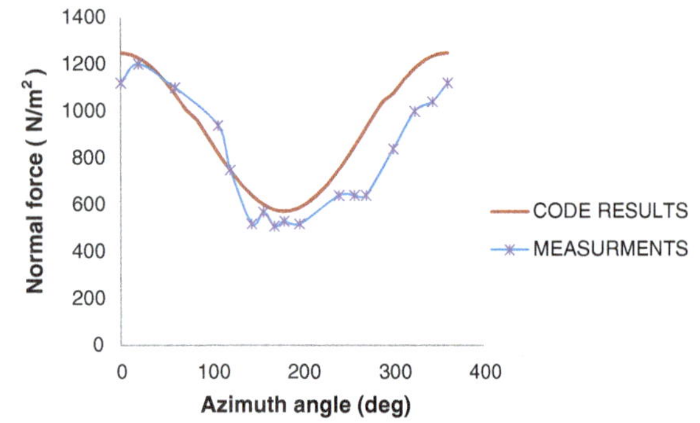

Figure 13 Normal force distribution at 95% span, for wind speed of 15 m/s.

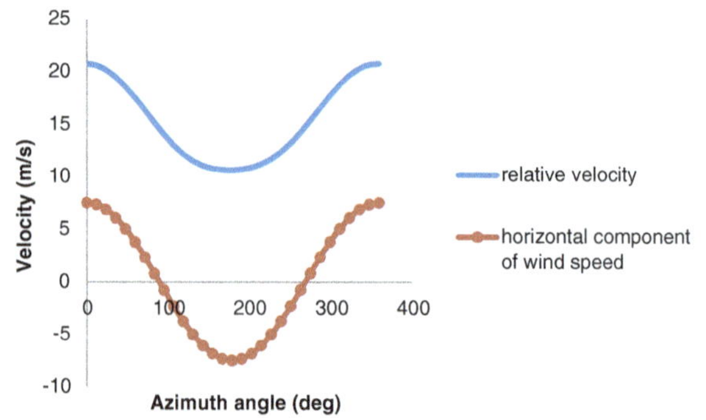

Figure 15 Azimuthal variation of relative velocity and horizontal component of wind velocity for wind speed of 15 m/s.

is reflected in the azimuthal variation of the normal force illustrated in Figures 12 and 13. We underline a correlation between these distributions.

We conclude that at high wind speeds, the variation of the normal force is mainly influenced by the advancing and retreating blade effect. This can be explained by the predominance of the advancing and retreating blade effect compared with the skewed wake effect.

Conclusion

We have developed a numerical code for a yaw-misaligned wind turbine. The numerical procedure is obtained by combining the Beddoes near-wake model and the free-wake method. The Beddoes-Leishman dynamic stall model was implemented in the calculation, and results of the complete model were compared with the NASA Ames wind tunnel measurements. This comparison demonstrates the capability of the models used to reproduce a physically realistic feature with acceptable tolerance. However, we have observed the existence of a difference in amplitude at high wind speed arising from the different assumptions adopted in this work.

The load fluctuation highlighted in this work is caused by the skewed wake effect and the advancing and retreating effect. The former plays a major role in the aerodynamic loads calculation at low wind speed. However, the latter is important at high wind speed and can provide a strong dynamic stall effect. This behavior explains the difference found in the amplitude between experiment and theory at high wind speed despite the implementation of the Beddoes-Leishman dynamic stall model in the calculation.

Abbreviations

α: angle of attack [deg]; β: global pitch angle [deg]; φ_r: azimuth angle [deg]; φ_y: yaw angle [deg]; ρ: air density [kg/m^3]; Ω: rotor speed [rpm]; Γ: circulation [m^2/s]; A_0: angle of attack at which $CL_1 = 0$; for all AR (deg); ACL_1: angle of attack at maximum pre-stall lift (deg); ACD_1: angle of attack at maximum pre-stall drag (deg); B: blade number; BEM: blade element momentum theory; C: chord [m]; Cd: drag coefficient [dimensionless]; CD_0: minimum drag coefficient, at $\alpha = A_0$, for all aspect ratios; CD_1: drag coefficient in the pre-stall regime [dimensionless]; CD_2: drag coefficient in the post-stall regime [dimensionless]; CD_{1max}: maximum pre-stall drag coefficient; at α = ACD1 [dimensionless]; CD_{2max}: maximum post-stall drag coefficient [dimensionless]; Cl: lift coefficient [dimensionnelss]; CL_1: lift coefficient in the pre-stall regime [dimensionless]; CL_{1max}: maximum pre-stall lift coefficient; at $\alpha = ACL_1$ [dimensionless]; CL_2: lift coefficient in the post-stall regime [dimensionless]; DS: dynamic stall; r: radial position on the blade [m]; R: rotor radius [m]; S1: slope of linear segment of pre-stall lift curve (1/deg); U: wind velocity [m/s]; V_{ind}: induced velocity [m/s]; V_{rel}: relative wind speed velocity [m/s].

Competing interest

The authors declared that they have no competing interests.

Authors' contributions

AB, AA, and NB have participated in the development and validation of the calculation code. All authors have read and approved the final manuscripts.

Authors' information

All authors are currently university professors at ENSAM Meknes. AB is assigned to the Department of Mechanical Engineering. His research is in the field of renewable energy, mainly wind energy. AA and NB work and conduct their research at the energy Department in the energetic field and fluid mechanics.

Acknowledgments

The authors are grateful to the Director of ENSAM Meknes and the Director of doctoral studies center.

Author details

[1]Department of Mechanical Engineering, Moulay Ismail University, ENSAM-MEKENS, PO. BOX. 4024, Meknes 50003, Morocco. [2]Department of Energetic, Moulay Ismail University, ENSAM-MEKENS, P.O. BOX. 4024, Meknes 50003, Morocco.

References

1. Lanzafame, R, Messina, M: Fluid dynamics wind turbine design: critical analysis, optimization and application of BEM theory. Renew Energ. **32**, 2291–2305 (2007)
2. Afjeh, AA, Keith Jr, TG: A vortex lifting line method for the analysis of horizontal axis wind turbines. Trans. ASME J. Sol. Energy Eng. **108**, 303–9 (1986)
3. Chkir, S: Unsteady loads evaluation for a wind turbine rotor using free wake method. Energy Procedia. **6**, 777–785 (2011)
4. Beddoes, TS: A near wake dynamic model. In: Aerodynamic and Acoustics National Specialist Meeting, pp. 1–9., Arlington, TX (1987)
5. Wang, T, Coton, FN: A modified near wake dynamic model for rotor analysis. Aeronaut J. **103**, 143–146 (1999)
6. Vermeer, LJ, Sorensen, JN, Crespo, A: Wind turbine wake aerodynamics. Prog Aerosp Sci **39**, 467–510 (2003)
7. Sebastien, T, Lackner, MA: Development of a free vortex wake method code for offshore floating wind turbines. Renew Energ. **46**, 269–275 (2012)
8. Hansen, MOL, Sørensen, JN, Voutsinas, S, Sørensen, N, Madsen, HA: State of the art in wind turbine aerodynamics and aeroelasticity. Prog Aerosp Sci. **42**, 285–330 (2006)
9. Leishman, JG: Principles of Helicopter Aerodynamics. Cambridge Universiy Press, Cambridge (2006)
10. Bhagwat, J, Leishman, J: Free-vortex filament methods for the analysis of helicopter rotor wakes. J Aircraft. **39**, 759–775 (2002)
11. Spera, DA: Models of lift and drag coefficients of stalled and installed airfoils in wind turbines and wind tunnels. Report NASA/CR—2008-215434. NASA, Cleveland, OH (2012)
12. Larsen, JW, Nielsen, SRK, Krenk, S: Dynamic stall model for wind turbine airfoils. J Fluid Struct. **23**, 959–982 (2007)
13. Leishman, JG, Beddoes, TS: A semi-empirical model for dynamic stall. J. Am Helicopter Soc. **34**(3), 3–17 (1989)
14. Galvanetto, U, Peiro, J, Chantharasenawong, C: Remarks on the nonlinear dynamics of a typical aerofoil section in dynamic stall. Aeronaut J. **111**, 731–739 (2007)
15. Dai, JC, Hu, YP, Liu, DS, Long, X: Aerodynamic loads calculation and analysis for large scale wind turbine based on combining BEM modified theory with dynamic stall model. Renew Energ. **36**, 1095–1104 (2011)
16. Huyer, H: Effects of forced unsteady separated flow fields on a rotating wind turbine blade. NREL Tech. Rep. 442–4864. Natl. Renewable Energy Lab, Golden, CO (1993)
17. Simms, D, Schreck, S, Hand, M, Fingersh, LJ: NREL unsteady aerodynamics experiment in the NASA-Ames wind tunnel: a comparison of predictions to measurements. National Renewable Energy Laboratory/TP-500-29494. Natl. Renewable Energy Lab, Golden, CO (2001)
18. Shepers, JG: IEA annex XX: comparison between calculations and measurements on a wind turbine in yaw in the NASA-Ames wind tunnel. ECN-E-07-072. Energy Research Center of the Netherlands, Petten (2005)
19. Leishman, JG: Challenges in modeling the unsteady aerodynamics of wind turbine. In: 21st ASME Wind Energy Symposium and 40th AIAA Aerospace Sciences Meeting., Reno, NV (2002)

10

Performance correlation for single-basin double-slope solar still

Shanmugasundaram Kandasamy*, Manikandan Vellingiri, Shanmugan Sengottain and Janarthanan Balasundaram

Abstract

In this study, an attempt has been made to develop a new correlation for the productivity of a single-basin double-slope solar still using experimental results obtained for different brine depths (0.01, 0.02, 0.03, 0.04, 0.05, and 0.06 m) in the basin in Coimbatore, Tamilnadu, India. The proposed correlation has been validated with the experimental results for one of the typical days in Coimbatore, Tamilnadu, India. A concluding correlation has been developed by including the results mentioned in the literature regarding double-slope solar still alone and the proposed correlation. The concluding correlation is compared with the correlation developed by the researcher based on the results in the literature regarding both single- and double-slope solar stills. The results indicate that the correlation coefficient R^2 is slightly higher as compared with that of previous results obtained for both single- and double-slope single-basin solar stills.

Keywords: Distillation, Single-basin solar still, Brine water depths, Correlation

Background

Water is very much essential for human beings. Supplying drinkable water is one of the major problems in developing countries. A solar still is a simple device which is used in converting brackish/saline water into potable/drinkable water. Many researchers have found the productivity for different brine depths in the basin of a single-slope single-basin solar still [1-15].

Further attempts have been made by researchers to investigate the effect of water depth in double-slope single-basin solar still. A digital simulation method has been done for the productivity of a complicated still by Cooper [16], and the influence of more common variables such as water depth, wind velocity, still insulation, double glass cover, cover slope, and daily variability on the productivity has been investigated. Results of the simulation indicate that water depth and thermal insulation have little effect on the productivity at shallow water depth and that a double glass cover and high cover slopes are not justifiable. Garg and Mann [17] have investigated the effect of climatic, operational, and design parameters for single- and double-slope solar stills under the Indian arid zone regions and to find the effect of all possible variables on the performance of the single-effect solar still. Akash et al. [18] have studied the basin-type solar still under local climatic conditions and concluded that lower water depth and salinity of the brine produced higher distillate yields for the double-slope passive solar still. The effect of wind speed on daily productivity of some active and passive solar stills using computer simulation for different solar still designs has been studied by El-Sebaii [19]. It has been shown that for the active and multi-effective passive stills, daily productivity increases with the increase of wind velocity up to a typical velocity; beyond that, the productivity becomes insignificant for single-effect passive stills. Al-Hinani et al. [20] have studied the effect of climatic, design, and operational parameters such as solar intensity, wind velocity, ambient temperature, glass cover slope angle, feedwater temperature water depth on the basin, and basin material on the distillate output of the basin-type solar still. They found that the 23° cover tilt angle, 0.1-m insulation thickness, and asphalt coating of solar still of the shallow water basin are the optimum parameters for producing higher yields. Al-Hayeka and Badran [21] have found the productivity of freshwater of two different designs of stills, which are asymmetric greenhouse with mirror and symmetric greenhouse solar stills, under

* Correspondence: sspriyashanmu2@gmail.com
Department of Physics, Karpagam University, Coimbatore, Tamilnadu 641 021, India

the climatic conditions of Jordan. It was inferred that the distilled water output of the asymmetric greenhouse type was 20% higher than that of the symmetric greenhouse type. Mowla and Karimi [22] have simulated a mathematical model for a solar still; the rate of production of freshwater has been calculated, and experiments have been carried out in Shiraz. It has been found that the experimental and calculated results are in good agreement. The effect of water depth on the transient performance of a double-basin solar still has been investigated by Tiwari and Tiwari [23].

Siddiqui and Bajpai [24] have made an attempt to develop a correlation between the thickness of dust collected and the difference in efficiencies for all seasons in Lucknow, India. The equation derived is very good for the measurement of correlation with experimental results for all the months. Pragya et al. [25] have made an attempt to investigate the advantages of a two-layer solar panel for the enhancement of solar energy absorption. Results have shown an increase of 70% in the output. Hitesh and Shah [26] have aimed to evaluate the effect of various energy-absorbing plates in solar still to increase the daily distillate yield. It has been concluded that the utilization of aluminium floating plate has a significant impact on the productivity of the still.

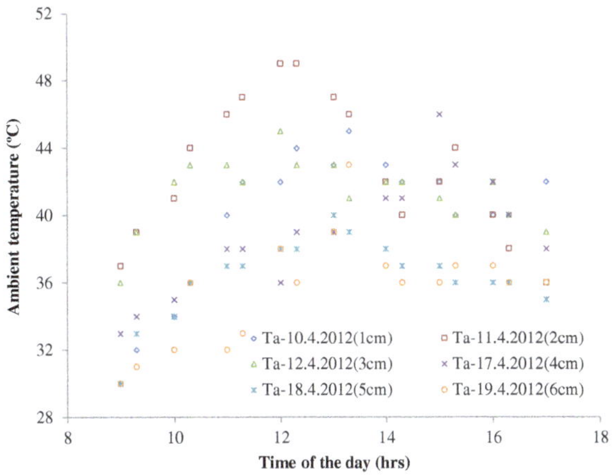

Figure 2 Variation of ambient temperature for experimental days with different brine depths.

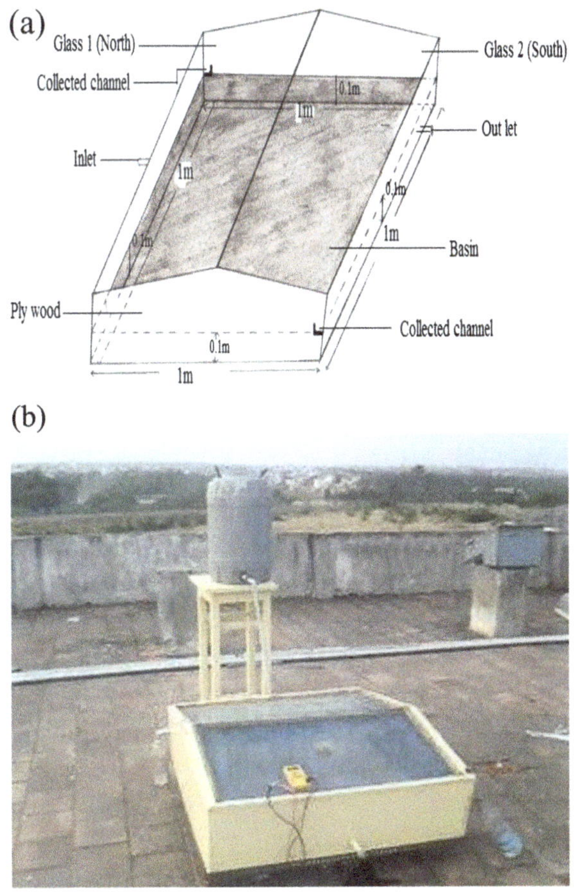

Figure 1 Schematic diagram of double-slope single-basin solar still (a) and photograph of experimental still (b).

The main objective of this work is to propose a correlation to find the productivity of the still based on the experimental results for different brine depth in the basin of the double-slope solar still. Experimental studies were carried out to validate the developed correlation. The correlation has been compared with the correlation developed by Khalifa and Hamood [27].

The decrease in productivity depends on the design parameter, location, and range of depth of brine water in the basin. Moreover, the existing models lack the inclusion of factors influencing the working of the still. These inconveniences paved the way for the present investigation to propose a correlation for a double-slope single-basin solar still alone.

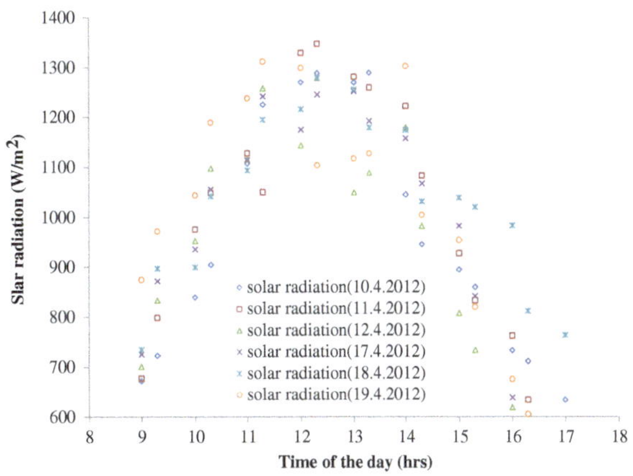

Figure 3 Variation of intensity of solar radiation for experimental days with different brine depths.

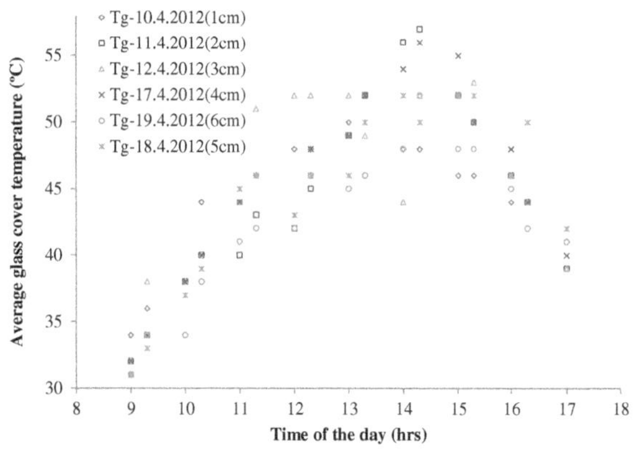

Figure 4 Variation of average glass cover temperature for experimental days with different brine depths.

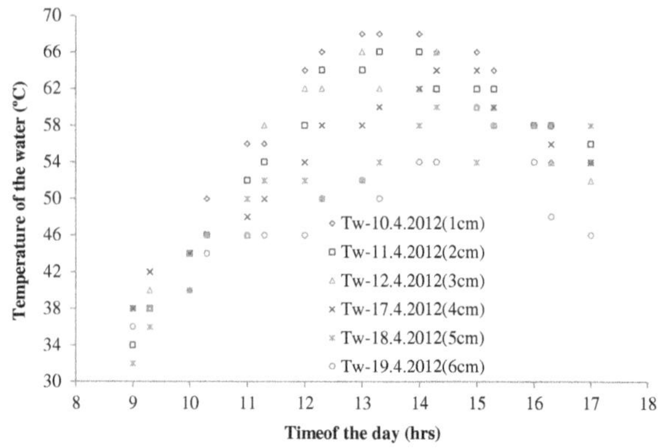

Figure 6 Variation of water temperature for the experimental days with different brine depths.

Methods

Experimental setup and observations

A schematic diagram of the double-slope single-basin solar still is shown in Figure 1a, and a photograph of the experimental setup of double-slope single-basin solar still with an inclination angle of the condensing glass cover (11°) is shown in Figure 1b. The experiments were conducted for many days with different brine depths in the basin from January to May 2012. Among the observations of experimental days, typical days have been selected for the proposed correlation based on the output and average solar radiation. Hence, in the month of April 2012, experimental observations for six different brine depths, namely 0.01, 0.02, 0.03, 0.04, 0.05, and 0.06 m, as the output and average solar radiation are high.

The experimental still is made of galvanized iron sheet with a basin area of 1 m^2. A glass cover with a 0.004-m thickness was used as condensing surface with an inclination angle of 11°, which is equal to the latitude of the location (latitude = 11°N). The bottom surface of the still basin is painted black to absorb a large amount of solar radiation. The yield of the solar still is collected using a graduated measuring jar. The solar still was oriented in the north–south direction to receive solar radiation throughout the working hours of the day. Experiments were conducted from 9 a.m. to 5 p.m. from the month of January to May 2012. Solar radiation, ambient temperature, distillate output, and basin, water, and cover glass temperatures have been measured for an interval of 30 min. The intensity of solar radiation is measured using a solar radiation monitor, and a digital thermometer is used to measure ambient temperature. Calibrated iron-constantan thermocouples are used to measure the basin, water, and glass cover temperatures.

Results and discussion

The variation of ambient temperature and solar radiation for experimental days are shown in Figures 2 and 3. From Figure 2, it is clear that for all experimental days, the

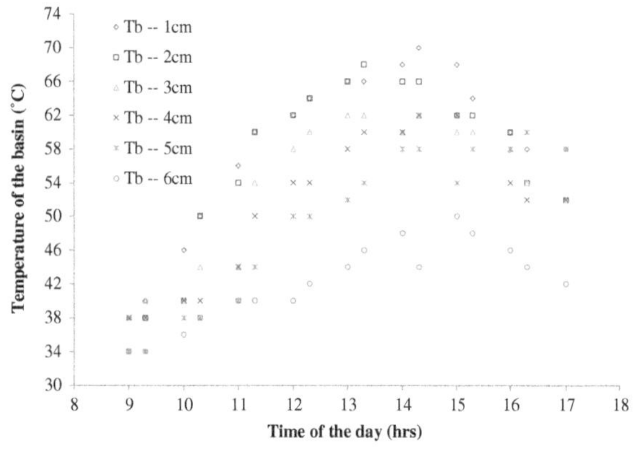

Figure 5 Variation of basin temperature for experimental days with different brine depths.

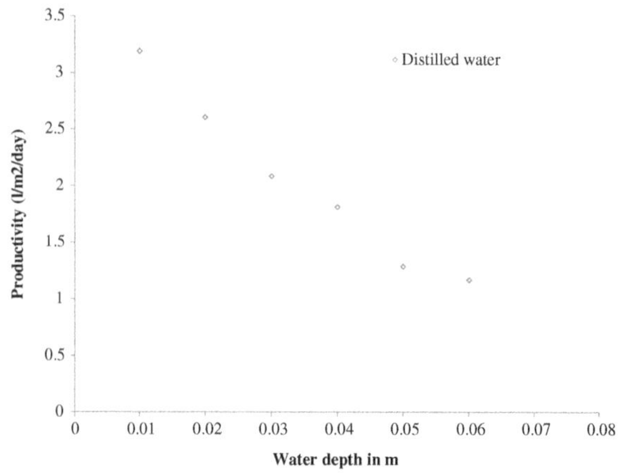

Figure 7 Variation of distilled water output between 9 a.m. to 5 p.m. for different brine depths.

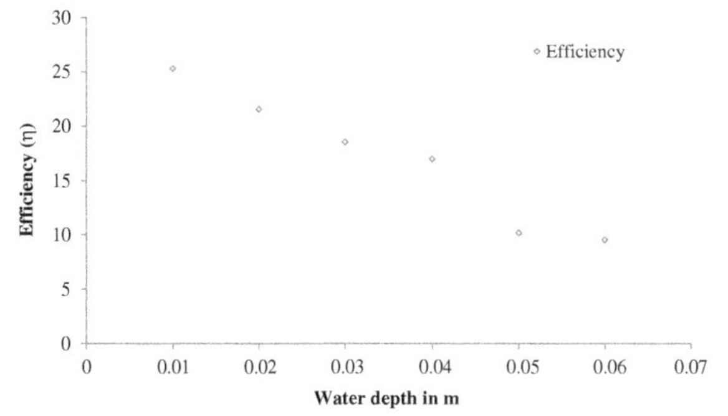

Figure 8 Variation of average efficiency for different depths.

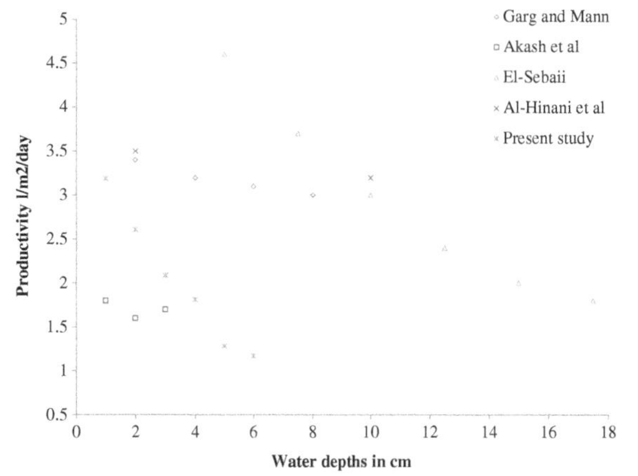

Figure 10 The scattered correlation of the literature for double-slope single-basin solar still.

ambient temperature is higher than 30°C from the morning hours. The maximum ambient temperature recorded was 49°C in one of the hours of the experimental day, and days are seen to be clear sunny days. The variation of solar radiation in Figure 3 shows that the intensity gradually increases from morning, reaches a maximum value at 1 p.m., and gradually decreases in the evening hours. The maximum intensity of solar radiation in one of the hours in experimental days is found to be 1,347 W/m^2.

The experiments were performed on a single-basin double-slope solar still for different brine depths of 0.01, 0.02, 0.03, 0.04, 0.05, and 0.06 m on 10 to 12 and 17 to 19 April 2012. Figure 4 shows the variation of the still's average glass cover temperature for different brine depths maintained in the basin. The average glass cover temperature gradually increases until 1 p.m. and decreases gradually during the evening hours. For the brine depth of 0.02 m, the maximum temperature of the glass cover is found to be 57°C due to the large convection and radiation heat transfer from water to the glass cover.

The variation of water and basin temperatures of the proposed still for different brine depths has been depicted in Figures 5 and 6. From Figures 5 and 6, it is observed that the water and basin temperatures for the brine depth of 0.01 m reached a maximum of 68°C and 70°C, respectively. Since the thermal capacity of the water in the basin is low when compared with the other brine depths, water and basin temperatures reached a maximum. Moreover, the temperature between the glass cover and water is high in order to produce more distillate yields as expected.

The amount of distillate yield from 9 a.m. to 5 p.m. has been measured for the considered brine depths and is presented in Figure 7. It is clear from the figure that for the least water depth, the distillate yield decreases every 30 min. Thus, for the water depth of 0.01 m, the distillate yield for the proposed still for every 30 min is higher than the other brine depths considered for the experiment. This is in accordance with previous results obtained by the researchers [1-15].

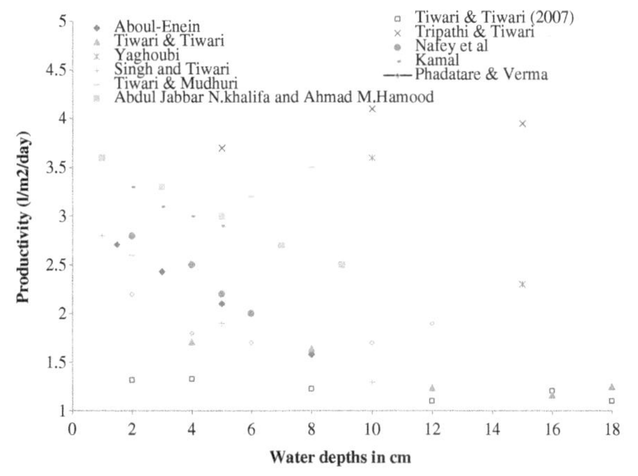

Figure 9 The scattered correlation of the literature for single-slope single-basin solar still.

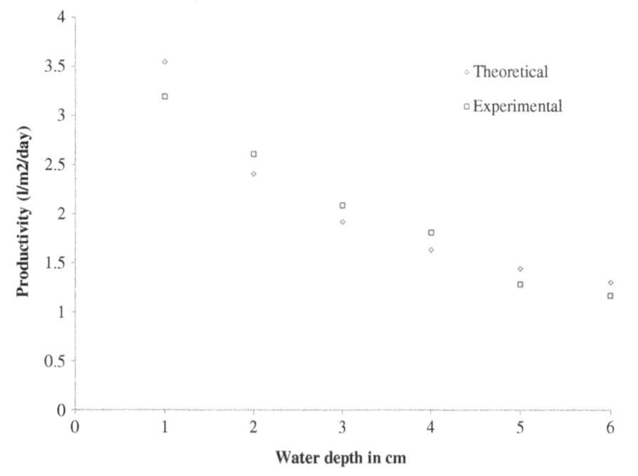

Figure 11 Variation of experimental and theoretical productivity for different depths.

For the 0.01-m brine depth, a maximum of 350 ml for a 30-min interval from 1 p.m. to 1:30 p.m. has been obtained.

Figure 8 shows the variation of average efficiency of the still for brine depths considered in the study. It is observed that the average efficiency of 25% has been obtained for 0.01 m of brine depth.

The result confirmed that the performance of the still is better for the least water depth, and it also inferred that when the brine depth increases, the efficiency decreases.

Developed correlation

Regression lines were drawn, and R^2 value had been found for the data collected from the literature regarding single- and double-slope basin-type solar stills separately by means of a least square method. A concluding correlation has been developed for single-basin double-slope solar still with literature data and correlation developed from the present investigation. A power regression in the form $y = C(d)^n$ was chosen so that no output is found at zero depth, where the productivity is in liters per square meter per day and the brine depth is in centimeters.

Figures 9 and 10 shows the correlations developed using the result of data from literature for single-slope solar still and double-slope solar still separately. The correlations for single-slope solar still and concluding correlation for double-slope solar still are as follows:

$$y = 2.833d^{-0.20}, R^2 = 0.829, \tag{1}$$

and

$$y = 5.885d^{-0.292}, R^2 = 0.869, \tag{2}$$

where R^2 is the correlation coefficient.

Performance of the correlation

A concluding correlation has been proposed with the experimental results of the present study and literature data for double-slope single-basin solar still. The concluding correlation developed has been compared with the concluding correlation developed by Khalifa and Hamood [27]. Also, a correlation has been developed for single-slope single-basin solar still separately from the data available in the literature. It has been found that for single-slope single-basin solar still alone, it has the regression coefficient of 0.829 which is higher than the correlation developed by Khalifa and Hamood [27] including both single- and double-slope basin-type solar stills.

The concluding correlation proposed which includes the present investigation and data in the literature for double-slope single-basin solar still has the regression coefficient of 0.869, which is higher than the correlation for the single-slope single-basin solar still. The concluding correlation has been validated for one of the typical days with different brine depths and is depicted in Figure 11. From the figure, it is observed that the theoretical results are in close agreement with the experimental observations with least error.

Conclusions

The correlations are developed separately for single-slope single-basin solar still using data from the literature, and the concluding correlation includes data from the literature and the present study. Results have shown that the R^2 value of 0.829 obtained for the single-slope single-basin solar still and the R^2 value of 0.869 for the concluding correlation are higher than previous correlations developed by Khalifa and Hamood [27]. Hence, it is confirmed that correlation should be done separately for single- and double-slope single-basin solar stills instead of combining the data of both single- and double-slope single-basin solar stills. The regression coefficient of the concluding correlation approaches to unity, and the correlation may be a perfect positive correlation with least error.

Competing interests

The authors declare that they have no competing interests.

Authors' contributions

SK developed the correlation for the proposed system. MV did the literature review regarding the correlation for solar still. SS compared the developed correlation with the existing correlation and finalized the correlation for single-basin double-slope solar still. JB went through the text of the paper and critically checked the correlation for correctness. All authors read and approved the final manuscript.

Acknowledgements

The authors would like to thank the University Grand Commission (UGC), The Government of India, New Delhi, for the award of Rajiv Gandhi National Research Fellowship for SC/ST.

References

1. Tiwari, GN, Madhuri: Effect of water depth on daily yield of the still. Desalination **61**, 67–75 (1987)
2. Kamal, WA: A theoretical and experimental study of the basin type solar still under the Arabian Gulf climatic conditions. Solar and Wind Technology **5**, 147–157 (1988)
3. Farid, M, Hamad, F: Performance of single-basin solar still. Renew. Energy **3**, 75–83 (1993)
4. Yaghoubi, MA, Sabzevari, A: Calculation of the hourly output of a solar still for various cities in Iran. Renew. Energy **7**, 427–435 (1996)
5. Toure, S, Meukam, P: A numerical model and experimental investigation for a solar still in climatic conditions in Abidjan (Cote D'ivoire). Renew. Energy **11**, 319–330 (1997)
6. Aboul-Enein, S, Eli-Sebaii, AA, El-Bialy, E: Investigation of a single-basin solar still with deep basins. Renew. Energy **14**, 299–305 (1998)
7. Nafey, AS, Abdelkader, M, Abdelmotalip, A, Mabrouk, AA: Parameters affecting solar still productivity. Energ. Convers. Manage. **41**, 1797–1809 (2000)
8. Nafey, AS, Abdelkader, M, Abdelmotalip, A, Mabrouk, AA: Enhancement of solar still productivity using floating perforated black plate. Energ. Convers. Manage. **43**, 937–946 (2002)
9. Singh, HN, Tiwari, GN: Monthly performance of passive and active solar still for different Indian climatic conditions. Desalination **168**, 145–150 (2004)

10. Tripathi, R, Tiwari, GN: Effect of water depth on internal heat and mass transfer for active solar distillation. Desalination **173**, 187–200 (2005)
11. Abu-Arabi, M, Zurigat, Y: Year round comparative performance study of solar desalination units. Desalination **172**, 137–143 (2005)
12. Tripathi, R, Tiwari, GN: Thermal modeling of passive and active solar stills for different depths of water by using the concept of solar fraction. Solar Energy **80**, 956–967 (2006)
13. Phadatare, MK, Verma, SK: Influence of water depth on internal heat and mass transfer in a plastic solar still. Desalination **217**, 267–275 (2007)
14. Tiwari, AK, Tiwari, GN: Thermal modeling based on solar fraction and experimental study of the annual and seasonal performance of a single slope passive solar still: the effect of water depths. Desalination **207**, 184–204 (2007)
15. Badran, OO: Experimental study of the enhancement parameters on a single slope solar still productivity. Desalination **209**, 136–143 (2007)
16. Cooper, PI: Digital simulation of transient solar still process. Solar Energy **12**, 313–331 (1969)
17. Garg, HP, Mann, HS: Effect of climatic, operational and design parameters on the year round performance of single slope and double slope solar still under Indian arid zone conditions. Solar Energy **18**, 159–164 (1976)
18. Akash, BA, Mohsen, MN, Nayfeh, W: Experimental study of the basin type solar still under local climatic conditions. Energ. Convers. Manage. **41**, 883–890 (2000)
19. El-Sebaii, AA: Effect of wind speed on active and passive solar stills. Desalination **45**, 1187–1204 (2004)
20. Al-Hinani, H, Al-Nassri, MS, Jubran, BA: Effect of climatic, design and operational parameters on the yield of a simple solar still. Energ. Convers. Manage. **43**, 1639–1650 (2002)
21. Al-Hayeka, I, Badran, OO: The effect of using different designs of solar stills on water distillation. Desalination **169**, 121–127 (2004)
22. Mowla, D, Karimi, G: Mathematical modeling of solar stills in Iran. Solar Energy **55**, 389–393 (1995)
23. Tiwari, AK, Tiwari, GN: Effect of water depths on heat and mass transfer in a passive solar still: in summer climatic condition. Desalination **195**, 78–94 (2006)
24. Siddiqui, R, Bajpai, U: Correlation between thickness of dust collected on photovoltaic module and difference in efficiencies in composite climate. Int. J. Energ. Res. Environ. Eng. **3**(26), 1–7 (2012)
25. Pragya, S, Tirumalachetty, H: Enhancement of energy generation from two layer solar panel. Int. J. Energ. Res. Environ. Eng. **3**(12), 1–9 (2012)
26. Hitesh, P, Shah, PK: Investigation on solar stills having floating plates. Int. J. Energ. Res. Environ. Eng. **3**(8), 1–5 (2012)
27. Khalifa, AJN, Hamood, AM: On the verification of the effect of water depth on the performance of basin type solar stills. Solar Energy **83**, 1312–1321 (2009)

Energy and exergy analyses of the Nigerian transportation sector from 1980 to 2010

Ismaila Badmus[1*], Ajiboye Saheeb Osunleke[2], Richard Olayiwola Fagbenle[3] and Miracle Olanrewaju Oyewola[3]

Abstract

This paper analyses energy utilisation in the transportation sector of Nigeria using exergy methods. The sector is dominated by the road subsector, with a share ranging from 70.65% in 1990 to 97.51% in the year 2005 and a mean of 88.44%. The road subsector is still the most efficient, with energy efficiency values that range from 9.93% in 1990 to 19.19% in 1986. The corresponding exergy values are 9.28% and 17.93%, respectively. Following the road subsector is the aviation subsector with its least energy consumption share of 0.10% in 2005 and biggest share of 26.25% in 1990. The subsector energy efficiency values range from 0.03% in 2005 to 7.35% in 1990, with corresponding exergy values of 0.026% and 6.87%, respectively. The overall mean energy efficiency in the Nigerian transportation sector for the three decades is 17.11%, while the overall mean exergy efficiency is 15.97%. The road subsector performance has been adversely affected by the massive importation of used vehicles into the country.

Keywords: Energy analysis, Exergy efficiency, Nigerian transportation sector

Background

Nigeria's transport system consists of rail, road, water, and air subsectors. Ninety-six per cent of the total quantity of premium motor spirit (gasoline) consumed and 40% of the automotive gas oil (diesel oil) consumed go to road transportation. Twenty-four per cent of the dual purpose kerosene consumed goes to aviation, and it is the sole fuel. Ten per cent of the fuel oil consumed and 6% of the automotive gas oil consumed go to water transportation [1].

Road transport is the most commonly used mode of transportation in Nigeria today. Road traffic depends on the pattern of human settlements, accounting for more than 90% of the subsector's contribution to the gross domestic product (GDP). Road transport activities involve the en-masse conveyance of passengers or in small numbers, the transportation of animals, farm products and merchandise and the rendering of mobile services (clinics, libraries and banks). The optional use of motor cars for pleasure, which can be distinguished from the three uses listed above, also contributes tremendously to the importance of road transport in Nigeria. This is more predominant in Nigeria than in most other African countries because of the poor state of alternative means of transportation by which journeys could have been made and also due to the psychological satisfaction offered by the possession of a car. According to Obih [2], the Nigerian vehicle fleet size is 2.2 million, gasoline engines represent 60% of vehicle fleet and road transportation is the main transportation mode, accounting for 90% of passenger/freight movements. There are also low turnover rates for replacement of older cars with newer models. Nearly, all vehicles are imported from overseas, often used cars and trucks. Potential savings are achievable using roadworthy vehicles and changing policies. Vehicles tend to have low fuel efficiency. The average fuel efficiency in Nigeria is estimated to be about 18 l of gasoline/100 km [3].

Water transport occupies a distant second position to road transport, with an average share of about 1.6% of Nigeria's gross domestic product. Water transport is slow and, therefore, unsuitable for passenger movement, except for holiday and tourist traffic where time is not a constraint or where other forms of transport are not available. Water transport has the following three components: ocean, coastal water and inland water transports. Inland water transport is only advantageous in terms of costs of moving heavy traffic, especially

* Correspondence: ismail.badmus@gmail.com
[1]Mechanical Engineering Department, Yaba College of Technology, Yaba, Lagos, 6457, Nigeria
Full list of author information is available at the end of the article

Table 1 Lower heating value, chemical exergy and exergy factor for different fuels (at 25 °C and 1 atm)

Fuel	Lower heating value (kJ/kg)	Exergy factor, φ_f	Fuel chemical exergy, ε_f (kJ/kg)
Motor gasoline	44,300	1.07	47,401
Aviation kerosene	44,100	1.07	47,187
Diesel oil	43,000	1.07	46,010
Fuel oil	40,400	1.07	43,228

where speed is less important than cost. Water transport parameters are at present dominated by port statistics, statistics of inland water transport and the operations of ferries in ocean and coastal waters especially in Lagos and Niger delta areas of the country. The Petroleum Products Marketing Company, a subsidiary of the Nigerian government owned Nigerian National Petroleum Corporation, also keeps a fleet of chartered marine vessels which are used to transport products from the refineries to water-fed facilities [4].

Air transport has a unique advantage over all other modes of transport if speed, time and distance are major considerations. Air transport is of high value in relation to weight. It is also preferred where accessibility by other modes is a problem. Its share of the GDP is, however, still negligible in Nigeria [5].

Rail transport is usually the most suitable mode of transportation for heavy traffic flows when speed is also an advantage because of the lower cost per person per load as the train load increases. Nigeria's single-narrow-gauge railway line constructed in the colonial period was, for many years, the only mode of freight movement between the northern and southern parts of the country.

Contribution of rail transport to the gross domestic products of the transport sector is less than a half per cent. Although rail has always contributed a tiny proportion of value added in transportation, its share of value added continues to decline because road transport (freight and passenger) has virtually taken over all the traffic previously conveyed by rail. The relegated status of the Nigerian Railways is a classic illustration of a transportation policy which has sidelined an important

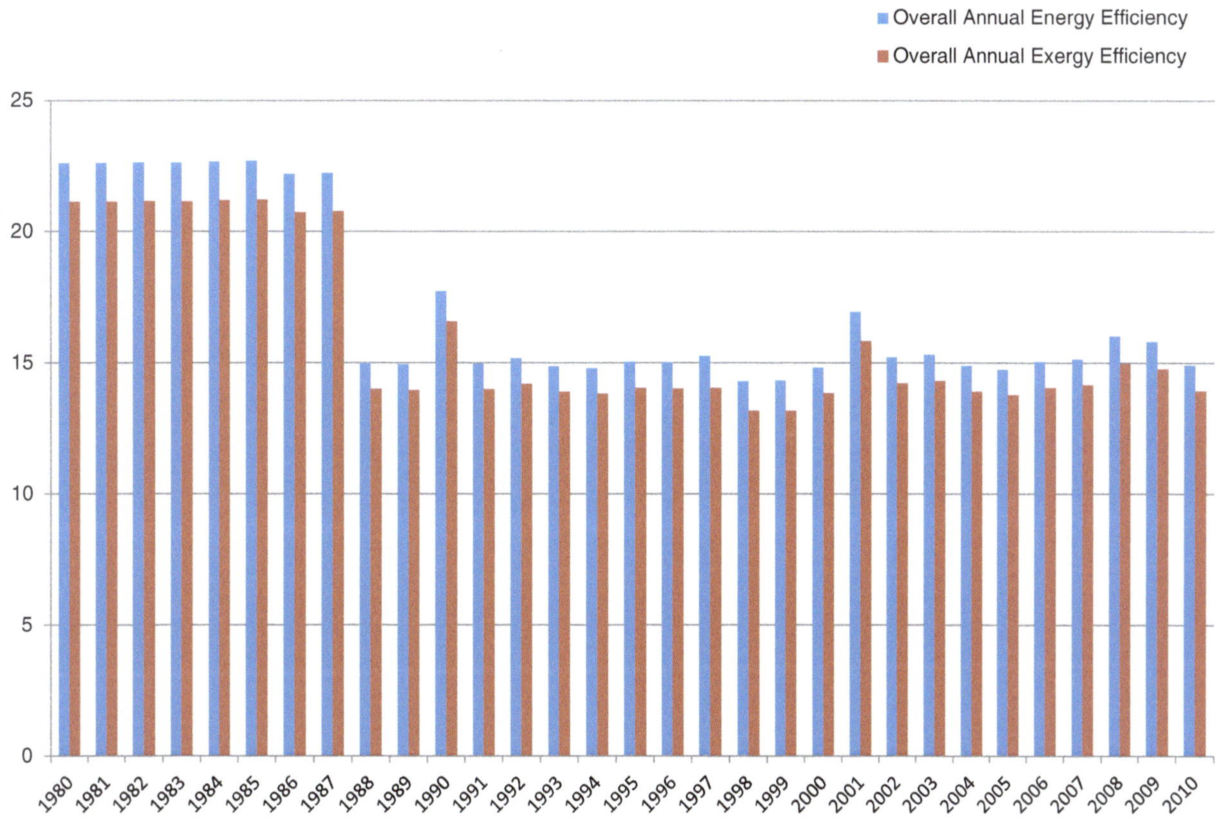

Figure 1 Overall energy and exergy efficiencies of the Nigerian transportation sector (1980 to 2010).

Table 2 Energy consumption values in the Nigerian transportation sector from 1980 to 2010

Year	Land transportation PMS (PJ)	Percentage of TEC	Land transportation AGO (PJ)	Percentage of TEC	Total land transportation (%)	Aviation turbine kerosene (PJ)	Percentage of TEC	Marine transportation FO (PJ)	Percentage of TEC	Marine transportation AGO (PJ)	Percentage of TEC	Total marine transportation (%)	Railway AGO (PJ)	Percentage of TEC	Annual total (PJ)
1980	121.7725	63.24	33.43869	17.37	80.60	29.09347	15.11	3.240969	1.68	5.015804	2.60	4.29	N/A	N/A	192.5615
1981	152.9379	63.93	39.31714	16.44	80.37	36.53939	15.27	4.530618	1.89	5.89757	2.47	4.36	N/A	N/A	239.2226
1982	171.9794	64.63	41.96783	15.77	80.40	41.08871	15.44	4.77431	1.79	6.295174	2.37	4.16	N/A	N/A	266.1054
1983	177.8283	64.61	43.31493	15.74	80.35	42.48611	15.44	5.1045	1.85	6.49724	2.36	4.22	N/A	N/A	275.2311
1984	169.3456	65.19	40.37992	15.54	80.74	40.45947	15.58	3.526152	1.36	6.056989	2.33	3.69	N/A	N/A	259.7682
1985	169.1236	66.22	37.06686	14.51	80.74	40.40643	15.82	3.225455	1.26	5.560029	2.18	3.44	N/A	N/A	255.3824
1986	154.016	72.29	31.8384	14.94	87.23	17.81905	8.36	4.618205	2.17	4.775761	2.24	4.41	N/A	N/A	213.0674
1987	155.5185	72.63	29.60361	13.83	86.45	19.4459	9.08	5.122074	2.39	4.440541	2.07	4.47	N/A	N/A	214.1306
1988	165.428	74.30	32.69032	14.68	88.98	14.26591	6.41	5.371018	2.41	4.903548	2.20	4.61	N/A	N/A	222.6588
1989	187.5791	76.18	34.40722	13.97	90.16	14.86699	6.04	4.209882	1.71	5.161084	2.10	3.81	N/A	N/A	246.2243
1990	185.69	57.87	40.98399	12.77	70.65	84.22042	26.25	3.8117	1.19	6.147598	1.92	3.10	N/A	N/A	320.8537
1991	185.7925	73.50	41.00136	16.22	89.72	16.03079	6.34	3.8117	1.51	6.150204	2.43	3.94	N/A	N/A	252.7866
1992	187.1253	75.83	32.13304	13.02	88.85	19.40576	7.86	3.287308	1.33	4.819956	1.95	3.29	N/A	N/A	246.7714
1993	226.9443	72.15	57.92496	18.42	90.57	17.59855	5.60	3.372632	1.07	8.688743	2.76	3.83	N/A	N/A	314.5292
1994	239.8583	78.83	39.73802	13.06	91.89	15.43368	5.07	3.277086	1.08	5.960703	1.96	3.04	N/A	N/A	304.2678
1995	175.6138	72.42	38.98208	16.08	88.49	16.21381	6.69	5.843981	2.41	5.847312	2.41	4.82	N/A	N/A	242.501
1996	169.48	71.48	38.95989	16.43	87.91	15.32255	6.46	7.501553	3.16	5.843984	2.46	5.63	N/A	N/A	237.108
1997	168.4867	67.00	45.59488	18.13	85.13	18.51314	7.36	10.00649	3.98	6.839232	2.72	6.70	2.021	0.80	251.4614
1998	150.1318	77.07	31.13056	15.98	93.05	1.154097	0.59	6.385382	3.28	4.669583	2.40	5.67	1.333	0.68	194.8045
1999	134.116	78.12	27.371	15.94	94.07	1.430163	0.83	3.273578	1.91	4.10565	2.39	4.30	1.376	0.80	171.6724
2000	190.2686	80.11	38.30822	16.13	96.23	0.434826	0.18	0.959548	0.40	5.746234	2.42	2.82	1.806	0.76	237.5235
2001	250.76	68.41	46.13529	12.59	81.00	59.91069	16.34	0.93609	0.26	6.920293	1.89	2.14	1.892	0.52	366.5544
2002	275.4001	81.56	41.35251	12.25	93.81	12.09123	3.58	1.062656	0.31	6.202876	1.84	2.15	1.548	0.46	337.6574
2003	271.5041	80.02	39.35402	11.60	91.62	14.31883	4.22	6.722852	1.98	5.903103	1.74	3.72	1.505	0.44	339.3079
2004	258.2866	88.05	24.72426	8.43	96.47	3.17961	1.08	2.252768	0.77	3.708639	1.26	2.03	1.204	0.41	293.3559
2005	307.2292	87.55	34.92473	9.95	97.51	0.345229	0.10	2.22513	0.63	5.23871	1.49	2.13	0.946	0.27	350.909
2006	376.2422	90.69	23.36103	5.63	96.32	10.71834	2.58	1.032407	0.25	3.504154	0.84	1.09	N/A	N/A	414.8581
2007	328.5612	92.89	10.77207	3.05	95.93	12.11762	3.43	0.647006	0.18	1.61581	0.46	0.64	N/A	N/A	353.7137
2008	306.4878	82.64	21.89909	5.90	88.54	37.05917	9.99	2.143436	0.58	3.284864	0.89	1.46	N/A	N/A	370.8743
2009	292.4471	87.29	11.15277	3.33	90.62	28.11002	8.39	1.630987	0.49	1.672915	0.50	0.99	N/A	N/A	335.0138
2010	386.5995	91.77	22.98541	5.46	97.23	7.251688	1.72	0.967965	0.23	3.447812	0.82	1.05	N/A	N/A	421.2524

AGO, automotive gas oil; FO, fuel oil; N/A, not applicable; PMS, premium motor spirit; TEC, total energy consumed.

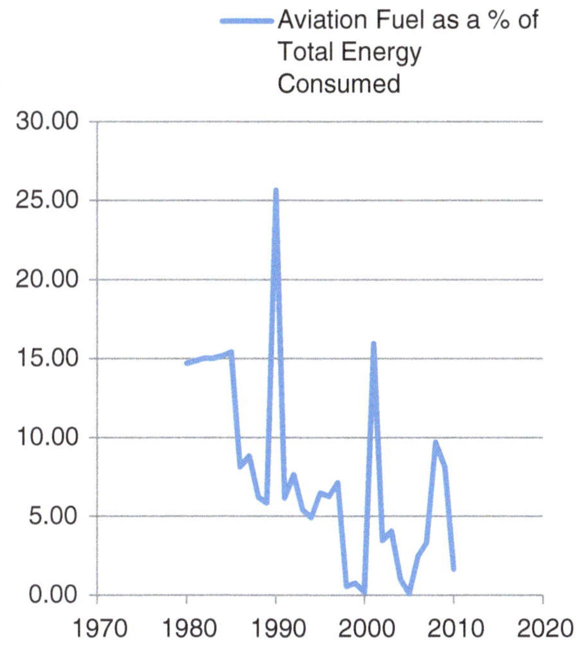

Figure 2 Trend in annual aviation fuel consumption in the Nigerian transportation sector.

and cheap means of transport to foster the growth of privately owned long haulage transport services. This policy has engendered the following:

(a) the Nigerian Railway Corporation (NRC) which has become a lame duck with total reliance on the government for subvention and
(b) a disorganised, unregulated private sector-owned road transport system providing freight and passenger services.

The effects of these are as follows:

(a) traffic congestion on urban roads,
(b) increasing rate of fatal road accidents emanating from bad roads, poorly maintained vehicles and careless driving, and
(c) worsening environmental pollution.

The sharp devaluation of the Nigerian local currency has also aggravated the situation as an increasing number of private car owners who would have travelled in their cars now have to join the pool of public transport commuters. The most comprehensive items of data on rail transport are published in the Annual Report issued by the Nigerian Railway Corporation. It contains a detailed account of all the activities of the corporation in the year covered. In one of the tables captioned 'General Statistical Returns', the corporation publishes time-series data of 54 observed and derived variables on the following items: receipts and expenditure, passenger statistics, freight statistics and combined passenger and freight train and engine statistics. Most of these publications have been discontinued in recent years.

Hence, from the foregoing, railway is a very small division of Nigeria's transport subsector. It once had the most comprehensive set of data which described its activities, but unfortunately, the Nigeria Railway Corporation (which is the sole authority in charge of rail transport) is just undergoing resuscitation and general overhaul presently. During the years of operation within the period covered by this work, the corporation only had functional diesel engines.

While countries like Turkey [6-8], China [9,10], Norway [11,12], Saudi Arabia [13] and Malaysia [14] have their transportation sectors analysed from the exergetic point of view in the open literature, there seems to be none analysing that of Nigeria, despite its importance in sub-Saharan Africa. Besides, the country transportation sector typifies a peculiar one, dominated by used and ill-maintained motor vehicles.

Methods

Device efficiencies

Virtually, all authors who have worked on this topic (like [9,14,15]) assume device efficiencies used by Reistad [16] to evaluate energy and exergy utilisation efficiencies. This work differs from theirs for one main reason: the present work is on a developing African country where cars and trucks that have been used elsewhere before being imported into the country and brand new ones compete for dominance. For instance, according to the Nigerian Customs and Excise sources, in 2005, 18,391 new vehicles and 33,134 used vehicles were imported into the country, while in 2006, 31,010 new vehicles of all types were imported as against 43,606 used vehicles of all types [17]. According to Oritse [18], for December 2011, out of the 3,725 vehicles arriving or expected at Lagos ports, about 1,000 were new while the remaining 2,725 are used vehicles. Specifically, more than 85% of motor vehicles being imported into the country annually are used vehicles [19].

In fact, the average vehicle fuel efficiency in the early 90s was 13.07 mpg [2] at a time when the average fuel efficiency in the US was 24.09 mpg for light duty vehicles [20]. In fact, the average new car in 2005 worldwide had a fuel economy level of about 29.4 mpg [21]. Fuel efficiency is low in Nigeria because the vehicle fleet is old and poorly maintained, there is traffic congestion in most urban centres and the driving habits are bad [22]. However, according to Ajayi and Dosunmu [23], there were no records of importation of used vehicles into the country prior to 1988. Hence, from 1980 to 1987, this work will be based on the assumptions of previous

Table 3 Nigerian transportation subsector's energy and exergy efficiencies

Year	Overall highway energy efficiency	Airway energy efficiency	Overall waterway energy efficiency	Railway energy efficiency	Overall highway exergy efficiency	Airway exergy efficiency	Overall waterway exergy efficiency	Railway exergy efficiency
1980	17.73	4.23	0.64	N/A	16.57	3.95	0.60	N/A
1981	17.68	4.28	0.65	N/A	16.52	4.00	0.61	N/A
1982	17.69	4.32	0.62	N/A	16.53	4.04	0.58	N/A
1983	17.68	4.32	0.63	N/A	16.52	4.04	0.59	N/A
1984	17.76	4.36	0.55	N/A	16.60	4.08	0.52	N/A
1985	17.76	4.43	0.52	N/A	16.60	4.14	0.48	N/A
1986	19.19	2.34	0.66	N/A	17.93	2.19	0.62	N/A
1987	19.02	2.54	0.67	N/A	17.78	2.38	0.63	N/A
1988	12.50	1.79	0.69	N/A	11.68	1.68	0.65	N/A
1989	12.67	1.69	0.57	N/A	11.84	1.58	0.53	N/A
1990	9.93	7.35	0.47	N/A	9.28	6.87	0.44	N/A
1991	12.61	1.78	0.59	N/A	11.78	1.66	0.55	N/A
1992	12.48	2.20	0.49	N/A	11.67	2.06	0.46	N/A
1993	12.73	1.57	0.58	N/A	11.89	1.46	0.54	N/A
1994	12.91	1.42	0.46	N/A	12.07	1.33	0.43	N/A
1995	12.43	1.87	0.72	N/A	11.62	1.75	0.68	N/A
1996	12.35	1.81	0.84	N/A	11.54	1.69	0.79	N/A
1997	11.96	2.06	1.00	0.23	11.18	1.93	0.94	0.21
1998	13.07	0.17	0.85	0.19	12.22	0.16	0.80	0.18
1999	13.22	0.23	0.64	0.22	12.35	0.22	0.60	0.21
2000	14.13	0.05	0.42	0.21	13.20	0.05	0.40	0.20
2001	11.89	4.58	0.32	0.14	11.11	4.28	0.30	0.14
2002	13.77	1.00	0.32	0.13	12.87	0.94	0.30	0.12
2003	13.45	1.18	0.56	0.12	12.57	1.10	0.52	0.12
2004	14.16	0.30	0.30	0.11	13.24	0.28	0.28	0.11
2005	14.31	0.03	0.32	0.08	13.38	0.03	0.30	0.07
2006	14.14	0.72	0.16	N/A	13.22	0.68	0.15	N/A
2007	14.08	0.96	0.10	N/A	13.16	0.90	0.09	N/A
2008	13.00	2.80	0.22	N/A	12.15	2.61	0.21	N/A
2009	13.30	2.35	0.15	N/A	12.43	2.20	0.14	N/A
2010	14.27	0.48	0.16	N/A	13.34	0.45	0.15	N/A

N/A, not applicable.

Table 4 Energy and exergy efficiencies of the transportation sector of some countries in the year 2000

Country	Energy efficiency	Exergy efficiency	Reference
Greece	22.69	21.23	Koroneos and Nanaki [26]
Jordan	22.83	22.45	Jaber et al. [15]
Turkey	23.71	23.65	Utlu and Hepbasli [7,8]
Saudi Arabia	22.24	22.24	Dincer et al. [13]
Malaysia	22.78	22.44	Saidur et al. [14]
Nigeria	14.81	13.85	This work

authors for land transportation engine efficiencies (22%), while the thermal efficiencies for subsequent years will be evaluated as in Equation 1:

$$\eta_{\text{th,Nigeria}} = \frac{\eta_{\text{fuel,Nigeria}}}{\eta_{\text{fuel,US}}} \times \eta_{\text{th,US}} \quad (1)$$

For the remaining modes of transportation, we rely on the values used by Reistad [16] as used by several other authors. These are 28%, 15% and 28% for railways, waterways and airways, respectively. Equation 1 yields 14.05% from 1988 up to the end of the 90s and 14.68% for the year 2000 to 2010.

Energy and exergy analyses

The expressions for energy efficiency (η) and exergy efficiency (ψ) for the processes in this paper are as follows:

$$\eta = \frac{\text{Energy in products}}{\text{Total energy input}} \quad (2)$$

$$\psi = \frac{\text{Exergy in products}}{\text{Total exergy input}}. \quad (3)$$

Exergy of a hydrocarbon fuel

The specific exergy of the fuel (ϵ_f) at environmental conditions reduces to chemical exergy, which can be written as follows:

$$\varepsilon_f = \phi_f \text{LHV}. \quad (4)$$

LHV is the lower heating value of the fuel and ϕ_f is the exergy factor of Szargut and Styrylska as enunciated in Szargut et al. [24]. The fuels and the liquid fuel exergy factor used in this work are in Table 1.

All the heating values have been taken from Garg et al. [25]. The exergy factor for liquid fuels has been taken from Table 3.4 of Szargut et al. [24].

Exergy of work

From the definition of exergy, mechanical work, W, is identical to the physical work exergy, E^W. Hence,

$$E^W = W. \quad (5)$$

From Equations 2, 3, 4 and 5, the energy η_m and exergy ψ_m efficiencies for the fossil fuel-driven kinetic energy production process which produces shaft work, W, from fuel mass m_f, can be expressed as follows:

$$\eta_m = \frac{W}{m_f \text{LHV}} \quad (6)$$

$$\psi_m = \frac{E^W}{m_f \phi_f \text{LHV}} = \frac{\eta_m}{\phi_f} \quad (7)$$

The weighted mean overall energy and exergy efficiencies for the transportation sector in each year are calculated following the method explained in Dincer et al. [13].

Results and discussion

The overall consumption efficiencies are generally between 22.19% and 22.71% for energy and between 20.74% and 21.22% for exergy before 1988, when mass importation of used vehicles into the country started. From 1988 to 2010, the efficiencies are generally around 15% and below (Figure 1). This is because the transportation sector has been dominated by land transportation (Table 2). The only exceptions to this general trend are prominently the years 1990 and 2001 with energy and exergy efficiencies 17.74% and 16.58% as well as 16.93% and 15.82%, respectively, and mildly the years 2008 and 2009 with energy and exergy efficiencies 16.02% and 14.97% as well as 15.80% and 14.77%, respectively. These are years with comparatively high annual percentages of aviation energy utilisation (Figure 2). They are also years of comparatively high aviation subsectoral annual efficiencies (Table 3). This positive influence is expected since aviation has a device part load thermal efficiency of 28% compared with that of land, which is 22%. A comparison of the overall sectoral efficiencies of various countries is shown in Table 4 where it is seen that Nigeria has the least values of energy utilisation efficiencies for the year 2000.

Conclusions

The overall mean energy efficiency in the Nigerian transportation sector for the three decades is 17.11%, while the overall mean exergy efficiency is 15.97%. This study also shows that airway transportation contribution should be increased to improve the overall energy and exergy efficiencies of the Nigerian transport sector. It is very necessary in the light of the results of this work that Nigerian policy makers review the issue of mass importation of used vehicles into the country. All efforts should also be made to ease air transportation and make it affordable for Nigerians. The Nigerian railway system should also be resuscitated for cheap and affordable transportation of goods and passengers.

Competing interests

The authors declare that they have no competing interests.

Authors' contributions

IB carried out the preliminary studies and data survey and prepared the draft of the manuscript. ASO together with IB did the data analysis. ROF and MOO both provided technical guide and review of the manuscript. All authors read and approved the final manuscript.

Authors' information

IB is a principal lecturer in the Department of Mechanical Engineering, Yaba College of Technology, Yaba, Lagos, Nigeria. He received his B.Sc. (Hons.) and M.Sc. in Mechanical Engineering from the University of Lagos in 1988 and 1996, respectively. He is a corporate member of both the Nigerian Society of Engineers and Nigerian Institution of Mechanical Engineers as well as a registered mechanical engineer of the Council for the Regulation of Engineering in Nigeria. His area of research is energy integration and exergy analysis. He has several publications in peer-reviewed international and local journals.

ASO holds B.Sc. and M.Sc. (Hons.) in Chemical Engineering from the Department of Chemical Engineering, Obafemi Awolowo University (OAU), Ile-Ife, Nigeria. He obtained his Ph.D. degree in Intelligent Adaptive Systems from the Division of Industrial Innovation Sciences, Okayama University, Japan. He is a prominent member of PSE Research group and an academic member of the Faculty of Technology, OAU, Nigeria. His areas of research include process modeling and simulation, linear and nonlinear intelligent adaptive control. He is a member of the Nigerian Society of Chemical Engineers (NSChE) and a registered engineer of the Council for the Regulation of Engineering in Nigeria (COREN) among others. He has published a number of research articles in journals and conference proceedings.

ROF is a retired professor and former head of the Mechanical Engineering Department at the University of Ibadan, Nigeria and is currently a contract

professor of Mechanical Engineering at the Obafemi Awolowo University, Ile-Ife, Osun State, Nigeria. Among other schools in which he has taught engineering courses are the University of Illinois (Urbana-Champaign, IL, USA), Iowa State University (Ames, IA, USA), Kwame Nkrumah University of Science and Technology (Kumasi, Ghana), South Bank University (London, UK) and Covenant University (Ota, Ogun State, Nigeria). As lead consultant on the 6.4 kW Solar PV street lighting and Central Administration lighting project of the University of Ibadan in 2008, he designed and supervised the entire project, the first of its type in Nigeria and Africa. He has been and is still involved in several national and international energy sector activities. He has consulted for numerous international agencies, and he was a key member of the consulting team, which developed the Niger State Renewable Energy Policy in 2010. He was also the chair of the Infrastructure Working Group of the National Adaptation Strategy and Plan of Action (NASPA) Project and on the Nigerian Environmental Study Action Team's (NEST) Building Nigeria's Resilience to Climate Change (BNRCC) from 2010 to 2011, and technology transfer lead researcher on the HBF-funded COP-17 Nigeria's National Position Paper (September/November 2011). Prof. ROF has about 50 publications in peer-reviewed international and local journals, and he has supervised many students at both the undergraduate, master's and doctoral levels. He continues to mentor students at the postgraduate level at the University of Ibadan, Obafemi Awolowo University and Covenant University, Nigeria, and has recently supervised students at Johns Hopkins University, MD, USA (2012) and Brandenburg University of Technology, Cottbus, Germany (2011). He is a fellow of both Nigerian Society of Engineers and Solar Energy Society of Nigeria.
MOO obtained B.Sc. (First Class Honours) in Mechanical Engineering from the OAU, Ile-Ife in 1997 and Ph.D. in Mechanical Engineering from the University of Newcastle, NSW, Australia in 2004. He is an editorial board member of several engineering journals. He has been a member of the American Society of Engineers (ASME) since 2003. He has been a reviewer of Experiments in Fluid Journal, since 2005, ASME Heat Transfer/Fluids Engineering Summer Conference, USA, 2004 and Mini-symposium on Energy and Thermal-Fluid Systems of CANCAM Conference, Canada, 2007. He was an invited visiting faculty in the Department of Mechanical Engineering, Howard University, USA, 2010. Currently, Dr. MOO is a reader in the Mechanical Engineering Department, University of Ibadan, Nigeria as well as the acting head of the department. He specialises in energy analyses and thermofluids. He has scores of publications in learned journals and conference proceedings.

Acknowledgements
The authors wish to acknowledge the Corporate Planning and Development Division of the Nigerian National Petroleum Corporation for providing information on the fuel consumption over the years. This work has been partly funded by the Education Trust Fund grant for researchers in Nigerian tertiary institutions.

Author details
[1]Mechanical Engineering Department, Yaba College of Technology, Yaba, Lagos, 6457, Nigeria. [2]Chemical Engineering Department, Obafemi Awolowo University, Ile-Ife, 220005, Nigeria. [3]Mechanical Engineering Department, University of Ibadan, Ibadan, 200284, Nigeria.

References

1. Oniwon, A: Oil and gas in Nigeria's national development: an assessment. A presentation at the National Defence College, Abuja (2011)
2. Obih, H: Fuel distribution and logistics. Sub-regional conference on the phase-out of leaded gasoline in Nigeria and neighbouring countries, Abuja, Nigeria (2001)
3. Adegbulugbe, AO: The energy-environment nexus: the role of energy efficiency. In: Energy issues in Nigeria: today and tomorrow. Proceedings of a conference held by the Nigerian National Committee of The World Energy Council at the Nigerian Institute of International Affairs, Victoria Island, Lagos (1991)
4. Adenikinju, AF, Falobi, N: Macroeconomic and distributional consequences of energy supply shocks in Nigeria. AERC Research Paper 162. African Economic Research Consortium, Nairobi (2006)
5. Nigerian Bureau of Statistics (NBS): Transport and Communications. Office of the Director, Abuja, Nigeria (2008)
6. Ediger, VS, Camdali, U: Energy and exergy efficiencies in Turkish transportation sector, 1988–2004. En. Pol. **35**, 1238–1244 (2007)
7. Utlu, Z, Hepbasli, A: Turkey's sectoral energy and exergy analysis between 1999 and 2000. Int. J. En. Res. **28**(13), 1177–1196 (2004). doi:10.1002/er.1023
8. Utlu, Z, Hepbasli, A: Assessment of the energy utilization efficiency in the Turkish transportation sector between 2000 and 2020 using energy and exergy analysis method. En. Pol. **34**, 1611–1618 (2006)
9. Ji, X, Chen, GQ: Exergy analysis of energy utilization in the transportation sector in China. En. Pol. **34**, 1709–1719 (2006)
10. Zhang, M, Li, G, Mu, HL, Ning, YD: Energy and exergy efficiencies in the Chinese transportation sector. 1980–2009. En **36**(2), 770–776 (2011). 10.1016/j.energy.2010.12.044
11. Ertesva°g, IS, Mielnik, M: Exergy analysis of the Norwegian society. En **25**, 957–973 (2000)
12. Ertesva°g, IS: Energy, exergy, and extended-exergy analysis of the Norwegian society 2000. En **30**, 649–675 (2005)
13. Dincer, I, Hussain, MM, Al-Zaharnah, I: Energy and exergy utilization in the transportation sector of Saudi Arabia. Appl. Therm. Engn. **24**, 525–538 (2004)
14. Saidur, R, Sattar, MA, Masjuki, HH, Ahmed, S, Hashim, U: An estimation of the energy and exergy efficiencies for the energy resources consumption in the transportation sector in Malaysia. En. Pol. **35**, 4018–4026 (2007)
15. Jaber, JO, Al-Ghandoor, A, Sawalha, SA: Energy analysis and exergy utilization in the transportation sector of Jordan. En. Pol. **36**(8), 2995–3000 (2008)
16. Reistad, G: Available energy conversion and utilization in the United States. J. Engn. Power. **97**, 429–434 (1975)
17. Ochonma, M: Vehicle importation. BusinessDay., (2012). www. businessdayonline.com (2007). Accessed 1 April
18. Oritse, G: Fuel, vehicles top Nigeria's import for month of December. Sweetcrude Newspaper, (2012). www.sweetcrudereporters.com (2011). Accessed 1 April
19. Agbo, COA: A critical evaluation of motor vehicle manufacturing in Nigeria. Nig. J. Tech. **30**(1), 8–16 (2011)
20. Research and Innovative Technology Administration Bureau of Transportation Statistics (RITA/BTS): Average fuel efficiency of US light duty vehicles. http: www.bts.gov (2011). Accessed 26 February 2012
21. Elzinga, D, Fulton, L, Heinen, S, Wasilik, O: Advantage energy: emerging economies, developing countries and the private-public sector interface. Information paper prepared by the International Energy Agency in support of the United Nations private sector forum 2011. International Energy Agency, Paris, France (2011)
22. Jochem, E: Energy end-use efficiency. In: Jochem, E (ed.) World energy assessment: energy and the challenge of sustainability, pp. 174–217. United Nations Development Programme, New York (2000)
23. Ajayi, AB, Dosunmu, OO: Environmental hazards of importing used vehicles into Nigeria. In: Proceedings of international symposium on environmental pollution control and waste management, pp. 521–532. EPCOWM'2002, Tunis (2002)
24. Szargut, J, Morris, DR, Steward, FR: Exergy analysis of thermal, chemical, and metallurgical processes. Hemisphere Publishing Corporation, Washington DC, USA (1988)
25. Garg, A, Kazunari, K, Pulles, T: Volume 2: energy. In: IPCC guidelines for national greenhouse gas inventories. IGES, Japan (2006)
26. Koroneos, CJ, Nanaki, EA: Energy and exergy utilization assessment of the Greek transport sector. Res. Cons. & Recyc. **52**, 700–706 (2008)

12

Correlation between thicknesses of dust collected on photovoltaic module and difference in efficiencies in composite climate

Rahnuma Siddiqui* and Usha Bajpai

Abstract

An effort has been made to develop an equation with the given data for all seasons for a location (Lucknow, India) consisting of composite climate, which is further helpful in developing a relation between difference in efficiencies of module with respect to thicknesses of dust collected on the module. This equation that is developed mathematically is in good correlation with the measured data. Here, data are shown for a whole year (from 2010 to 2011); the study done gives us a broad view of finding out the difference in efficiencies of module when dust collects on it.

Keywords: (Atmospheric) dust, regression analysis, efficiency of solar photovoltaic module

Background

In solar cells, lots of power get lost due to various reasons such as reflection losses at the top surface, incomplete absorption of photon energy due to limited cell thickness, series and shunt resistance loss, curve or fill factor loss, etc. Results obtained show that there is a direct proportionality between solar flux, output current, and efficiency of the photovoltaic module as given by Omubo-Pepple et al. [1].

Dust contributes to as much as 40% degradation in peak power of photovoltaics; there is surprisingly little scientific work done on the subject. Since no information about the type of dust, density, and rate of accumulation of dust was noted, no general understanding of the underlying physical principles could be deduced. Dust accumulation on the photovoltaic (PV) panel surface depends on different parameters such as PV panel inclination and kind of installation as given by Del Cueto [2].

The research done previously include degradation analysis of silicon photovoltaic modules and effective efficiency of PV modules under field conditions as given [3,4]. Some studies have also been done on corrosion effects in thin film photovoltaic modules as given by Carlson et al. [5]. The airborne particles in the atmosphere affect the amount and properties of the radiation finally reaching the collectors. The accumulation of dust particles on the surface of PV module greatly affects its performance especially in desert areas. Some correlation studies have done direct beam solar radiation received by photovoltaic panel with sand dust accumulated on its surface as given by Al-Hasan [6].

Desert countries are of course best suited to photovoltaic power generation due to the abundant availability of sunlight throughout the year. Nowadays, the ideas of setting up vast solar arrays in desert countries and exporting the power to other countries are being discussed. In a bigger PV solar plant, more work force and machines will be needed to help make the rounds and clean the panels, especially after a stand storm.

Many research results discuss about performance of panel with dust concentration on the surface, but for a common PV user, it is important to know how frequently the panel has been cleaned. In case if frequent cleaning is not possible, it is important to know the performance loss due to dust for additional estimation to compensate the loss.

It has been concluded in many studies that dust accumulation considerably deteriorates the performance of photovoltaic cells. However, in carrying out the investigation on the effect of dust and particulate pollution, the

* Correspondence: rmsphy@gmail.com
Renewable Energy Research Laboratory, Department of Physics, University of Lucknow, Lucknow 226007, India

physical characteristics of dust must be determined and correlated to the observed effects [7].

Research done also include characteristic distribution of total, diffuse and direct solar radiation at given locations, qualification testing of modules, etc. [8-11]. Different analysis, performance, and reliability testing of photovoltaic modules and arrays have been done as given [12-14].

As we know that module performance is greatly affected by the amount of dust collected on it [15], sometimes when we want to get the amount of difference value in efficiency of modules, the instruments for measuring it at that instant are not found. The main aim of this paper is to solve these kinds of problems. Here, an attempt has been made to obtain a relation between 'thicknesses of dust collected and difference in efficiencies' for all seasons so that difference in efficiencies of modules can be easily estimated by measuring the thicknesses of dust collected using simple devices.

Here, a statistical analysis has been done in finding out the correlation between thicknesses of dust collected on photovoltaic module and the difference in efficiencies of the module for a whole year considering all seasons of a composite climate.

Methods

Instrumentation

The SPV module (number 2007.20.685) which is manufactured in the year 2007 by M/s Rajasthan Electronics and Instruments Ltd, Jaipur, India is studied. The module has been connected to a rheostat ammeter digital multimeter (DT9205A, Agronic Ltd.). Other instruments used are the digital thermometer (SE-221-P-K, Agronic Ltd.) to measure cell temperature, the Thermo-Hygro Clock (288-CTH, Agronic Ltd.) to measure ambient temperature and humidity and digital luxmeter (range 0 to 50,000 lux in three ranges of 0 to 2,000, 0 to 20,000, and 0 to 50,000 lux; accuracy ±5%, ±2 least count) to measure solar flux. Data are taken for one whole year from 2010 to 2011.

The commercial photovoltaic module used has the following components:

- Solar type is polycrystalline silicon.
- Cell area = 0.216 m^2.
- Electrical efficiency STP = 12% with a power peak of 36 W.
- Electrical specifications is at standard test conditions of 100 MW/cm^2, AM 1.5, and at 25°C cell temperature within normal production of tolerance ±3%.
- Area of the panel = 0.340 m^2

Development of correlation between thicknesses of dust collected on photovoltaic module and the difference in efficiencies

As stated earlier, we have tried to verify the data taken for the whole year while considering different values, depending on the variation of climate. This fluctuation resulted from different factors affecting the performance of solar photovoltaic module. For example, these factors may be temperature, wind velocity, humidity, cloud cover, and dust.

The aim of this study is to find out the relation of the difference in efficiencies with respect to thicknesses of dust collected. Here, reading is taken three times for each month after every 10 days, and finally, the average of the three readings is taken as measured value of the data for both the variables (difference in efficiencies and thicknesses of dust collected on the module) for that particular month. In this way, each month has its individual reading taken throughout the year. In Table 1,

Table 1 Difference in efficiencies of solar photovoltaic module and thicknesses of dust collected

Months	Monthly average of difference in efficiencies (before and after collection of dust)($\Delta\eta$) (%)	Monthly average of thicknesses of dust collected on the module (t) (mm)
January	1.5666	0.0002400800
February	1.9574	0.0010534300
March	1.3707	0.0007545513
April	1.9563	0.0007006547
May	1.6332	0.0014258080
June	1.8172	0.0005536642
July	0.9202	0.0003919747
August	Data unavailable(rainy month)	Data unavailable
September	Data unavailable(rainy month)	Data unavailable
October	1.8790	0.0009505386
November	1.5544	0.0006369589
December	2.0198	0.0006516579

both variables are taken as monthly average value of difference in efficiencies and monthly average value of thicknesses of dust collected on the module.

Thus, the efficiencies of the module and their corresponding thicknesses of dust particles are taken into account. Efficiency of module has been found with the help of the maximum power, solar flux, and area of solar panel. On the other hand, thickness of dust particles collected on the module is calculated using the following the basic formula:

Thickness of dust collected = Volume of dust collected/Area of solar panel,

where

Volume of dust collected = Mass of dust collected/Density of dust collected on module.

The mass of dust collected is measured for every month using a chemical balance. The area of solar panel = 0.340 m^2, and the density of dust collected = 3g/cm^3.

As we know that density of dust is governed by the diameter of dust particles, dust consists of multiple composite particles ranging from 0.1 to 80 μm in diameter; sand consists of loose siliceous particles whose size range is approximately 0.08 to 1 mm in diameter (full text is found in the handbook [16]); thus, knowing the average characteristic of dust, density is known to be the above given value).

General variation in the difference in efficiencies and thicknesses of dust has been recorded which has been observed to be different for different months. Here, readings have been taken consecutively for 12 months, as given in columns 2 and 3 in Table 1.

Calculation

For calculations of the relation, we have used the process of linear regression as it is given by the method of least squares solved for two variables, as shown in Table 2. Problems involving more than two variables can be treated in a manner analogous to that for two variables.

$$\text{Now, } \Delta\eta = a_o + a_1 t, \tag{1}$$

$$\sum \Delta\eta = n\, a_o + a_1 \sum t, \tag{2}$$

$$\sum \Delta\eta.t = a_o \sum t + a_1 \sum t^2. \tag{3}$$

Substituting the values of $\sum t$, $\sum \Delta\eta$, $\sum t^2$, and $\sum \Delta\eta.t$ in Equations 2 and 3 when n =10, we get the following:

$$16.6748 = 10\, a_o + a_1 73.593183 \times 10^{-4}, \tag{4}$$

$$126.30879920 \times 10^{-4} = 73.593183 \times 10^{-4} a_o + a_1 645.4632795 \times 10^{-8}. \tag{5}$$

Now, solving Equations 4 and 5 to get the values of a_o and a_1,

$$1.66748 = a_o + a_1 7.3593183 \times 10^{-4}, \tag{6}$$

$$1.71631 = a_o + a_1 8.770693877 \times 10^{-4}, \tag{7}$$

$$a_1 = 345.9471.$$

Keeping the value of a_1 to find out a_o in Equation 6,

$$a_1 = 345.9471.$$

Therefore, the equation comes out as follows:

$$\Delta\eta = 1.4128 + 345.9471t. \tag{8}$$

The corresponding relation between the 'monthly average of thicknesses of dust collected and monthly

Table 2 Method of least squares solved for two variables

Monthly average of thicknesses of dust collected (t) (mm)	Monthly average of difference in efficiencies ($\Delta\eta$) (%)	t^2	$\Delta\eta.t$
0.0002400800	1.5666	5.7638406 × 10^{-8}	3.76109328 × 10^{-4}
0.0010534300	1.9574	110.9714765 × 10^{-8}	20.61983882 × 10^{-4}
0.0007545513	1.3707	56.9347664 × 10^{-8}	10.34263460 × 10^{-4}
0.0007006547	1.9563	49.0917008 × 10^{-8}	13.70690790 × 10^{-4}
0.0014258080	1.6332	203.2928453 × 10^{-8}	23.28629600 × 10^{-4}
0.0005536642	1.8172	30.6544046 × 10^{-8}	10.06118580 × 10^{-4}
0.0003919747	0.9202	15.3644165 × 10^{-8}	3.60714717 × 10^{-4}
0.0009505386	1.8790	90.3523630 × 10^{-8}	17.86062029 × 10^{-4}
0.0006369589	1.5544	40.5716640 × 10^{-8}	9.90088914 × 10^{-4}
0.0006516579	2.0198	42.4658018 × 10^{-8}	13.16218620 × 10^{-4}
$\sum t$ = 0.0073593183	$\sum \Delta\eta$ = 16.6748	$\sum t^2$ = 645.4632795 × 10^{-8}	$\sum \Delta\eta.t$ = 126.30879920 × 10^{-4}

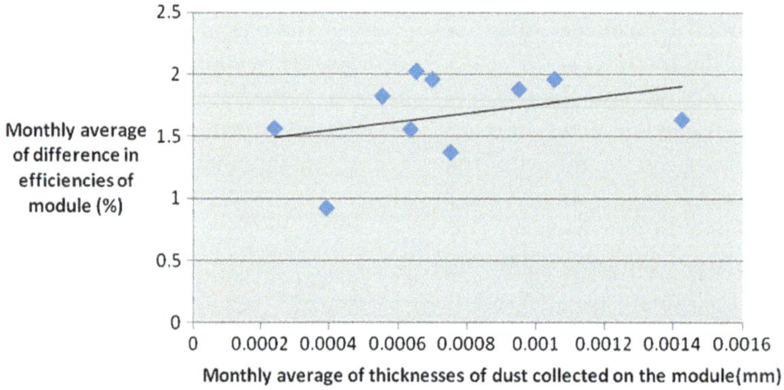

Figure 1 Correlation between thicknesses of dust and difference in efficiencies. Correlation between the monthly average of thicknesses of dust collected on photovoltaic module and the monthly average of difference in efficiencies for a whole year (2010 to 2011).

average of difference in efficiencies' graph for the data are given in Figure 1.

Here, t and $\Delta\eta$ are two variables; t is considered as an independent variable, and $\Delta\eta$ is considered as an dependent variable (t = monthly average of thicknesses of dust collected) and ($\Delta\eta$ = monthly average of difference in efficiencies). In statistics, linear regression is an approach to model the relationship between the two variables. Problems involving more than two or three variables can be treated in a manner analogous to that for two variables.

To further reconfirm the relation between thicknesses of dust collected and the difference in efficiencies, equations to the lines of regression and regression coefficients have been found with the help of the deviation method using Pearson's formula as given in Table 3. We have taken the estimated value of thicknesses of dust collected as 0 mm and therefore accordingly estimated the value of difference in efficiencies of solar photovoltaic module used as 0%, respectively.

Substituting the above values of Table 3 in the formula, we get the following:

$$\text{Regression coefficient } (\Delta\eta \text{ on } t) = \frac{\sum uv - \sum u \sum v/n}{\sum u^2 - \left(\sum u\right)^2/n},$$

$$\begin{aligned} b_{\Delta\eta.t} &= 3.5936 \times 10^{-4}/103.8674 \\ &= 0.0346 \times 10^{-4}. \end{aligned} \tag{9}$$

$$\text{Regression coefficient } (t \text{ on } \Delta\eta) = \frac{\sum uv - \sum u \sum v/n}{\sum v^2 - \left(\sum v\right)^2/n},$$

$$\begin{aligned} \mathrm{b}_{t\Delta\eta} &= 3.5936 \times 10^{-4}/1.0292, \\ &= 3.4916 \times 10^{-4}. \end{aligned} \tag{10}$$

$$\begin{aligned} t &= \text{estimated value} + \sum u/n = 73.593183 \times 10^{-4}/10, \\ &= 7.3593183 \times 10^{-4}. \end{aligned} \tag{11}$$

$$\begin{aligned} \Delta\eta &= \text{estimated value} + \sum v/n = 16.6748/10, \\ &= 1.66748. \end{aligned} \tag{12}$$

Substituting the above values in the regression equation of t on $\Delta\eta$, we get the following:

$$t - t = b_{t\Delta\eta}(\Delta\eta - \Delta\eta).$$

Similarly, with the equation of the line of regression of $\Delta\eta$ on t, we get the following:

$$\Delta\eta - \Delta\eta = b_{\Delta\eta.t}(t - t).$$

Therefore, the final two equations come out to be as follows:

$$t = 3.4916 \times 10^{-4}\Delta\eta + 1.5371 \times 10^{-4}, \tag{13}$$

$$\Delta\eta = 1.4128 + 346\,t. \tag{14}$$

Using Pearson's formula, coefficient of correlation has been found out as follows:

$$r = \frac{\sum uv - \sum u \sum v/n}{\surd\left(\sum u^2 - \left(\sum u\right)^2/n\right)\surd\left(\sum v^2 - \left(\sum v\right)^2/n,\right.}$$

$$\begin{aligned} &= 3.5936 \times 10^{-4}/\surd 103.8674 \times 10^{-8}\surd 1.0292 \\ &= 0.34756839. \end{aligned}$$

Note. We have seen that there are two lines of regression (Equations 13 and 14): one of $\Delta\eta$ on t and the other of t on $\Delta\eta$. In the first one, $\Delta\eta$ is the dependent variable and t is the independent variable, while in the other one, $\Delta\eta$ is the independent variable and t is the dependent variable. Since the assumptions for deriving the equations of two lines of regression are different, the two equations are not reversible. In other words, one equation gives the

Table 3 Equations to the lines of regression and regression coefficients

Monthly average of thicknesses of dust collected on the module			Monthly average of difference in efficiencies			Product of deviation
Monthly average of thicknesses 2of dust for a year	Deviation from estimated value	Square of deviation	For a whole year	Deviation from estimated value	Square of deviation	
t	u	u^2	$\Delta\eta$	v	v^2	uv
0.0002400800	0.0002400800	$5.76384064 \times 10^{-8}$	1.5666	1.5666	2.4542	$3.76109328 \times 10^{-4}$
0.0010534300	0.0010534300	$110.9714765 \times 10^{-8}$	1.9574	1.9574	3.8314	$20.61983882 \times 10^{-4}$
0.0007545513	0.0007545513	$56.93476643 \times 10^{-8}$	1.3707	1.3707	1.8788	$10.34263460 \times 10^{-4}$
0.0007006547	0.0007006547	$49.09170081 \times 10^{-8}$	1.9563	1.9563	3.8271	$13.70690790 \times 10^{-4}$
0.0014258080	0.0014258080	$203.2928453 \times 10^{-8}$	1.6332	1.6332	2.6673	$23.28629600 \times 10^{-4}$
0.0005536642	0.0005536642	$30.6544046 \times 10^{-8}$	1.8172	1.8172	3.3022	$10.06118580 \times 10^{-4}$
0.0003919747	0.0003919747	$15.3644165 \times 10^{-8}$	0.9202	0.9202	0.8468	$3.60714717 \times 10^{-4}$
0.0009505386	0.0009505386	$90.3523630 \times 10^{-8}$	1.8790	1.8790	3.5306	$17.86062029 \times 10^{-4}$
0.0006369589	0.0006369589	$40.5716640 \times 10^{-8}$	1.5544	1.5544	2.4161	$9.90088914 \times 10^{-4}$
0.0006516579	0.0006516579	$42.4658018 * 10^{-8}$	2.0198	2.0198	4.0796	$13.16218620 \times 10^{-4}$
$\sum t = 73.593183 \times 10^{-4}$	$\sum u = 73.593183 \times 10^{-4}$	$\sum u^2 = 645.46327 \times 10^{-8}$	$\sum \Delta\eta = 16.6748$	$\sum v = 16.6748$	$\sum v^2 = 28.8341$	$\sum uv = 126.30879920 \times 10^{-4}$

Table 4 Values of $\Delta\eta$ measured and calculated

$\Delta\eta$ measured (from measured values)	$\Delta\eta$ calculated (calculated from equation)	Difference of the two $\Delta\eta$ values	Difference^2
1.5666	1.4959	0.0707	0.0049
1.9574	1.7772	0.1802	0.0324
1.3707	1.6738	−0.3031	0.0918
1.9563	1.6552	0.3011	0.0906
1.6332	1.9060	−0.2728	0.0744
1.8172	1.6043	0.2129	0.0453
0.9202	1.5480	−0.6277	0.3940
Data unavailable	Data unavailable	-	-
Data unavailable	Data unavailable	-	-
1.8790	1.7416	0.1374	0.0188
1.5544	1.6331	−0.0787	0.0062
2.0198	1.6382	0.3816	0.1456
$\sum\Delta\eta$measured = 16.6748	$\sum\Delta\eta$calculated = 16.6733	$\sum$difference = 0.0016	$\sum$difference^2 = 0.9040

estimate of $\Delta\eta$, while the other gives the estimate of t. However, if r is equal to (±) 1, the two lines of regression can be reduced to the same form, but from the above value of r, it is clear that $r = 0$, i.e., the estimated value of $\Delta\eta$ (or t) is the same for all values of t (or $\Delta\eta$). Thus, the degree of correlation between the two variables is much greater but not perfect; there are other climatic variables influencing the correlation between the two.

Results and discussion

We have made an effort to test the validity of the equation deduced. Table 4 gives the value of $\Delta\eta$ measured and calculated from Equation 8 and also from Equation 14.

While keeping manual error in mind, the values obtained ($\Delta\eta$ measured and $\Delta\eta$ calculated) are quite in agreement with each other as seen from the table and from calculating the error of prediction given as follows:

$$\begin{aligned}\text{Estimate of error}\\ &= (\mathring{a}\Delta\eta\text{ measured} \;-\; \mathring{a}\Delta\eta\text{ calculated})^{1/2}/n\\ &= (0.9040)^{1/2}/10 = 0.0950789,\end{aligned}$$

where the value 0.0950789 signifies the amount of deviation of the predicted value from the observed value, which is considerably very small.

Thus, the equation obtained (Equation 8) may be taken as a valid equation, and using it, $\Delta\eta$ (difference in efficiencies of module) can be calculated using the known values of t (thicknesses of dust collected on module) for a location.

Conclusions

Silicon solar cell operating efficiency is at optimum level in the absence of dust which is an ideal condition. Practically, it is seen that there is in average a reduction in output, near 10% to 20%, when heavy layers of dust are accumulated on terrestrial modules [17].

Dust consists of multiple composite particles ranging approximately from 0.1 to 80 μm in diameter. Dust particles may be electrically conducting and are usually soluble in water. Sand dust is most severe in low humidity regions. Dust becomes airborne with slight winds and may remain suspended for hours as dust clouds [16].

During wind storms, dust particles penetrate almost any enclosure which is not hermetically sealed, and it may be hygroscopic. Its presence on metallic surfaces may aggravate corrosion.

It is reported that a small amount of dust on silicon solar photovoltaic module covers has a negligible effect on the sunlight transmission to the silicon photovoltaic module. An attempt has been made to develop an equation to calculate difference in efficiencies from thicknesses of dust particles collected for a whole year. The standard equation obtained is very good to measure the correlation with obtained results for almost every month. This is clear from the calculated percentage error from Table 4. Therefore, the standard equation helps to evaluate the difference in efficiencies of module while knowing thicknesses of dust collected on it for any climatic conditions of the area.

This result has been obtained with a particular module studied the whole year; a combined study can be done by taking different modules with their respective climatic conditions, and difference in output of solar photovoltaic modules can be easily evaluated.

Abbreviations

$\Delta\eta$: monthly average of difference in efficiencies of solar photovoltaic module (before and after collection of dust) (percentage); a_0: first constant; t: monthly average of thicknesses of dust collected (thickness in millimeter); u: deviation from the estimated value of thicknesses of dust collected; v: deviation from the estimated value of difference in efficiencies; $\Delta\eta$: arithmetic average of the thicknesses of dust collected on module; $\Delta\eta_{measured}$: measured value of difference in efficiencies of module; $\Delta\eta_{calculated}$: calculated value of difference in efficiencies of module, $b_{t\Delta\eta}$, regression coefficient (t on $\Delta\eta$); $b_{\Delta\eta.t}$: regression coefficient ($\Delta\eta$ on t); t: arithmetic average of the difference in efficiencies of photovoltaic module.

Competing interests

The authors declare that they have no competing interests.

Authors' contributions

UB supervised the whole work, drafted the manuscript, and corrected the manuscript. RS conducted the experimental analysis, studying the performance of solar photovoltaic module under different climatic parameters, and conducted the calculation according to the instruction provided by the supervisor. All authors read and approved the final manuscript.

Acknowledgments

The authors would like to thank the Ministry of New and Renewable Energy (MNRE), New Delhi, India for providing financial support for current research work.

References

1. Omubo-Pepple, V.B., Israel-Cookey, C., Alaminokuma, G.I.: Effects of temperature, solarflux and relative humidity on the efficient conversion of solar energy to electricity. European J Sci Res **35**(2), 173–180 (2009)
2. Del Cueto, J.A.: Comparison of energy production and performance from flat plate photovoltaic module technologies deployed at fixed tilt. Proceedings of the 29th IEEE Photovoltaic Specialists Conference, Piscataway (2002)
3. Meyer, E.L., Van Dyk, E.E.: Degradation analysis of silicon photovoltaic modules. Proceedings of the 16th European Photovoltaic Solar Energy Conference, Glasgow (2000)
4. Topic, M., Breel, K., Sites, J.: Effective efficiency of PV modules under field conditions. Progress in Photovoltaics: Res Appl **15**, 19–26 (2007)
5. Carlson, D.E., Romero, R., Willing, F., Meakin, D.H., Gonzalez, L., Murphy, R.: Corrosion effects in thin film photovoltaic modules. Progress in Photovoltaics: Res Appl **11**, 377 (2003)
6. Al-Hasan, A.Y.: A new correlation for direct beam solar radiation received by photovoltaic panel with sand dust accumulated on its surface. Solar Energy **63**(5), 323–333 (1998)
7. Sukhatme, S.P.: Solar Energy: Principles of Thermal Collection and Storage. Tata McGraw-Hill, New Delhi (2003)
8. Armstrong, S., Hurley, W.G.: A new methodology to optimize solar energy extraction under cloudy conditions. Renew Energy **35**, 780–787 (2010)
9. Abentin, J., Cueli, A.B., Diaz, J., Moracho, J., Lagunas, A.R.: New accredited laboratory for PV module testing. Proceedings of the 20th European Photovoltaic Solar Energy Conference, Barcelona (2005)
10. Akpabio, L.E., Udoimuk, A.B.: Characteristics distribution of total, diffuse and direct solar radiation at Calabar. Glob J Pure Appl Sci **9**(1), 45–49 (2003)
11. Osterwald, C.R., McMohan, T.J.: History of accelerated and qualification testing of terrestrial photovoltaic modules: a literature review. Progress in Photovoltaics: Res Appl **2017**(1), 11–33 (2008)
12. Malik, A.Q., Lim Chee, M., Tan Kha, S., Blundell, M.: Influence of temperature on the performance of photovoltaic polycrystalline silicon module in the Bruneian climate. AJSTD **26**(2), 61–72 (2010)
13. Arya, R.R., Carlson, D.E.: Amorphous silicon PV module manufacturing at BP solar. Progress in Photovoltaics: Res Appl **10**(2), 69–76 (2002)
14. Wohlgemuth, J.H., Conway, M., Meakin, D.H.: Reliability and performance testing of photovoltaic modules, pp. 15–22. 28th IEEE PVSC, Egan Convention Center, Anchorage (2000)
15. Shaharin, A., Sulaiman, S.A., Hussain, H.H.: Nik Leh, NSH, Mohd SI, R: Effects of dust on the performance of PV panels. World Academy of Science, Engineering and Technol **58**, 588–592 (2011)
16. TRW Systems Group, Power Sources Engineering Department, Rauschenbach HS, Jet Propulsion Laboratory, United States National Aeronautics and Space Administration: Solar Cell Array Design Handbook, 1st edn. NASA, Washington, DC (1976)
17. Monto, M., Rohit, P.: Impact of dust on solar photovoltaic (PV) performance: research status, challenges and recommendations. Renew Sustain Energy Rev **14**(9), 3124–3131 (2010)

Dilute H_2SO_4-catalyzed hydrothermal pretreatment to enhance enzymatic digestibility of *Jatropha curcas* fruit hull for ethanol fermentation

Ahmad Marasabessy[1,2,3*], A Maarten J Kootstra[1,4], Johan PM Sanders[1] and Ruud A Weusthuis[1]

Abstract

Dilute sulfuric acid pretreatment of the *Jatropha curcas* fruit hull at high temperatures (140°C to 180°C) performed in a 110-mL stainless steel reactor was investigated to enhance the enzymatic digestibility of its lignocellulosic components. Carbohydrates accounted for 43% of the dry matter of the *J. curcas* fruit hull biomass. The goal of the study was to optimize the pretreatment conditions (acid concentration, time, and temperature) in order to obtain the highest sugar yield after subsequent enzymatic hydrolysis. A Box-Behnken design was applied to the experimental setup in order to reduce the number of experiments. The optimal pretreatment conditions are 30-min incubations at a temperature of 178°C with a sulfuric acid concentration of 0.9% (*w/v*). Using these pretreatment conditions for a fruit solid loading of 9.52% followed by a 24-h enzymatic hydrolysis resulted in a liberation of 100% of all pentoses present (71% yield and 29% degradation to furfural) and 83% of the hexoses (78% yield and 5% degradation to 5-hydroxymethylfurfural). The simultaneous saccharification and fermentation experiment showed that acid-pretreated fruit hull can be used as a substrate for *Saccharomyces cerevisiae* to produce ethanol.

Keywords: Box-Behnken, *Jatropha curcas*, Fruit hull, Hemicellulose, Cellulose, Pentose, Hexose, Ethanol

Background

Jatropha curcas has become widely known as a biofuel crop due to the high oil content of the oilseeds harvested from this plant [1]. The biodiesel produced from *J. curcas* oil via transesterification compares well to petrochemical diesel and meets the latest biodiesel standards [2]. However, whereas palm oil is widely used as feedstock for biodiesel production in Indonesia, *Jatropha* oil has not been traded at price levels that satisfactorily reward growers/farmers. Low oil productivity and high labor cost for fruit harvesting are the most important reasons. Many potential Indonesian farmers in the subsistence sector are now reluctant to invest time and money in planting *J. curcas* since the profit from *Jatropha* products is much less compared to other agricultural commodities. Hence, one of the strategies to improve the economics of this potentially profitable plant is to increase the value of the side streams of oil production (trimmed stems/branches, seed or kernel cake, seed shells, and fruit hulls).

J. curcas bears fruit from the second year after crop establishment, but seed production becomes economically feasible from the fifth year onward [3,4]. In Indonesia, *J. curcas* is planted at a 2 × 2-m distance, yielding 2,500 plants/ha. A productivity of 2 kg seeds/plant/year on a *J. curcas* plantation in Indonesia has been reported [5]. One hectare of the *Jatropha* plant therefore would yield 5 tons of seeds annually (containing 25% to 33% oil), with 1 ton/year of fruit hull biomass (dry weight) as side stream. Fruit hull is composed of lignocellulose (39.3% cellulose and 14.2% lignin) with 5.3% protein [4], making this biomass a potential raw material for the production of bioethanol, enzymes, organic acids, and other fermentative products, or as animal feed [6,7]. The use of fruit hulls so far focuses on applications such as fertilizer and substrate for co-firing installations or for biogas

* Correspondence: madbessy@gmail.com
[1]Agrotechnology and Food Sciences Group, Biobased Commodity Chemicals, Wageningen University, P.O. Box 17, Wageningen 6700 AA, The Netherlands
[2]Department of Chemical Engineering, Rijksuniversiteit Groningen, Nijenborgh 4, Groningen 9747 AG, The Netherlands
Full list of author information is available at the end of the article

production [1,3,4,8]. Another way to valorize the fruit hull side stream is to convert it to fermentation products such as ethanol and lactic acid. In that case, pretreatment is required to enhance the enzymatic digestibility of the lignocellulosic biomass. Pretreatment aims at a partial dislocation of the inter- and intra-fibrillic structure of lignocellulose with a reasonable energy input [9]. From an economic point of view, pretreatment must improve fermentable sugar yield, maintain carbohydrate content, minimize formation of degradation products that are inhibitory to subsequent fermentation processes, and be cost-effective [10]. Various methods, from physical to chemical pretreatments or combinations, have been widely studied; one of which is hot dilute sulfuric acid pretreatment. Dilute acid pretreatment of lignocellulosic biomass conducted at high temperature (more than 160°C) results in high xylan conversion yields and low levels of sugar degradation, thus resulting in an improved overall cellulose hydrolysis [10]. There are no reports on pretreatment of the fruit hull of *J. curcas* yet.

The subject of this study is the pretreatment of the *Jatropha* fruit hull by sulfuric acid at an elevated temperature in order to improve the enzymatic digestibility of this lignocellulosic byproduct. We study the influence of varying sulfuric acid concentration, pretreatment time, and temperature on the following three factors:

1. Pentose and hexose degradations after fruit hull pretreatment, calculated from furfural and 5-HMF formed, respectively
2. Pentose and hexose yields after subsequent enzymatic hydrolysis of the pretreated fruit hull
3. Total sugar yield after subsequent enzymatic hydrolysis of the pretreated fruit hull

A Box-Behnken design was applied to the experimental setup in order to reduce the number of experiments. The main objective is to obtain an optimum sulfuric acid concentration, time, and temperature of pretreatment that give a low level of sugar degradation and a high level of sugar yield in the subsequent enzymatic hydrolysis. Simultaneous saccharification and fermentation (SSF) of *J. curcas* fruit hull hydrolysates by *Saccharomyces cerevisiae* for ethanol production is also studied.

Methods

Experimental design and setup

Design-Expert 8.0.3 software (Stat-Ease, Inc., MN, USA) was used for the experimental design, model fitting, and statistical data analysis. In order to reduce the number of experiments, a Box-Behnken design (BBD) [11] was applied. Experimental data of each response factor (Y) were expressed in a second order mathematical model:

$$Y = \beta_o + \sum_{i}^{n} \beta_i X_i + \sum_{i}^{n} \beta_{ii} X_i^2 + \sum_{i}^{n} \sum_{j}^{n} \beta_{ij} X_i X_j \quad (1)$$

where $i = 1$ to 3, $j = 2$ to 3, and $X =$ input variables. The number of runs (N) required to measure the responses in BBD is defined as $N = 2k(k-1) + C_o$, where k is the number of input variables and C_o is the number of central points. Since there were three input variables tested (sulfuric acid concentration, time of pretreatment, and temperature) and four replicates in the central point of the design to calculate the experimental error, a total of 16 experimental runs were carried out. The response factors are pentose degradation, hexose degradation, pentose yield, hexose yield, sugar degradation, and sugar yield. The significant effects and two variable interactions were estimated by ANOVA. The relationship between the response factor and input variables was further elucidated using response surface plots.

Materials

All chemicals were of analytical grade and used as received. Termamyl and amyloglucosidase were purchased from Novo Nordisk, Bagsvaerd, Denmark. Cellulase of GC220 was purchased from Genencor, Rochester, NY, USA. *S. cerevisiae* CBS 8066, maintained on YPD agar medium, was used for ethanol fermentation.

Preparation and analysis of fruit hull

Ripened *J. curcas* fruits (with yellow color only, harvested in November 2009, Serpong, Indonesia), were peeled. The hulls were collected and dried at 60°C for 48 h. The dried hulls were milled in a home blender (Philips HR 2071, Royal Philips Electronics, Amsterdam, The Netherlands) and sieved through a strainer with a 0.8-mm hole diameter. The dry matter of the milled hull was 95.32% (*w*/*w*) (24 h, 105°C). The milled hull material was kept in a sealed plastic container at 4°C until used.

The chemical composition of the hull (Table 1) was analyzed in triplicate according to the following methods. The organic solvent extractives, the total water extractives, and the non-extractives (except protein) were measured by TAPPI methods [12-17], with minor modifications as described previously [18]. The protein content of the fruit hull was determined using the Kjeldahl method, which consists of a destruction unit (Gerhardt Kjeldahlterm) and distillation unit (Gerhardt Vapodest). The amounts of amino acids in the water extractives were determined using a Dionex Ultra-HPLC instrument (Dionex Corporation, Sunnyvale, CA, USA) as described previously [19]. For measuring monomeric sugars and water-soluble oligosaccharides in the fruit hull, the samples were prepared as

Table 1 ***J. curcas*** **fruit hull composition (% (*w/w*) dry matter)**

Constituents	Content (% (*w/w*) dry matter)
Organic solvent extractives	
- Ethanol-toluene extractives	4.1 ± 0.02
- Ethanol extractives	2.0 ± 0.02
Water extractives	
- Glucose	4.0 ± 0.01
- Galactose, mannose, rhamnose, and arabinose	0.3 ± 0.00
- Glucan	7.2 ± 0.26
- Other oligosaccharide (galactan and mannan)	0.6 ± 0.00
- Amino acids	0.1 ± 0.00
- Others	24.0
Non-extractives	
- Glucan	20.4 ± 0.01
- Xylan	5.7 ± 0.01
- Galactan	2.0 ± 0.01
- Mannan	1.1 ± 0.01
- Rhamnan	0.8 ± 0.00
- Arabinan	0.8 ± 0.01
- Protein	4.9 ± 0.35
- Uronic acid	3.2 ± 0.16
- Acid-soluble lignin (ASL)	0.8 ± 0.00
- Acid-insoluble lignin (AIL)	11.7 ± 0.00
- Others	6.3
Total	100.0
Total pentose sugars[a]	6.5
Total hexose sugars[b]	35.6

[a]For response surface analysis, the data of pentose sugars (xylose and arabinose) were grouped to calculate pentose yield. [b]Similarly, the data of hexose sugars (glucose, galactose, mannose, rhamnose) were grouped to calculate hexose yield.

shown in Figure 1. Monomeric sugars were measured by high-performance anion-exchange chromatography with pulsed amperometric detection (HPAEC-PAD) as described earlier [18].

Fruit hull pretreatment

Milled fruit hull (5.00 g; 4.76 g dry matter) was mixed with 45.0 mL of sulfuric acid solution (0.1%, 0.5% or 0.9% (*w/v*)), resulting in 9.52% (*w/w*) dry hull solid loading. The hull/acid mixture was soaked for 20 to 24 h at room temperature and then transferred to a 316-SS (stainless steel) reactor (inner height × diameter, 90.0 × 40.0 mm; 5.0-mm wall), fitted with thermocouples. Four reactors were heated at a time in an oil bath (Haake B bath filled with silicon oil of DC 200 fluid, 100 cSt, Dow Corning, Midland, MI, USA), equipped with a Haake N3 temperature controller (Thermo Fisher Scientific, Waltham, MA, USA). Sample core temperature was digitally recorded using a Picotech data collector and software (Picotech, Neots, Cambridgeshire, UK). Pretreatments were conducted at 140°C, 160°C, and 180°C. Holding time was 30, 45, or 60 min, starting from when the desired core temperature was reached. The heating bath oil was preheated between 1°C and 5°C above the desired sample core temperature; by this way, the time until the desired core temperature was reached ranged between 16 and 20 min. During the holding time, the reaction temperature in the reactors oscillated at a maximum of 1°C from the desired temperature. The reactors were cooled after the incubation to 25°C in less than 10 min by quenching in ice water. After the pretreatment, the resulting material was transferred to pre-weighed 250-mL baffled shake flasks. The pH was adjusted to 5.0 with 10 M NaOH, and sodium azide solution (0.025% (*w/w*) final concentration; Merck KGaA, Darmstadt, Germany) was added. A 1-mL sample was taken for analysis. Monomeric sugars were measured by HPAEC-PAD method [18]. Furfural, 5-hydroxymethylfurfural (5-HMF), and acetic acid were measured by high-performance liquid chromatography (HPLC) [18]. The monomeric sugar yield was calculated as follows:

$$Yield(\%) = \frac{S_L}{S_S} * 100(\%) \tag{2}$$

where S_L is the amount of sugar (g) in the aqueous phase, and S_S is the amount of monomeric sugar present in the sample of dry hull (g sugar equivalents in polymeric sugar). The sugar degradation was calculated in a similar way in which S_L is accounted as the amount of sugar equivalents (g) to the amount of furfural or 5-HMF present in the aqueous phase, calculated in mole basis.

Enzymatic hydrolysis of pretreated fruit hull

To start enzymatic hydrolysis, GC220 (0.4 g (*w/w*) dry matter hull) was added into the baffled shake flask containing the pH-adjusted (pH 5.0) pretreated fruit hull. GC220 is a cellulase enzyme mixture (batch 4900759148, 7608 IU/mL cellulase activity). The amount of GC220 used corresponded to 46 FPU/g original dry matter hull. We used GC220 in excess (in the plateau region of the dose-effect curve of the enzyme mixture) to ensure that the effect of pretreatment on the sugar yield was measured, not the effect of the enzyme concentration. The weight of the material plus the flask was determined; after which, the flasks were closed with airtight plugs and placed in an Innova 44 incubator shaker (50°C, 150 rpm, 2-in. stroke; NBSC, NJ, USA). Samples of 1.5 mL were taken at $t = 0$, 24, and 72 h. GC220 was inactivated by

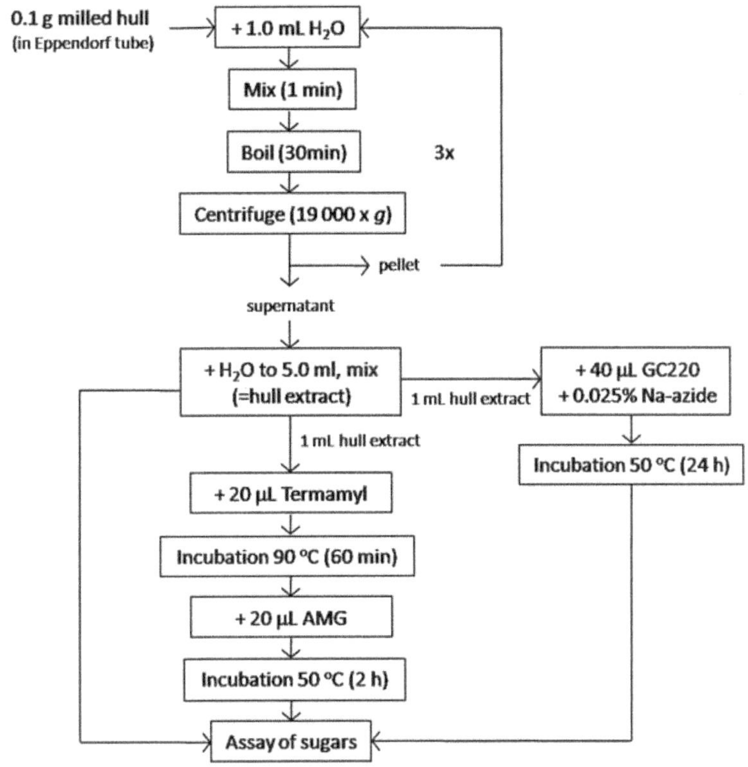

Figure 1 A scheme for determination of water-soluble sugars in fruit hull. A Hermle Z33M2 (HERMLE Labortechnik GmbH, Wehingen, Germany) was used in centrifugation. The amount of soluble oligosaccharides in the fruit hull was calculated by subtracting the initial monosaccharide present in the fruit hull from the total soluble sugar measured after enzymatic digestion.

incubation at 90°C for 10 min. Samples were stored at −20°C until analysis. Monomeric sugars were measured by HPAEC-PAD method [18].

SSF of acid-pretreated hull hydrolysate into ethanol

Ten grams of milled fruit hull pretreated with 90 mL of sulfuric acid solution at the optimum conditions was used for ethanol fermentation tests with the yeast *S. cerevisiae* CBS 8066.

To prepare inoculums, yeast cells were grown in a 100-mL liquid medium of YPD (Merck) at pH 5.0 in a 250-mL Erlenmeyer flask. The culture was incubated for 24 h in an Innova 44 incubator shaker (37°C, 150 rpm, 2-in. stroke; NBSC, NJ, USA). The yeast cake was prepared by centrifugation at 17,000 × *g* for 15 min using the SORVALL RC6+ centrifuge (Thermo Fisher Scientific, Waltham, MA, USA).

Prior to fermentation, the pH of the acid-pretreated hull was adjusted to 5.0 using 10 M NaOH. Pre-hydrolysis was conducted to reduce the viscosity of the pretreated hull suspension and was performed by the addition of 0.95 g GC220 (equivalent to 0.1 g/g dry matter hull) followed by incubation for 6 h in an Innova 44 incubator shaker (37°C, 150 rpm, 2-in. stroke; NBSC, NJ, USA). After pre-hydrolysis, the pre-hydrolysed substrate was enriched with the following (per liter): 4.2 g $(NH_4)_2SO_4$, 2.5 g KH_2PO_4, 0.42 g $MgSO_4 \cdot 7H_2O$, trace elements (consisting of 15.0 mg Na_2EDTA, 4.5 mg ZnSO4·$2H_2O$, 0.84 mg $MnCl_2 \cdot 2H_2O$, 0.3 mg $CoCl_2 \cdot 6H_2O$, 0.3 mg $CuSO_4 \cdot 5H_2O$, 0.4 mg $Na_2MoO_4 \cdot 2H_2O$, 4.5 mg $CaCl_2 \cdot 2H_2O$, 3 mg $FeSO_4 \cdot 7H_2O$, 1 mg H_3BO_3, 1 mg KI, 0.05), vitamins (consisting of 0.05 mg biotin, 1.0 mg calcium panthotenate, 1.0 mg nicotinic acid, 25 mg inositol, 1.0 mg thiamine, 1.0 mg pyridoxine HCl, 0.2 mg para-aminobenzoic acid), and fatty acids (10.0 mg ergosterol and 420 mg Tween 80 dissolved in 1.25 mL ethanol). Trace element solution, vitamin solution, and fatty acid solution were prepared separately in stocks of 1,000×, 1,000×, and 800×, respectively, as described previously [20]. A 2.86-g enzyme mixture (GC220), equivalent to 0.3 g/g dry matter hull, was added to the substrate. The total volume of the substrate just before fermentation was circa 100 mL. An initial yeast cake concentration between 0.50 and 0.55 g/100-mL substrate was applied in the SSF experiments [21]. Fermentation was conducted at 37°C and 70 rpm for 72 h. All SSF experiments were done in duplicate, and homogenous samples of 1.5 mL were withdrawn at 0 (just after yeast addition), 6, 11, 24, 35, 48, and 72 h of incubation for the analysis of monosaccharides, organic acids, and ethanol. For the analysis of ethanol and organic acids in SSF samples, the supernatant of centrifuged samples (5 min at 17,400 × *g*) was diluted 1:1 with 6 mM sulfuric acid and filtered in order to remove solids and precipitated proteins. Samples were analyzed using a HPLC system (Waters Corp., Milford, MA, USA) using the organic acid column IOA-1000 (Alltech, Deerfield, IL, USA) with 3.0 mM sulfuric acid as the mobile phase at 90°C and a flow rate of 0.4 mL/min. Peaks were detected by using a RI detector (Waters 2414, Waters Corporation, Milford, MA, USA). SSF results of sugars, organic acids, and ethanol were reported in grams of product formed per liter fermentation medium. The theoretical ethanol yield was calculated from the total glucose consumed by the yeast, taking into account a fermentation yield of 0.51 [21,22].

Results and discussion

Raw material composition

Table 1 shows the composition of the *J. curcas* fruit hull. Carbohydrates account for 42.9% of the dry weight. This value is lower than those reported for other lignocellulosic materials such as wheat straw (59%) and cardoon biomass (52%) [22,23]. A glucan (as glucose) content of 31.6% is comparable to those reported for agricultural residues such as cardoon, sunflower, or *Brassica* [22,24,25] but slightly lower than those for wheat straw and rice straw (36%) [18,26]. Unlike wood and straw, the hull contains about 12% water-soluble sugars. The total lignin value is 12.5% of which acid-soluble lignin (ASL) accounts for 0.8% only. The lignin value of the hull is less than 50% of that of wood [27,28]. ASL will solubilize during acid pretreatment, while acid-insoluble lignin remains in solid form. Hemicellulosic sugars (xylose, galactose, mannose, rhamnose, and arabinose) account for 10.4% of the hull with xylose (5.7%) as the main hemicellulosic carbohydrate. No xylose was found after incubation of the water-extractable fraction with the enzyme mixture GC220 (data not shown), indicating that the material does not contain soluble xylan. Together, the organic solvent extractables and water extractables account for 30% of the total dry matter, which may include non-structural components of fruit hulls such as waxes, fats, tannins, some resins, and soluble pectins [28,29]. In comparison, the values for glucan and lignin content obtained are lower than those of the *J. curcas* fruit hull (39.3% and 14.2%, respectively) previously reported [4], which could be due to differences in cultivar, cultivation conditions, or ripeness of the fruit.

Pretreatment and enzymatic hydrolysis of fruit hull

The experimental conditions for the 16 runs according to the BBD can be seen in Table 2. Table 2 also summarizes the overall experimental results of sugar liberation from milled *Jatropha* fruit hull after pretreatment and subsequent enzymatic hydrolysis for 24 h.

Both the highest pentose yield (34.4%) and pentose degradation (42.4%) after pretreatment were achieved at 0.9% sulfuric acid, 45 min, and 180°C, in which 57.4% pentose and 80.6% hexose were recovered after subsequent enzymatic hydrolysis (Table 2). After the subsequent 24-h enzymatic hydrolysis, the highest pentose yield (84.7%) was obtained from the biomass pretreated at less severe conditions (0.9% sulfuric acid, 30 min, and 160°C) in comparison to the highest hexose yield (80.6%) which was obtained at more severe conditions (0.9% sulfuric acid, 45 min, and 180°C).

In terms of total sugar liberation (Table 2), both the highest sugar yield (21.0%) and sugar degradation (24.4%) after pretreatment were achieved at 0.9% sulfuric acid, after 45 min, and at 180°C. After the subsequent 24-h enzymatic hydrolysis, the highest sugar yield of 77.0% was reached at this pretreatment condition.

Statistical analysis of sugar degradation and sugar yield

In the experimental Box-Behnken design, the influence of the three input variables (sulfuric acid concentration, time, and temperature of pretreatment) on pentose degradation, hexose degradation, pentose yield, and hexose yield were determined. The Box-Behnken model used to describe the experimental results is based on the effect of the three input variables studied, extended with parameters for interactions and squared factors. A square root transformation of the response factors was applied for improved model fit. To describe the interactive effects of the input variables on responses in the statistical analysis, one variable is set constant while the other two variables are varying in the design space. The input variables of X_1, X_2, and X_3 are the coded factors of sulfuric acid concentration, time, and temperature of pretreatment, respectively. The coded variables are defined as follows: $X_1 = (S_A - S_{A,C})/S_{A,S}$, $X_2 = (t - t_C)/t_S$, and $X_3 = (T - T_C)/T_S$, in which S_A = concentration of sulfuric acid (%), t = pretreatment time (min), and T = pretreatment temperature (°C); subscript C = center value and subscript S = step value; $S_{A,C} = 0.5\%$, $S_{A,S} = 0.4\%$, $t_C = 45$ min, and $t_S = 15$ min, $T_C = 160°C$ and $T_S = 20°C$.

Pentose and hexose degradation after pretreatment

The starting point was a quadratic model which was found significant for the degradation of pentose into furfural and the degradation of hexose into 5-HMF during the pretreatment (both with $P < 0.0001$). The quadratic model was then adjusted by backward elimination: taking out terms that had no significant contribution ($P > 0.05$) one by one and then recalculating the model with the remaining terms. The adjusted regression models fit the data with the R^2_{adjusted} of 0.99 for pentose degradation (Equation 3) and 0.97 for hexose degradation (Equation 4) as follows:

Table 2 Experimental conditions according to Box-Behnken design and experimental results of sugar liberation from milled *J. curcas*

	Run number															
	1	2	3	4	5	6	7	8	9	10	11	12	13	14	15	16
Pretreatment conditions																
Sulfuric acid (% *w/v*)	0.1	0.5	0.5	0.9	0.1	0.1	0.9	0.9	0.5	0.5	0.5	0.5	0.1	0.5	0.5	0.9
Time (min)	45	30	60	45	30	60	30	60	45	45	45	45	45	30	60	45
Temperature (°C)	140	140	140	140	160	160	160	160	160	160	160	160	180	180	180	180
Response factors																
A. Pentose (% *w/w*)																
Pretreatment																
Pentose yield	1.2	5.4	7.8	16.1	2.9	2.8	26.5	33.5	10.2	11.1	10.9	10.1	2.5	10.4	9.6	34.4
Pentose degraded	0.2	0.6	1.1	2.0	0.5	0.9	7.7	12.7	3.9	3.8	3.8	3.5	3.0	12.1	20.2	42.4
Total pentose liberated	1.3	5.9	8.9	18.2	3.4	3.7	34.1	46.2	14.1	15.0	14.7	13.6	5.5	22.4	29.8	76.8
After 24-h enzymatic hydrolysis																
Pentose yield	29.3	35.1	40.4	56.6	47.7	54.4	84.7	80.7	69.5	66.0	67.7	69.3	57.6	70.5	46.8	57.4
B. Hexose (% *w/w*)																
Pretreatment																
Hexose yield	12.5	14.0	14.5	15.2	8.8	6.6	17.8	18.1	13.7	12.8	13.0	13.9	1.5	8.3	6.0	18.6
Hexose degraded	0.1	0.5	0.6	0.9	0.2	0.3	1.8	2.5	1.3	1.3	1.3	1.3	0.4	2.7	3.6	6.4
Total hexose liberated	12.6	14.5	15.1	16.1	9.0	6.9	19.6	20.6	15.0	14.1	14.3	15.2	1.9	11.0	9.6	25.0
After 24-h enzymatic hydrolysis																
Hexose yield	57.6	58.7	57.6	57.6	54.0	51.2	65.6	69.9	59.9	58.3	59.0	59.6	51.2	64.8	63.7	80.6
C. Total sugars (% *w/w*)																
Pretreatment																
Sugar yield	10.7	12.7	13.4	15.3	7.8	6.1	19.1	20.5	13.1	12.5	12.7	13.3	1.7	8.6	6.5	21.0
Sugar degraded	0.1	0.5	0.9	1.5	0.4	0.6	4.7	7.6	2.6	2.6	2.5	2.4	1.7	7.4	11.9	24.4
Total sugar liberated	10.8	13.2	14.3	16.8	8.2	6.7	23.8	28.1	15.7	15.1	15.2	15.7	3.4	16.0	18.4	45.4
After 24-h enzymatic hydrolysis																
Sugar yield	53.3	55.1	55.0	57.4	53.0	51.7	68.6	71.5	61.4	59.5	60.4	61.1	52.2	65.7	61.1	77.0

Only data of the 24-h enzymatic hydrolysis are presented because no increase of sugar liberation was observed between 24 and 72 h of hydrolysis (data not shown).

$$\left(Y_{PD,Pr}\right)^{0.5} = 1.97 + 1.31X_1 + 0.29X_2 + 1.57X_3 + 0.94X_1X_3 + 0.19X_2X_3 + 0.51X_3^2 \quad (3)$$

$$\left(Y_{HD,Pr}\right)^{0.5} = 1.13 + 0.56X_1 + 0.50X_3 + 0.30X_1X_3 - 0.15X_1^2 + 0.13X_3^2 \quad (4)$$

in which $Y_{\mathrm{PD,Pr}}$ (% *w/w*) is the pentose degradation, and $Y_{\mathrm{HD,Pr}}$ (% *w/w*) is the hexose degradation after pretreatment, respectively.

Figure 2 shows the response curves of pentose and hexose degradation as three-dimensional surfaces. As can be seen in Figure 2a and Equation 3, minimizing the degradation of pentose into furfural depends, for a large part, on sulfuric acid concentration and temperature. Degradation of pentose increases rapidly at higher sulfuric acid concentration and temperature. Because the degradation of pentose is, for a small part, influenced also by time, conducting the pretreatment step in a shorter time (30 min) is preferable than a longer pretreatment time if high pentose recovery is the target. A pentose degradation of 34% is obtained at 30 min, 0.9% sulfuric acid, and 180°C (Figure 2a) in which 33% of the pentose is simultaneously recovered (data not shown), resulting in a total pentose liberation from the biomass of 67%. A corresponding result has been reported in the pretreatment of corn stover at 20% (*w/w*) solid loading by using sulfuric acid (0.5% to 1.4% (*w/w*)) at 165°C to 195°C for 3 to 12 min, in which pentose degradation ranged between 5% and 31% [30].

In case of hexose degradation, from the three input variables studied, the pretreatment time has no significant contribution on the degradation of hexose (Equation 4). The

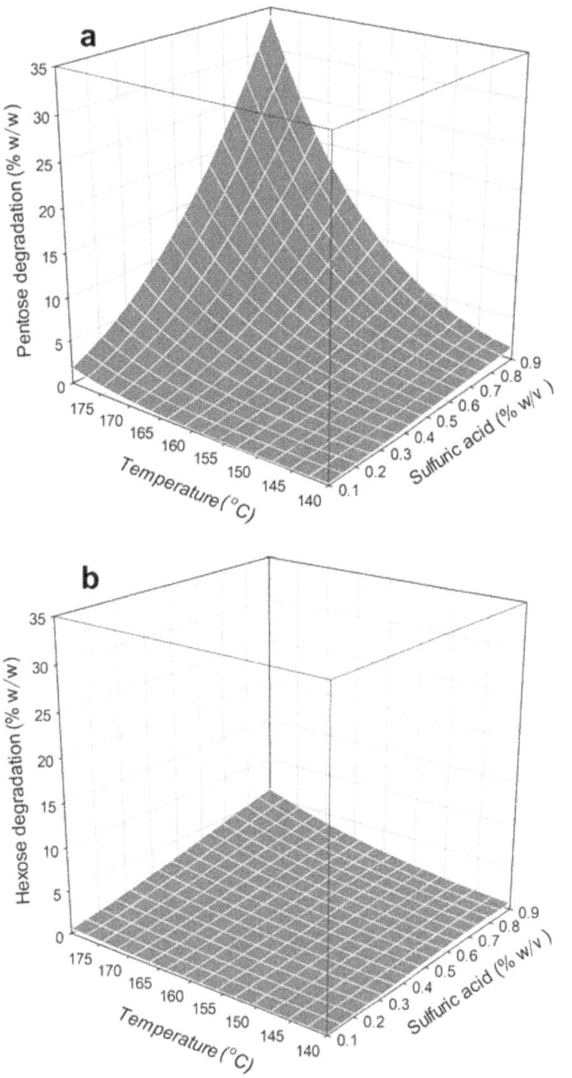

Figure 2 Three-dimensional response surfaces for (a) pentose degradation and (b) hexose degradation of *J. curcas* fruit hull. The response surfaces were graphed after dilute sulfuric acid pretreatment. Pretreatment time is 30 min.

influence of the sulfuric acid concentration and temperature causes less than 6% hexose degradation in the design space as shown in Figure 2b. The low hexose degradation at pretreatment demonstrates that only a small fraction of the six-carbon polysaccharides (mostly cellulose) of the biomass was completely depolymerized and degraded during the pretreatment, a positive property of a pretreatment, assuming the treatment's ability to facilitate the subsequent enzymatic hydrolysis.

Pentose and hexose yields after enzymatic hydrolysis

The starting point was a quadratic model which was found significant for the pentose yield and the hexose yield after the subsequent 24-h enzymatic hydrolysis of pretreated fruit hulls (both with $P < 0.0001$). The quadratic model was then adjusted by backward elimination: taking out terms that had no significant contribution ($P > 0.05$) one by one and then recalculating the model with the remaining terms. The adjusted regression models fit the data with the $R^2_{adjusted}$ of 0.91 for pentose yield (Equation 5) and 0.99 for hexose yield (Equation 6) as follows:

$$(Y_{PY,24h})^{0.5} = 8.19 + 0.75X_1 + 0.65X_3 - 0.53X_1X_3 - 0.50X_2X_3 - 1.23X_3{}^2 \quad (5)$$

$$(Y_{HY,24h})^{0.5} = 7.69 + 0.47X_1 + 0.22X + 0.11X_1X_2 + 0.45X_1X_3 + 0.11X_3{}^2 \quad (6)$$

in which $Y_{PY,24h}$ (% w/w) is the pentose yield, and $Y_{HY,24h}$ (% w/w) is the hexose yield after the 24-h enzymatic hydrolysis, respectively. Figure 3 shows the response analysis of pentose and hexose yields as three-dimensional surfaces.

As can be seen in Figure 3a and Equation 5, maximizing the pentose yield is, for a large part, dependent on sulfuric acid concentration and temperature. The time of pretreatment on its own has no direct significant contribution on the pentose yield, but its interaction with the temperature has. In comparison to our results, it was reported that the yield of xylose (a major component of pentose) after a dilute sulfuric acid (0.5% to 2%) pretreatment of rapeseed straw at high temperature (180°C) is dependent on acid concentration and time of pretreatment [31].

Figure 3 shows that pretreatment of the fruit hull at the conditions studied results in a significant increase of the pentose yield and the hexose yield in the hydrolysis step. Approximately 84% pentose yield after enzymatic hydrolysis can be achieved at 0.9% sulfuric acid (30-min pretreatment at 160°C) in which approximately 10% pentose is degraded (Figure 2a). Increasing the temperature above 160°C or the time of pretreatment for more than 30 min (data not shown) does not improve the pentose yield due to the degradation of pentose into furfural. The 30-min pretreatment results in a higher pentose yield than the longer pretreatment periods.

We found that without enzymatic hydrolysis, pretreatment alone of the fruit hull with 0.1% sulfuric acid at a temperature between 140°C and 180°C resulted in pentose yields of less than 4% (data not shown). However, high overall pentose yields obtained after enzymatic hydrolysis of the fruit hull pretreated with 0.1% sulfuric acid at 140°C to 180°C (Figure 3a) demonstrate that the hydrothermal treatment itself contributes to hemicellulose digestibility, thereby enhancing the enzymatic hydrolysis. It has been reported that heating lignocellulosic biomass in aqueous media at a temperature above 150°C to 180°C will solubilize parts of the biomass, firstly the hemicellulose and shortly after the lignin [32].

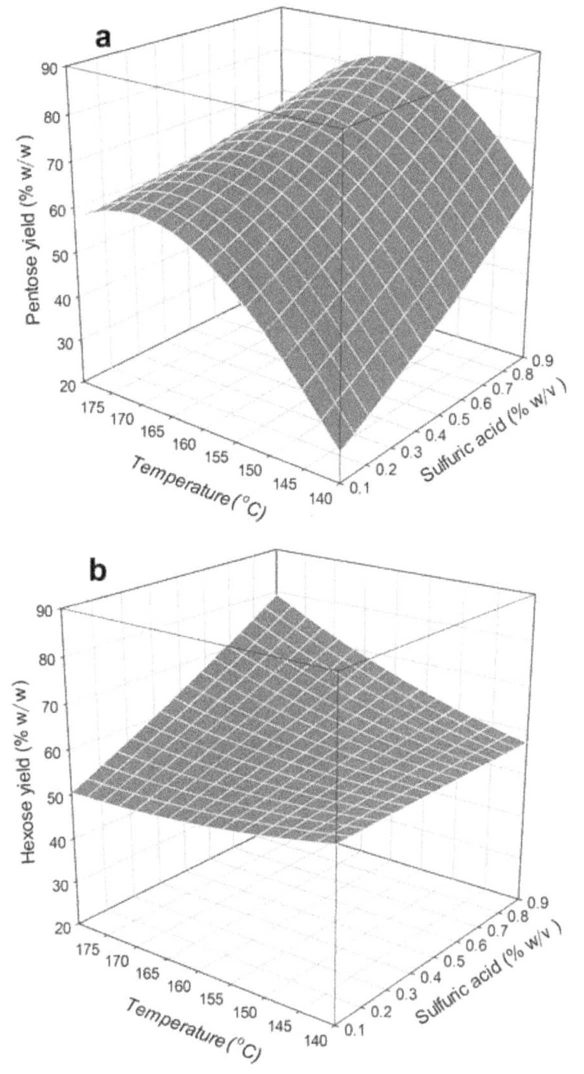

Figure 3 Three-dimensional response surfaces for (a) pentose yield and (b) hexose yield. The response surfaces were graphed after the 24-h enzymatic hydrolysis, respectively, of the pretreated *J. curcas* fruit hull. Pretreatment time is 30 min.

From Figure 3b and Equation 6, the hexose yield ranged from approximately 52% at the least severe pretreatment (0.1% sulfuric acid, 30 min, 140°C) to approximately 78% at the more severe conditions (0.9% sulfuric acid, 30 min, 180°C).

Total sugar yield after enzymatic hydrolysis

For the total sugar degradations (pentose plus hexose degradation) after pretreatment and the total sugar yield (pentose plus hexose yield) after the subsequent 24-h enzymatic hydrolysis of the pretreated fruit hull, the starting point of a quadratic model was found significant ($P < 0.0001$). The quadratic model was then adjusted by backward elimination: taking out terms that had no significant contribution ($P > 0.05$) one by one and then recalculating the model with the remaining terms. The adjusted regression models fit the data with the R^2_{adjusted} of 0.98 for the total sugar degradation (Equation 7) and 0.96 for the total sugar yield (Equation 8) as follows:

$$\left(Y_{SD,Pr}\right)^{0.5} = 1.58 + X_1 + 0.21X_2 + 0.15X_3 + 0.69X_1X_3 + 0.38{X_3}^2 \quad (7)$$

$$\left(Y_{SY,24h}\right)^{0.5} = 7.79 + 0.51X_1 + 0.28X_3 + 0.32X_1X_3 - 0.09{X_3}^2 \quad (8)$$

in which $Y_{\text{SD,Pr}}$ (% *w*/*w*) is the total sugar degradation after pretreatment, and $Y_{\text{SY,24h}}$ (% *w*/*w*) is the total sugar yield after the subsequent 24-h enzymatic hydrolysis.

Figure 4 shows the response curve of the total sugar yield and the total sugar degradation as three-dimensional surfaces. Statistical analysis of combined sugars (pentose plus hexose) showed that the time of pretreatment has no influence on the total sugar yield (Equation 8). Sulfuric acid concentration and temperature as well as their interaction have positive influences on the total sugar yield, to the increase of overall sugar yield as sulfuric acid concentration and temperature increase. Because of the negative squared term of the temperature (Equation 8), an optimum point somewhere in the experimental space might be expected; however, the temperature coefficient seems too small to level off the curve. A yield of 78% was the maximum achieved value at the

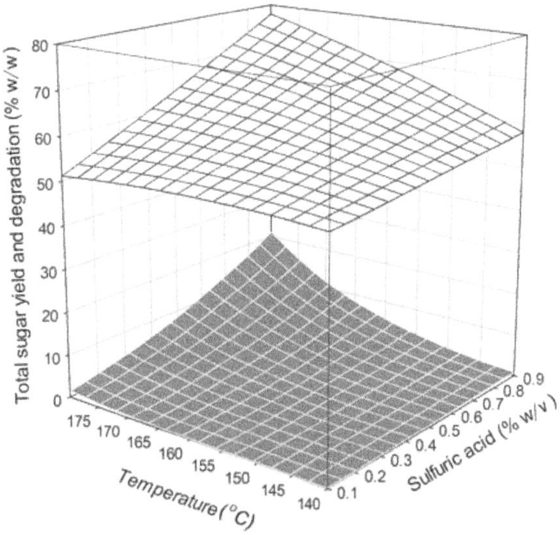

Figure 4 Three-dimensional response surfaces for total sugar degradation (dark) and total sugar yield (light). The response surface for total sugar degradation was graphed after pretreatment, and that for total sugar yield was graphed after the 24-h enzymatic hydrolysis of the pretreated *J. curcas* fruit hull. Pretreatment time is 30 min.

temperature of 180°C using 0.9% sulfuric acid, in which 21% sugar is degraded into furfural and 5-HMF (Figure 3).

Optimization and confirmation test

It has been shown that the maximum sugar yield as well as the hexose yield, respectively, after enzymatic hydrolysis can be achieved only if pretreatment of the fruit hull (9.52% solid loading) was performed using 0.9% sulfuric acid at a temperature around 180°C. Using the Design-Expert 8.0.3 software, optimization was conducted to obtain the highest hexose/sugar yields with minimum sugar degradations. The optimal conditions were as follows: sulfuric acid equals 0.9%; time equals 30 min, and temperature equals 178°C. Under these conditions, the model predicts the responses as follows: pentose degradation of 31%, hexose degradation of 6%, a pentose yield (after 24-h enzymatic hydrolysis) of 72%, and a hexose yield (after 24-h enzymatic hydrolysis) of 77%.

A confirmation test was performed under the pretreatment conditions described above, in duplicate. The experimental results with the deviations from the average were as follows: pentose degradation of $29.4 \pm 1.27\%$ (equivalent to 1.5 ± 0.05 g/L furfural), hexose degradation of $5.2 \pm 0.14\%$ (equivalent to 1.2 ± 0.03 g/L 5-HMF), a pentose yield of $70.8 \pm 1.20\%$ (equivalent to 4.4 ± 0.08 g/L pentose), and a hexose yield (24 h) of $78.3 \pm 0.70\%$ (equivalent to 26.5 ± 0.25 g/L hexose). These experimental sugar values agree well to those predicted by the model with deviations less than 1%. In addition, we found that acetic acid liberation was 1.5 ± 0.02 g/L.

SSF of hydrolysed fruit hull into ethanol

Fruit hull of 9.52% solid loading pretreated at optimum conditions (0.9% sulfuric acid, 178°C, 30 min) was applied in a SSF process using *S. cerevisiae* at 37°C. In a SSF process, the glucose liberated by enzymatic hydrolysis is simultaneously converted into ethanol, thus reducing product inhibition of the enzyme by glucose. Figure 5 shows the course of ethanol formation for the 72-h fermentation in the standard medium containing 30 g/L glucose (Figure 5a) in comparison to the medium containing the pretreated fruit hull (Figure 5b).

In the standard medium, glucose is depleted within 12 h. In that period, ethanol was rapidly produced and peaked at about 10.7 g/L. This ethanol concentration corresponds to 74% of the maximum theoretical yield.

In the pretreated fruit hull medium (Figure 5b), a 6-h pre-hydrolysis to reduce the viscosity of the hydrolysate suspension resulted in 18 g/L glucose initially available in the medium before the fermentation started. As can be seen in Figure 5b, glucose depletion in the fruit hull medium was slower than in the standard medium, reaching 2 g/L after 36 h and remaining constant from that point; the maximum ethanol concentration was 8.4 g/L at 24 h (corresponding to 71% of the maximum theoretical yield). This trend of ethanol formation, peaking at 24 h, corresponds well to the results previously reported [21]. It is clearly seen that ethanol formation was slightly inhibited in the pretreated fruit hull biomass, leading to a lower ethanol yield (71% compared to 74%) and longer fermentation time (24 h compared to 12 h) at peak points. In biomass hydrolysis, limiting concentrations of byproducts are important when the hydrolysate is to be used as a fermentation medium. For ethanol fermentation, it was reported that the presence of acetic acid (5 g/L), furfural (1.2 g/L), and 5-HMF (1.3 g/L) slightly decreased the ethanol yield in the fermentation process; however, a low acetic acid concentration (about 1 g/L) was found to have a positive effect on the ethanol production yield [33]. Due to pretreatment of the fruit hull at optimum conditions, 1.5 g/L furfural, 1.2 g/L 5-HMF,

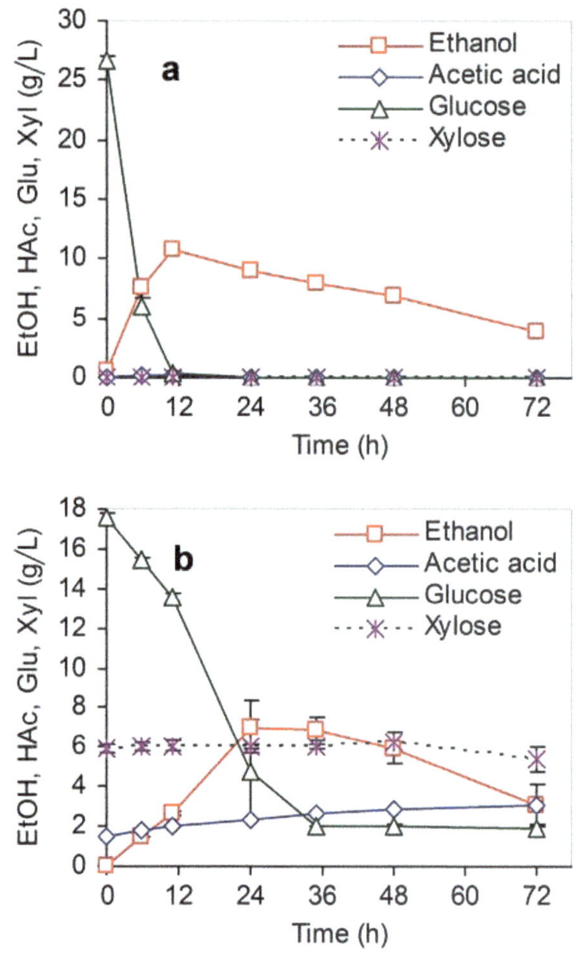

Figure 5 Course of ethanol formation. Formation of ethanol, liberation of acetic acid, and the consumption trends of glucose and xylose during fermentation by *S. cerevisiae* CBS 8066 of (**a**) standard medium containing glucose as a sole carbon source in comparison to (**b**) SSF of the pretreated *J. curcas* fruit hull.

and 1.5 g/L acetic acid were also formed in the fermentation substrate. Therefore, inhibition of ethanol formation in the fruit hull substrate was probably caused by the presence of these compounds.

Conclusions

It is demonstrated that the model equations developed using the Box-Behnken design with the three input variables studied (sulfuric acid, time, and temperature) can be used to predict liberation and degradation of sugars from the *J. curcas* fruit hull after dilute sulfuric acid-catalyzed hydrothermal pretreatment and subsequent enzymatic hydrolysis. A pretreatment at optimum conditions (0.9% sulfuric acid, 30 min, 178°C) followed by a 24-h enzymatic hydrolysis liberates nearly all of the sugars present, consisting of 71% pentose, 78% hexose, 29% pentose degradation into furfural, and 5% hexose degradation into 5-HMF. Therefore, higher sugar yields cannot be expected within the space studied. The SSF experiment of ethanol production showed that the sulfuric acid-pretreated fruit hull can be used to produce ethanol by *S. cerevisiae* in the simultaneous saccharification and fermentation process.

Competing interests

The authors declare that they have no competing interests.

Authors' contributions

AM carried out the design of the study and laboratory experiments, performed statistical analysis and drafted the manuscript. AMJK participated in the design of the study and involved in HPLC measurements. JPMS conceived the study and participated in its design. RAW participated in the design of the study and coordination. All authors read, edited, and eventually approved the final manuscript to be submitted.

Acknowledgments

The authors gratefully acknowledge the Koninklijke Nederlandse Akademie van Wetenschappen, Scientific Programme Indonesia - Netherlands (SPIN-KNAW) for the financial support.

Author details

[1]Agrotechnology and Food Sciences Group, Biobased Commodity Chemicals, Wageningen University, P.O. Box 17, Wageningen 6700 AA, The Netherlands. [2]Department of Chemical Engineering, Rijksuniversiteit Groningen, Nijenborgh 4, Groningen 9747 AG, The Netherlands. [3]Agency for the Assessment and Application of Technology (BPPT), Jl. MH Thamrin 8, Jakarta 10340, Indonesia. [4]Bioprocess Engineering Group, Wageningen University, P.O. Box 17, Wageningen 6700 AA, The Netherlands.

References

1. Gubitz, GM, Mittelbach, M, Trabi, M: Exploitation of the tropical oil seed plant *Jatropha curcas* L. Bioresour Technol **67**, 73–82 (1999)
2. Tiwari, AK, Kumar, A, Raheman, H: Biodiesel production from jatropha (*Jatropha curcas*) with high free fatty acids: an optimized process. Biomass Bioenergy **31**, 569–575 (2007)
3. Foidl, N, Eder, P: Agro-industrial exploitation of *J. curcas*. In: Gubitz, GM, Mittelbach, M, Trabi, M (eds.) Biofuels and Industrial Products from *Jatropha curcas*, pp. 88–91. The Austrian Ministry of Foreign Affairs, Managua, Nicaragua (1997)
4. Gunaseelan, VN: Biomass estimates, characteristics, biochemical methane potential, kinetics and energy flow from *Jatropha curcus* on dry lands. Biomass Bioenergy **33**, 589–596 (2009)
5. Purwaamijaya, IM, Masri, RM, Tukimin, S: Social economic analysis of seeding and cultivation of *Jatropha curcas*, Linnaeus as alternative energy sources (bioediesel) which environmental friendly. Indonesia Education of University, Bandung (2007)
6. Wyman, CE: Ethanol from lignocellulosic biomass: technology, economics, and opportunities. Bioresour Technol **50**, 3–15 (1994)
7. Gonzalez-Garcia, S, Moreira, MT, Feijoo, G: Comparative environmental performance of lignocellulosic ethanol from different feedstocks. Renew Sustain Energy Rev **14**, 2077–2085 (2010)
8. Openshaw, K: A review of *Jatropha curcas*: an oil plant of unfulfilled promise. Biomass Bioenerg **19**, 1–15 (2000)
9. Abatzoglou, N, Bouchard, J, Chornet, E, Overend, RP: Dilute acid depolymerization of cellulose in aqueous phase: experimental evidence of the significant presence of soluble oligomeric intermediates. Can J Chem Eng **64**, 781–786 (1986)
10. Sun, Y, Cheng, JY: Hydrolysis of lignocellulosic materials for ethanol production: a review. Bioresour Technol **83**, 1–11 (2002)
11. Box, GEP, Behnken, DW: Simplex-sum designs: a class of 2nd order rotatable designs derivable from those of 1st order. Ann Math Stat **31**, 838–864 (1960)
12. TAPPI: Test Method T 418 cm-97. Organic Nitrogen in Paper and Paperboard, TAPPI, Georgia (2004)
13. TAPPI: Test Method T 211 om-02. Ash in Wood, Pulp, Paper and Paperboard: Combustion at 525°C, TAPPI, Georgia (2004)
14. TAPPI: Test Method T 222 om-02. Acid-Insoluble Lignin in Wood and Pulp, TAPPI, Georgia (2004)
15. TAPPI: Test Method T 249 cm-00. Carbohydrate Composition of Extractive-Free Wood and Wood Pulp by Gas–Liquid Chromatography, TAPPI, Georgia (2004)
16. TAPPI: Test Method T 412 om-02. Moisture in Pulp, Paper and Paperboard, TAPPI, Georgia (2004)
17. TAPPI: Test Method T 204 cm-97. Solvent Extractives of Wood and Pulp, TAPPI, Georgia (2004)
18. Kootstra, A.M.J., Beeftink, H.H., Scott, E.L., Sanders, J.P.M.: Optimization of the dilute maleic acid pretreatment of wheat straw. Biotechnol Biofuel **2**, 1–14 (2009)
19. Teng, YL, Scott, EL, van Zeeland, ANT, Sanders, JPM: The use of L-lysine decarboxylase as a means to separate amino acids by electrodialysis. Green Chem **13**, 624–630 (2011)
20. Verduyn, C, Postma, E, Scheffers, WA, Vandijken, JP: Effect of benzoic acid on metabolic fluxes in yeasts: a continuous-culture study on the regulation of respiration and alcoholic fermentation. Yeast **8**, 501–517 (1992)
21. Cuevas, M, Sanchez, S, Bravo, V, Garcia, JF, Baeza, J, Parra, C, Freer, J: Determination of optimal pre-treatment conditions for ethanol production from olive-pruning debris by simultaneous saccharification and fermentation. Fuel **89**, 2891–2896 (2010)
22. Ballesteros, I, Ballesteros, M, Manzanares, P, Negro, MJ, Oliva, JM, Saez, F: Dilute sulfuric acid pretreatment of cardoon for ethanol production. Biochem Eng J **42**, 84–91 (2008)
23. Kootstra, AMJ, Beeftink, HH, Scott, EL, Sanders, JPM: Comparison of dilute mineral and organic acid pretreatment for enzymatic hydrolysis of wheat straw. Biochem Eng J **46**, 126–131 (2009)
24. Ballesteros, I, Oliva, JM, Negro, MJ, Manzanares, P, Ballesteros, M: Enzymic hydrolysis of steam exploded herbaceous agricultural waste (*Brassica carinata*) at different particule sizes. Process Biochem **38**, 187–192 (2002)
25. Ruiz, E, Cara, C, Ballesteros, M, Manzanares, P, Ballesteros, I, Castro, E: Ethanol production from pretreated olive tree wood and sunflower stalks by an SSF process. Appl Biochem Biotechnol **130**, 631–643 (2006)
26. Bak, JS, Ko, JK, Choi, IG, Park, YC, Seo, JH, Kim, KH: Fungal pretreatment of lignocellulose by *Phanerochaete chrysosporium* to produce ethanol from rice straw. Biotechnol Bioeng **104**, 471–482 (2009)
27. Yang, B, Boussaid, A, Mansfield, SD, Gregg, DJ, Saddler, JN: Fast and efficient alkaline peroxide treatment to enhance the enzymatic digestibility of steam-exploded softwood substrates. Biotechnol Bioeng **77**, 678–684 (2002)
28. Cara, C, Ruiz, E, Ballesteros, I, Negro, MJ, Castro, E: Enhanced enzymatic hydrolysis of olive tree wood by steam explosion and alkaline peroxide delignification. Process Biochem **41**, 423–429 (2006)
29. Winkler, E, Foidl, N, Gubitz, GM, Staubmann, R, Steiner, W: Enzyme-supported oil extraction from *Jatropha curcas* seeds. Appl Biochem Biotechnol **63–65**, 449–456 (1997)
30. Schell, DJ, Farmer, J, Newman, M, McMillan, JD: Dilute-sulfuric acid pretreatment of corn stover in pilot-scale reactor:investigation of yields, kinetics, and enzymatic digestibilities of solids. Appl Biochem Biotechnol **105**, 69–85 (2003)

31. Lu, XB, Zhang, YM, Angelidaki, I: Optimization of H_2SO_4-catalyzed hydrothermal pretreatment of rapeseed straw for bioconversion to ethanol: focusing on pretreatment at high solids content. Bioresour Technol **100**, 3048–3053 (2009)
32. Hendriks, A, Zeeman, G: Pretreatments to enhance the digestibility of lignocellulosic biomass. Bioresour Technol **100**, 10–18 (2009)
33. Erdei, B, Barta, Z, Sipos, B, Reczey, K, Galbe, M, Zacchi, G: Ethanol production from mixtures of wheat straw and wheat meal. Biotechnol Biofuel **3**, 9 (2010)

Box-Behnken design application to study enhanced bioremediation of soil artificially contaminated with spent engine oil using biostimulation strategy

Samuel E Agarry* and Oladipupo O Ogunleye

Abstract

This work studies the biodegradation of spent engine oil in soil using Box-Behnken design under response surface methodology. NPK fertilizer (inorganic nutrient), Tween 80 (nonionic surfactant), and pig manure (organic nutrient) concentrations were used as independent biostimulant variables, while total petroleum hydrocarbon (TPH) and hexavalent chromium (Cr (VI)) reductions as dependent variables (response) in a 42-day remediation period. A statistically significant second-order quadratic regression model for TPH and Cr (VI) removal was obtained. The coefficient of determination ($R^2 = 0.9995$ for TPH and 0.9988 for Cr (VI)) and probability value ($P < 0.0001$) demonstrated significance for the regression model. Numerical optimization technique based on desirability function was carried out for initial spent engine oil concentration of 10% *w/w* to optimize the biodegradation process. The optimum values for biostimulation agents to achieve a predicted maximum TPH and Cr (VI) removal of 67.20% and 53.20%, respectively, were found to be as follows: NPK fertilizer, 4.22 g; Tween 80, 10.69 mg/l; and pig manure, 47.76 g. At this optimum point, the observed TPH and Cr (VI) reductions were found to be 66.47% and 52.33%, respectively. The statistical analyses and the closeness of the experimental results and model predictions show the reliability of the regression model, and thus, biostimulation of indigenous microbial density and activity can reduce remediation period of petroleum hydrocarbon and heavy metal-contaminated environment and subsequently the cost of remediation.

Keywords: Biodegradation, Biostimulation, Nutrients, Second-order quadratic regression model, Spent engine oil

Background

Spent engine oil (SEO) which is also known as used motor oil is produced when new engine oil (or motor oil) is subjected to high temperature and high mechanical strain [1]. It is a brown-to-black liquid mixture of heavy metal contaminants such as zinc, lead, and chromium that come from engine parts as they wear down, including low to high molecular weight (C_{15} to C_{18}) aliphatic and aromatic hydrocarbons, polychlorinated biphenyls, chlorodibenzofurans, lubricative additives, and decomposition products [1,2]. There are improper disposals of SEO during manual oil changing operation which is not recycled but spilled and dumped by automobile and generator mechanics into runoff, gutters, water drains and open vacant plots, and farmlands used as automobile workshops [3-5], thereby polluting both soil and water. In addition, the oil is also released into the environment from the exhaust system during engine use and due to engine leaks [6]. In Nigeria and some developing countries, about 20 million gallons of SEO are generated annually from mechanic workshops and discharged carelessly into the environment [7].

Chromium which is a heavy metal component of SEO has been designated as a priority pollutant by US EPA [8] and found to be toxic, carcinogenic, and teratogenic when present at high concentration [9]. Chromium (Cr) does not occur naturally in elemental form, only in compounds. Chromium exists in nine valence states ranging

* Correspondence: sam_agarry@yahoo.com
Biochemical Engineering and Biotechnology Research Laboratory, Department of Chemical Engineering, Ladoke Akintola University of Technology, Ogbomoso-Ilorin Road, Ogbomoso, Oyo State 260001, Nigeria

from 2 to +6; nevertheless, trivalent chromium (Cr (III)) and hexavalent chromium (Cr (VI)) are of major environmental significance because of their stability in the natural environment [10]. Hexavalent chromium (chromate) compounds are highly water soluble and, therefore, can overcome the cellular permeability barrier, entering via sulfate transport pathways since it bears structural similarity with SO_4. It has a high tendency to bind with oxygen, and unless it is rapidly reduced, it can oxidatively damage the DNA via the production of free radicals [9]. It has been reported that hexavalent chromium causes lung cancer [11], chromate ulcer, perforation of nasal septum, and kidney damage in humans, and it is also toxic to other organisms as well. Trivalent chromium is an essential micronutrient for many microorganisms, relatively insoluble in water, and 100 times less toxic than the hexavalent form. Thus, reduction of Cr (VI) to Cr (III) represents a potentially useful detoxification process. Therefore, prolonged exposure to high SEO concentration may cause the development of liver or kidney disease, possible damage to the bone marrow, and an increased risk of cancer [12,13].

Soil is a key component of natural ecosystems because environmental sustainability depends largely on a sustainable soil ecosystem [14]. When soil is polluted, the ecosystem is altered, and agricultural activities are affected. Contamination of the soil by SEO creates an unsatisfactory condition for life in the soil, which is due to the poor aeration it causes in the soil, immobilization of soil nutrients, loss of water-holding capacity, lowering of soil pH, and reduction in soil catalase enzyme activity [15-17], as well as inhibitory effect on the nitrate reductase activities of plants [18]. The soil remediation processes leading to the eventual removal of petroleum hydrocarbon and heavy metal pollutants from the environment involve the trio of physical, chemical, and biological alternatives [19]. However, due to limitations of the physicochemical technologies which are expensive to implement at full scale, not environmentally friendly, their technologies are complex, can lead to destruction of soil texture and characteristics, and do not always result in complete neutralization of pollutants [20,21], bioremediation technology through the mechanism of biodegradation has been recognized to be a valuable alternative for the detoxification and disposal of toxic substance. This is because of their cost-effectiveness, environmental friendliness, simplicity in technology, and conservation of soil texture and characteristics [21-25].

Bioremediation is an ecologically acceptable technology that employs the use of microorganisms to efficiently degrade pollutants. Nevertheless, bioremediation is an emerging technology and still has its limitations or disadvantages which are as follows: partial degradation of pollutants to metabolites that can potentially be more toxic and highly mobile than the parent compound [26,27], it is not always possible to obtain complete pollutant removal because there might be a threshold concentration below which rates of biodegradation are slow or negligible [28], bioavailability of pollutants may decrease as biodegradation proceeds, and thus, recalcitrant compounds may persist [29], retardation or inhibition of the bioremediation process may occur when there are high variations in environmental conditions and concentration of the pollutant, microorganisms may be susceptible to toxins or natural predators in the environment [30], and it requires longer treatment time. The rate of hydrocarbon biodegradation can be limited by many factors which are physicochemical, biological, and environmental such as microorganism type, nutrients, pH, temperature, moisture, oxygen, soil type and properties, and concentration, chemical structure, and type of contaminant [31-33]. The basic principle of bioremediation is superficially to optimize the environmental conditions so that microbial degradation can occur rapidly and as completely as possible [34]. A number of bioremediation strategies have been developed to treat waste and contaminated sites [35]. However, selecting the most appropriate strategy to treat a specific site can be guided by considering three fundamental principles [29]: the amenability of the pollutant to be biologically transformed to less toxic products (biochemistry), the accessibility of the pollutant to microorganism (bioavailability), and the opportunity for optimization of biological activity (bioactivity).

Biostimulation is an effective bioremediation strategy that can improve or accelerate pollutant degradation [36-38] by optimizing conditions such as bioavailability, addition of nutrients, aeration, pH, and temperature control [39]. Biostimulation can be considered as an appropriate remediation technology for SEO removal in soil and requires the evaluation of the environmental parameters involved in the kinetics of the *in situ* process. Three of these parameters are bioavailability, nutrients, and aeration which can be improved in bioremediation systems by the use of surfactants and inorganic/organic nutrients. Biostimulation in ecological systems where different physical and chemical factors were controlled has been well reported in the literature [40,41]; however, only very few works have been reported in the literature on the bioremediation of soil contaminated with used motor oil through the use of biostimulation [42,43]. Abdulsalam et al. [43] focused on the use of NPK fertilizer (biostimulation) and mixture of *Pseudomonas aeruginosa* and *Bacillus subtilis* (bioaugmentation), while Abioye et al. [42] studied the enhanced bioremediation of used motor oil using organic wastes that comprised banana skin, brewery spent grain, and spent mushroom compost. Nevertheless, in developing countries, fertilizers are not sufficient for agriculture, let alone for cleaning oil spills. It therefore necessitates the search for cheaper and environmentally friendly options of enhancing petroleum hydrocarbon degradation. One of such option is the use of pig manure which acts as both bulking agents and bacterial biomass suppliers. The addition of organic waste materials such as pig manure to the soil facilitates aeration through small pores and increases the water-holding capacity of the soil, thus enhancing bioremediation [44,45]. The potential of this manure in enhancing the cleanup of soil contaminated with petroleum hydrocarbons has been investigated by very few workers [46,47]; however, there is a lack of information of its use in accelerating the degradation of petroleum hydrocarbon and reduction in heavy metal components of SEO. The information on *ex situ* uptake of heavy metal component of SEO (for example chromium) contaminated soil is rarely reported [5]. Hence, further works are still needed to fully establish the potential of pig manure in the enhancement of soil bioremediation. Knaebel et al. [48] had observed that microbial mineralization of organic chemicals added to soil is controlled by the interactions between the chemicals present in the soil. While Okolo and Amadi [49] had also observed that there are conflicting results obtained from the addition of biostimulating or amendment agents in soil treatment and that this may probably be due to the heterogeneity of soil and oil sample as well as the variability and possible interactions between the soil amendments and the natural soil constituents. As a result of this, there is a dearth of information in the literature on the interaction effects of animal manure (organic nutrient), inorganic fertilizer (inorganic nutrient), and surfactant in enhancing bioremediation of soil contaminated with mixtures of petroleum hydrocarbons and heavy metals such as SEO. Response surface methodology is a statistical experimental design that provides information about direct effects, pair-wise interaction effects, and curvilinear variable effects. This statistical technique has been successfully applied in many fields [50-53].

Therefore, the purpose of this work is to study through the use of response surface methodology (RSM) the bioremediation of a tropical agricultural soil artificially contaminated with SEO using NPK fertilizer (inorganic nutrient), Tween 80 (surfactant), and pig manure (organic nutrient) as biostimulating agents. Also, in this study, 2^3 full-factorial Box-Behnken designs of experiment were implemented in order to evaluate the interaction effects of the biostimulating agents (variables) on the biodegradation rate of petroleum hydrocarbons and reduction (or uptake) of hexavalent chromium components of SEO as well as to optimize the variables for maximum SEO removal.

Methods

Soil sample and characterization

An un-impacted soil samples from Ladoke Akintola University of Technology (LAUTECH) Agricultural Farms, Ogbomoso were collected from the surface layer of the vadose zone 15 to 30 cm below the land surface. The soil samples were air-dried, homogenized, passed through a 2-mm (pore size) sieve, and stored in a polyethylene bag at room temperature. Soil samples were characterized for their physicochemical and microbial parameters according to standard methods. Total organic carbon and total nitrogen of soil were determined using Walkley-Black and Macro-Kjeldahl methods, respectively [54,55]. Soil pH was determined using pH meter fitted with a combined glass pH and reference electrode [55]. Soil moisture content was determined by evaporation on Whatman filter paper No. 1 at 103°C to 105°C in an electrical oven. Available phosphorus was determined using Bray No. 1 method [54,55].

The total hydrocarbon-degrading bacteria (THDB) population was determined by the pour plate method [56]. In this method, 10 g of soil was transferred into 250-ml Erlenmeyer flasks containing 100 ml of mineral salt medium (MSM). The MSM [57] consists of 2.13 g Na_2HPO_4, 0.5 g NH_4Cl, 0.2 g $MgSO_4{\cdot}7H_2O$, 1.3 g KH_2PO_4, 0.01 g $FeSO_4{\cdot}7H_2O$, 0.1 g NaCl, 20 g agar, and 1% spent engine oil in 1-L de-ionized water, pH 7.4. The flasks were shaken for 3 days at 180 rpm and 28 ± 2°C. After the third day, 10 ml of the supernatant was transferred to 125 ml of fresh medium. The incubation transfer to the fresh medium was repeated three times. Thereafter, serially diluted samples (0.1 ml) were plated on nutrient agar medium (Oxoid) supplemented with 50 μg/ml nystatin to suppress the growth of fungi. The oil agar plates were incubated at 30°C for 5 days, and the colonies were counted and randomly picked.Pure isolates were obtained by repeated sub-culturing on nutrient agar (Oxoid). The bacterial isolates were characterized using microscopic techniques and biochemical tests. The identities of the isolates were determined by comparing their characteristics with those of known taxonomy as described in Bergey's manual of determinative bacteriology [58].

The determined soil parameter values are as follows: moisture content $5.95 \pm 0.05\%$, total nitrogen $0.25 \pm 0.04\%$, available phosphorus $0.12 \pm 0.02\%$, potassium $0.31 \pm 0.05\%$, total organic carbon $1.21 \pm 0.03\%$, pH 5.9 ± 0.2, and THDB3.7×10^5cfug^{-1}. The hydrocarbon-degrading bacterial types isolated from the soil samples were mainly *Acinetobacter*, *Bacillus*, and *Pseudomonas* species. The soil characterization showed that the soil did not fulfill the nutrient (NPK) requirements for an efficient biodegradation process. Therefore, these elements were added in the form of NPK inorganic fertilizer (20:10:10) to provide the proper nutrients required for the biodegradation process.

Chemicals

The spent engine oil was obtained from an automobile mechanic workshop located in Ogbomoso, Nigeria. This study employed a biodegradable nonionic surfactant Tween 80 manufactured by Sigma-Aldrich, St. Louis, MO, USA, which has an average molecular weight of 1,310 and a critical micelle concentration value of 15 mg/l. NPK fertilizer (20:10:10) was purchased from an agro-chemical store in Ogbomoso, Nigeria. Hexane solvent (BDH Chemicals Ltd., Poole, England) used for extraction of oil from soil was bought from a chemical store in Lagos, Nigeria.

Microcosm preparation and bioremediation experimentation

To optimize the range of experimentation for 2^3 full-factorial Box-Behnken design, the following experiments were performed in earthen pots (used as bioreactors) maintained at room temperature. Soil samples (200 g) were placed in earthen pots (microcosm) and were artificially contaminated with SEO to a level of 10% *w*/*w*. The SEO-contaminated soil in each microcosm was amended with different amounts of NPK fertilizer (2 to 10 g), Tween 80 (5 to 25 mg/L), and pig manure (20 to 100 g), respectively. Soil used as control was not amended with any nutrient or surfactant. In total, 16 microcosms were settled and incubated for 42 days. All microcosms were mixed manually once per week to enhance oxygenation and kept moist during the 42-day experimental period. Samples were withdrawn at intervals of 1 week for residual total petroleum hydrocarbon (TPH) and chromium analysis.

Experimental design and data analysis

The Box-Behnken factorial experimental design employed had three independent variables *viz.*, NPK (20:10:10) fertilizer, Tween 80 (surfactant), and pig manure (Table 1). Each of the independent amendment variables was studied at three levels (1, 0, and +1), with 17 experimental runs (Table 2). The levels were selected based on the above preliminary study results. The variables optimized were NPK fertilizer in the range of 2 to 6 g; Tween 80, 5to 15 mg/l; and pig manure, 20 to 60 g, respectively. Change in TPH (i.e., percentageof TPH reduction) and change in chromium (Cr (VI)) were considered as experimental responses. Efficiency of SEO removal was assessed after 42 days. Table 3 shows the coded and actual values of factors and levels used in the experimental design. SEO-contaminated soil without biostimulation was also assayed as a control. The statistical software Design Expert6.0.8, (Stat-Ease Inc., Minneapolis, MN, USA) was used to evaluate the analysis of variance ($P < 0.05$) to determine the significance of each term in the fitted equations and to estimate the goodness of fit in each case.

Table 1 Experimental range and the levels of the variables

Dependent variable	Low level	Medium level	High level
	(−1)	(0)	(+1)
NPK fertilizer (*A*), g	2.0	4.0	6.0
Tween 80 (*B*), mg/l	5.0	10	15
Pig manure (*C*), g	20	40	60

Table 2 Coded and uncoded Box-Behnken design for the three independent variables

Run	NPK fertilizer (*A*)		Tween 80 (*B*)		Pig manure (*C*)	
	Code	Value (g)	Code	Value (mg/l)	Code	Value (g)
1	−1	2	−1	5	0	40
2	+1	6	−1	5	0	40
3	−1	2	+1	15	0	40
4	+1	6	+1	15	0	40
5	−1	2	0	10	−1	20
6	+1	6	0	10	−1	20
7	−1	2	0	10	+1	60
8	+1	6	0	10	+1	60
9	0	4	−1	5	−1	20
10	0	4	+1	15	−1	20
11	0	4	−1	5	+1	60
12	0	4	+1	15	+1	60
13	0	4	0	10	0	40
14	0	4	0	10	0	40
15	0	4	0	10	0	40
16	0	4	0	10	0	40
17	0	4	0	10	0	40
18 (control)	-	-	-	-	-	-

Estimation of total petroleum hydrocarbon

The TPH in the treatments was determined spectrophotometrically following the methods of Mishra et al. [13] and Vu-Duc et al. [59] for 30 min. Thereafter, the mixtures were allowed to settle, and extracts were decanted into a volumetric flask and plug. The above procedure was repeated two times using 20 ml of hexane each time. The total extracts were combined and was then filtered using a Whatman filter paper, and the filtrate was dried at room temperature by evaporation of the hexane solvent under a gentle nitrogen stream in a fume hood. After evaporation of the solvent, the extracted TPH in the filtrate (liquid phase extract) was detected spectrophotometrically at an absorbance of 400 nm using visible range spectrophotometer (model 6100, PYE UNICAM Ltd., England, UK). The TPH concentration was estimated from the standard curve of

Table 3 Experimental design and results for bioremediation of spent engine oil

Run	Percentage of TPH reduction		Percentage of Cr (VI) reduction	
	Observed value	Predicted value	Observed value	Predicted value
1	43.92	44.10	31.88	32.01
2	45.25	45.40	32.65	33.11
3	44.65	44.50	32.22	31.76
4	47.84	47.66	35.44	35.31
5	45.38	45.49	34.25	34.41
6	43.68	43.82	32.93	32.75
7	48.94	48.80	36.12	36.30
8	55.04	54.93	42.76	42.60
9	50.80	50.51	40.55	40.27
10	49.29	49.33	39.62	39.92
11	55.25	55.21	45.12	44.82
12	58.78	59.07	46.83	47.11
13	66.89	66.65	52.65	52.46
14	66.51	66.65	52.40	52.46
15	66.92	66.65	52.25	52.46
16	66.15	66.65	52.11	52.46
17	66.78	66.65	52.88	52.46
18 (control)	28.80	-	10.4	-

absorbance against concentration derived from fresh SEO that has been diluted with hexane. Values of TPH concentration obtained were multiplied by a dilution factor to give the actual concentration.

Extraction and analysis of hexavalent chromium (Cr (VI)) and total chromium

For the extraction of Cr (VI) and total chromium from soil, an alkaline digestion method and nitric acid/sulfuric acid digestion method as per standard methods [55] were used, respectively. The hexavalent chromium was measured colorimetrically at 540 nm by reaction with diphenylcarbazide in acidic conditions [55]. In the case of Cr (III), potassium permanganate was used to oxidize Cr (III) to Cr (VI). Cr (III) concentration was determined by taking the difference between total chromium concentration and Cr (VI) concentration.

Results and discussion

Natural bioattenuation and enhanced bioremediation

After performing 17 runs of the Box-Behnken design (BBD) and one control, the results of the statistical experiment were analyzed with regard to the coded design matrix. The regression equation shows that the SEO degradation rate was an experimental function of test variables in coded units. Table 3 shows that on day 42 (6th week), SEO content had decreased in all the soil microcosms. In control, natural biodegradation (natural bioattenuation) removed 28.8% of petroleum hydrocarbons and 10.4% of Cr (VI), respectively. The removal of 10.4% hexavalent chromium from the SEO-contaminated soil indicated that the intrinsic *Acinetobacter, Bacillus*, and *Pseudomonas* species in the soil have great potentials in accumulating or removing chromium heavy metal associated with SEO. Bader et al. [60] studied the potential of aerobic reduction of Cr (VI) by an indigenous soil microbial community and found that Cr (VI) in the soil was reduced by 33% within 21 days and suggested that Cr (VI)-reducing microbial populations is widespread in soil. Moreover, the respective reduction in petroleum hydrocarbon (TPH) and Cr (VI) content of soil microcosms containing amendments was much higher (Table 3) in the same period. This result indicated that the addition of biostimulant increased the rate of TPH biodegradation and hexavalent chromium uptake, respectively. This is in agreement with the report of Odokuma and Akponah [61] who observed an increase in crude oil biodegradation and heavy metal uptake from fresh and brackish waters impacted with crude oil when NPK fertilizer and poultry litters were used as nutrient supplements. A considerable decrease in SEO reduction was observed in runs 13 to 17, with a relatively high amount of NPK fertilizer, surfactant (Tween 80), and pig manure; the residual TPH and Cr (VI) reached 33.08% to 33.85% and 47.12% to 47.89%, respectively of the initial SEO concentration. Agarry et al. [62] in their study of enhanced bioremediation of soil artificially contaminated with kerosene have shown that natural attenuation removed 38.5% of kerosene oil after 42-day incubation, and when the soil was amended with NPK fertilizer, hydrogen peroxide, and Tween 80, 77.89% of the kerosene oil was removed. In comparing this study with the results of Agarry et al. [62], it could be seen that the percentage of hydrocarbon removal was lower in both the natural attenuation and biostimulation treatments. This may be due to the presence of heavy metals in the SEO which tends to inhibit the metabolic activities of the autochthonous hydrocarbon-degrading bacteria in the soil [63,64]. The comparison of SEO-enhanced biodegradation and natural bioattenuation for each run is shown in Figure 1.

At 10% (*w*/*w*) SEO concentrations, runs number 1 and 3, and 11 and 12 had the same remediation condition with different concentrations of Tween 80 (surfactant); the results show that the addition of surfactant can enhance SEO degradation. Similar observations have been reported for the use of nonionic surfactant as well as biosurfactant for the remediation of environment contaminated with petroleum hydrocarbons [62,65-69]. The

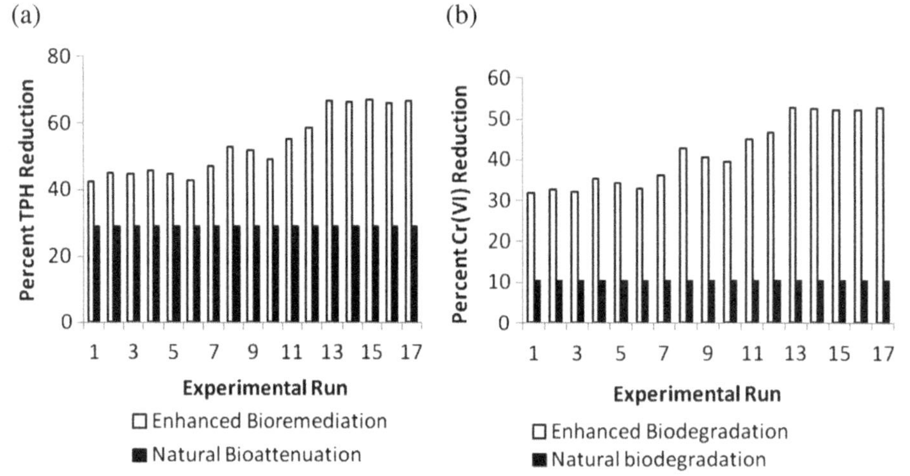

Figure 1 Comparison of enhanced and natural biodegradations of SEO for (a) TPH and (b) Cr (VI) reductions.

observed enhancement as a result of the added surfactant may be due to the fact that surfactant might have played a role in emulsifying the SEO which made it readily available for degradation by the intrinsic hydrocarbon-degrading bacterial populations. However, one of the major factors that limit hydrocarbon biodegradation is their low availability to microbial cells [70].

Effect of different concentrations of pig manure were investigated at the same condition of NPK and Tween 80 (run numbers 6 and 8, and 10 and 12), and the findings demonstrated that the addition of pig manure can enhance the biodegradation process of SEO contaminant in soil. Yakubu [46], Agarry et al. [47], and Agarry and Owabor [71] in their respective work have demonstrated the positive effect of pig manure on enhanced biodegradation of petroleum hydrocarbons. Abioye et al. [42] have also reported the positive effect of organic waste (brewery spent grain, spent mushroom compost, and banana skin) on enhanced biodegradation of used motor oil. Relatively, run numbers 3 and 4 and run numbers 7 and 8 had the same condition with different concentrations of NPK; the results show that extra addition of NPK fertilizer (nutrient) improved SEO biodegradation. A similar observation has been reported for SEO biodegradation [43] as well as for kerosene degradation [62].

In addition, efficiency of bioremediation is a function of the microbial viability in the natural environment [72]. Factors, such as nitrogen, phosphorus, and microorganism presence, have been reported to affect bioremediation, for example, limiting nitrogen is reported to enhance bioremediation [73,74]. Abdulsalam et al. [43] and Abioye et al. [42] showed that natural attenuation removed 50% of oil and grease and 68% of TPH in SEO after 70- and 84-day incubations, respectively. When the soil was supplemented with nutrients (nitrogen and phosphorus), 66% and 92% to 95% of the SEO was respectively removed. The results suggest that high dose of nutrient amendment can accelerate the initial SEO degradation rate and may shorten the period to clean up contaminated environments. The accelerating effect of amendment is stronger when nutrient availability is a limiting factor in the biodegradation of oil [51].

Second-order polynomial regression model and statistical analysis

The experimental data were fitted to a second-order polynomial regression model (Equation 1) containing three linear, three quadratic, and three interaction terms [75] using the same experimental design software to derive the equation for SEO removal from contaminated soil.

$$Y = \beta_0 + \beta_1 A + \beta_2 B + \beta_3 C + \beta_{11} A^2 + \beta_{22} B^2 + \beta_{33} C^2 + \beta_{12} AB + \beta_{13} AC + \beta_{23} BC, \quad (1)$$

where β_0 is the value of the fixed response at the center point of the design; β_1,β_2, and β_3, the linear coefficients; β_{11},β_{22}, and β_{33}, the quadratic coefficients; β_{12}, β_{13}, and β_{23}, the interaction effect coefficient regression terms, respectively; and A, B, and C, the levels of independent amendment variables. The significance of each coefficient in the equation was determined by F test and P values. F test indicated that all the factors and interactions considered in the experimental design are statistically significant ($P < 0.05$) at the 95% confidence level. The regression equation obtained after analysis of variance gives the level of total petroleum hydrocarbon reduction and Cr (VI) reduction (SEO removal) as a function of the different amendment variables: NPK, Tween 80, and pig manure. All terms regardless of their significance are included in Equations 2 and 3:

$$Y_{\mathrm{TPH}} = 66.65 + 1.11A + 0.67B + 3.61C - 13.25A^2 - 7.98B^2 - 5.14C^2 + 0.47AB + 1.95AC + 1.26BC \quad (2)$$

$$Y_{\mathrm{Cr(VI)}} = 52.46 + 1.16A + 0.49B + 2.94C - 12.96A^2 - 6.45B^2 - 2.98C^2 + 0.61AB + 1.99AC + 0.66BC, \quad (3)$$

where A is NPK concentration; B, Tween concentration; and C, pig manure concentration.

To test the fit of the model, the regression equation and determination coefficient (R^2) were evaluated (Table 4). The model F value of 1,431.99 and 656.30 implies that the model is significant for TPH and Cr (VI), respectively. There is only a 0.01% chance that a model F value could occur due to noise alone for TPH and Cr (VI), respectively. The low probability value (<0.0001) for both TPH and Cr (VI) indicates that the model is significant. The value of the determination coefficient ($R^2 = 0.9995$ for TPH and 0.9988 for Cr (VI)) being a measure of goodness of fit to the model indicates a high degree of correlation between the observed value and predicted values. The determination coefficient ($R = 0.9997$ for TPH and 0.9994 for Cr (VI)) suggests that more than 99.97% and 99.94% of the variance are attributable to the variables and indicated a high significance of the model for TPH and Cr (VI), respectively. Thus, 0.03% and 0.06% of the total variance cannot be explained by the model for TPH and Cr (VI), respectively. The fitted model is considered adequate if the F test is significant ($P < 0.05$). The analysis of variance (ANOVA) quadratic regression model demonstrated that the model was highly significant for both TPH and Cr (VI) as was evident from the very low probability ($P < 0.0001$) of the F test and insignificant result from the lack-of-fit model (Table 4). The lack-of-fit test is performed by comparing the variability of the current residual model to the variability between observations at replicate settings of the factors.

The lack of fit is designed to determine whether the selected model is adequate to describe the observed data or whether a more complicated model should be used. The predicted R^2 values of 0.9957 and 0.9861 for TPH and Cr (VI), respectively, are correspondently in reasonable agreement with the adjusted R^2 values of 0.9988 and 0.9973. Adequate precision measures the signal-to-noise ratio. A ratio >4 is desirable. The ratio of 96.000 and 62.746 obtained for TPH and Cr (VI), respectively, in this research indicates an adequate signal. This model can be used to navigate the design space. The coefficient of variation (CV) as the ratio of the standard error of estimate to the mean value of the observed response is a measure of reproducibility of the model; generally, a model can be considered reasonably reproducible if its CV is not greater than 10%. Hence, the low variation coefficient value (CV = 0.61% for TPH and 1.03% for Cr (VI)) obtained indicates a high precision and reliability of the experiments.

Table 4 ANOVA for the quadratic response surface model fitting to the bioremediation data of spent engine oil

Source	SS	DF	MS	*F* value	Probability > *F*(*P* value)
TPH					
Residual model	1,366.679	9	151.85	1,417.10	<0.0001
Lack of fit	0.33	3	0.11	1.07	0.4573
Pure error	0.42	4	0.10		
Total correlation	1,367.42	16			$R^2 = 0.9995$
					Adjusted $R^2 = 0.9987$
					Predicted $R^2 = 0.9956$
					Adequate precision = 90.943
Cr (VI)					
Residual model	1,092.66	9	121.41	656.30	<0.0001
Lack of fit	0.91	3	0.30	3.18	0.1466
Pure error	0.38	4	0.096		
Total correlation	1,093.95	16			$R^2 = 0.9988$
					Adjusted $R^2 = 0.9973$
					Predicted $R^2 = 0.9861$
					Adequate precision = 62.746

SS sum of square, *DF* degrees of freedom, *MS* mean square.

The coefficient of the model (parameter estimation) and the corresponding *P* values are presented in Table 5. The significance of regression coefficients was considered at a significance level of 95%. The *P* values of the regression coefficients suggest that among the independent test variables, linear, quadratic, and interaction effects of NPK fertilizer (inorganic nutrient), Tween 80 (surfactant), and pig manure (organic nutrient) are highly significant. In this study, A, B, C, A^2, B^2, C^2, AB, AC, and BC are significant model terms for TPH and Cr (VI), respectively. Thus, statistical analysis of all the experimental data showed that NPK fertilizer, Tween 80, and pig manure had a significant effect on TPH and Cr (VI) reductions in this study.

Moreover, it is observed that pig manure and NPK fertilizer concentrations exerted more pronounced linear effect (higher coefficient values) on TPH and Cr (VI) reductions, respectively, that is, TPH and Cr (VI) reductions were mostly and positively influenced by the addition of pig manure followed by NPK fertilizer and Tween 80, respectively. A higher positive linear effect of NPK fertilizer over Tween 80 in the biostimulation treatment of soil contaminated with kerosene had been reported in the previous work of Agarry et al. [62]. Furthermore, the pig manure exerted the highest positive linear effect (due to higher coefficient) than the interaction effect between the amendment variables. The strong influence of animal-derived organic manure on petroleum hydrocarbon degradation has been shown in the previous works of Agarry et al. [47] and Agarry and Owabor [71]. The quadratic effect of the independent variables on the rate of TPH and Cr (VI) reductions was negative.

Influence of variable interaction on spent engine oil removal

Table 5 showed that SEO removal (i.e., TPH and Cr (VI) reductions) was influenced positively by the interaction of NPK fertilizer (A) and Tween 80 (B); NPK fertilizer (A) and pig manure (C); and Tween 80 (B) and pig manure (C), respectively. The interaction effect of NPK fertilizer (inorganic nutrient) and pig manure (organic nutrient) exerted more pronounced positive influence (due to higher coefficient) on SEO removal (i.e., TPH and Cr (VI) reductions) than the linear effect of NPK fertilizer. The interaction effect of NPK fertilizer and Tween 80 exerted less positive influence (due to lower coefficient) on SEO removal than the linear effect of NPK fertilizer and Tween 80, respectively. A similar observation has been reported [62]. On the other hand, Thavasi et al. [69] have reported that the biodegradation

Table 5 Coefficient of the model for bioremediation of spent engine oil

Factor	Coefficient estimate	Standard error	*F* value	*P* value	Remarks
TPH					
β_0	66.65	0.15	1,417.10	<0.0001	Significant
β_1	1.11	0.12	92.82	<0.0001	Significant
β_2	0.67	0.12	33.26	0.0007	Significant
β_3	3.61	0.12	971.59	<0.0001	Significant
β_{11}	−13.25	0.16	6,900.98	<0.0001	Significant
β_{22}	−7.98	0.16	2,503.76	<0.0001	Significant
β_{33}	−5.14	0.16	1,037.10	<0.0001	Significant
β_{12}	0.47	0.16	8.07	0.0250	Significant
β_{13}	1.95	0.16	141.94	<0.0001	Significant
β_{23}	1.26	0.16	59.26	0.0001	Significant
Cr (VI)					
β_0	52.46	0.19	656.30	<0.0001	Significant
β_1	1.16	0.15	58.57	0.0001	Significant
β_2	0.49	0.15	10.33	0.0148	Significant
β_3	2.94	0.15	372.53	<0.0001	Significant
β_{11}	−12.96	0.21	3,824.64	<0.0001	Significant
β_{22}	−6.45	0.21	946.27	<0.0001	Significant
β_{33}	−2.98	0.21	202.16	<0.0001	Significant
β_{12}	0.61	0.22	8.11	0.0248	Significant
β_{13}	1.99	0.22	85.63	<0.0001	Significant
β_{23}	0.66	0.22	9.42	0.0181	Significant

of crude oil in soil was more enhanced by the addition of biosurfactant + fertilizer than the addition of biosurfactant alone, but the difference between them was minimal. Thus, they concluded that biosurfactants alone are capable of promoting biodegradation process without using fertilizers. Furthermore, the interaction effect of Tween 80 and pig manure exerted higher positive influence on SEO removal than the linear effect of Tween 80. Generally, the interaction effects of NPK fertilizer and pig manure as well as Tween 80 and pig manure exerted more pronounced positive effects on SEO removal than the individual linear effects of NPK fertilizer and Tween 80, respectively. Okolo and Amadi [49] have reported the positive influence of the mixture of cow dung + poultry manure + Corexit (surfactant) on the biodegradation of crude oil.

The graphical representation of the response shown in Figures 2,3,4 helped to visualize the effect of NPK fertilizer (A), Tween 80 (B), and pig manure (C) on percentage of TPH and Cr (VI) reductions (SEO removal). The interaction effect of NPK fertilizer and Tween 80 on SEO biodegradation is illustrated in Figure 2. The plot shows that higher rate of TPH reduction was attained with increase in NPK fertilizer (nutrient) and Tween 80 (surfactant) concentrations (Figure 2a). The maximum TPH degradation yield of 66.68% was obtained with 10 mg/l of Tween 80 surfactant and 4 g/l (or 0.02 μg/kg) of NPK fertilizer at a fixed pig manure concentration of 40 g/l (0.2 μg/kg). A similar effect was observed for Cr (VI) percentage reduction (Figure 2b), where it can be seen that increasing NPK fertilizer and Tween 80 concentrations in the soil also caused an increase in Cr (VI) reduction. This may be due to better bioavailability of substrate for the intrinsic microorganisms. However, availability of hydrocarbon and heavy metal-utilizing microorganism is a key issue in SEO biodegradation [76]. The positive influence of NPK fertilizer and Tween 80 interaction on petroleum hydrocarbon degradation in SEO is higher (due to high coefficient value of 0.47) though with lower maximum TPH yield (66.85%) as compared to the same interaction effect of NPK fertilizer and Tween 80 on petroleum hydrocarbon in kerosene (coefficient value of 0.27) with maximum degradation yield of 74.95% at a fixed concentration of hydrogen peroxide as presented in the previous work of Agarry et al. [62] where they used Box-Behnken design of experiment to study the effect of NPK fertilizer, Tween 80, and hydrogen peroxide as biostimulating agents on the degradation of kerosene in soil. The lower maximum TPH degradation value obtained in this study may probably be due to the presence of heavy metals in SEO which tends to inhibit the metabolic activities of the autochthonous hydrocarbon-degrading bacteria in the soil [63,64] as well as due to the heterogeneity of soil and SEO sample [49]. Metals may interfere with the biodegradation of pollutants through either interaction with enzymes directly involved in biodegradation or interaction with enzymes involved in general metabolism. In either case, inhibition is mediated by the ionic form of the metal [64,77].

Figure 3 shows the 3D response surface plot of the interaction effect between NPK fertilizer and pig manure concentrations. This three dimensional plot explains that both NPK fertilizer and pig manure have individual

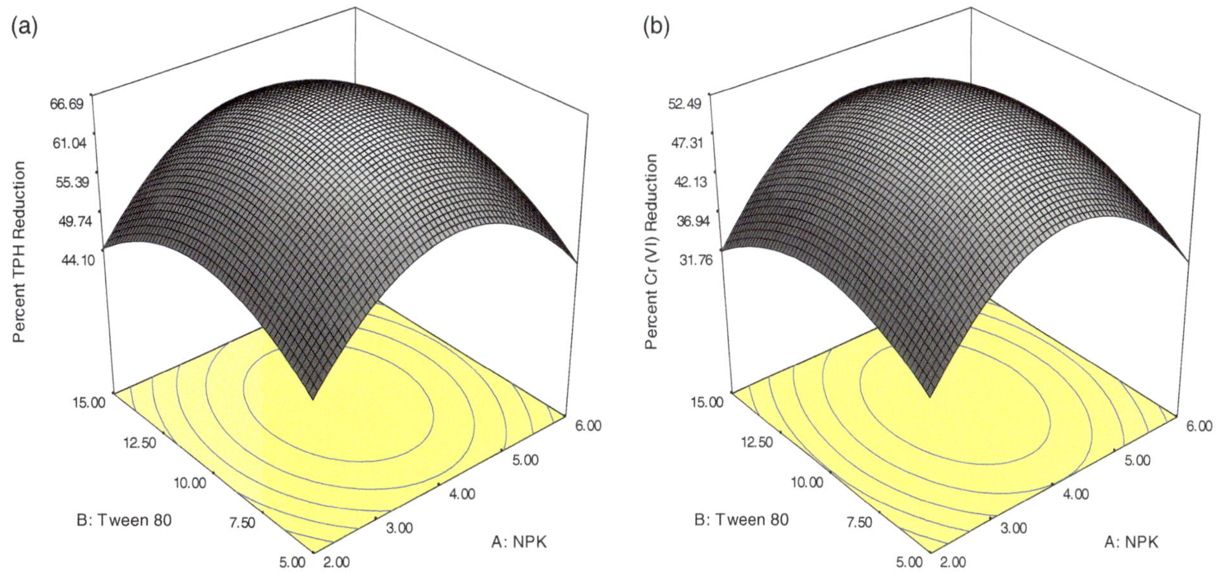

Figure 2 Response surface plot. Indicating the interaction effect of NPK fertilizer and Tween 80 on SEO biodegradation for (**a**) TPH and (**b**) Cr (VI) reductions.

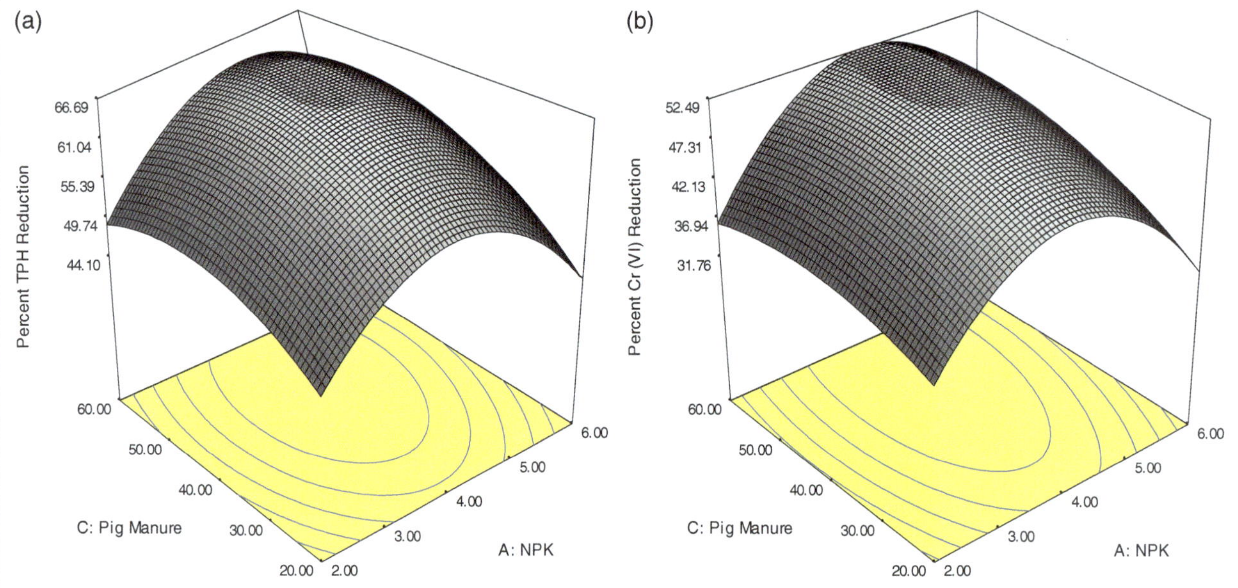

Figure 3 Response surface plot. Indicating the interaction effect of NPK fertilizer and pig manure on SEO biodegradation for (**a**) TPH and (**b**) Cr (VI) reductions.

impact on SEO removal as the individual coefficient of both NPK fertilizer and pig manure is positive, and their interaction effect is positive. However, the impact of pig manure is more than the NPK fertilizer concentration as the individual coefficient value is higher for pig manure than for NPK fertilizer. It is seen that a higher percentage of TPH reduction was obtained at a higher pig manure concentration and relatively high amount of NPK fertilizer (Figure 3a). A similar effect was observed for Cr (VI) reduction in the contaminated soil (Figure 3b). The maximum removal yield of 67.34% TPH and 53.27% Cr (VI) reductions were achieved, respectively, with around 40 g (or 200,000 mg/kg) of pig manure and 4 g (or 20,000 mg/kg) of NPK fertilizer. Further increase in the amount of NPK fertilizer (>4.0 g), a significant decrease in bioremediation yields occurred. This suggests that at a fixed concentration of Tween 80, the amount of NPK fertilizer can be decreased and that of the pig manure concentration has to be increased for higher reduction in TPH and Cr (VI) yields. Nevertheless,

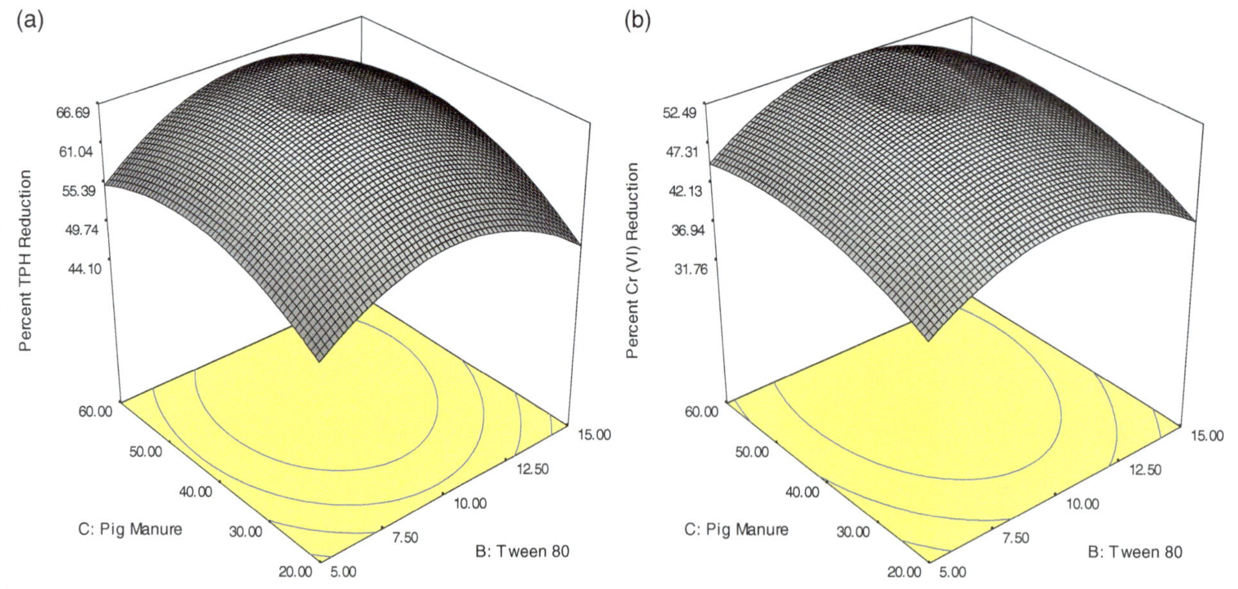

Figure 4 Response surface plot. Indicating interaction effect of Tween 80 and pig manure on SEO biodegradation for (**a**) TPH and (**b**) Cr (VI) reductions.

eutrophication and harmful algal blooms in the aquatic environment may occur due to excessive nutrient concentration [55,78,79]. The positive influence of NPK fertilizer and pig manure interaction on petroleum hydrocarbon degradation in SEO is higher (due to high coefficient value of 1.96) though with lower maximum TPH yield (67.34%) as compared to the interaction effect of NPK fertilizer and hydrogen peroxide on the biodegradation of petroleum hydrocarbon in kerosene (coefficient value of 1.02) with a maximum degradation yield of 75.07% at a fixed concentration of Tween 80 as presented in the previous work of Agarry et al. [62]. Though the maximum degradation yield may be low due to the earlier suggested reasons, however, the comparisons imply that pig manure (animal organic waste material) can perform the purpose of improving aeration process for the soil microorganisms just as the hydrogen peroxide (an oxygen-releasing compound) serves to do.

Finally, Figure 4 shows the 3D response surface plot of the effect of interaction between Tween 80 (surfactant) and pig manure (organic nutrient) concentrations. This plot demonstrates that both Tween 80 and pig manure have positive mutual impact on the biodegradation process. At a fixed concentration of NPK fertilizer, it was observed that increase in Tween 80 and pig manure concentrations resulted in higher TPH and Cr (VI) reductions. The maximum reductions in TPH (67.32%) and Cr (VI) (53.21%) were obtained with 10 mg/l of Tween 80 and 4 g (i.e., 20,000 mg/kg) of NPK fertilizer.

Factor plot

The factor effect function plot (Figure 5) was used to assess the effect of each factor graphically. From the trace plot as shown in Figure 5, it can be seen that each of the three variables used in the present study has its individual effect on SEO (TPH and Cr (VI)) removal by the intrinsic microbial populations in the soil. Gradual increase in NPK fertilizer (*A*), Tween 80 (*B*), and pig manure (*C*) concentrations from low level (coded value −1) to a higher level (coded value +1) resulted in both increase and decrease of TPH (Figure 5a) and Cr (VI) (Figure 5b), respectively. Moreover, it is also to be noted from Figure 5a,b that over the range of −1 (2 g) to +1 (6 g) of NPK fertilizer, the TPH and Cr (VI) degradation changed in a wide range. This was also the case for Tween 80; however, for pig manure, it did not change over a wide range. This clearly indicates that keeping pig manure at the optimum level, a change in NPK fertilizer (nutrient) and Tween 80 (surfactant) will affect the biodegradation process more severely than done otherwise.

Optimization and validation

Numerical optimization technique based on desirability function was carried out to determine the workable optimum conditions for the SEO bioremediation process. In order to provide an ideal case for biodegradation, the goal for NPK fertilizer, Tween 80, and pig manure was set in range based upon the requirements of the oil biodegradation, and TPH and Cr (VI) removal was set to maximum. The predicted optimum (uncoded) values of NPK fertilizer, Tween 80, and pig manure were correspondingly found to be 4.22 g (or 20,100 mg/kg), 10.69 mg/l, and 47.76 g (or 238,800 mg/kg) to achieve 67.20% and 53.20%maximum TPH and Cr (VI) reductions, respectively, while desirability was 1.000 for the experiment (Figure 6). Nevertheless, validation experiment was conducted to determine the optimum SEO removal when the amendment variables were set at the favorable optimum levels established above, through BBD and RSM. Standard deviation and percentage error were investigated for validation of the experiments. Errors between predicted and actual values were calculated according to Equation 4:

$$\text{Error} = \frac{\text{Actual value} - \text{Predicted value}}{\text{Actual value}} \times 100 \qquad (4)$$

In the optimized condition for initial SEO of 10% (*w*/*w*) concentration, 66.47% and 52.40% TPH and Cr (VI) reductions were obtained, respectively. The percentage error between the predicted and actual values was found to be −1.10 and −1.53 for TPH and Cr (VI), respectively. The results clearly indicated that no significant difference was observed.

Conclusions

RSM is a reliable and powerful tool for modeling and optimization of spent engine oil bioremediation processes. The variations in total petroleum hydrocarbon and hexavalent chromium contents of spent engine oil bioremediation pattern with respect to NPK fertilizer, Tween 80, and pig manure contents were observed to be very significant. The results indicated that bioremediation of spent engine oil-contaminated soil with the use of NPK fertilizer, Tween 80, and pig manure as biostimulating agents resulted in the acceleration or improvement of petroleum hydrocarbon and hexavalent chromium reductions. However, the presence of heavy metals in spent engine oil tends to reduce the degradation rate (percentage degradation) of its petroleum hydrocarbon component. Pig manure (organic nutrient) alone or in combination with either NPK fertilizer (inorganic nutrient) or Tween 80 (surfactant) can greatly and significantly enhance the biodegradation of petroleum hydrocarbon and heavy metal uptake or reduction in soil than the addition of either NPK fertilizer alone or Tween 80. Thus, based on the coefficient values of the second-order polynomial regression model, the order of biostimulation effectiveness among the biostimulating

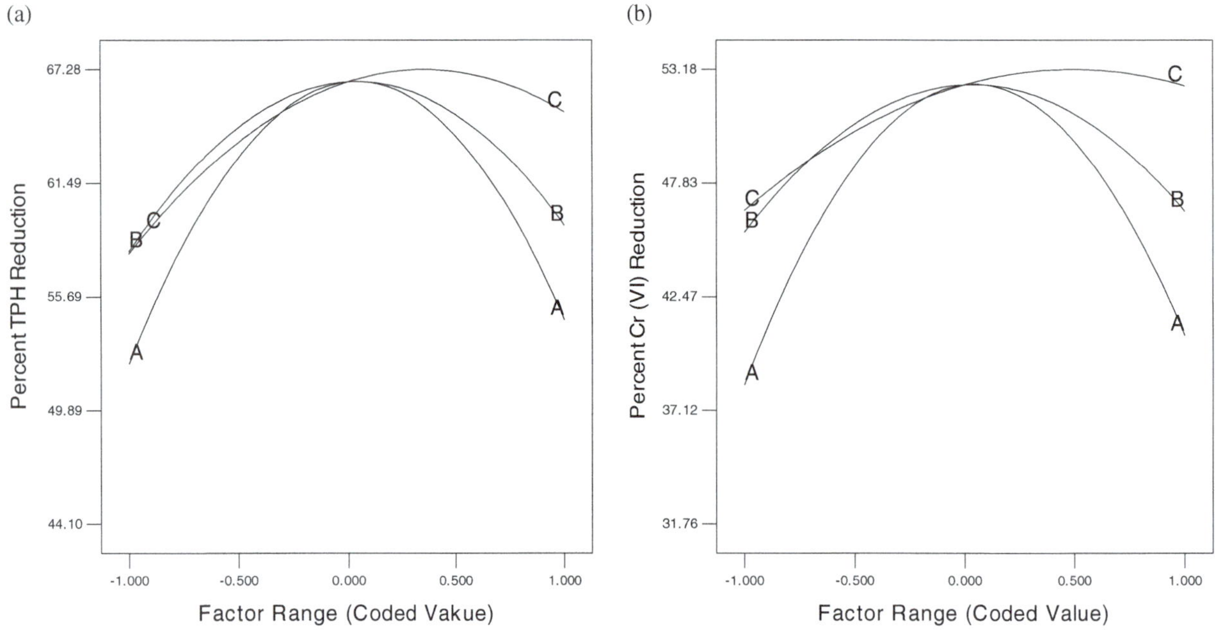

Figure 5 Factor plot representing the individual variable effect on SEO biodegradation. For (**a**) TPH and (**b**) Cr (VI) reductions during spent engine oil bioremediation.

agents and their interactions are as follows: pig manure > NPK fertilizer + pig manure > Tween 80 + pig manure > NPK fertilizer > Tween 80 > NPK + Tween 80. Therefore, the application of these biostimulation strategies can reduce the length of remediation time in the bioremediation of hydrocarbon and heavy metal-contaminated ecological system and subsequently the cost of remediation. The success and efficiency of these biostimulation techniques may vary considerably from one site to another; hence, there is no universal soil treatment regimen for the bioremediation of all petroleum hydrocarbon-contaminated soils. The effectiveness of any soil treatment applied for such purpose has to be evaluated on a case-specific basis. However, two important issues need to be addressed before

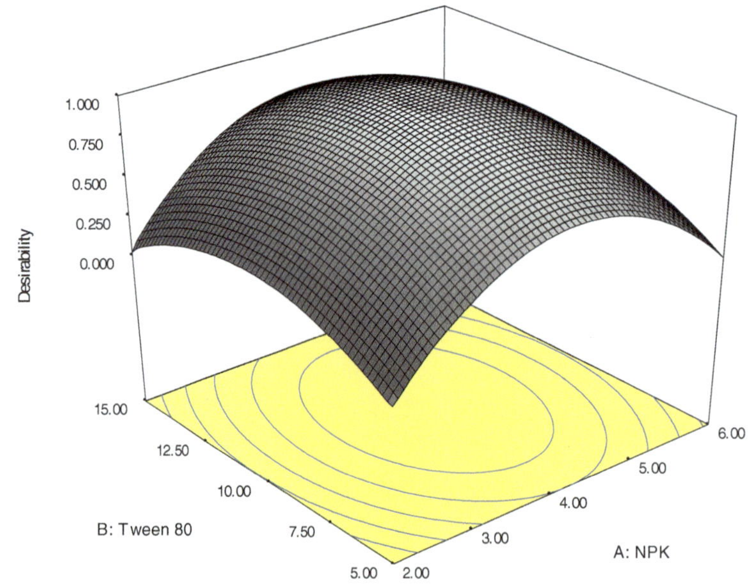

Figure 6 Desirability plot to optimize the biodegradation process of SEO.

taking such technology to the field (1) evaluation of the metabolic potential of the intrinsic microbial species and (2) evaluation of the environment where such bioremediation process is needed for lack of essential nutrients.

Competing interests

The authors declare that they have no competing interests.

Authors' contributions

SEA participated in the design and coordination of the study, interpreted the data, and drafted and corrected the manuscript. OOO participated in the interpretation of the data and corrected the draft manuscript. All authors read and approved the final manuscript.

Authors' information

SEA is a lecturer in the Department of Chemical Engineering, Ladoke Akintola University of Technology, Ogbomoso, Nigeria. His areas of specialization are biochemical engineering and environmental biotechnology. He has published over 20 papers in reputable journals. OOO is a senior lecturer in the Department of Chemical Engineering, Ladoke Akintola University of Technology, Ogbomoso, Nigeria. His areas of specialization are process optimization and process development. He has published over 20 papers in reputable journals.

Acknowledgments

The authors wish to thank Mr. J. Osakwe of Nigerian Institute of Science and Technology, Ibadan, Nigeria and Mr. W. Azeez of Ladoke Akintola University of Technology, Ogbomoso, Nigeria for their technical support. The source of funding for SEA and OOO and for the manuscript preparation is the salary received from the management of Ladoke Akintola University of Technology, Ogbomoso, Nigeria.

References

1. ATSDR (Agency for Toxic Substances and Disease Registry) Toxicology: Profile for used mineral base crankcase oil. Department of Health and Human Services, Public Health Service Press, Atlanta (1997)
2. Wang, J, Jia, CR, Wong, CK, Wong, PK: Characterization of polycyclic aromatic hydrocarbon created in lubricating oils. Water Air Soil Pollut. **120**, 381–396 (2000)
3. Anoliefo, GO, Edegbai, BO: Effect of spent engine oil as a soil contaminant on the growth of two eggplant species *Solanum melongena* (L) and *S. incanum* (L). J. Agric. Forestry Fisheries **1**, 21–25 (2000)
4. Odjegba, VJ, Sadiq, AO: Effect of spent engine oil on the growth parameters, chlorophyll and protein levels of Amaranthus hybridus L. Environmentalist **22**, 23–28 (2002)
5. Ikhajiagbe, B, Anoliefo, GO: Natural attenuation of a 14-month-old waste engine oil polluted soil. J. Soil Sci. Environ. Manage. **2**(7), 184–192 (2011)
6. Osubor, CC, Anoliefo, GO: Inhibitory effect of spent lubrication oil on the growth and respiratory function of Arachis hypogea L. Benin Sci. Dig. **1**, 73–79 (2003)
7. Faboya, OOP: Industrial pollution and waste management. In: Akinjide, O. (ed.) Dimensions of environmental problems in Nigeria, pp. 12-25. Friedrich Ebert Foundation, Washington D.C (1997)
8. Urvashi, T, Rasesh, P: Reduction of chromate by cell-free extract of Brucella sp. isolated from Cr VI contaminated sites. Bioresour. Technol. **98**, 1541–1547 (2007)
9. Poornima, K, Karthik, L, Swadhini, SP, Mythili, S, Sathiavelu, A: Degradation of chromium by using a novel strains of *Pseudomonas* species. J. Microbial Biochem. Technol. **2**, 095–099 (2010)
10. Schroeder, DC, Lee, GF: Potential transformations of chromium in natural waters. Water Air Soil Pollut. **4**, 355–365 (1975)
11. De Flora, S: Threshold mechanisms and site specificity in chromium(VI) carcinogenesis. J. Carcinog. **21**, 533–541 (2000)
12. Propst, TL, Lochmiller, RL, Qualis, CW, McBee Jr., K: In situ (mesocosm) assessment of immune-toxicity risks to small mammals inhabiting petrochemical waste sites. Chemosphere **38**, 1049–1067 (1999)
13. Mishra, SJ, Jyot, RC, Kuhad, BL: Evaluation of inoculum addition to stimulate in situ Bioremediation of oily-sludge-contaminated soil. Appl. Environ. Microbiol. **67**(4), 1675–1681 (2001)
14. Adriano, DC, Chlopecka, A, Kaplan, KI: Role of soil chemistry in soil and ecosystem conservation, pp. 361-386. Soil Science Society of America Special Publication, Madison (1998)
15. Sathiya-Moorthi, P, Deecaraman, M, Kalaichelvan, PT: Bioremediation of automobile oil effluent by *Pseudomonas sp*. Advanced Biotech. **31**, 34–37 (2008)
16. Atuanya, EJ: Effect of oil pollution on physical and chemical properties of soil: a case study of waste oil contaminated delta soil in Bendel State Nigeria. J. Appl. Sci. **55**, 155–176 (1987)
17. Achuba, FI, Peretiemo-Clarke, BO: Effect of spent engine oil on soil catalase and dehydrogenase activities. Int. Agrophysics **22**, 1–4 (2008)
18. Odjegba, VJ, Atebe, JO: The effect of used engine oil on carbohydrate, mineral content and nitrate reductase activity of leafy vegetable (Amaranthus hybridus L.). J. Appl. Sci. Environ. Manage. **11**(2), 191–196 (2007)
19. Okoh, AI: Biodegradation alternative in the cleanup of petroleum hydrocarbon pollutants. Biotechnol. Mol. Biol. Rev. **1**, 38–50 (2006)
20. Zhang, H, Dang, Z, Zheng, LC, Yi, XY: Remediation of soil co-contaminated with pyrene and cadmium by growing maize (Zea mays L.). Int. J. Environ. Sci. Tech. **6**(1), 69–76 (2009)
21. Yerushalmi, L, Rocheleau, S, Cimpoia, R, Sarrazin, M, Sunahara, G, Peisajovich, A, Leclair, G, Guiot, RS: Enhanced bioremediation of petroleum hydrocarbons in contaminated soil. Bioremed. J. **7**(1), 37–51 (2003)
22. Vidali, M: Bioremediation: an overview. J. Appl. Chem. **73**(7), 1163–1172 (2001)
23. Mandri, T, Lin, J: Isolation and characterization of engine oil degrading indigenous microorganisms in Kwazulu-Natal, South Africa. Afri. J. Biotechnol. **6**(1), 23–27 (2007)
24. Adams, RH, Guzmán-Osorio, FJ: Evaluation of land farming and chemico-biological stabilization for treatment of heavily contaminated sediments in a tropical environment. Int. J. Environ. Sci. Tech. **5**(2), 169–178 (2008)
25. Fouépé, TA, Kengni, TL, GurunadhaRao, WS, Ndam, NJR: Transfer of moisture through the unsaturated zone in the tropical forest using the neutron probe. Int. J. Environ. Sci. Tech. **6**(3), 379–388 (2009)
26. Sutherland, JB: Detoxification of polycyclic aromatic hydrocarbons by fungi. J. Ind. Microbiol. **9**, 53–62 (1992)
27. Bradley, PM: Microbial degradation of chloroethanes in groundwater systems. Hydrogeology J. **8**, 104–111 (2000)
28. Allard, AS, Neilson, AH: Bioremediation of organic waste sites: a critical review of microbiological aspects. Int. Biodeter. Biodegrad. **39**, 253–285 (1997)
29. Kim, S-J: Bioaugmentation for the remediation of pesticide contaminated soil with microorganisms directly enriched in soil or compost. The Ohio State University, (2003). Dissertation
30. Singleton, I: Microbial metabolism of Xenobiotics: fundamental and applied research. J. Chem. Technol. Biotechnol. **59**, 9–23 (1994)
31. Ting, YP, Hu, HL, Tan, HM: Bioremediation of petroleum hydrocarbons in soil microcosms. Resour. Environ. Biotechnol. **2**, 197–218 (1999)
32. Semple, KT, Reid, BJ, Fermor, TR: Impact of composting strategies on the treatment of soils contaminated with organic pollutants. Environ. Poll. **112**, 269–283 (2001)
33. Bundy, JG, Paton, GI, Campbell, CD: Microbial communities in different soil types do not converge after diesel contamination. J. Appl. Microbiol. **92**, 276–288 (2002)
34. Smith, J: Biotechnology, 3rd edn, p. 143. Press Syndicate Publishers, Great Britain (1996)
35. Agarry, SE, Owabor, CN, Yusuf, RO: Studies on biodegradation of kerosene in soil under different bioremediation strategies. Bioremediation J. **14**(3), 135–141 (2010)
36. Hollender, J, Althoff, K, Mundt, M, Dott, W: Assessing the microbial activity of soil samples, its nutrient limitation and toxic effects on contaminants using a simple respiration test. Chemosphere **53**, 269–275 (2003)
37. Semple, KT, Dew, NM, Doick, KJ, Rhodes, AH: Can mineralization be used to estimate microbial availability of organic contaminants in soil? Environ. Poll. **140**, 164–172 (2006)
38. Walworth, J, Pond, A, Snape, I, Rayner, J, Ferguson, S, Harvey, P: Nitrogen requirements for maximizing petroleum bioremediation in a sub-Antarctic soil. Cold Regions Sci. Technol. **48**, 84–91 (2007)
39. Margesin, R, Zimmerbauer, A, Schinner, F: Monitoring of bioremediation by soil biological activities. Chemosphere **40**, 339–346 (2000)
40. Elektorowicz, M: Bioremediation of petroleum-contaminated clayey soil with pretreatment. Environ. Technol. **15**, 373–380 (1994)

41. Piehler, MF, Swistak, JG, Pinckney, JL, Paerl, HW: Stimulation of diesel fuel biodegradation by indigenous nitrogen fixing bacterial consortia. Microbial Ecol. **38**, 69–78 (1999)
42. Abioye, PO, Abdul Aziz, A., Agamuthu, P: Enhanced biodegradation of used engine oil in soil amended with organic wastes. Water Air Soil Pollut. **209**, 173–179 (2009)
43. Abdulsalam, S, Bugaje, IM, Adefila, SS, Ibrahim, S: Comparison of biostimulation and bioaugmentation for remediation of soil contaminated with spent motor oil. Int. J. Environ. Sci. Tech. **8**(1), 187–194 (2011)
44. Jobson, AM, Mclaughlin, M, Cook, FD, Westlake, DWS: Effects of amendments on microbial utilization of oil applied to soil. Appl. Microbiol. **27**, 166–171 (1974)
45. Amadi, A, Ue Bari, Y: Use of poultry manure for amendment of oil-polluted soils in relation to growth of maize (*Zea mays* L.). Environ. Int. **18**, 521–527 (1992)
46. Yakubu, MB: Biodegradation of Lagoma crude oil using pig dung. Afr. J. Biotechnol. **6**, 2821–2825 (2007)
47. Agarry, SE, Owabor, CN, Yusuf, RO: Bioremediation of soil artificially contaminated with petroleum hydrocarbon mixtures: evaluation of the use of animal manure and chemical fertilizer. Bioremediation J. **14**(4), 189–195 (2010)
48. Knaebel, DB, Federie, TW, McAvoy, DC, Vestal, JR: Effects of mineral and organic soil constituents on microbial mineralization of organic compounds in natural soil. Appl. Environ. Microbiol. **60**, 4500–4508 (1994)
49. Okolo, JC, Amadi, EN: Optimising crude oil biodegradation in a sandy loam soil using a mixture of cow dung and poultry manure. Int. J. Agric. Rural. Dev. **5**, 69–76 (2004)
50. Ahmadi, M, Vahabzadeh, F, Bonakdarpour, B, Mofarrah, E, Mehranian, M: Application of the central composite design and response surface methodology to the advanced treatment of olive oil processing wastewater using Fenton's peroxidation. J. Hazard. Mater. **123**, 187–195 (2005)
51. Pala, DM, de Carvalho, D, Pinto, J, Sant'Anna Jr., G: A suitable model to describe of a petroleum-contaminated soil. Int. Biodeter. Biodegr. **58**, 254–260 (2006)
52. Rigas, F, Papadopoulou, K, Dritsa, V, Doulia, D: Bioremediation of a soil contaminated by lindane utilizing the fungus *Ganoderma australe* via response surface methodology. J. Hazard. Mater. **140**, 325–332 (2007)
53. Huang, L, Mab, T, Li, D, Liang, F, Liu, R, Li, G: Optimization of nutrient component for diesel oil degradation by *Rhodococcus erythropolis*. Mar. Pollut. Bull. **56**, 1714–1718 (2008)
54. Black, CA: Method of Soil Analysis II, pp. 573-590. American Society of Agronomy, Madison (1965)
55. APHA: Standard Methods for Examination of Water and Wastewater. American Public Health Association, Washington D.C (1985)
56. Colores, GM, Macur, RE, Ward, DM, Inskeep, WP: Molecular analysis of surfactant-driven microbial population shifts in hydrocarbon-contaminated soil. Appl. Environ. Microbiol. **66**, 2059–2964 (2000)
57. Kästner, M, Breuer-Jammali, M, Mahro, B: Enumeration and characterization of the soil microflora from hydrocarbon contaminated soil sites able to mineralize polycyclic aromatic hydrocarbons. Appl. Microbiol. Biotechnol. **41**, 267–273 (1994)
58. Krieg, NR, Holt, JG, Sneath, PHA, Stanley, JT, Williams, ST: Bergey's Manual of Determinative Bacteriology, 9th edn. Williams and Wilkins, Baltimore (1994)
59. Vu-Duc, T, Huynh, CK, Lafontaine, M, Bonnet, P, Binet, S: A spectrophotometric method for the determination of organic soluble matter in bitumen fumes. Appl. Occupational Environ. Hygiene **17**(7), 495–500 (2002)
60. Bader, JL, Gonzalez, G, Goodell, P, Ali, AM, Pillai, S: Aerobic reduction of hexavalent chromium in soil by indigenous microorganism. Biorem J. **3**, 201–212 (1999)
61. Odokuma, LO, Akponah, E: Effect of nutrient supplementation on biodegradation and metal uptake by three bacteria in crude oil impacted fresh and brackish waters of the Niger Delta. J. Cell and Animal Biol. **4**(1), 001–018 (2010)
62. Agarry, SE, Owabor, CN, Yusuf, RO: Enhanced bioremediation of soil artificially contaminated with kerosene: optimization of biostimulation agents through statistical experimental design. J. Pet. Environ. Biotechnol. **3**, 120 (2012)
63. Al-Saleh, ES, Obuekwe, C: Inhibition of hydrocarbon bioremediation by lead in a crude oil. Int. Biodeter. Biodegr. **56**(1), 1–7 (2005)
64. Al-Mahin, A, Chowdhury, AZ, Alam, MK, Aktar, Z, Fakhruddin, ANM: Phenol biodegradation by two strains of *Pseudomonas putida* and effect of lead and zinc on the degradation process. Int. J. Environ. **1**, 27–33 (2011)
65. Wang, S, Mulligan, CN: An evaluation of surfactant foam technology in remediation of contaminated soil. Chemosphere **57**, 1079–1089 (2004)
66. Tsai, TT, Kao, CM, Surampalli, RY, Chien, HY: Enhanced bioremediation of fuel-oil contaminated soils: laboratory feasibility study. J. Environ. Eng. **135**(9), 845–853 (2009)
67. Calvo, C, Manzanera, M, Silva-Castro, GA, Uad, I, González-López, J: Application of bioemulsifiers in soil oil bioremediation processes. Future prospects. Sci. Total. Environ. **407**, 3634–3640 (2009)
68. Rahman, KSM, Banat, IM, Thahira, J, Thayumanavan, T, Lakshmanaperumalsamy, P: Bioremediation of gasoline contaminated soil by bacterial consortium amended with poultry litter, coir pith and rhamnolipid biosurfactant. Bioresour. Technol. **81**, 25–32 (2002)
69. Thavasi, R, Jayalakshmi, S, Banat, IM: Effect of biosurfactant and fertilizer on biodegradation of crude oil by marine isolates of *Bacillus megaterium,* Corynebacterium kutscheri and Pseudomonas aeruginosa. Bioresour. Technol. **102**, 772–778 (2011)
70. Mohajeri, L, AbdulAziz, H, Isa, MH, Zahed, MA: A statistical experiment design approach for optimizing biodegradation of weathered crude oil in coastal sediments. Bioresour. Technol. **101**, 893–900 (2010)
71. Agarry, SE, Owabor, CN: Anaerobic bioremediation of marine sediment artificially contaminated with anthracene and naphthalene. Environ. Technol. **32**(12), 1375–1381 (2011)
72. Joo, H, Ndegwa, PM, Shoda, M, Phae, C: Bioremediation of oil-contaminated soil using Candida catenulata and food waste. Environ. Pollut. **156**, 891–896 (2008)
73. Odokuma, LO, Dickson, AA: Bioremediation of a crude oil polluted tropical mangrove environment. J. Appl. Sci. Environ. Manage. **7**(2), 23–29 (2003)
74. Mohan, SV, Reddy, BP, Sarma, PN: Ex situ slurry phase bioremediation of chrysene contaminated soil with the function of metabolic function: process evaluation by data enveloping analysis (DEA) and Taguchi design of experimental methodology (DOE). Bioresour. Technol. **100**, 164–172 (2009)
75. Montgomery, D.C.: Design and Analysis of Experiments, 7th edn. John Wiley, New York (2008)
76. Zhu, X, Venosa, AD, Suidan, MT, Lee, K: Guidelines for the Bioremediation of Marine Shorelines and Freshwater Wetlands. US Environmental Protection Agency Office of Research and Development National Risk Management Research Laboratory Land Remediation and Pollution Control Division, Cincinnati (2001)
77. Angle, JS, Chaney, RL: Cadmium resistance screening in nitrilotriacetate-buffered minimal media. Appl. Environ. Microbiol. **55**, 2101–2104 (1989)
78. Atlas, RM: Petroleum biodegradation and oil spill bioremediation. Mar. Pollut. Bull. **31**, 178–182 (1995)
79. Tam, NFY, Wong, YS, Wong, MH: Novel technology in pollutant removal at source and bioremediation. Ocean Coast. Manage. **52**, 368–373 (2009)

A simple and fast approach for allocation and size evaluation of distributed generation

Partha Kayal* and Chandan Kumar Chanda

Abstract

Penetration of distributed generation (DG) units in distribution network has increased rapidly stimulated by reduced network power loss, improved bus voltage profile, and better power quality. Appropriate size and allocation of DG units play a significant role to get beneficial effects. The objective of this study is to demonstrate a simple and fast technique to determine appropriate location and size of DG units. A voltage stability indicator (VSI) is derived which can quantify the voltage stability conditions of buses in distribution network. According to VSI, vulnerable buses of the network are arranged rank-wise to form a priority list for allocation of DG units. To determine the size of DG units, a feed forward artificial neural network is prepared in MATLAB environment (The MathWorks, Inc., Massachusetts, USA). The effectiveness of the proposed methodology has been tested on a 52-bus radial distribution network. After appropriate allocation of DG units, voltage profiles of most of the buses are increased significantly. The results also indicated that the total loss of the distribution network has reduced by nearly 76.39%, and voltage stability conditions of buses are improved considerably. Voltage stability conditions of bus-13, bus-36, and bus-44 are raised by 23.16%, 29.23%, and 37.64% respectively.

Keywords: Voltage stability condition, Distributed generation units, Priority list, Allocation, ANN

Background

Distributed generation (DG) is going to play a very important role in power systems worldwide. Penetration of DG units in distributed system presents a significant impact on system reliability, power flow, voltage profile, stability, and power quality. Renewable energy source-based DG can alleviate the energy poverty by utilizing abundant wind and solar photovoltaic available in the region [1]. Most of the radial distribution networks suffer with voltage instability at feeders. DG units can boost up the low voltage at the end of the feeders. The planned application of DG units can provide the transmission capacity release, reduction in network losses, and avoidance of high investment costs for network upgrades. But, unplanned uses of DG units may increase the problems. Therefore, some tools or techniques are needed to be examined for the allocation and sizing of the DG units.

A method for placement of DG units using continuation power flow analysis has been proposed by Hedayati et al. [2]. They have not studied about the size of DG units. Kashem et al. have discussed about optimal use of DG units to support voltage in distribution feeders [3]. They have applied sensitivity analysis to determine appropriate location of voltage support DG units. The effect of DG real and reactive power injections for the inclusion of DG units is also investigated. Rafidah et al. [4] have discussed about a methodology to evaluate appropriate DG size and its impact on power losses and voltage profile in distribution system. They have proposed complex artificial immune system optimization algorithm for sizing of DG units. Islam has presented a methodology to optimize the size of pumped hydro-storage system using optimization software HOMER (NREL, Washington DC, USA) [5]. Acharya et al. [6] have derived an expression to calculate size and location of DG units to minimize distribution losses. The size of DG units is calculated updating the loss coefficients of network loss formula. With proposed methodology by Acharya et al., appropriate size of DG units can be estimated after running a number of load flow iterations. A genetic algorithm-based optimal sizing and placement of DG units considering the system energy loss minimization in different load conditions have been

* Correspondence: partha_kayal@yahoo.co.in
Department of Electrical Engineering, Bengal Engineering & Science University, Shibpur, Howrah 711103, India

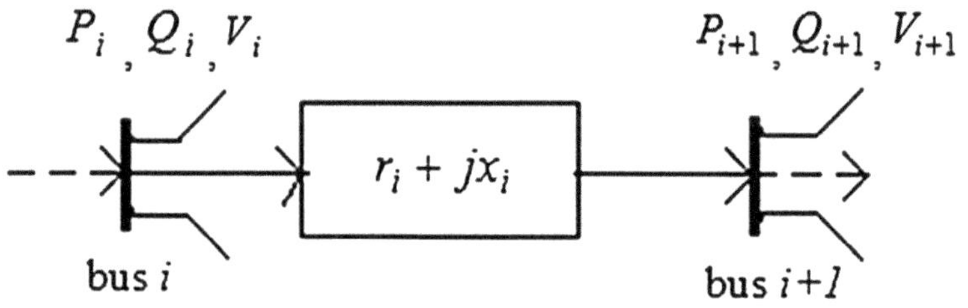

Figure 1 Two-bus power system network.

presented by Singh et al. [7]. However, this method needs extensive calculations. Analytical approaches to choose optimal location for DG units in radial distribution network to minimize loss have been presented by Caisheng et al. [8]. They have not discussed sizing issue of DG units in the literature. Recent researches focus on the selection of best places for the allocation of DG units in large distribution network, but appropriate size calculation of DG units using simple and fast techniques has not been emphasized yet.

The paper is organized as follows: At first, a voltage stability indicator (VSI) is developed from conventional power flow equation to determine the stability condition of buses. Then, a priority list is set up using VSI to allocate DG units. In the next section, artificial neural network (ANN) technique is used to determine the proper size of the DG units to ensure the permissible static voltage of each bus. After that, a proposed methodology is tested on a 52-bus radial distribution network, and the impact of the DG units on static voltage profile is illustrated. In the last section, some relevant conclusions are given.

Methods

Allocation of DG units

In developing countries, most of the distribution networks are operated with radial structure, leading to

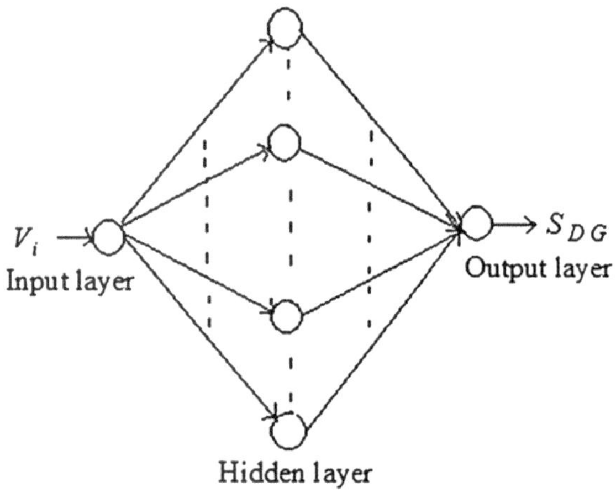

Figure 2 Architecture of the three-layered feed forward ANN.

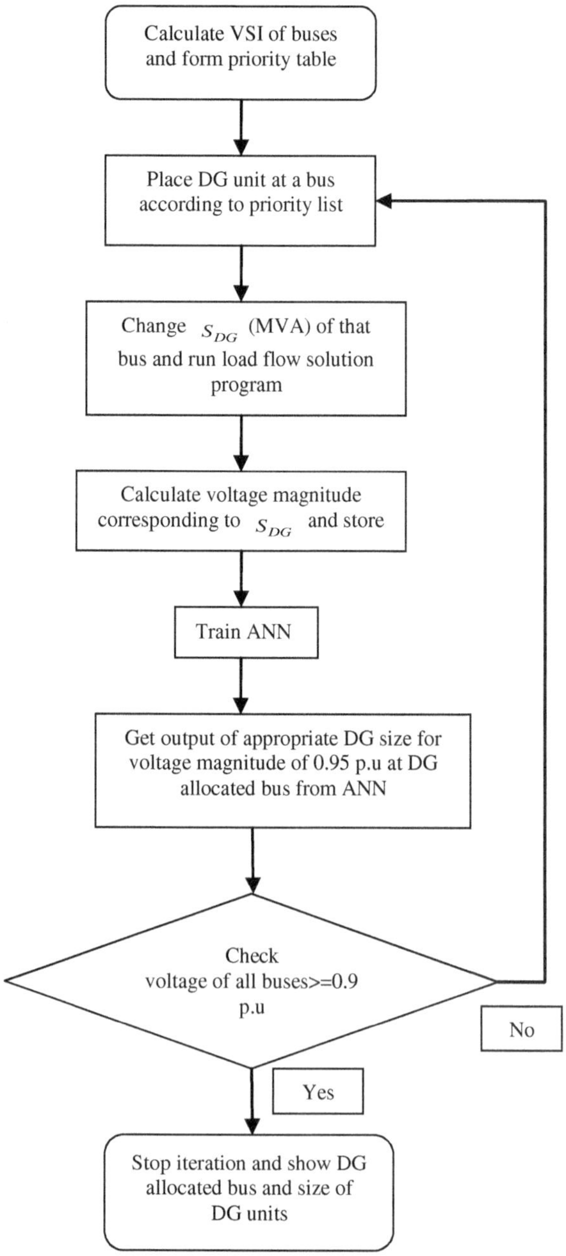

Figure 3 Flow chart for allocation and sizing of DG units.

larger voltage drop. Inefficient utilization of energy is another reason to get low quality power at customer end [9]. Considering the fact that most DG units are located at the distribution level, determination of the best locations for DG units to maximize their benefits is very important in system design and expansion. By analyzing the voltage sensitivity of lines, weakness of network voltage may be identified and opportunities for improvement with active or reactive power compensation via DG units can be examined.

Proposed VSI

Power networks are becoming heavily stressed to meet ever-increasing load demand. This situation has resulted into deterioration of voltage magnitudes at buses. One of the major problems that may associate with such a stressed system is voltage collapse. From the necessity of accurate analysis of voltage stability, a number of analytical and computational tools have been discussed [10-12]. In this section, a VSI is formulated for radial distribution network to estimate the distance of voltage collapse. The indicator uses the bus voltage and network information provided by load flow program. Any branch $r_i + jx_i$ connected between bus i and bus $i + 1$ of the radial distribution system may be represented by an equivalent circuit model, shown in Figure 1. The conventional power flow equations of the receiving bus and branch connect the receiving bus can be written as follows:

$$I^2 = \left(P_{i+1}^2 + Q_{i+1}^2\right)/V_{i+1}^2 \quad (1)$$

$$I^2 = \frac{P_{L,i}^2 + Q_{L,i}^2}{\left(V_i - V_{i+1}\right)^2} \quad (2)$$

where $P_{L,i}$ and $Q_{L,i}$ are the active and reactive power loss of the line connected between two nodes.

$$\frac{P_{i+1}^2 + Q_{i+1}^2}{V_{i+1}^2} = \frac{P_{L,i}^2 + Q_{L,i}^2}{\left(V_i - V_{i+1}\right)^2} \quad (3)$$

Power balance equation is always maintained in each section of the network for steady operation.

$$P_{i+1} = P_i - P_{L,i} \quad (4)$$

$$Q_{i+1} = Q_i - Q_{L,i} \quad (5)$$

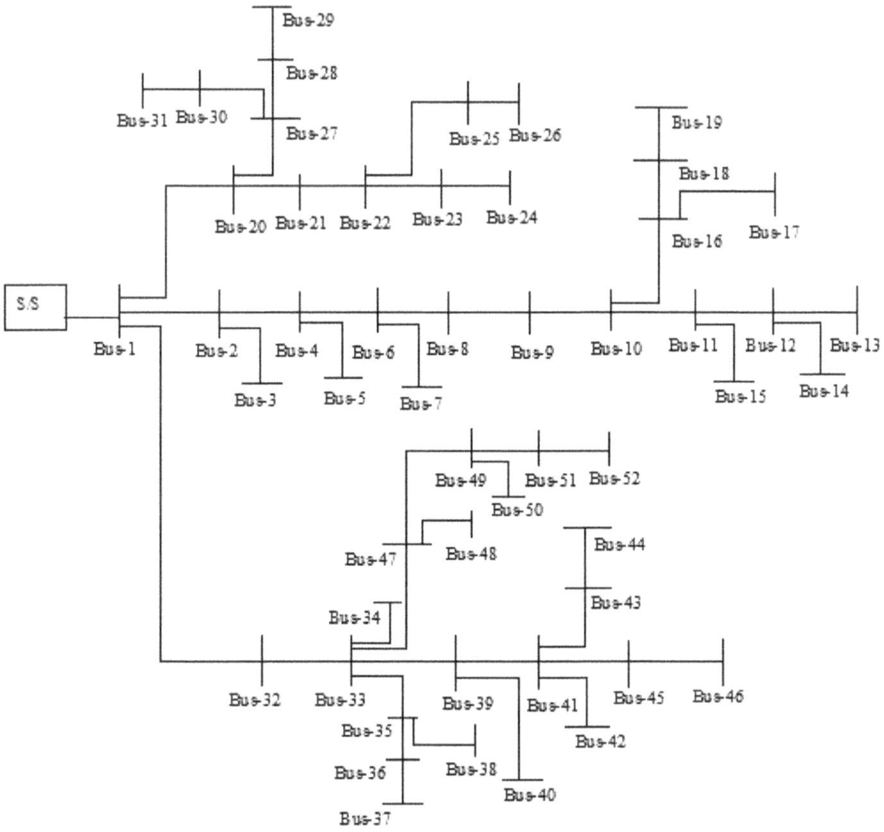

Figure 4 Single line diagram of the 52-bus radial distribution network [14].

Table 1 Priority list of test system

Rank	Bus number	Value of VSI
1	44	0.4323
2	36	0.3483
3	13	0.3433
4	18	0.3365
5	03	0.3102
6	33	0.2993
7	15	0.2969
8	47	0.2801
9	17	0.2537
10	23	0.2390
11	50	0.2230
12	19	0.2084
13	31	0.2061
14	46	0.1960
15	42	0.1927
16	10	0.1873
17	08	0.1628
18	39	0.1467
19	40	0.1415
20	14	0.1354
21	37	0.1338
22	25	0.1327
23	09	0.1243
24	24	0.1200
25	30	0.1148
26	28	0.0968
27	45	0.1040
28	22	0.1021
29	34	0.1019
30	02	0.0982
31	28	0.0968
32	35	0.0896
33	38	0.0875
34	43	0.0825
35	26	0.0724
36	05	0.0722
37	29	0.0718
38	04	0.0692
39	32	0.0671
40	41	0.0670
41	49	0.0653
42	16	0.0585
43	21	0.0556
44	07	0.0524
45	27	0.0419
46	20	0.0410
47	48	0.0394
48	52	0.0265
49	11	0.0235
50	12	0.0156
51	06	0.0119
52	1	–

$$P_{L,i} = r_i \left(\frac{P_{i+1}^2 + Q_{i+1}^2}{V_{i+1}^2} \right) \quad (6)$$

$$Q_{L,i} = x_i \left(\frac{P_{i+1}^2 + Q_{i+1}^2}{V_{i+1}^2} \right) \quad (7)$$

As the value of $P_{L,i}$ and $Q_{L,i}$ from Equations (6) and (7) have used in Equation (3),

$$\left(V_i.V_{i+1} - V_{i+1}^2\right)^2 = \left(P_{i+1}^2 + Q_{i+1}^2\right).\left(r_i^2 + x_i^2\right) \quad (8)$$

Equation (8) is solved according to the fact that the right hand side of Equation (8) is always positive for any value of P_{i+1}, Q_{i+1}, r_i, x_i.

$$V_{i+1}^2 - V_{i+1}.V_i + \sqrt{\left(\left(P_{i+1}^2 + Q_{i+1}^2\right).\left(r_i^2 + x_i^2\right)\right)} = 0. \quad (9)$$

Roots of Equation (9) are real if

$$V_i^2 - 4.\sqrt{\left(\left(P_{i+1}^2 + Q_{i+1}^2\right).\left(r_i^2 + x_i^2\right)\right)} \geq 0 \quad (10)$$

From Equation (10), the VSI is given as

$$L_{i+1} = \frac{4.\sqrt{\left(\left(P_{i+1}^2 + Q_{i+1}^2\right).\left(r_i^2 + x_i^2\right)\right)}}{V_i^2} \leq 1 \quad (11)$$

In practice, the distribution system operators always try to maintain the system within a given voltage stability margin so that small contingencies do not make the system unstable. Therefore, L_{i+1} must be less than 1.0 for maintaining stability at that bus. The smaller VSI, the more stable is the distribution network.

Priority list

Voltage stability level of each bus is calculated using the VSI. The buses are ranked in descending order according to their VSIs to form a priority list. The top-ranked buses are chosen first for allocation of DG units. Subsequently, lower-ranked buses are fed with power from the DG units. A number of DG units would be installed depending on the sizing issues of the DG units. After the installation of each DG unit, load flow solution is performed to monitor voltage magnitude of buses, network power losses, and therefore voltage stability condition of the system.

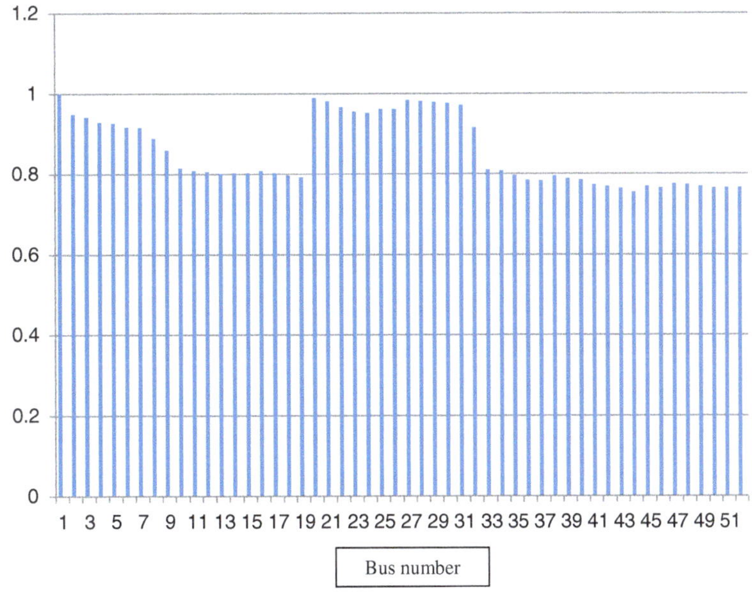

Figure 5 Voltage magnitude (p.u.) of all buses at normal condition.

Sizing of DG units

A DG unit is a small scale power generation ranging from multi-kilowatt to few megawatt and is usually connected to a distribution network. The impact of the DG unit on distribution network may be positive or negative, depending upon the size and operating condition of the DG unit. Usually, the DG unit size should be consumable within the distribution substation boundary. Further increase of the DG size can cause reverse flow of power through DG buses and then high system losses.

ANN model

For a particular bus, as the size of the DG unit is increased beyond its appropriate size, network losses start increasing rather than decreasing. So, appropriate size evaluation of DG is very significant. The use of ANN [13] is capable of indicating the best solution for a given distribution system. This is because of the advantage of high computation rate of three-layered feed forward ANN in approximating a complex nonlinear mapping. A feed forward ANN works on the basis of

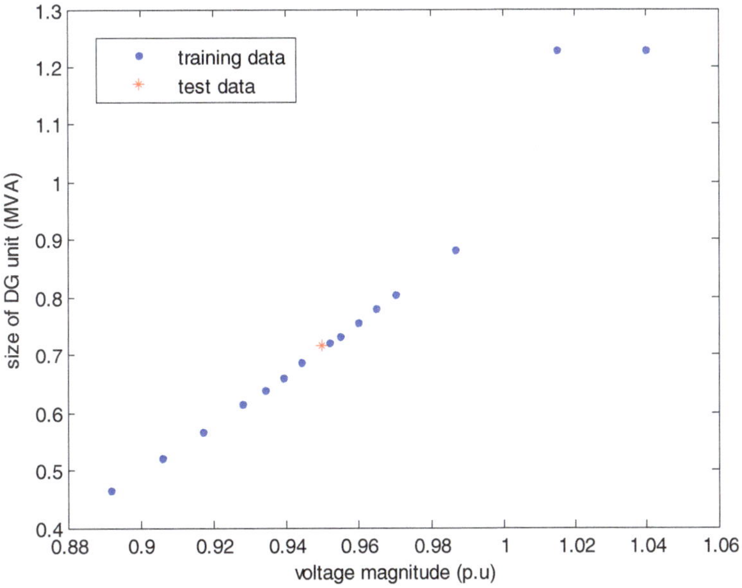

Figure 6 Graph of test output along with training data obtained from ANN for different voltage magnitudes at bus-44.

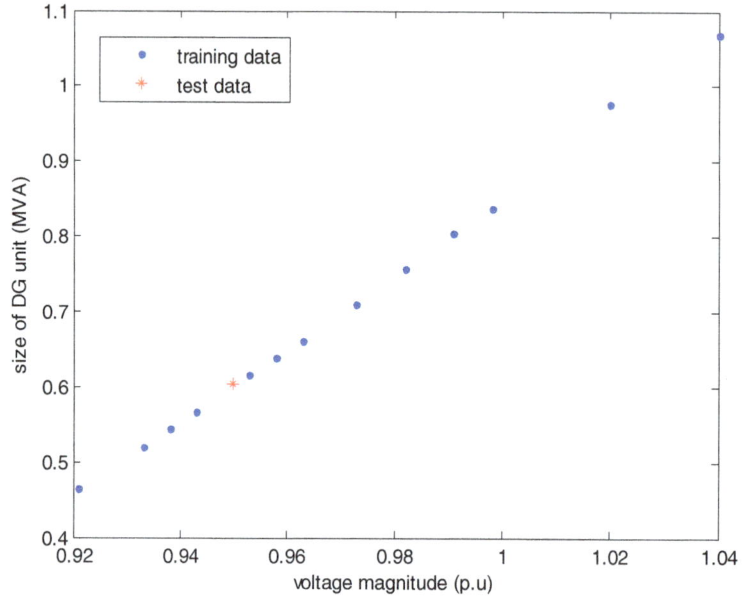

Figure 7 Graph of training and test output data obtained from ANN with respect to voltage magnitudes at bus-13.

propagation of signal in only one direction from an input stage to an output stage through intermediate neurons. The number of hidden layer is chosen to match the complexity of the function. Error backpropagation learning algorithm is used to train the ANN for the faster learning and reliable convergence. The error function chosen for the learning process is mean square error (MSE) of outputs. The architecture of the three-layered ANN is shown in Figure 2.

The appropriate size of a DG unit for a particular bus can be determined for desired voltage profile of that bus. With random change of DG unit at that bus, values of voltage magnitudes are determined from power flow solutions. Feed forward ANN is trained rigorously with DG size corresponding to voltage magnitudes at poor-voltage stable bus obtained from load flow solution. Then, for any required bus voltage, the appropriate size of the DG unit (MVA) for that particular bus is evaluated. The proposed methodology for the allocation and size evaluation of the DG units is shown in Figure 3. In the flowchart of Figure 3, the iteration starts with calculating the VSI of each bus at base case of the network. After iterations, on view point of voltage profile improvement of buses beyond 0.90 p.u., the appropriate size and location of the DG units are obtained at modified network.

Computational procedure

The installation of non-optimal size DG unit results in an increase of system losses, implying the reduction of voltage magnitudes of buses in the network. A fast systematic approach to allocate DG units and then determination of their sizes need step-by-step computational procedure. The procedure for proposed methodology is as follows:

1. Run the power flow solution at base case of the system.
2. Calculate the VSI of each bus and store.
3. Arrange the VSI of the buses in descending order to form a priority list.
4. Place the DG unit at top-ranked bus.
5. Change the size of the DG unit randomly within a certain limit and calculate the voltage magnitude of the highest priority bus to form the training data set of the proposed ANN.
6. Train the three-layered feed forward ANN properly with training data set.
7. Evaluate the MVA for the voltage profiles of 0.90 p.u. for buses using the ANN model.
8. Repeat steps 5 to 7 to allocate the DG units subsequently at other weak voltage stable buses and continue until all buses of the network reach to desired voltage.

Test network

The proposed technique has been tested on practical 11 kV, 52-bus radial distribution network. There are three main feeders supplying a total load of 4.184 +

Table 2 Size of DG units at three vulnerable buses

Bus number	Size of DG		MSE
	kW	kVAR	
44	641.4	309.8	0.00069
36	418.4	202.5	–
13	547.1	264.3	0.00033

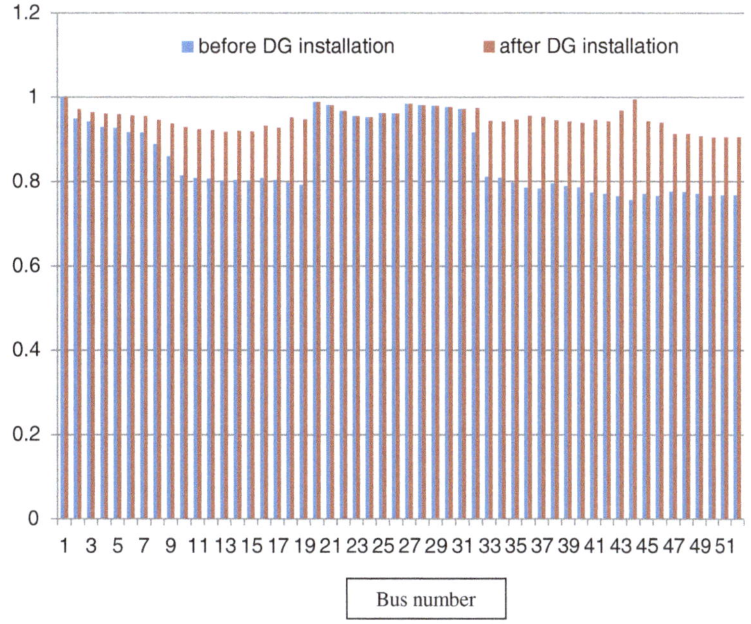

Figure 8 Comparison of voltage magnitude (p.u) of buses before and after DG installation.

j2.025MVA [14]. Only bus-1 is connected to the substation, and the rest of the buses in the network are load buses. The schematic diagram of the test network is shown in Figure 4. The line impedance of the system is 0.0086 + j0.0037 Ω/km. A Newton–Raphson algorithm is written in MATLAB 7 to solve the power flow problem.

Results and discussion

The method to allocate DG units to improve voltage profile and voltage stability condition of the distribution network is demonstrated through the simulation of the test network. The base power and the voltage for simulation of the system are taken as 10 MVA and 11 kV, respectively. The VSI of each bus is calculated with the network data of base configuration. The priority list is prepared and presented in Table 1.

According to the load flow, the active and reactive power losses of the system are obtained as 741 kW and 307 kVAR, respectively. Voltage magnitudes of the system are also noted and shown in Figure 5. From Figure 5,

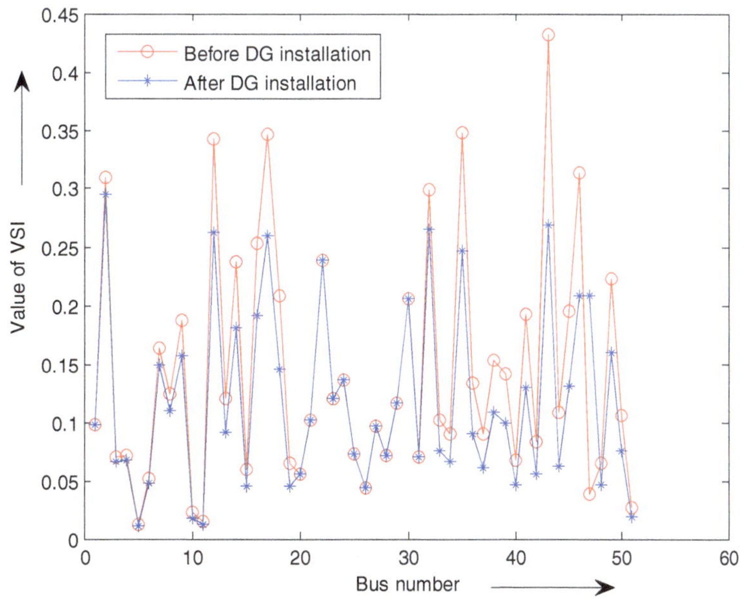

Figure 9 Comparison of voltage stability condition of different buses before and after DG installation.

Table 3 Comparative study of power losses before and after allocation of DG units

	Active power loss (kW)	Reactive power loss (kVAR)
Before allocation of DG units	741	307
After allocation of DG units	176	70
Reduction of loss after allocation of DG units	76.25%	77.20%

it is seen that many buses (bus-18, bus-19, bus-35, bus-36, etc.) suffer with low voltage problem and having a voltage lower than 0.8 p.u.

According to the priority list, the first DG unit is placed at bus-44 because it is with highest value of VSI. The size of the distribution system in term of loads plays an important role in selecting the size of the DG unit. To obtain a reasonable solution, the size of the DG unit should not be so small or so large with respect to load value. Therefore, the DG range is between 0.4648 MVA (10% of total load) and 1.3944 MVA (30% of total load) for this system. The optimal size of the DG unit is quantified through raising the voltage magnitude of low voltage stable bus to 0.95 p.u. The appropriate size of the DG unit at bus-44 is calculated using ANN. Figure 6 obtained from the simulation of ANN model shows training and target output data at bus-44. Very low MSE (0.00069) confirms the validation of the proposed model.

The appropriate DG size at bus-44 is 0.7123MVA, i.e., 15.06% of the total load. In the same way, two DG units are installed consecutively at bus-36 and at bus-13. The minimum size of the DG unit, i.e., 0.4648 MVA, is sufficient to raise the voltage beyond 0.95 p.u at bus-36. For bus-13, the size of DG unit is determined from ANN as shown in Figure 7. Appropriate sizes of DG units along with MSEs are tabulated in Table 2. From Table 2, it is to be noted that the appropriate size of DG unit is 12.82% of the total load of the test network.

After the allocation of the three DG units, it is seen that the voltage profile of all the buses are raised above 0.90 p.u. as shown in Figure 8. From Figure 8, it is seen that the voltage magnitudes of bus-18, bus-19, bus-35, and bus-36 are increased by 19.07%, 19.44%, 19.12%, and 21.78%. The voltage stability condition of all the buses has improved considerably compared to the initial condition and shown in Figure 9. It is noted from the figure that the voltage stability condition of bus-13, bus-36, and bus-44 are improved by 23.16%, 29.23%, and 37.64%. The active and reactive power losses of the network are reduced significantly as shown in Table 3. It is noted from Table 3 that the active and reactive power losses are reduced by 76.25% and 77.20%, respectively, after proper allocation of three DG units in the test distribution system.

Conclusions

In this study, a priority list is developed using VSI to accommodate DG units. The appropriate allocation of the DG unit has been done easily using the priority list. Proposed methodology for the determination of appropriate size of DG units for desired voltage profile using ANN technique has emerged as a very fast and efficient tool. The appropriate size of the DG units varies from bus to bus, depending on connected loads. The results reveal that the integration of DG units is highly effective in reducing power losses in the distribution network. It is shown that with appropriate sizing and allocation of DG units, voltage magnitudes of the poor-voltage buses can be raised above 0.90 p.u. Results also indicate that there is a significant improvement of voltage stability for most of the buses. The study may be further facilitated considering economical and geographical location factors of DG units.

Abbreviations

P_i: Active power demand at bus i; Q_i: Reactive power demand at bus i; V_i: Voltage magnitude at bus i; r_i: Resistance of branch i; x_i: Reactance of branch i; P_{L+i}: Active power loss of branch i; $Q_{L,i}$: Reactive power loss of branch i; P_{i+1}: Active power demand at bus $i+1$; Q_{i+1}: Reactive power demand at bus $i+1$; V_{i+1}: Voltage magnitude at bus $i+1$; L_{i+1}: VSI of bus $i+1$; S_{DG}: Size of DG unit (MVA).

Competing interest

The authors declare that they have no competing interests.

Authors' contributions

PK has carried out the survey and the simulation of the methodology. CKC analyzed the results and made the conclusion. Both authors have participated in drafting the manuscript. All authors read and approved the final manuscript.

Authors' information

PK has completed B. Tech. in Electrical Engineering in 2008 from West Bengal University of Technology, India. He received his masters degree in Electrical Engineering from Bengal Engineering and Science University, Shibpur, India in 2010. He joined the Future Institute of Engineering and Management, Kolkata, India as an Asst. Professor in 2010. Currently, he is working on his Ph.D. at Bengal Engineering and Science University, Shibpur, India. CKC has received his B.E. with honors from Regional Engineering College, Durgapur, (presently National Institute of Technology, Durgapur), India in 1983. He received M. Tech. from the Department of Electrical Engineering, Indian Institute of Technology, Kharagpur, India in 1989. He received his Ph.D. degree from the Bengal Engineering and Science University, Shibpur, India in 2003. Now, he is a professor at Bengal Engineering and Science University, Shibpur, India. He was formerly a lecturer in Regional Engineering College, Durgapur, India.

Acknowledgments

The authors are thankful to the editors and reviewers of the *International Journal of Energy and Environmental Engineering* for their valuable suggestions.

References

1. Lal, S, Laturi, A: Techno-economic analysis of a hybrid mini-grid system for Fiji islands. Int. J. Energy. Environ. Eng. **3**, 10 (2012)
2. Hedayati, H, Nabaviniaki, SA, Akbarimajd, A: A method for placement of DG units in distribution networks. IEEE T. Power Deliv. **23**(3), 1620–1628 (2008)

3. Kashem, MA, Ledwich, G: Multiple distributed generators for distribution feeder voltage support. IEEE T. Energy Conver. **20**(3), 676–684 (2005)
4. Rafidah, S, Rahim, A: Implementation of DG for loss minimization and voltage profile in distribution system. In the 4th international power engineering and optimization conference, Shah Alam, Malaysia (2010)
5. Islam, SM: Increasing wind energy penetration level using pumped hydro storage in island micro-grid system. Int. J. Energy. Environ. Eng. **3**, 9 (2012)
6. Acharya, A, Mahat, P, Mithulananthan, N: An analytical approach for DG allocation in primary distribution network. Electr. Pow. and Energy Syst. **28**, 669–678 (2006)
7. Singh, D, Singh, D, Verma, KS: GA based energy loss minimization approach for optimal sizing and placement of distributed generation. Int. J. of Knowledge-based and Intelligent Eng. Syst. **12**, 147–156 (2008)
8. Caisheng, W, Nehrir, MH: Analytical approaches for optimal placement of distributed generation sources in power system. IEEE T. Pow. Syst. **19**(4), 2068–2076 (2004)
9. Oyedepo, SO: Efficient energy utilization as a tool for sustainable development in Nigeria. Int. J. Energy. Environ. Eng. **3**, 11 (2012)
10. Dey, S, Chanda, CK, Chakraborti, A: Development of global voltage security indicator (VSI) and role of SVC on it in longitudinal power supply (LPS) system. Electr. Pow. Syst. Res. **68**, 1–9 (2004)
11. Gubina, F, Strmcnik, B: A simple approach to voltage stability assessment in radial network. IEEE T. Pow. Syst. **12**(3), 1121–1128 (1997)
12. Shin, JR: A new optimal routing algorithm for loss minimization and voltage stability improvement in radial power systems. IEEE T. Pow. Syst. **22**(2), 648–657 (2007)
13. Ismail, KS: Neural network based load balancing and reactive power control by static VAR compensator. Int. J. of Computer and Electr. Eng. **1**(1), 1793–8198 (2009)
14. Thukaram, D, Khincha, HP, Vijaynarasimha, HP: Artificial neural network and support vector machine approach for locating faults in radial distribution system. IEEE T. Pow. Deliv. **20**(2), 710–721 (2005)

16

Application of energy spilling mechanism by *para*-nitrophenol in biological excess sludge reduction in batch-activated sludge reactor

Afshin Takdastan[1*] and Azadeh Eslami[2]

Abstract

Reduction of biomass production coefficient is an ideal solution for the reduction of excess sludge especially in industrial wastewater treatment plants. Studies were carried out in two sequencing batch reactors, which were controlled online. After presenting a stable situation in reactors, during 24 months of the study, sampling and examining of chemical oxygen demand (COD), biochemical oxygen demand, pH, sludge volume index (SVI), specific oxygen uptake rate (SOUR), remaining *p*NP, and biomass yield (*Y*) were implemented.

Results have shown that among different retention times (5, 10, 15, 25 days), maximum COD removal efficiency (95%) was achieved in 10 days, without bulking and foaming problems. In 10 days of sludge retention time, average *Y* and kinetic coefficient (K_d) were calculated: 0.58 mg biomass/mg COD and 0.058 1/day, respectively, and correlation coefficient (R^2) was 0.98. Different concentration of *p*NP were used due to energy spilling effect in the reactor and the results show that injection of 100 mg/L *p*NP to the reactor can reduce synthetic coefficient *Y* from 0.58 to 0.27 mg biomass/mg COD without *p*NP injection, so that the excess sludge was reduced by 0.56%. Although, an increase of 193 mg/L soluble COD in the effluent was observed. On the other hand, in this concentration of *p*NP, SOUR rate reached 31 mg O_2/h/g volatile suspended solids, and SVI rate reached less than 48 mL/g. In the concentration of 150 mg/L *p*NP, no sludge was produced, but COD rate of the effluent increased to 480 mg/L. Otherwise, *p*NP rate is an environmental limitation in effluent and sludge disposal.

Keywords: Sequencing batch reactor; Biological sludge; Biomass yield; COD; *Para*-nitrophenol

Background

Removal of organic materials by biological oxidation is a core technology in wastewater treatment process. New cells (sludge), carbon dioxide, soluble microbial products, and water are the end products for this process. The activated sludge process is widely used for municipal and industrial wastewater treatment and generates a large quantity of excess sludge daily. So far, the ultimate disposal of excess sludge has been and continues to be one of the most expensive problems faced by wastewater utilities, e.g., the treatment of the excess sludge may account for up to 65% of the total plant operation cost. So in recent years, increased attention has been given to the minimization of waste sludge in wastewater treatment process [1-6].

Reduction of biomass production in wastewater treatment was discussed when costs and difficulties of treatment and disposal of the sludge were considered. On the other hand, new rules and severe standards of reuse and disposal of the sludge about different organic and inorganic pollutants and pathogens forced wastewater treatment experts to invent aerated biological treatment methods which produce less sludge amount. In other words, if the problem of excess sludge production is solved, most of the problems would also be solved considerably in the treatment and disposal of the biomass [7,8]. Due to severe problems and heavy costs of operations of sludge treatment, mechanisms of biomass reduction were considered in the recent years. A collection of effective determinations in this field are:

* Correspondence: afshin_ir@yahoo.com
[1]Department of Environmental Health and Environmental Technology Research Centre, Ahvaz Jundishapur University of Medical Sciences, Ahvaz, Iran
Full list of author information is available at the end of the article

- Self-destructive process [9-12]
- Uncoupled metabolism using OSA process [5,12-18]
- Increasing soluble oxygen of aeration pond [19]
- Oxidation of a part of sludge by chlorine or ozone [2,5,7,8,18,20-23]
- Increasing temperature in returned sludge to the reactor [6,24-26]
- Energy spilling by compounds resistant to degradation and toxicant [3-5,17,27-30]
- pH changes [11,24]
- Using electrical pulse in returned sludge [31]
- Using ultrasonic waves in returned sludge [32]
- Using bacteriophages such as protozoa and metazoan [33,34]

For most of the aerobic bacteria, adenosine-5′-triphosphate (ATP) is generated by oxidative phosphorylation, in which process electrons are transported through the electron transport system from a source of electrons at elevated energy levels (substrate) to a final electron acceptor (oxygen). The chemiosmotic theory shows that the oxidative phosphorylation is driven by proton gradient built up across cell membrane [11]. However, the tight coupling of respiration and phosphorylation can be disturbed by molecules known as metabolic uncouplers. In the presence of metabolic uncouplers, the energy generated from the oxidation of organic substrate would be lost as heat rather than being captured in ATP. As a result, the growth efficiency is much lowered in uncoupler-containing microbial culture. Metabolic uncouplers include a diverse group of molecule-structures, but they are all lipophilic weak acids [5,29,35,36] many of which have been used to reduce excess sludge production from the activated sludge processes, such as nitrophenol, chlorophenol, 3;3′,4′,5-tetrachlorosalicylanilide (TCS), 2,4,5-trichlorophenol (TCP), carbonyl cyanide-p-trifluoromethoxyphenyl hydrazone, cresol, aminophenol, and so on [3,4,20,27-29,37,38].

In 1998, Mayhew and colleagues found that injecting 35 mg/L of 2,4 dinitrophenol to the activated sludge reactor in 20°C, 0.3 mg mixed-liquor suspended solids (MLSS)/g chemical oxygen demand (COD) in pH = 7, and 2.5 g/L MLSS, sludge retention time (SRT) = 15 days and hydraulic retention time (HRT) = 5.5 h can reduce biological excess sludge, but the only problem is a 3.7% COD increase in the effluent [38]. In 1998, Low et al., found that continuous injection of 100 mg/L *para*-nitrophenol (*p*NP) to a culture media containing pseudomonas at 30°C and pH = 6.2 to 7 can reduce biological excess sludge from 62% to 70% [28]. In 1999, Strand and colleagues found that a 2 to 2.5 mg/L increase of TCP to the activated sludge continuously cultivated at 20°C and pH = 7, MLSS = 2.5 g/L, SRT = 5 days, HRT = 3.5 h can reduce biological excess sludge about 50% [39]. In 2000, Low reported that in a lab-scale-activated sludge system, 49% of the biomass production would be reduced by injecting 100 mg/L *p*NP in the culture media. The COD removal efficiency was reduced about 25%. No sludge production was observed when 120 mg/L *para*-nitrophenol was added [29].

In 2003, Yang et al., found that 20 mg/L m-chlorophenol (mCP) injection to non-continuous activated sludge reactor at 25°C, pH = 7 can reduce biological excess sludge 86.9%, but there is the problem of 13.5% increasing of COD in effluent [40].

In 2003, Chen et al., found that efficiency of cellular growth or sludge production can reduce excess sludge about 78% in a non-continuous medium by injecting 0.8 mg/L of TCS at 20°C and pH = 7 but is ineffective in COD removal. In fact, most of the organic substrates would be oxidized by energy spilling mechanism, and cell synthesis rarely happens in the presence of TCS combined within the medium [15]. In 2007, Chase et al. indicated that at the dosage of 100 mg/L *p*NP in the activated sludge stage, the reduction of excess sludge was 16% and 27% of TSS in the case of wastewater handling units fed with raw and settled wastewater, respectively [41].

Although the reduction of sludge production happens by adding degradation-resistant materials and toxicants, reduction of pollutants to an acceptable value in effluent is necessary due to environmental standards. This method increases the cost of biological treatment system operation but is an acceptable way for a stronger industrial wastewater treatment. Therefore, this study was conducted to investigate and analyze the influence of energy spilling mechanism by

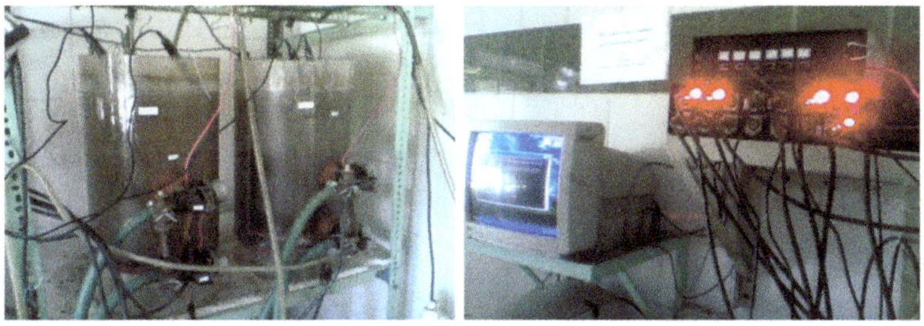

Figure 1 Sequencing batch reactor (SBR) pilot plant.

Table 1 Effect of injected *p*NP on *Y*, percentage of sludge reduction, and remaining *p*NP after reactions

Sludge reduction (%)	*p*NP in effluent (mg/L)	*p*NP in sludge (mg/L)	$Y\ \frac{mg\ Biomas}{mg\ COD}$	*p*NP injection to the reactor (mg/L)
-	0	0	0.63	0
14.3	0.06	0.05	0.54	20
33.4	0.07	0.06	0.42	50
42.85	31	0.09	0.36	70
57.15	36.7	0.09	0.27	100
73	70	0.1	0.17	120
87.3	-	-	0.08	130
100	103	0.5	0	150

para-nitrophenol in biological excess sludge reduction on the activated sludge in a batch culture process.

Methods

Collection of samples and wastewater characteristics

Two non-continuous cylinder-shaped reactors made of Plexiglas were used for this study with 25-cm internal diameter, 60-cm height, 20-L working volume, and 10-L treatment capacity in each cycle. Figure 1 shows a pilot plant view of non-continuous sequencing reactors. Planning system operation was done by software. The software was able to control and save all operations of the system.

Pilot sewage input was made of 40 gr of industrial milk powder and 100 L drinking water. In this study, quality profile of the synthetic wastewater was as below:

COD = 600 mg/L
Biological oxygen demand (BOD_5) = 420 mg/L
Nitrous nitrogen concentration = 4.7 mg/L nitrogen
Ammoniac nitrogen concentration = 0.7 mg/L
Organic nitrogen concentration = 30 mg/L
Kejeldal nitrogen concentration = 30.7 mg/L
Phosphorus concentration= 10.5 mg/L

Pilot start-up

For setting up and starting up the sequencing batch reactor (SBR), reverse activated sludge of Ekbatan wastewater treatment plant was used as seed - which did not have any bulking, foaming, and pinpoint flock - in about 2 L for each SBR pilot reactors with a total volume of 20 L, and synthetic wastewater with COD of 600 ± 20 mg/L was added to the reactor.

According to the type and characteristics of the used wastewater in the present work, time duration was the same in both the reactors. Filling ends in 3 min, aeration in 4 h, sedimentation in 1 h 45 min, and draining in 12 min, and in fact, the required time for filling was shorter and approximately took place in 1 min and 10 s.

Aeration and reaction took place approximately 2 weeks for flock or biomass formation in such a way that only the reaction was done, but nutrients were added to the reactor daily. After 2 weeks, the pilot system of SBR was launched by 5 cycles which are filling, reaction, discharge, sludge discharge, and rest. COD, suspended solids (SS), pH of the effluent in different turns were very close, which showed that reactor start-up came to its end. After having stable conditions in the reactors during 12 months of study, sampling, and testing parameters such as COD, MLSS, MLVSS, dried sludge solid percentage, sludge volume index (SVI), F/M, *p*NP, and finally, *Y* were determined. Sampling methods and examinations were all under guidance of the standard methods for examination of water and wastewater.

Changing conditions

While sludge age and injected *para*-nitrophenol concentration to the reactor were changing, at least 2 weeks (42 cycles) were considered for the system to be adopted with the new situation. It is assigned 3 days adaptation for changing influent COD, for determining kinetic coefficient (K_d, Y) in various sludge detention times (5, 10, 15, 20, 25 days), 1 week for changing sludge detention time (equal to 21 cycles), and 1 week for changing *p*NP concentration. All the measurements were carried out after stabilization; it is considered, and in normal condition (without adding

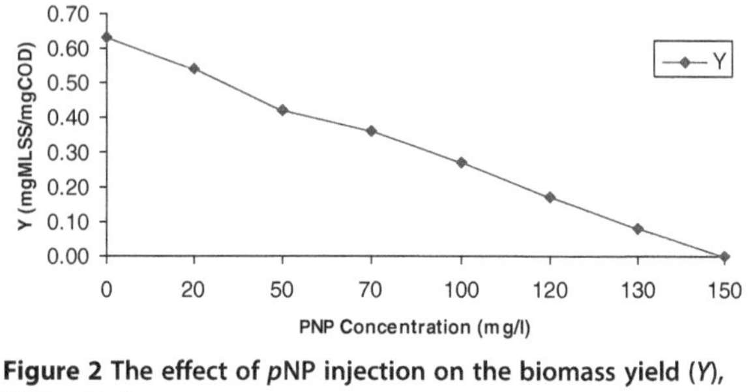

Figure 2 The effect of *p*NP injection on the biomass yield (*Y*), SRT = 10 days.

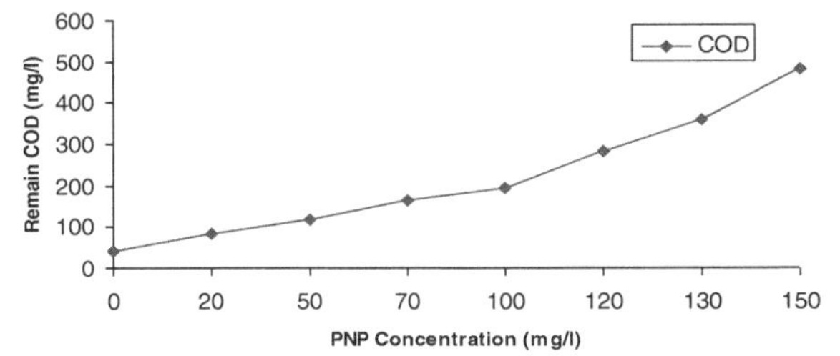

Figure 3 Effect of *p*NP ratio on COD removal efficiency, SRT = 10 days.

*p*NP to the reactor), reactor operation was examined in various SRT (5, 10, 15, 20, 25 days). The SBR system worked very well without any bulking and foaming problems in 10 days retention time; so, it was selected as the optimum detention time, and all the experiments were carried out in this time. Maximum COD removal efficiency was achieved in SRT = 10 days without any bulking and foaming problems.

Concentration of suspended solids in the reactor and COD of effluent were considered as stabilization indicators. For more accuracy and certitude, all the samples were examined three times.

Results and discussions

Determination of *Y* and remaining COD in different amounts of injected *p*NP to the reactor

The rate of biomass changes per time, to COD changes per time are used for determining the biosynthetic coefficients especially *Y*. *Y* and sludge reduction in different concentrations of injected *p*NP to the reactor are shown in Table 1.

As it can be seen in Figure 2, in 10 days, sludge retention time (maximum efficiency of COD removal), *Y* per gr used COD is 0.63. By adding *p*NP to the reactor, biomass yield will be decreased, as in 100 mg/L and 130 mg/L of injected *p*NP, the coefficient of biomass production rate was calculated, respectively, as 0.27 and 0.08 gr produced biomass per 1 gr used COD. Many researchers achieved similar results with other uncoupling combinations in the reduction of produced excess biological sludge by decreasing biomass production coefficient [27,40,42]. The reason of decreasing sludge production coefficient was energy spilling because of uncoupled *para*-nitrophenol in the reactor as the result of less electron transportation from electron donor [42].

As it can be seen in Figure 3, removal efficiency of COD will be 95% without increasing *p*NP and with input COD of 600 mg/L, but *Y* reduced by adding *p*NP to the reactor, so that *Y* coefficient reached to 0.27 mg biomass/mg COD by adding 100 mg/L *p*NP to the reactor, which caused decreasing production of sludge. However, the problem was the increasing soluble COD in effluent by 193 mg/L (COD removal efficiency reached 75%). By increasing *p*NP to 130 mg/L, COD removal efficiency decreased by 56%, and effluent COD reached 360 mg/L. Table 2 shows results of the *p*NP chemical uncoupling compared with other researchers in minimization of biological excess sludge.

Table 2 Comparing *p*NP uncoupling results with other literature data in biomass minimization

Uncoupling chemicals and operating condition	Sludge reduction %	COD reduction %	References
35 mg/L of 2,4-Dinitrophenol to the continuously activated sludge reactor, *T* = 20°C, 0.3 g MLSS/g COD, pH = 7, MLSS = 2.5 g/L, SRT = 15 days, HRT = 5.5 h	0.3	Decreased by 3.7	[38]
2 to 2.5 mg/L 2,4,5-Trichlorophnol in continuously activated sludge culture, *T* = 20°C, pH = 7, MLSS = 2.5 g/L, SRT = 5 days, HRT = 3.5 h	50	-	[39]
0.8 to 1 mg/L 3,3,4,5-Tetrachlorosalicylanilide in continuously activated sludge reactor, *T* = 20°C, pH = 7, HRT = 8 h, SRT = 7 days, MLSS = 2 g/L	40	Nearly unaffected	[15,42]
Adding 20 mg/L of m-chlorophenol (mCP), batch-activated sludge culture, *T* = 25+1°C, pH = 7	86.9	Decreased by 13.5	[40]
Continuous addition of 100 mg/L *para*-nitrophenol in continuous mono culture of *P. putida*, *T* = 30°C, pH = 6.2/7.0	62 to 77	-	[28]
Continuous addition of 100 mg/L *p*NP, continuous activated sludge culture, dilution rate = 0.29/h, sludge discharge rate = 0.02 l/h, MLSS = 0.71 g/L, *T* = 20°C, pH = 7.7	49	Decreased by 25	[29]
Pilot plant scale, synthetic wastewater, application of TCS in conventional activated system at 1% g TCS/g MLSS	80	Nearly unaffected	[30]
Para-nitrophenol (*p*NP), sequencing batch reactor culture, *T* = 25°C, pH = 7.6, influent COD = 600 mg/L, addition *p*NP by:	-	COD in effluent (mg/L)	In current research
1. 50 mg/L	34	117	
2. 100 mg/L	57	165	
3. 120 mg/L	73	284	
4. 150 mg/L	100	480	

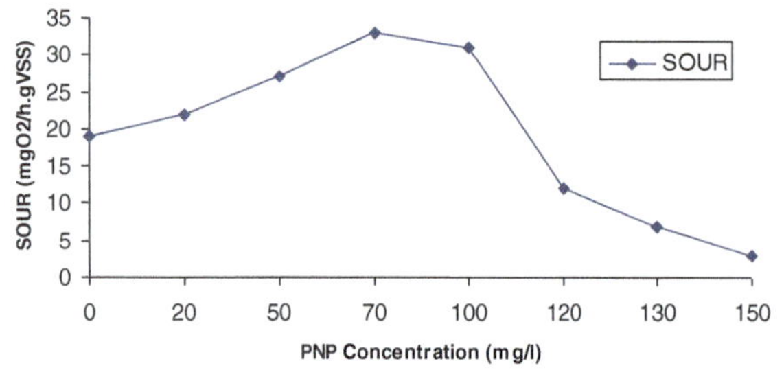

Figure 4 The effect of *p*NP injection on SOUR, SRT = 10 days.

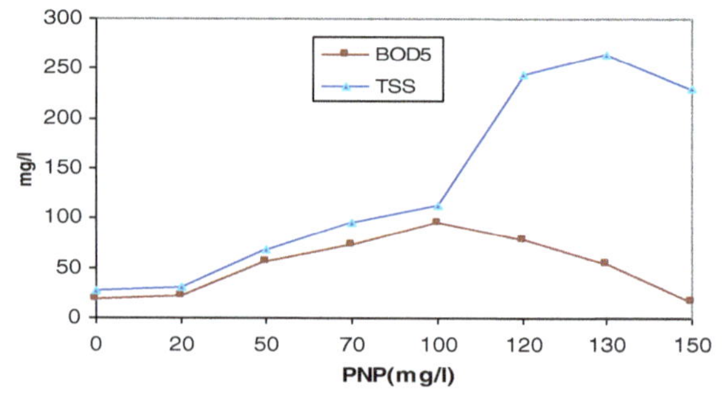

Figure 6 BOD_5 and TSS in the effluent in various *p*NP concentrations.

Effluent of various rates of injected *p*NP on SOUR and SVI

Figures 4 and 5 show that without *p*NP injection and addition of 100 and 130 mg/L *p*NP, SOUR was calculated with results 19, 31, and 7 mg used oxygen per h per gr MLVSS and SVI was 95, 48, and 13 mL/gr, respectively. It appears from Figure 3 that by adding *p*NP in the reactor, microorganisms have more activity up to a certain range, and consequently, SOUR rate increases up to 70 mg/L of *p*NP and decreases in more concentration of *p*NP because of its toxicity.

Mayhew and Stephenson in 1998 also reported the similar result by adding 2,4-dinitrophenol. They found out that uncoupled compounds caused break downs and energy loss in the culture; thus, SOUR increasing is not related to the growth [15]. Chen in 2002 found that in the presence of 0.8 mg/L TCS uncoupled compounds, SOUR increases 50% in the reactor, and increasing microbial activity happened in the presence of TCS in operating duration. High oxygen consumption indicates energy loss for metabolic conduction which ends to sludge minimization [43]. With increasing *p*NP concentration, as the result of toxicity of these metabolism uncoupling combinations on microorganism, SOUR will be reduced and so most of the microorganisms will get deactivated (except for few capsular ones). Microbial activity will be reduced and SOUR (breath rate) as a proper indicator, shows microbial death. SOUR reduction shows reduction of microorganism as the result of *p*NP presence in the system.

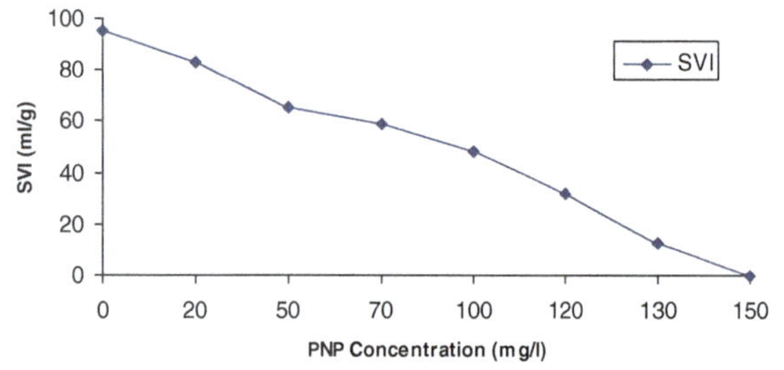

Figure 5 The effect of *p*NP injection on SVI concentration, SRT = 10 days.

As it can be seen in Figure 5, SVI reduced by increasing injected *p*NP which caused low biomass production coefficient. When injecting *p*NP was increased more than 100 mg/L to the reactor, SVI reached to 48 mL/gr, and when *p*NP was increased to 130 mg/L, SVI reached 13 mL/gr, but in 150 mg/L of *p*NP injection, no sludge was produced.

BOD_5 and TSS in the effluent with various *p*NP injections

Figure 6 shows BOD_5 and TSS in effluent when different *p*NP concentrations are injected to the reactor. As it can be seen in the reactor, by increasing *p*NP up to 100 mg/L, BOD_5 and TSS of the effluent increased. When no *p*NP was injected, BOD_5 and TSS in the effluent were 18 and 28 mg/L, respectively. However, with the 100 mg/L *p*NP adding to the reactor, effluent BOD_5 and TSS will be 96 and 112 mg/L, respectively. Also, by adding 150 mg/L *p*NP to the reactor, effluent BOD_5 and TSS will be 17 and 230 mg/L, respectively. The cause of such a BOD_5 reduction is microorganism death as the result of *p*NP increasing concentration.

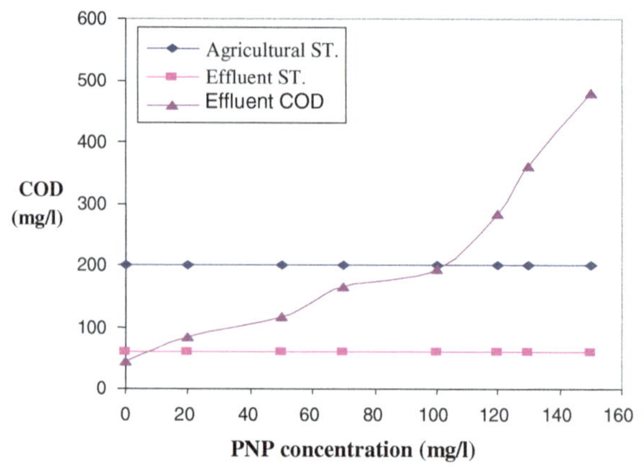

Figure 7 The comparison of effluent COD in the reactor with standards in various *p*NP injection.

Comparing effluent COD with disposal and reuse standards in different *p*NP injection

As it can be seen in Figure 7, in 10 days cell retention time (maximum COD removal efficiency) only *p*NP concentration lower than 20 mg/L can be accepted by receptive water environmental rules, but effluent standards used in agricultural works, in *p*NP dosages higher than 100 mg/L, cannot estimate standards of irrigation and agriculture.

Conclusion

Metabolic uncoupling is effective in reducing sludge production from the activated sludge process. In the 10 days of cellular retention, time of *Y* was 0.63 per gr used COD, but this amount was 0.27 and 0.08 gr biomass/gr COD if injected *p*NP increases to 100 and 130 mg/L. COD removal efficiency was 95% if the influent COD was 600 mg L-1, without *p*NP injection. However, biomass yield reduced by injecting *p*NP to the reactor: *Y* reached to 0.27 mg biomass/mg COD by injecting 100 mg/L *p*NP. This fact may cause sludge minimization, but its problem was a small increase in effluent COD which may be about 193 mg/L (COD removal efficiency will be 75%). SOUR amounts were 19, 31, and 7 mg used oxygen per h per gr of MLVSS, when amounts of injected *p*NP were 0, 100, and 130 mg/L, respectively. It means that in high concentrations, *p*NP behaves like a bacterial growth limiter. SVI rate reduced by the increase of injected *p*NP because biomass production reduced then. SVI was 48 mg/L when the injected *p*NP was more than 100 mg/L, and it was 13 mg/L if *p*NP was 130 mg/L, but there was no sludge made if injected *p*NP amount was 150 mg/L. Also, sludge minimization happened by increasing degradation resistant compounds such as *para*-nitrophenol, but it is necessary to reduce pollutant concentration in the effluent to an acceptable environmental standard amount.

Abbreviations

SBR: Sequencing batch reactor; COD: Chemical oxygen demand; BOD: Biochemical oxygen demand; SVI: Sludge volume index; SOUR: Specific oxygen uptake rate; VSS: Volatile suspended solids; *Y*: Biomass yield; SRT: Sludge retention time.

Competing interests

The authors declare that they have no competing interests.

Authors' contributions

AT designed and conceived of the study; generated, assembled, analyzed, and interpreted the data; and drafted and revised the manuscript. AE designed and conceived of the study; collected, assembled, and interpreted the data; and drafted the manuscript. Both authors read and approved the final manuscript.

Acknowledgements

Finally, we gratefully thank the Development of Water and Wastewater Resources Company for their financial support, and labs of Tehran Environment School for the examinations.

Author details

[1]Department of Environmental Health and Environmental Technology Research Centre, Ahvaz Jundishapur University of Medical Sciences, Ahvaz, Iran. [2]Department of Water and Wastewater Engineering, Sazabpardazan Consultant Engineering, Ahvaz, Iran.

References

1. Tokumura, M, Sekine, M, Yoshinari, M, Znad, HT, Kawase, Y: Photo-Fenton process for excess sludge disintegration. Process Biochem. **42**, 627–633 (2007)
2. Takdastan, A, Mehrdadi, N, Azimi, AA, Torabian, A, Nabi Bidhendi, G: Investigation of the excess sludge reduction in SBR by oxidizing some sludge by ozone. Iran. J. Chem. Chem. Eng. **28**(4), 95–104 (2009)
3. Xie, WM, Ni, BJ, Sheng, GP, Yu, HQ, Yang, M: Substrate consumption and excess sludge reduction of activated sludge in the presence of uncouplers: a modeling approach. Appl. Microbiol. Biotechnol. **85**, 2001–2008 (2010)
4. Guowei, C, Pengge1, X, Deqian1, X, Hanqing, Y: Comparison between inhibitor and uncoupler for minimizing excess sludge production of an activated sludge process. Environ Sci Engin China **1**(1), 63–66 (2007)
5. Perez-Elvira, SI, Diez, PN, Polanco, FF: Sludge minimization technologies. Reviews in Environmental Science and Bio/Technology **5**, 375–398 (2006)
6. Yan, S, Miyanagab, K, Xing, XH, Tanjib, Y: Succession of bacterial community and enzymatic activities of activated sludge by heat-treatment for reduction of excess sludge. Biochemical Engineering **39**, 598–603 (2008)
7. Takdastan, A, Mehrdadi, N, Azimi, A.A., Torabian, A., Nabi Bidhendi, G: Intermittent chlorination system in biological excess sludge reduction by sequencing batch reactor. Iran J. Environ. Health. Sci. Eng. **6**(1), 53–60 (2009)
8. Takdastan, A, Azimi, AA, Jaafarzadeh, N: Biological excess sludge reduction in municipal wastewater treatment by chlorine. Asian Journal of Chemistry **22**(3), 1665–1674 (2010)
9. Liu, Y, Tay, JH: Strategy for minimization of excess sludge production from the activated sludge process. Biotech. Adv. **19**(2), 97–107 (2001)
10. Low, EW, Chase, HA: Reducing production of excess biomass during wastewater treatment. Wat. Res **33**(5), 1119–1132 (1999)
11. Liu, Y: Chemically reduced excess sludge production in the activated sludge process. Chemosphere **50**, 1–7 (2003)
12. Canales, A, Pareilleux, A, Rols, J1, Goma, C, Huyard, A: Decreased sludge production strategy for domestic waste water treatment. Wat. Sci. Thecno **30**(8), 96–106 (1994)
13. Chen, GH, An, KJ, Saby, S, Brois, E, Djafer, M: Possible cause of excess sludge reduction in an oxic-settling-anaerobic activated sludge process (OSA process). Water Res. **37**, 3855–3866 (2003)
14. Sabya‹, S, Djafera, M, Hao Chenb, G: Effect of low ORP in anoxic sludge zone on excess sludge production In OSA activated sludge process. Water Res **37**(1), 11–20 (2003)
15. Chen, G, Saby, S: New approaches to minimize excess sludge in activated sludge system. Water Sci. Techno. **44**(10), 203–208 (2003)
16. Wang, J, Zhao, Q, Jin, W, Lin, J: Mechanism on minimization of excess sludge in oxic-settling-anaerobic (OSA) process. Environ Sci. Engin. **2**(1), 36–43 (2008)
17. Khursheed, A, Kazmi, AA: Retrospective of ecological approaches to excess sludge reduction. Water Res. **45**, 4287–4310 (2011)
18. Torregrossa, M, Di Bella, G, Di Trapani, D: Comparison between ozonation and OSA process: analysis of excess sludge reduction and biomass activity in two different pilot plants. Water Sci. Technol. **66**(1), 185–192 (2012)
19. Abbassi, B, Dullstein, S, Rabiger, N: Minimization of excess sludge production by increase of oxygen concentration in activated sludge flocks: experimental and theoretical apperoch. Water Res. **34**(1), 139–146 (2000)
20. Saby, S, Djafer, M, Chen, GH: Feasibility of using a chlorination step to reduce excess sludge in activated sludge process. Water Res. **36**(3), 656–666 (2002)
21. Dytczaka, MA, Londryb, KL, Siegristc, H, Oleszkiewicza, JA: Ozonation reduces sludge production and improves denitrification. Water Res. **41**, 543–550 (2007)
22. Chua, L, Yanb, S, Xingb, XH, Sunc, X, Jurcik, B: Progress and perspectives of sludge ozonation as a powerful pretreatment method for minimization of excess sludge production. Water Research **43**, 1811–1822 (2009)

23. Fazelipour, M, Takdastan, A, Sekhavatjou, MS: Survey on chlorine application in sequencing batch reactor waste sludge in order to sludge minimization. Asian J. Chem. **23**(7), 2994–2998 (2011)
24. Rocher, M, Roux, G, Goma, G, Begue, AP, Louvel, L, Rols, JL: Excess sludge reduction in activated sludge processes by integrating biomass alkaline heat treatment. Water Sci. Technol. **44**(2–3), 437–444 (2001)
25. Mahmoudi, P, Takdastan, A, Nadali Alavi, A, Mosavi, AJ, Kaydi, N: Study of excess sludge reduction in conventional activated sludge process by heating returned sludge. Asian J. Chem. **25**(5), 2627–2630 (2013)
26. Pazoki, M, Takdastan, A, Jaafarzadeh, N: Investigation of minimization of excess sludge production in sequencing batch reactor by heating some sludge. Asian J. Chem. **22**(3), 1751–1759 (2010)
27. Liu, Y, Tay, JH: A kinetic model for energy spilling-associated product formation in substrate-sufficient continuous culture. Journal of Apply Microbiology **88**, 663–668 (2000)
28. Low, EW, Chase, HA: The use of chemical uncouplers for reducing biomass production during biodegradation. Water Sci. Technol. **7**, 399–402 (1998)
29. Low, EW, Chase, HA, Milner, MG, Curtis, TP: Uncoupling of metabolism to reduce biomass production in the activated sludge process. Water Res. **34**, 3204–3212 (2000)
30. Hassani, AH, Nejaei, A, Torabian, A: Excess sludge minimization in conventional activated sludge pilot plant by three chemical matters. Int. J. Environ. Res **5**(4), 981–988 (2011)
31. Koners, U, Toepfl, S, Heinz, V, Camacho, P, Ginestel, P, Knorr, D: Application of pulsed electric field treatment for sludge reduction on wastewater treatment plants. In: Proceedings of the 2nd European Pulsed Power Symposium (EPPS), pp. 47–52. International Society on Pulsed Power Applications, Shaker, Aachen (2004)
32. Hoon Yoon, S, Soo Kim, H, Lee, S: Incorporation of ultrasonic cell disintegration into a membrane bioreactor for zero sludge production. Process Biochem. **39**, 1923–1929 (2004)
33. Lee, NM, Welander, T: Use of protozoa and metazoa for decreasing sludge production in aerobic wastewater treatment. Biotechnol. Lett **18**(4), 429–434 (1996)
34. Ramakrishna, DM: Strategies for sludge minimization in activated sludge process-a review. Fresenius Environmental Bulletin **14**(1), 2–12 (2005)
35. Wei, YS, Van Houten, RT, Borger, AR, Eikelboom, DH, Fan, Y: Minimization of excess sludge production for biological wastewater treatment. Water Res. **37**, 4453–4467 (2003)
36. Hazrati, H, Shaygan, J: Excess sludge reduction methods in activated sludge systems- a review. Iranian Chemical Engineering Journal **10**(54), 67–75 (2011)
37. Okey, RW, Stensel, DH: Uncouplers and activated sludge: the impact on synthesis and respiration. Toxicol. Environ. Chem. **40**, 235–254 (1993)
38. Mayhew, M, Stephenson, T: Biomass yield reduction: is biochemical manipulation possible without affecting activated sludge process efficiency? Water Sci. Technol. **38**, 137–144 (1998)
39. Strand, SE, Harem, GH, Stensel, HD: Activated sludge yield reduction using chemical uncouplers. Water Environ. Res. **71**, 454–458 (1999)
40. Yang, XF, Xie, ML, Liu, Y: Metabolic uncouplers reduce excess sludge production in an activated sludge process. Process Biochem **38**(9), 1373–1377 (2003)
41. Chase, HA, Dennis, JS, Woodgate, J: In: Ginestet, P. (ed.) Scaling up of uncoupling route in comparative evaluation of sludge reduction routes. IWA Publishing, London, UK (2007)
42. Chen, GH, Mo, HK, Saby, S, Yip, WK, Liu, Y: Minimization of activated sludge production by chemically stimulated energy spilling. Water Sci. Technol. **42**(12), 189–200 (2000)
43. Chen, GH, Mo, HK, Liu, Y: Utilization of a metabolic uncoupler, 3,30,40,5-tetrachlorosalicylanilide (TCS) to reduce sludge growth in activated sludge culture. Water Res. **36**, 2077–2083 (2002)

17

Study of a low-temperature Stirling engine driven by a rhombic drive mechanism

Hind El Hassani[1*], Nour-eddine Boutammachte[2], Juergen Knorr[3] and Mohamed El Hannaoui[1]

Abstract

In this paper, a parametric study of the rhombic drive mechanism of a solar low-temperature Stirling engine was conducted. The goal is to find out the relationship between the rhombic drive mechanism parameters and the performance of a low-temperature differential Stirling engine. The results indicate that the offset distance from the gear center and gear radius of the rhombic drive mechanism have to be maximized, in order to increase the indicated power, while other rhombic drive parameters should be reduced for the same reason. Another result is that for realizing the optimized phase angle for beta-type Stirling engines, one should act on working piston-related bars and rhombic drive gear radius.

Keywords: Rhombic drive; Stirling engine; Indicated power; Phase angle

Background

The Stirling engine is an externally heated machine which runs according to a reversible closed cycle. It was invented in 1816 by the Stirling brothers [1]. This engine is known by high heat conversion efficiency, reliability, low noise operation, and ability to use many fuels. These characteristics make the Stirling engines adapted to the demand of the effective use of energy and environmental preservation [2,3]. In fact, reducing environmental impacts of conventional energy resources and meeting the growing energy demand of the global population had motivated considerable research attention in a wide range of environmental and engineering application of renewable form of energy [4], and among all possible alternative energy options, solar energy is becoming more popular in the world. This is mainly due to the availability of plenty of sunlight in many countries [5].

The low-temperature differential Stirling engine (LTD-SE) is a kind of Stirling engine that can run with a small temperature difference between the hot and the cold sources. In 1983, Kolin [6] demonstrated the first LTD-SE. Senft [6,7] developed the Ringbom engine using Kolin's conclusions. Many other prototypes of LTD-SE were manufactured and tested. In 2006, Kongtragool and Wongwises [8] constructed twin power pistons and four power pistons, gamma-configuration LTD Stirling engines. In 2007, Martaj et al. [9] presented a thermodynamic analysis of a low-temperature Stirling engine at steady state operation. In 2012, Boutammachte and Knorr [10] studied a γ-type LTD-SE. They presented several experimental measurements made under laboratory and field-test conditions in Morocco [11]. Measurements with flat plate cooler and discontinuous motion of the displacer were conducted to verify some recommendations of Kolin in expectation of power output improvement. They concluded that the assumption of Kolin related to the proportionality of speed to temperature difference is justified, but the expectation about improvement of the machine performance with discontinuous motion of the two pistons needs to be reviewed.

As a result of a scientific cooperation between Moulay Ismail University in Morocco and the University of Technology of Dresden in Germany, an LTD γ-type Stirling engine was manufactured. This engine was tested and studied under real conditions [11]. Based on the conclusions of this study, a new prototype of LTD-SE with rhombic drive mechanism is being developed. This new prototype is designed to be simple, robust, cheap, and stable to be used for water pumping in poor rural African areas. In fact, this engine is manufactured more

* Correspondence: hind_elhassani@yahoo.fr
[1]Department of Physics, Faculty of Sciences, Moulay Ismail University, Meknès 50050, Morocco
Full list of author information is available at the end of the article

than 90% from mechanical cheap parts available in poor areas, so that it could be easily manufactured and maintained right there. Besides, it is a solar engine, fact that lowers significantly its use price. In addition to that, the new prototype is a beta type which has a symmetric geometry that gives the engine more stability compared to γ-type Stirling engines, and the drive mechanism used, which is the rhombic drive, allows the coaxial position of the engine pistons, fact that improves also the engine stability.

Since this machine is developed to be as low expensive as possible, it was designed without any concentrator mechanism; consequently, its power output is relatively low. Therefore, a special attention should be kept for each construction parameter in order to optimize the machine performance.

In this optic, this paper aims to study the drive mechanism adopted in the studied machine, which is the rhombic drive mechanism. Thus, a parametric study based on Schmidt theory is conducted to find out the relationship between the rhombic drive parameters and the performance of low-temperature Stirling engines. A thermodynamic analysis of the prototype is also conducted. Moreover, as the relationship between rhombic drive mechanism parameters is not enough studied and available for researchers, a study of these parameters is conducted, and a methodology of choosing their values is suggested.

Methods

Prototype description

As a result of a scientific cooperation between Dresden University of Technology and Moulay Ismail University in Morocco, an LTD-SE (beta type) is under construction. The mechanical arrangement of the manufactured LTD-SE and its specifications are shown in Figure 1 and Table 1.

Kinematic relations

Figure 2 shows the parameters of the rhombic drive mechanism. The position and velocity vectors of each link are expressed as follows:

$$\vec{oa} = \left(r_g + e\cos\emptyset\right)\vec{i} + e\sin\emptyset\,\vec{j} \tag{1}$$

$$\vec{v_a} = \frac{d\,\vec{oa}}{dt} = -ew\ \sin\emptyset\,\vec{i} + ew\ \cos\emptyset\,\vec{j} \tag{2}$$

$$\vec{ob} = \left(r_g + e\cos\emptyset + d_1\cos\emptyset_1\right)\vec{i} + \left(e\sin\emptyset + d_1\sin\emptyset_1\right)\vec{j} \tag{3}$$

$$\vec{v_b} = \frac{d\,\vec{ob}}{dt} = \left(-ew\ \sin\emptyset - d_1w_1\sin\emptyset_1\right)\vec{i} + \left(ew\ \cos\emptyset + d_1w_1\cos\emptyset_1\right)\vec{j} \tag{4}$$

$$\vec{ob'} = \vec{oa} + \vec{ab'} = \left(r_g + e\cos\emptyset + d_2\cos\emptyset_2\right)\vec{i} + \left(e\sin\emptyset + d_2\sin\emptyset_2\right)\vec{j} \tag{5}$$

$$\vec{v_{b'}} = \frac{d\,\vec{ob'}}{dt} = \left(-ew\sin\emptyset - d_2w_2\sin\emptyset_2\right)\vec{i} + \left(ew\ \cos\emptyset + d_2w_2\cos\emptyset_2\right)\vec{j}\,. \tag{6}$$

From the geometrical constraints,

$$r_g + e\cos\emptyset + d_1\cos\emptyset_1 = d_3/2 \tag{7}$$

$$r_g + e\cos\emptyset + d_2\cos\emptyset_2 = d_4/2 \tag{8}$$

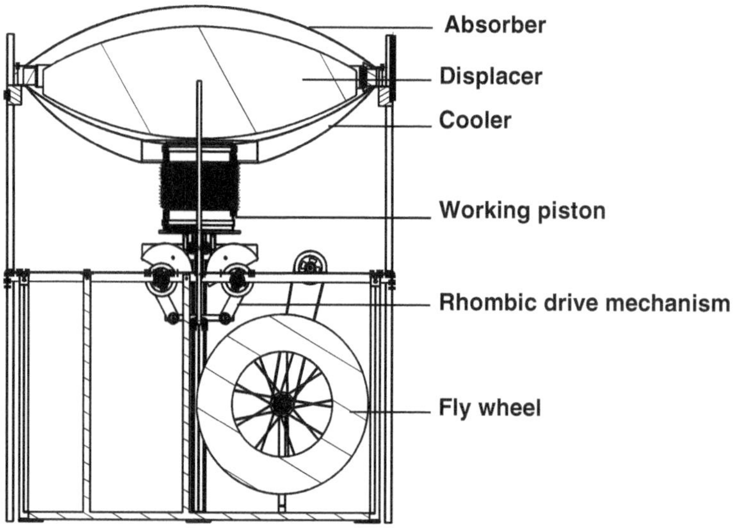

Figure 1 Schematic illustration of the prototype.

Table 1 Technical specifications of the prototype (baseline case)

Parameters	Description/value
Engine type	β
Working fluid	Air
Cooling system	Water cooled
Displacer diameter	1.35 m
Working piston diameter	0.29 m
Absorber diameter	1.51 m
Speed	30 rpm
Absorber temperature	70°C
Cooler temperature	20°C
Regenerator volume	0.063 m^3

$$\emptyset_1 = \cos^{-1}\left[\frac{1}{d_1}\left(\frac{d_3}{2} - r_g - e\cos\emptyset\right)\right] \quad (9)$$

$$\emptyset_2 = \cos^{-1}\left[\frac{1}{d_2}\left(\frac{d_4}{2} - r_g - e\cos\emptyset\right)\right]. \quad (10)$$

Then, the y position of the working piston and the displacer are as follows:

$$Y\ (\text{working piston}) = \text{Ld} + e\sin\emptyset + d_1\sin\emptyset_1 \quad (11)$$

$$Y(\text{displacer}) = \text{Lp} + e\sin\emptyset + d_2\sin\emptyset_2. \quad (12)$$

Based on the prototype data (baseline case), available in Tables 1 and 2, one can get the vertical positions of the working piston and displacer (Figure 3) and variation of engine spaces versus crank angle (Figure 4).

Engine work and power

Schmidt [12] presented a mathematically exact expression for calculating the indicated work of Stirling engines. The Schmidt theory adopts many ideal assumptions, like perfect regeneration, isothermal compression and expansion, and harmonic motion of the displacer and the working piston. The Schmidt theory is widely used due to its mathematical simplicity, and it is considered by many researchers, suitable for studying Stirling engines in designing stage, as it is the case for this study.

Schmidt first calculated the gas mass in different parts of the machine and used the perfect gas equation. Applied on our prototype (Figure 5, Table 3), we obtain the following equations:

$$p = mr\left(\frac{Vc}{Tc} + \frac{V_f}{Tf} + \frac{VTr}{Tf} + \frac{Vr}{Tr} + \frac{V_{mc}}{Tc} + \frac{V_{mf}}{Tf} + \frac{VmTr}{Tf}\right)^{-1}. \quad (13)$$

Schmidt adopted the assumption of harmonic motion:

$$Vc = \frac{V_{SC}}{2}(1 + \cos(\theta + \alpha)) \quad (14)$$

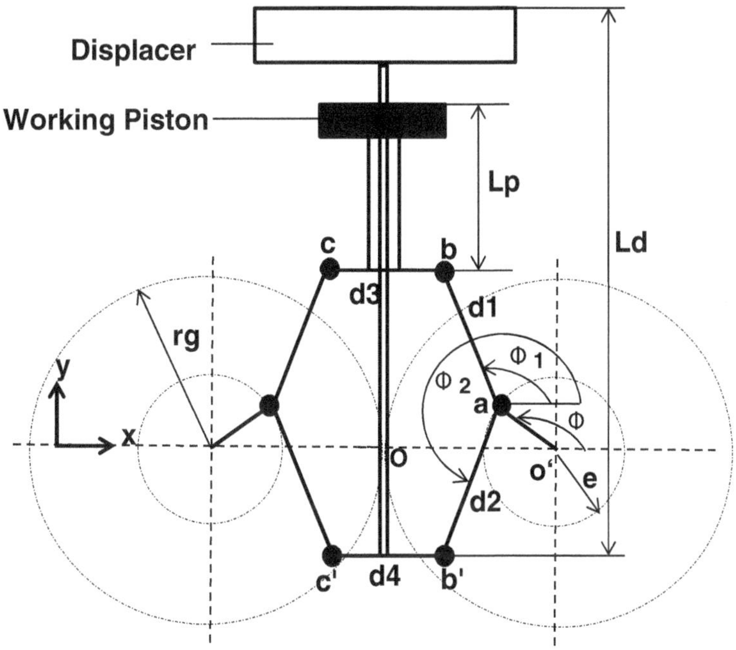

Figure 2 Schematic illustration of the rhombic drive mechanism parameters.

Table 2 Geometrical variables of the rhombic drive mechanism (baseline case)

	Value (m)
e	0.035
d_1	0.125
d_2	0.127
d_3	0.228
d_4	0.226
rg	0.18
Ld	0.808
Lp	0.112

$$Vf = \frac{V_{\mathrm{SC}}}{2}(1 - \cos(\theta + \alpha)) \tag{15}$$

$$V_{\mathrm{TR}} = \frac{V_{\mathrm{STR}}}{2}(1 + \cos\theta). \tag{16}$$

Table 3 Nomenclature and indices

	Description
Nomenclature	
C_p	Specific heat at $p = \mathrm{cst}$ (J/kg K)
C_v	Specific heat at $v = \mathrm{cst}$ (J/kg K)
e	Offset distance from the gear center (m)
h	Specific enthalpy (J/kg)
Ld	Displacer length (m)
Lp	Working piston length (m)
n	Engine speed (Hz)
P	Pressure (Pa)
Q	Heat (J)
r	Specific gas constant (J/kg K)
rg	Rhombic drive gear radius (m)
T	Temperature (K)
V	Volume (m^3)
V_{SC}	Maximal volume of hot space (m^3)
V_{STR}	Maximal volume of working space (m^3)
W	Work (J)
α	Phase angle
Φ	Crank angle
η	Efficiency
Indices	
c	Hot space
f	Cold space
m	Dead (volume)
r	Regenerator
th	Thermal
TR	Working space

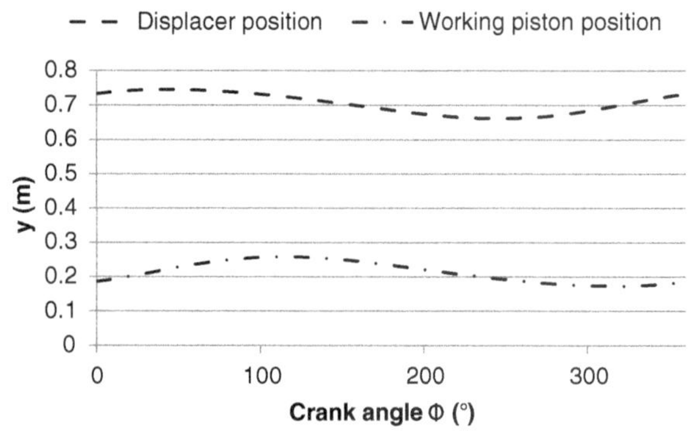

Figure 3 **Working piston and displacer positions versus crank angle.**

The pressure is expressed as follows:

$$P = \frac{mr/k}{1 + \left(\frac{A}{k}\right)\cos(\theta - \beta)} \tag{17}$$

$$k = \frac{V_{\mathrm{SC}}}{2Tc} + \frac{V_{\mathrm{SC}}}{2Tf} + \frac{V_{\mathrm{STR}}}{2Tf} + \frac{2V_r}{Tf + Tc} + \frac{V_{mc}}{Tc} + \frac{V_{mf}}{Tf} + \frac{V_{m\mathrm{TR}}}{Tf} \tag{18}$$

$$A = \frac{1}{2}\left[\left(\frac{V_{\mathrm{STR}}}{Tf}\right)^2 + \left(\frac{V_{\mathrm{SC}}}{Tc} - \frac{V_{\mathrm{SC}}}{Tf}\right)^2 - 2\frac{V_{\mathrm{STR}}}{Tf}\left(\frac{1}{Tf} - \frac{1}{Tc}\right)V_{\mathrm{SC}} \times \cos\alpha\right]^{\frac{1}{2}} \tag{19}$$

$$\tan\beta = \frac{\left(\frac{1}{Tf} - \frac{1}{Tc}\right)\sin\alpha}{\frac{V_{\mathrm{STR}}}{Tf \times V_{\mathrm{SC}}} - \left(\frac{1}{Tf} - \frac{1}{Tc}\right)\cos\alpha}. \tag{20}$$

Cyclic work and power are expressed as follows:

$$W_{\mathrm{tot}} = W_{\mathrm{TR}} = \int_0^{2\pi} P dV_{\mathrm{TR}} = -\frac{1}{2}V_{\mathrm{STR}}\int_0^{2\pi} P\sin\theta\, d\theta \tag{21}$$

$$W_{\mathrm{tot}} = V_{\mathrm{STR}}\frac{\pi mR}{A}\left(\frac{1}{\sqrt{1 - {k_1}^2}} - 1\right)\sin\beta \tag{22}$$

$$P = \frac{n \times W_{\mathrm{tot}}}{2\pi} = V_{\mathrm{STR}}\frac{nmr}{2A}\left(\frac{1}{\sqrt{1 - {k_1}^2}} - 1\right)\sin\beta, \tag{23}$$

where *n* is the engine speed and *m* is the total gas mass.

Thermodynamic analysis

Applying the thermodynamic analysis proposed by Martaj et al. [9] to the particular studied prototype, we obtain the following results.

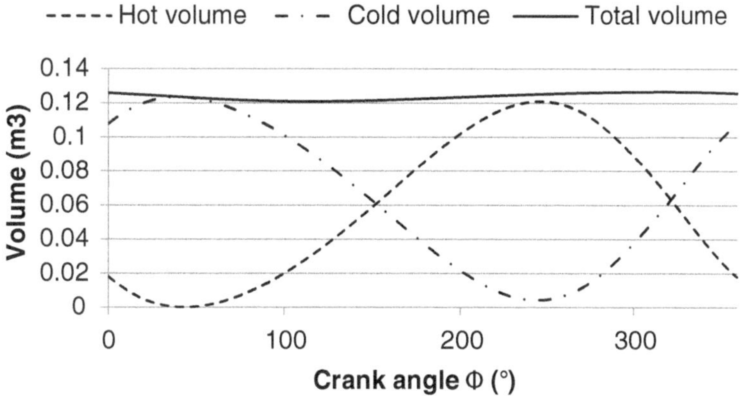

Figure 4 Variation of engine spaces versus crank angle.

Regenerator analysis

The regenerator is used to store and release the heat exchanged with the fluid during its movement from the cold cell towards the hot cell and conversely. Schmidt assumes that the regeneration is perfect. But in practice, it is difficult to realize a perfect regenerator. In this thermodynamic analysis, it is supposed an imperfect regeneration. Thus, the fluid temperature at the exit of the regenerator towards the cold cell $T_{f'}$ is higher than T_f, and the fluid temperature at the exit of the regenerator towards the hot cell $T_{c'}$ is lower than T_c (Figure 6).

The regenerator efficiency is defined by

$$\eta_r = \frac{Tc - Tf'}{Tc - Tf} = \frac{Tc' - Tf}{Tc - Tf} = 1 - \frac{\Delta Tr}{Tc - Tf}, \tag{24}$$

where ΔTr represents the temperature pinch in the regenerator, assumed identical at the two extreme orifices of the regenerator:

$$\Delta Tr = Tf' - Tf = Tc - Tc'. \tag{25}$$

Since the regenerator volume is constant, the work exchanged is null, and the average temperature Tr is supposed to be constant.

$$Tr = \frac{Tf + Tc}{2}. \tag{26}$$

The first thermodynamic principle for an open system is written as

$$\delta Q + \delta W + \sum h_i dm_i = Cv\, d(mT). \tag{27}$$

In the regenerator cell, Equation 27 can be written as

$$\delta Qr = C_p T_{rf} dm_f + C_p T_{rc} dm_c + \frac{C_v}{r} V_r dp, \tag{28}$$

where dm_i is positive when it enters the volume i. Two cases arise for each opening:

- Flow from cold side towards the regenerator: $dm_f < 0$, then $T_{rf} = T_f$, $T_{rc} = T_{c'}$.
- Flow from hot side towards the regenerator: $dm_c < 0$, then $T_{rc} = T_c$, $T_{rf} = T_{f'}$.

The mass in this cell is

$$m_r = \frac{PV_r}{rT_r}. \tag{29}$$

The mass variation in the regenerator is

$$dm_r = -dm_c - dm_h = \frac{V_r}{rT_r} dp. \tag{30}$$

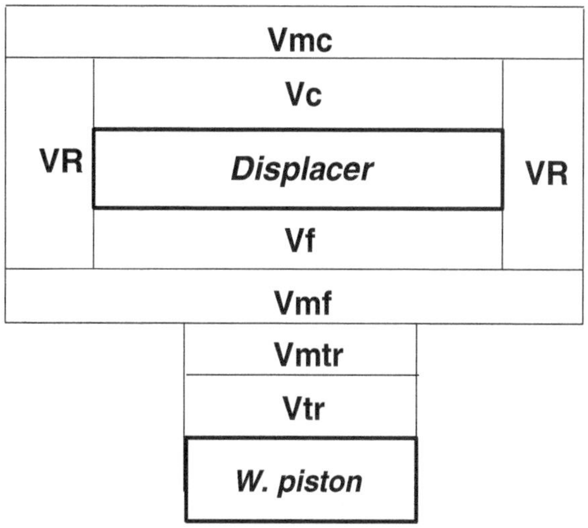

Figure 5 Different prototype spaces.

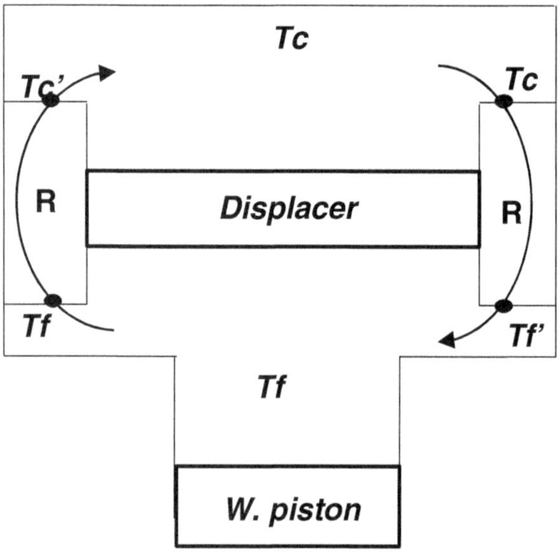

Figure 6 Schematic illustration of the prototype temperatures.

Heating cell

The heating cell has a single communication orifice; Equation 27 could be written in the following form:

$$\delta Q_c + \delta W_c + C_p T_{rc} dm_c = C_v \cdot d(mT)_c. \quad (31)$$

As stated before, if elementary mass is entering the hot volume ($d_{mc} > 0$), then $T_{rc} = T_{c'}$, else $T_{rc} = T_c$.

The differential form of the perfect gas state equation is

$$\frac{dp}{p} + \frac{dV_c}{V_c} = \frac{dm_c}{m_c} + \frac{dT_c}{T_c}. \quad (32)$$

The hot cell is supposed to be isothermal; thus, Equation 33 becomes

$$\frac{dp}{p} + \frac{dV_c}{V_c} = \frac{dm_c}{m_c}. \quad (33)$$

According to the perfect gas state equation,

$$m_c = \frac{PV_c}{rT_c}$$

and

$$dm_c = \frac{V_c\, dp}{T_c\, r} + \frac{P\, dV_c}{T_c\, r} = \frac{1}{rT_c}(V_c dp + PdV_c). \quad (34)$$

Equation 31 becomes

$$\delta Q_c = -C_p T_{rc} dm_c + \frac{C_v}{r} V_c dp + \frac{C_p}{r} PdV_c \quad (35)$$

$$\delta Q_c = V_c \frac{C_p}{r}\left(\frac{1}{\gamma} - \frac{T_{rc}}{T_c}\right)dP + P\frac{C_p}{r}\left(1 - \frac{T_{rc}}{T_c}\right)dV_c. \quad (36)$$

Cooling cell

Using the same reasoning as for the previous cell, Equation 27 becomes

$$\delta Q_f = -C_p\, T_{rf} dm_f + \frac{C_v}{r} V_f\, dp + \frac{C_p}{r} P\, dV_f. \quad (37)$$

If $dm_f > 0$, then $T_{rf} = T_{f'}$, else $T_{rf} = T_f$.

As for the heating cell, we obtain also

$$dm_f = \frac{V_f\, dp}{T_f\, r} + \frac{P\, dV_f}{T_f\, r} = \frac{1}{rT_f}(V_f dp + PdV_f). \quad (38)$$

Thus, Equation 37 becomes

$$\delta Q_f = V_f \frac{C_p}{r}\left(\frac{1}{\gamma} - \frac{T_{rf}}{T_f}\right)dP + P\frac{C_p}{r}\left(1 - \frac{T_{rf}}{T_f}\right)dV_f. \quad (39)$$

Engine balance

During a crankshaft rotation, the heat exchanged in the cooling and heating cells are $Q_c = \oint \delta Q_c$ and $Q_f = \oint \delta Q_f$ obtained by integration of Equations 36 and 39. Since the regenerator is supposed imperfect, an additional quantity of heat Q_r to bring by the hot source, is essential:

$$\oint \delta Q_r = \oint C_p T_{rf} dm_f + \oint C_p T_{rc} dm_c. \quad (40)$$

The overall engine balance is written as

$$W + Q_f + Q_c + Q_r = 0 \qquad W = -\left(Q_f + Q_c + Q_r\right). \quad (41)$$

The work is carried out by the working piston during compression and expansion. It can also be calculated by the following expression:

$$W = W_f + W_c. \quad (42)$$

This work is also

$$W = -\oint PdV, \quad (43)$$

where dV is the total fluid variation and P is the internal pressure. The thermal engine efficiency is expressed by

$$\eta_{\text{th}} = \frac{W}{Q_c + Q_r}. \quad (45)$$

Rhombic drive study

The rhombic drive mechanism was proposed by the Philips Company [13] and widely used for beta-type Stirling engines [14]. The major advantage of the rhombic drive mechanism is that the coaxial movement of the piston and the displacer is quiet and requires no serious lubrication [15].

As the relationship between rhombic drive mechanism parameters is not studied enough and available for researchers, a study of these parameters was conducted. Here are some conditions for good functioning of the rhombic drive mechanism:

1. At $\Phi = 0°$, the rigid rods d_1 and d_2 should be tall enough in order to reach the most far point during the cycle:

$$d_1 > \text{rg}-\frac{d_3}{2}+e \quad (46)$$

$$d_2 > \text{rg}-\frac{d_4}{2}+e. \quad (47)$$

2. At $\Phi = 180°$, gears should be far enough from each other:

$$\text{rg} > e+\frac{d_3}{2} \quad (48)$$

$$\text{rg} > e+\frac{d_4}{2}. \quad (49)$$

As a consequence, during the design of the rhombic drive mechanism, we propose the following order to choose the rhombic drive parameters:

Choosing the crossbar lengths d_3 and d_4 and the offset distance from the gear center e

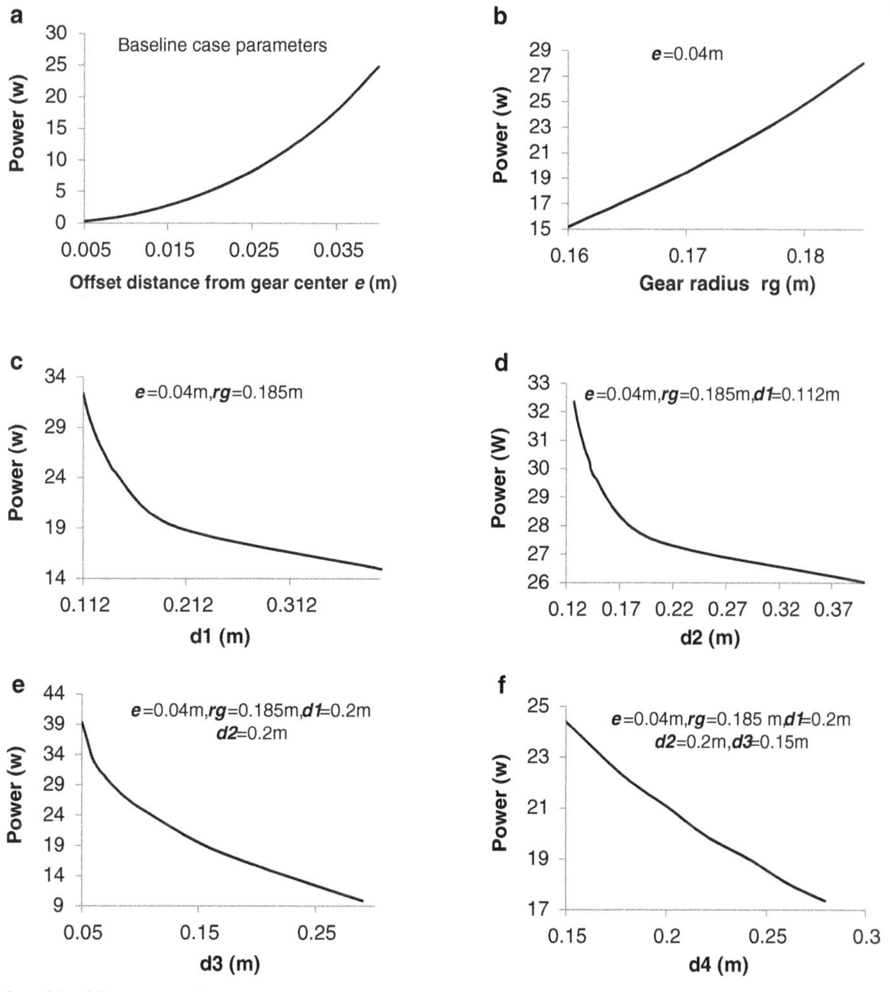

Figure 7 Effect of rhombic drive parameters on power.

1. Calculating the interval of values allowed to gear radius as follows:

$$\min\left(e+\frac{d_3}{2}, e+\frac{d_3}{2}\right) < rg < \max\left(d_2+\frac{d_4}{2}-e, d_1+\frac{d_3}{2}-e\right). \quad (50)$$

2. Choosing the connecting bar lengths d_1 and d_2, with respect to the conditions below:

$$d_1 > rg-\frac{d_3}{2}+e \quad (51)$$

$$d_2 > rg-\frac{d_4}{2}+e. \quad (52)$$

A parametric study based on the theoretical analysis is performed in order to evaluate the effect of each rhombic drive parameter on the machine characteristics. Note that Schmidt calculating assumptions are adopted. At each study, only one parameter is changed in order to find out its single effect. Results are found using a computer program on the Microsoft EXCEL. For a complete revolution of the crank and at a step of 1°, the program computes the instantaneous positions of the displacer and the working piston, the volume of each space of the machine, and the total mass, work, and power of the cycle. The program has the ability to change power calculation parameters in order to show their single effect.

The geometrical and the operating variables which are regarded as the baseline case in this study are listed in Tables 1 and 2. In this case, the rotational speed is set at 30 rpm; the absorber and the cooler temperatures are set at 70°C and 20°C, respectively.

Results and discussion

Figure 7 displays the effect of rhombic drive parameters on indicated power. Figure 7a,b shows that any increase in offset distance from the gear center (e) or gear radius

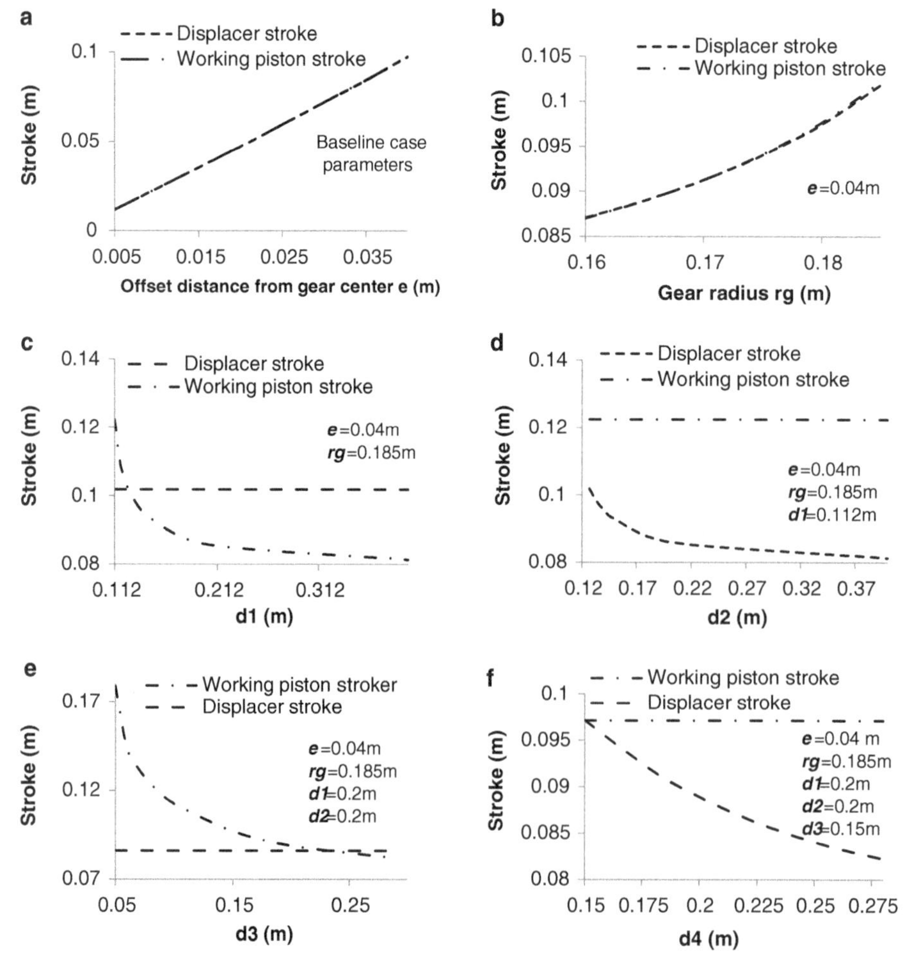

Figure 8 **Effect of rhombic drive parameters on piston strokes.**

(rg) has a positive influence on power. Figure 8a,b gives the explanation of the former fact; when increasing e and rg, strokes of displacer and also working piston increase, then dead volumes are reduced, and total volume variation is increased. All of these have direct positive influence on indicated power. It is noticed that for the same variation of e and rg ($\Delta e \approx \Delta rg \approx 0.025$ m), the offset distance from gear e has larger positive effect on the indicated power than rg ($\Delta P(e) = 18$ W, $\Delta P(rg) = 13$ W). Then, considering only power, the optimal values of e and rg are their maxima values, but one should keep in mind that we can increase the stroke and power consequently - just as much as the machine volume allows it. Maximum value of rg is

$$rg = \min\left(d_2 + \frac{d_4}{2} - e, d_1 + \frac{d_3}{2} - e\right). \quad (53)$$

It is noticed that strokes of displacer and working piston are equal and vary in the same way while changing e and rg. Shown in Figure 7c,d,e,f are the respective effects of the rhombic drive parameters d_1, d_2, d_3, and d_4 on indicated power. These figures indicate that increasing the value of any of these parameters has negative effect on power. The same reasoning as for e and rg could be done. Figure 8c,d,e,f gives the explanation of the former fact; any increase in d_1, d_2, d_3, and d_4 values decreases strokes of both displacer and working piston, then dead volumes will increase, and total volume variation will be decreased, and consequently, the indicated power will be reduced. It is noticed that for the studied machine, for the same variation of d_1 and d_2 ($\Delta(d_1) \approx \Delta(d_2) \approx$ 0.273 m), the negative effect of d_1 is greater three times than that of d_2 ($\Delta P(d_1) = -18$ W, $\Delta P(d_2) = -6$ W). This is due to the fact that the working piston is related to bar d_1; thus, its effect is much more important than that of bar d_2, which is related to the displacer which has no other role but moving gas from one side to another, whereas the working piston has an effect on total volume variation, directly related to indicated work and power. Also, for bars d_3 and d_4, similar remarks and conclusion could be done. For the same variation, the effect of d_3 is greater two times than that of d_4 for the

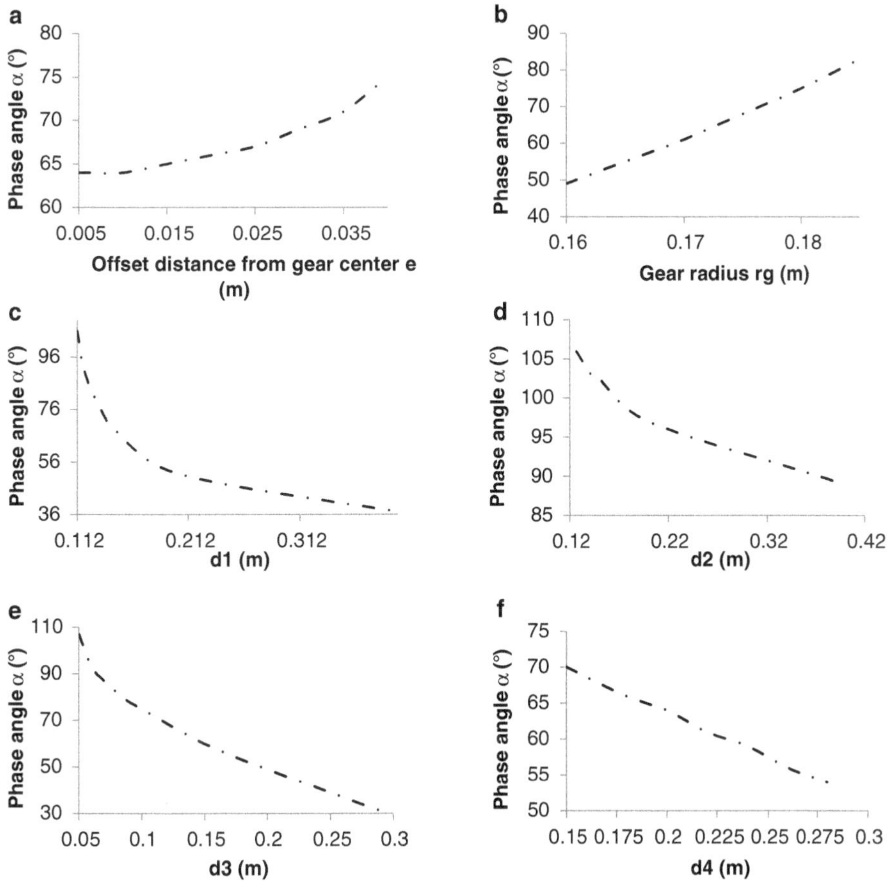

Figure 9 Effect of rhombic drive parameters on phase angle.

studied prototype. Therefore, for increasing the machine power, minima values of these parameters should be adopted:

$$d_1 = e + \text{rg}-\frac{d_3}{2} \tag{54}$$

$$d_2 = e + \text{rg}-\frac{d_4}{2} \tag{55}$$

$$d_3 = 2(e + \text{rg}-d_1) \tag{56}$$

$$d_4 = 2(e + \text{rg}-d_2). \tag{57}$$

Figure 8 indicates that parameters d_1 and d_3 have an effect only on working piston stroke, but no influence on displacer stroke, while parameters d_2 and d_4 have an influence only on displacer stroke and no effect on working piston stroke. For parameters *e* and rg, they have simultaneous equal effect on both displacer and working piston strokes. This gives a directive about adjusting the stroke of one of the pistons (for example, for reducing dead volume) without changing the stroke of the other. Many investigations about optimal phase angle for beta-type Stirling engines show that the optimal value is 90° [16].

Figure 9 shows the influence of different parameters of the rhombic drive mechanism on phase angle. As noticed for power, the increase in values of the offset distance from the gear center (*e*) or gear radius (rg) increases the phase angle, whereas increasing the rest of the rhombic drive parameter values decreases the phase angle. It is noticed that for the same variation of *e* and rg, the effect of gear radius rg is greater four times than the effect of *e*. Also, the effect of parameter d_1 is greater three times than that of d_2 on the phase angle of the studied prototype, and the effect of d_3 is greater two times than that of d_4. This fact is due to the tightness of working space versus displacer-related space, which makes the working piston parameters having more considerable influence on the phase angle than that of the displacer-related parameters. As a conclusion, for obtaining a special phase angle, one should pay a special attention to parameters rg, d_1, and d_3.

Conclusions

Theoretical analysis has been performed to study the rhombic drive mechanism parameter effect on a low-temperature solar Stirling engine. The variation of the machine spaces was investigated, and a thermodynamic analysis and a calculation of power based on Schmidt theory were conducted. A parametric study was then performed to evaluate the effect of each rhombic drive mechanism parameter on some machine characteristics, like power, piston and displacer strokes, and phase angle. In accordance with the obtained results of this study, offset distance from the gear center and gear radius have to be maximized in order to maximize the indicated power, while other rhombic drive parameters should be reduced for the same reason. Their minima values were given relatively to each other. Directives about realizing the optimal phase angle of beta-type Stirling engines were also given. As a conclusion, based only on rhombic drive optimization, one can increase the studied prototype power about 50%.

Competing interests

The authors declare that they have no competing interests.

Authors' contributions

HEH performed the rhombic drive and the thermodynamic studies. JK designed and manufactured the studied prototype. NB supervised the research, performed the Schmidt calculation, and helped JK on constructing the studied machine. MEH supervised the research. All authors read and approved the final manuscript.

Acknowledgements

The authors are grateful to the staff of the Energy Institute of the Technical University of Dresden for their support.

Author details

[1]Department of Physics, Faculty of Sciences, Moulay Ismail University, Meknès 50050, Morocco. [2]Department of Energy, Ecole Nationale Supérieure d'Arts et Métiers, Moulay Ismail University, Meknès 50050, Morocco. [3]Institute of Energy, Faculty of Mechanical Engineering, University of Technology, Dresden 01062, Germany.

References

1. Kolin, I: Stirling Motor: History-Theory-Practice. Zagreb University Publications, Zagreb (1991)
2. Bahrami, M, Hamidi, A, Porkhial, S: Investigation of the effect of organic working fluids on thermodynamic performance of combined cycle Stirling-ORC. Int. J. Energ. Environ. Eng. **4**, 12 (2013)
3. Thombarea, DG, Verma, SK: Technological development in the Stirling cycle engines. Renew Sustain Energy Rev. **12**, 1–38 (2008)
4. Sunday, O, Oyedepo, SO, Adaramola, MS, Paul, S: Analysis of wind speed data and wind energy potential in three selected locations in south-east Nigeria. Int. J. Energ. Environ. Eng. **3**, 7 (2012)
5. Sharma, P, Harinarayana, T: Enhancement of energy generation from two layer solar panels. Int. J. Energ. Environ. Eng. **3**, 12 (2012)
6. Senft, JR: An Introduction to low temperature differential Stirling engines. Moriya Press, River Falls (2004)
7. Robson, A, Grassie, T, Kubie, J: Modeling of a low-temperature differential Stirling engine. Proc. IMechE. Part C J. Mech. Eng. Sci. **221**, 927–943 (2007)
8. Kongtragool, B, Wongwises, S: Performance of low-temperature differential Stirling engines. Renew. Energ. **32**, 547–566 (2007)
9. Martaj, N, Grosu, L, Rochelle, P: Thermodynamic study of a low temperature difference Stirling engine at steady state operation. Int. J. Thermodynam. **10**, 165–176 (2007)
10. Boutammachte, N: Untersuchungen zur instationären Wärmeübergangs- und Strömungsvorgängen in einem Niedertemperatur-Stirlingmotor. PhD thesis. University of Technology of Dresden (2007)
11. Boutammachte, N, Knorr, J: Field-test of a solar low delta-T Stirling engine. Solar. Energ. **86**, 1849–1856 (2012)
12. Schmidt, G: Theorie der lehmannschen calorischen maschine. Zeit des Vereines deutsch Ing **15**, 97–112 (1871)
13. Hargreaves, CM: The Philips Stirling Engines. Elsevier, New York (1991)

14. Kerdchang, P, Win, M, Teekasap, S, Hirunlabh, J, Khedari, J, Zeghmati, B: Development of a new solar thermal engine system for circulating water for aeration. Sol. Energy **78**, 518–527 (2005)
15. Cheng, C, Yu, Y: Combining dynamic and thermodynamic models for dynamic simulation of beta-type Stirling engine with rhombic-drive mechanism. Renew. Energy **37**, 161–173 (2012)
16. El Hassani, H, Boutammachte, N, Knorr, Y, Hannaoui, M: Etudes des paramètres influençant la puissance d'un moteur Stirling solaire à basse température. In abstracts of the 11th Mechanics Congress, University Ibn Zohr, Agadir, 23-26 (2013)

Integration of evaporative cooling technique with solar chimney to improve indoor thermal environment in the New Assiut City, Egypt

Amr Sayed Hassan Abdallah[1,2*], Hiroshi Yoshino[2], Tomonobu Goto[2], Napoleon Enteria[2], Magdy M Radwan[1] and M Abdelsamei Eid[1]

Abstract

Cooling buildings in summer is one of the main environmental problems for architects and occupants in many hot dry countries. The summer temperature during these countries reaches peaks of more than 40°C in some. Mechanical air conditioners can solve the problem, but they put a heavy strain on the electricity consumption. Egypt in general has rich sunny and clear skies. Therefore, these conditions encourage to enhance evaporating with natural ventilation and save energy. This paper develops an integration of direct evaporative cooling tower with a solar chimney multi-zone thermal ventilation model. Simulation is done using commercial couple multi-zone airflow under COMIS-TRNSYS software (Madison, WI, USA) to assess natural ventilation and indoor thermal comfort. The results show that the system generates 130.5 m^3/h under the effect of solar radiation only and minimum 2 ACH without pressure coefficient which is considered the minimum requirement of ACH. The findings show that the new integrated system interacts with the building envelope and weather conditions to achieve a decrease in indoor temperatures that reach 10°C to 11.5°C compared to outdoor temperatures.

Keywords: Passive cooling; Solar chimney; Integration system; TRNSYS-COMIS; Multi-zone models

Background

Buildings contribute over 40% of the total global primary energy use corresponding to 24% of the world CO_2. Heating, ventilation, and air conditioning (HVAC) system is the technology of indoor and vehicular environmental comfort which is responsible for about half of the energy use in buildings [1]. The quest to reduce environmental impacts of conventional energy resources and, more importantly, to meet the growing energy demand of the global population had motivated considerable research attention in a wide range of environmental and engineering application [2]. Effective integration of passive features into the building design can significantly minimize the air conditioning demand while maintaining thermal comfort [3]. Passive evaporative cooling is one of the most efficient and long recognized ways of inducing thermal comfort in predominantly hot dry climates. Historically, evaporative cooling was used extensively in traditional architecture throughout the world's hot arid countries [4-6]. In order to extend the use of evaporative cooling, short wind tower cooling was integrated with inclined solar chimney to increase the pressure difference and the system efficiency. This was done after observing the indoor discomfort for one of the Egyptian houses built in the New Assiut City [7].

Many researches have been conducted on using natural ventilation and evaporative cooling strategies for producing cool air and the effect of using a solar chimney on thermal-induced ventilation in buildings. Three important researches were under taken to study these passive systems. Bahadori introduced the idea of capturing the wind within a tower and then pass it through wetted conduit walls [4]. Maerefat and Haghighi also put forward a new solar system employing a solar chimney together with an evaporative cooling cavity. The numerical calculation showed that this integrated system with

* Correspondence: arch_amrsayed@yahoo.com
[1]Department of Architecture, Assiut University, P.O. Box 71516, Assyut, Egypt
[2]Graduate School of Engineering, Tohoku University, Aoba 6-6-11-1201, Sendai 980-8579, Japan

the proper configurations was capable of providing good indoor conditions during daytime in the living room even at a poor solar intensity of 200 W/m^2 and a high ambient air temperature of 40°C [8]. Alemu et al. developed an integrated model incorporating passive airflow components into a coupled multi-zone ventilation and building thermal model. This model allows an assessment of a combination of passive features such as solar chimney and wind-induced earth-air tunnel for both natural and hybrid ventilation systems at the design stage [9]. A lot of awareness is ongoing worldwide on solar energy utilization [10].

The main objective of this paper is to develop an integrated solar chimney with passive cooling strategies into a coupled multi-zone ventilation and building thermal model. This was done by developing a mathematical model to incorporate passive features into a multi-zone ventilation model incorporating a simple thermal model using COMIS-TRNSYS software (Madison, WI, USA). This development system is used in a room of a single zone to study the performances and the advantages of an integrated system in the numerical model. Also, it focuses on the simulated performances and its impact on the comfort of occupants for the summer period with an integrated proposed system.

A theoretical study of the direct evaporative wet medium

An evaporative cooler (EC) is a device that cools air by passing it across a wet medium. As hot air passes over the wet medium, some of its energy goes to evaporate water from the medium. The air comes out of the other side of the device having given up some of its energy and consequently exits at a lower temperature with higher humidity and constant enthalpy. The minimum temperature that can be reached is the thermodynamic wet bulb temperature (TWBT) of the incoming air. It was reported that half of the conventional cooling device in the south west of the USA was replaced by simple direct evaporative air coolers; this would save 18 million barrels of oil per year [5,11].

In this paper, heat and mass transfer from the wet medium into the airflow are taken into account. Based on the theoretical study and initial system conditions, evaporative cooler performance is based on the concept of the adiabatic process. This means no heat is added to or removed from the system; the process of exchanging the sensible heat of the air for latent heat of evaporation from the water is an adiabatic process [12].

A description of the system

The performance of the system is to provide desired comfortable conditions and a suitable rate of ventilation, depending on several parameters such as the ambient conditions (temperature, solar radiation, relative humidity, wind speed, and pressure coefficient). The following dimensions and specifications are applied to the model: The system is located in the New Assiut City, Egypt. New Assiut City is located at 27.30°N latitude and 31.15°E longitude. The solar chimney is oriented towards the south. The calculations were carried out for a single zone and to study the integration of the new proposed system, having a dimension of 4.0 m × 4.0 m × 3.125 m. This system was developed from the reference system of inclined solar chimney attached to a room with an evaporative cooling cavity (ECC) [8]. The old model assumes zero pressure boundaries at the inlet and outlet except for the stack effect. Therefore, it is not taken under consideration the variety of the buildings' static pressure from the atmospheric pressure and the effect of wind pressure coefficient on buildings. Also, there are many reasons that make it very difficult to integrate this system into Egyptian housing. The ECC strategies consumed a lot of water because of water film and a lot of space. Figure 1 shows the diagram of the reference system and the new proposed system.

Nevertheless with the new proposed system, the inclination length of the chimney is 2 m on the inclined angle in order not to extend the vertical height (1.5 m) according to Egyptian Building Regulation Law, and the height of the cooling tower is 1 m. The chimney air gap is 0.2 m, and the inclination of the chimney is 50° based on the angle of the reference system. Table 1 shows the parameters of the solar chimney. Figure 2 shows the location of the room and its dimension in relation to a northern orientation. The performance of the system was monitored at a steady state of condition and then 5 days in the month of June (19, 20, 21, and 23) were selected for studying the performance of the solar chimney between sunrise and sunset in order to study the timewise to focus on the hottest day in the summer season (20 June). Then, parametric studies for solar chimney and wind tower are needed in the future studies in order to choose the best dimensions for a new proposed compact design. Some parameters used in the calculations of solar chimney appear in Table 1.

The evaporative pad provides a large water surface, and the pad is wetted by dripping water from the above source. The EC works by a concentric float valve, which opens when the water level is low in the collecting grill, allowing more water to enter. When the water level returns to the full level, the valve is shut automatically.

Methods

The system is being developed with a solar chimney and a small evaporative cooling wind tower. The experimentation would become time consuming and very expensive. Therefore, the numerical model is made and developed for the system. The developed model is implemented in

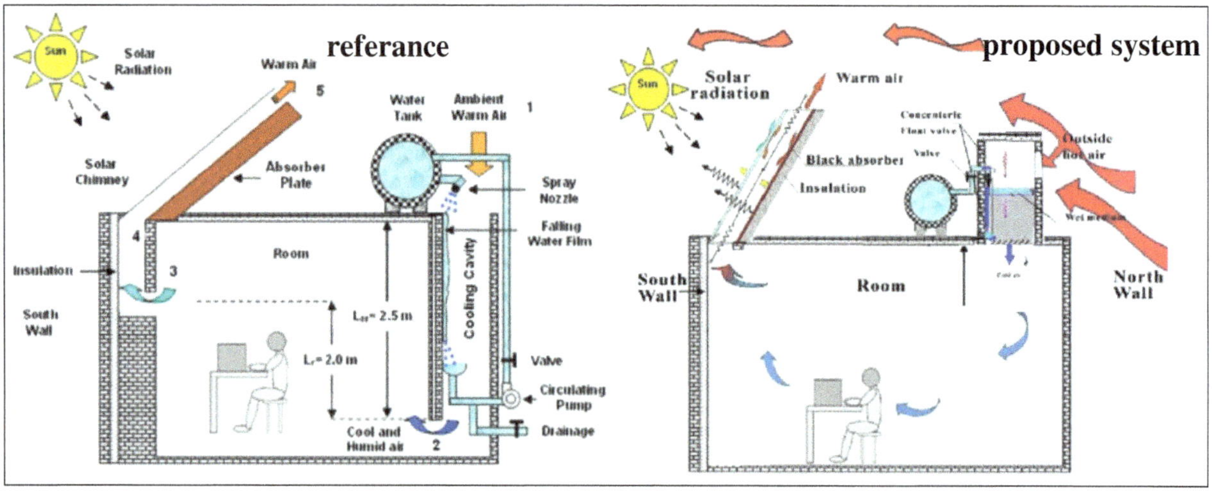

Figure 1 Diagram of solar chimney with evaporative cooler in the reference system and new proposed system.

the COMIS-TRNSYS simulation software. This direct evaporative cooling tower used a component 506d-TESS library which uses a wet medium at the top, and this component was assembled and validated by the Thermal Energy System Specialists, LLC (Madison, WI, USA) and was modified in 2004. The present model includes special airflow component which needs simultaneous prediction of temperature and airflow rates. In the developed multi-zone ventilation model, a building is idealized as a system of zones, openings, and ducts linked together by discrete airflow paths. Zones are represented by nodes. Chimney and evaporative cooling channels are represented by ducts for resistance and pressure drop calculations. A hydrostatic condition is assumed in zones, and the flow rate in each link is defined as a function of zone pressure, which results in a system of nonlinear equation solver, defined by the mass conservation for each zone [13]. Figure 3 shows the flow chart solving procedure of TRNSYS-COMIS.

The equations are solved numerically to predict zone pressures, and zone pressures are back substituted in the flow equations to predict the airflow rate at each link [13]. The detailed equations used for large openings and ducts are presented with the fundamentals of the multi-zone airflow model (COMIS, [14,15]. A description of the mathematical models used in each component can be found in the TRNSYS 16 manual handbook. Multi-zone airflow network models deal with the complexity of flows in a building by recognizing the effects of internal flow restrictions. They require extensive information about flow characteristics and pressure distributions [16]. The pressures are calculated in the previous steps as an initial guess for pressures in the next step. Initial zone densities are based on the pressures found in the last step. The solver updates zone densities as it updates zone pressures [13]. Figure 4 shows the flow chart solving procedure for COMIS-TRNSYS.

Table 1 The parameters used in the calculation [7]

Parameters	Value (dimensionless)
Transmissivity of glass	0.84
Absorptivity of glass	0.06
Emissivity of the glass	0.90
Absorptivity of absorber wall	0.95
Emissivity of the absorber surface	0.95

Pressure calculation and mass flow rate through openings

To estimate the amount of air flow through an opening, it is necessary to know the pressure difference across the opening and its effective flow area. The pressure at an opening can be due to wind, as well as buoyancy, and it is therefore determined by the location of the opening, as well as the internal and external environmental parameters such as wind velocity, pressure coefficient, and outdoor temperatures.

As solar radiation passes through the glazing and absorbed at the wall surface. The air in the chimney is then heated by convection and radiation from the absorber. The decrease in density experienced by the air causes it to rise, whereupon it is then replaced by air from below, i.e., from the attached room. The rate at which air is drawn through the room depends upon the buoyancy force experienced (i.e., depending upon the temperature differential), the resistance of flow through the chimney, and the resistance to the entry of fresh air from the room. While in a wind tower, the opening faces north towards the preferable wind. This compact design with dimensions of 1 m (width) × 0.7 m (depth) and a height of 1 m depends on solar radiation and solar chimney that sucks air from the room and increases the air flow inside the room.

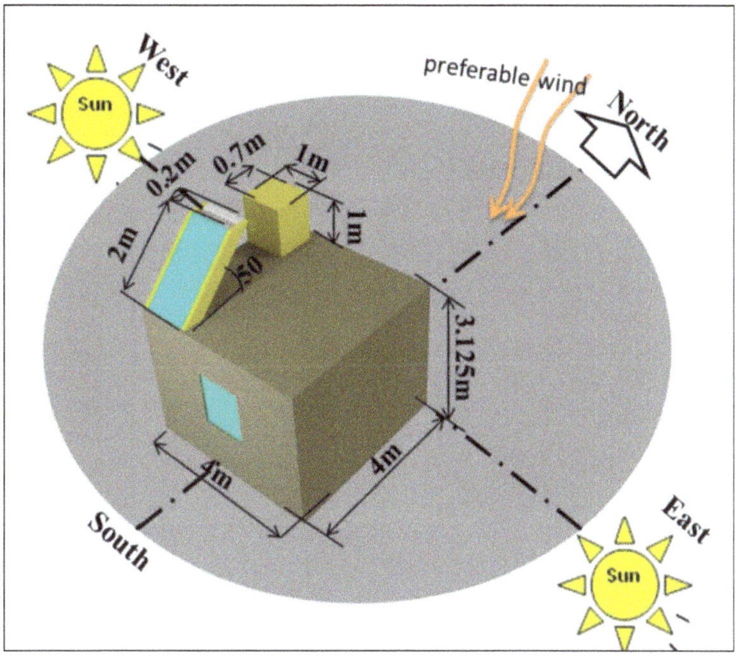

Figure 2 **The location of the room with the integrated system according to north orientation.**

The ventilation rate inside the system is calculated according to six variables (P_2, P_3, P_4, P_5, P_6, and P_7) which are solved from Equations 1 to 6, taking into account the base pressure on the ground ($P = 0$) as presented in Figure 3.

$$P_2 = \left(-\rho g h_2 + 0.5\rho_1 Cp_{\text{north}} v^2\right) - \frac{\rho_2}{2}\left(\frac{Q}{\alpha_1 A_1}\right)^2, \quad (1)$$

where $P_1 = (-\rho g h_2 + 0.5\rho_1 Cp_{\text{north}} v^2)$ is the pressure at point 1 N/m^2, α is the discharge coefficient of the opening in the wind tower, and A_1 is the area of the opening (1 m^2).

$$P_3 = P_2 - \left(\zeta_{\text{WetMediumInlet}} + \zeta_{\text{DuctOutlet}} + \zeta_{\text{DuctFriction}}\right)\frac{\rho_3}{2} v_A^2 + \rho_A g h_r, \quad (2)$$

where the local pressure losses in the inlet $\zeta_{WetMediumInlet}$ of the evaporative cooling (wet medium), the friction of

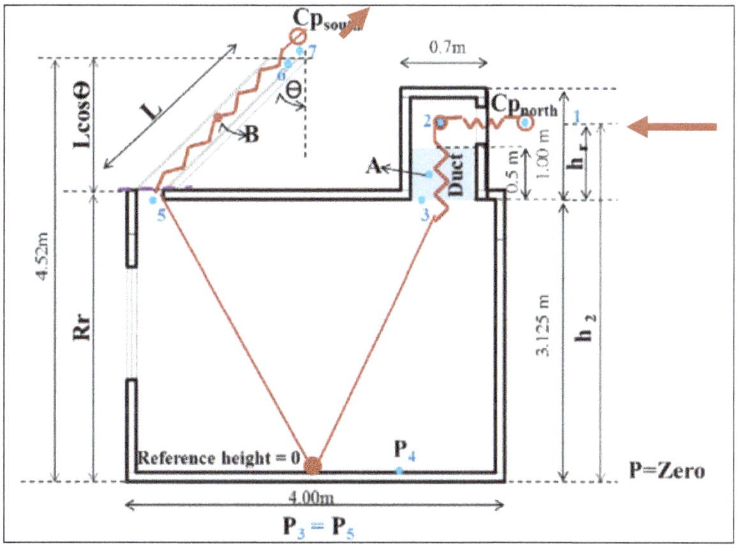

Figure 3 **A zone connected to a roof solar chimney with wind tower.**

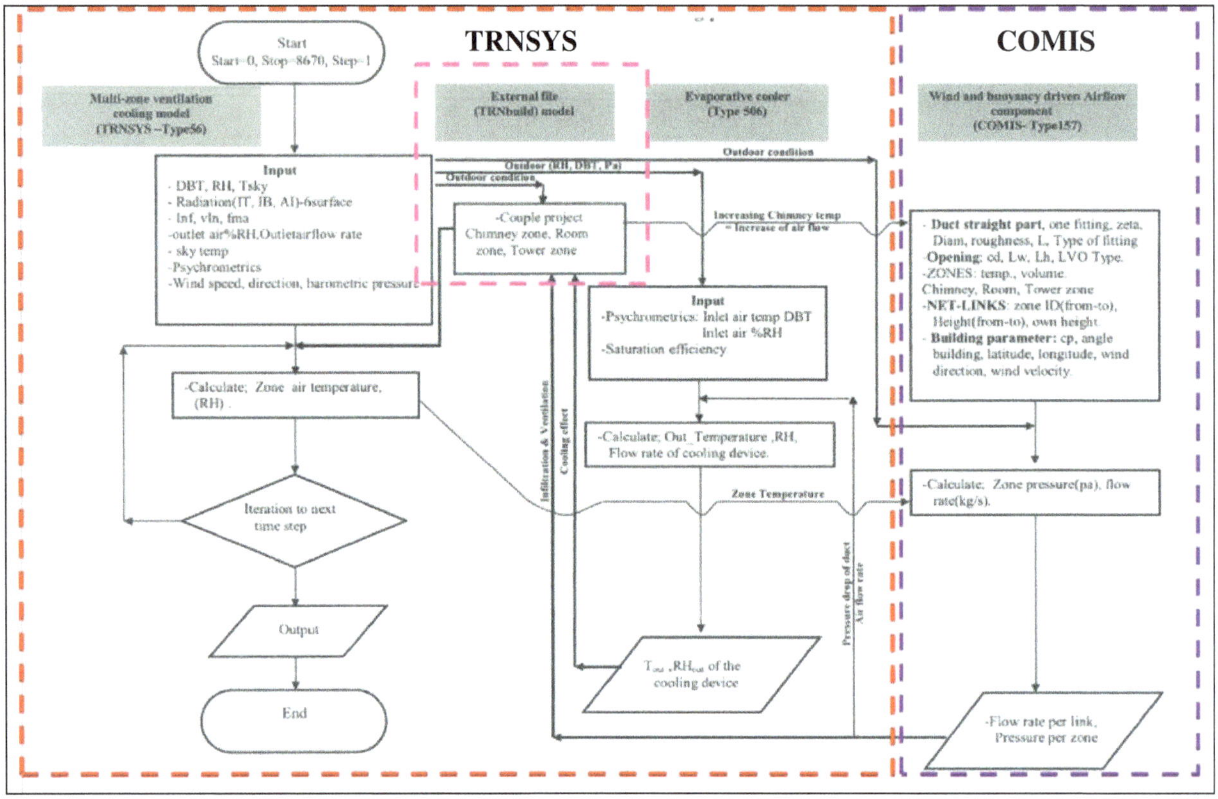

Figure 4 Flow chart solving procedure of TRNSYS-COMIS.

duct $\zeta_{DuctFriction}$, and local losses of the duct outlet $\zeta_{DuctOutlet}$ can be calculated according to the dimensions of dynamic losses due to duct fitting, *Re*, roughness, duct dimension, and λ. It can be determined in details from COMIS [13]. Also, the indoor velocity equals $\boldsymbol{V}_A = \frac{Q}{A}$ (m/s).

$$P_4 = P_3 - \rho_3 g R_r \tag{3}$$

$$P_5 = P_4 + \rho_4 g R_r \tag{4}$$

$$P_6 = P_5 - \left(\zeta_{\text{ChimneyInlet}} + \zeta_{\text{ChimneyOutlet}} + \zeta_{\text{ChimneyFriction}}\right)\frac{\rho_B}{2}v_B^2 + \rho_B g(L\cos\theta), \tag{5}$$

where the local pressure losses $\zeta_{ChimneyInlet}$ in the inlet of a solar chimney, the friction of chimney $\zeta_{ChimneyFriction}$, and local losses of the chimney outlet $\zeta_{ChimneyOutlet}$ can be calculated in a similar approach as used in evaporative cooling. The solar chimney is treated in the COMIS program as a rectangular duct connects to the room from one side and a node with volume that equals to zero from the other side to transfer the calculated temperature of the chimney from TRNSYS and then another rectangular duct to outside.

$$P_7 = P_6 - \frac{\rho_6}{2}\left(\frac{Q}{\alpha_6 A_6}\right)^2, \tag{6}$$

where $P7 = -\rho g(R_r L\cos\theta) + 0.5\rho_1 Cp_{\text{south}} v^2$, and $\rho = \frac{353.25}{T} = \frac{353.25}{\theta + 273.15}$ is the air density (kg/m^3). Then, the pressure differences across the opening due to buoyancy and wind are required and solved with each others in order to predict the pressure of the room and ventilation rate Q (m^3/s).

Temperature prediction in the solar chimney

The average air density in the chimney channel ρ_6 is predicted using the thermal resistance network model in TRNbuild-TRNSYS, based on the heat balances of each element of solar chimney. The model used by Ong [17] ang Bassiouny and Koura [18] modified by Alemu et al. [9] is adopted with some assumptions in order to solve the mathematical model. Flow through the chimney was considered in a steady state. The chimney inlet area and exit areas are equal. The thermal network for the physical model considered is shown in Figure 5.

The heat balance equations from the thermal network at each element of the solar chimney are described in Equations 7, 8, and 9:

$$T_g(\text{glass cover}):\; S_g + h_{rwg}(T_w - T_g) + h_g(T_f - T_g) = U_g(T_g - T_a) \quad (7)$$

$$T_{f(Air)}:\; h_w(T_w - T_f) = h_g(T_f - T_g) + q'' \quad (8)$$

$$T_w(\text{absorber}):\; S_w = h_w(T_w - T_f) + h_{rwg}(T_w - T_g) + U_w(T_w - T_a). \quad (9)$$

The energy balance on the air flowing through the chimney q'' with a length L and width of the air channel:

$$q^{WL''} = \dot{m}\, C_f(T_{airout} - T_{airin}), \quad (10)$$

where $m^{\bullet}$ is the mass flow rate of the air component.

The mean air temperature T_f is given by

$$T_{f(air)} = \gamma T_{airout} + (1 - \gamma) T_{airin}, \quad (11)$$

where γ (mean temperature weighting factor) = 0.75 [19].

Consider the air inlet to the chimney that has a temperature equal to the room average temperature, substitute in the above equation of q''_{WL}:

$$q''_{WL} = \dot{m}\, C_f \frac{T_{f(air)} - T_{airin}}{\gamma}. \quad (12)$$

The heat transfer coefficient from the glass to atmospheric air

$$U_g = h_{rs} + h_{wind}. \quad (13)$$

The radiation heat transfer coefficient between the glass and the sky is

$$h_{rs} \frac{\sigma_g \varepsilon_g g (T_g + T_{sky})(T_g^2 + T_{sky}^2)(T_g - T_{sky})}{T_g - T}, \quad (14)$$

where σ = solar absorptance, ε = emissivity, g = gravitational acceleration (9.81), $T_{sky} = 0.0552\, T_a^{1.5}$.

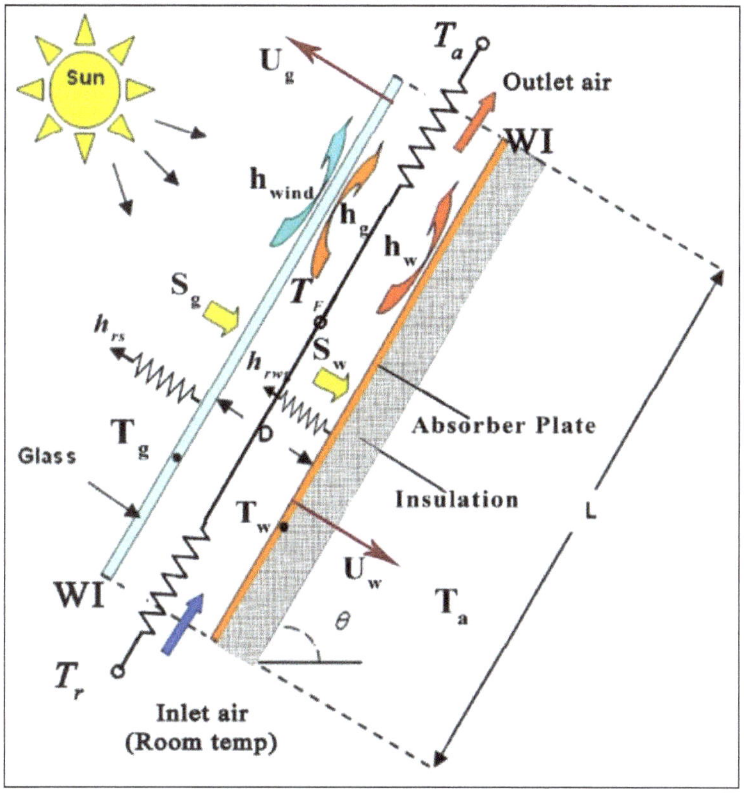

Figure 5 Thermal network for solar chimney with air flow.

The convective heat transfer outside the glass due to the wind is

$$h_{\text{wind}} = 5.7 + 3.8V_{(\text{wind speed at reference height})}. \quad (15)$$

The radiation heat transfer coefficient between wall and glass h_{rwg} is

$$h_{\text{rwg}} = \frac{\sigma(T_g + T_w)(T_g^2 + T_w^2)}{1/\varepsilon_g + 1/\varepsilon_w - 1}. \quad (16)$$

The heat transfer coefficient between the absorber and outside air is

$$U_w = \frac{1}{1/h_w + \Delta\omega_{\text{insulation}}/k_{\text{insulation}}}, \quad (17)$$

Where Δw is the thickness of insulation, $k_{\text{insulation}}$ is the thermal conductivity

The convective heat transfer coefficient between air and absorber plate or glass ($h_{\text{air.absorber}}, h_{\text{air.glass}}$)

$$h_{\text{w.x}} = \frac{Nu_{\text{air.}x} k_{\text{air}}}{d_{\text{chimney.gap}}}, \quad (18)$$

where x refers to absorber plate or glass, and d is the channel gap.

The Nusselt number correlation for a constant heat flux on one side of the channel [20] is

$$\text{Nu}_{\text{air.}x} = 0.9282\, Ra_{\text{air.}X}^{0.2035} \left(\frac{d_{\text{ChimneyGap}}}{Dl}\right)^{0.8972}, \quad (19)$$

where $Ra_{\text{air.}x}$ is the Rayleigh number for constant heat flux, Dl is the length.

The solar energy absorbed by the glass is

$$S_g = \sigma_g I_{\text{incident radiation}}. \quad (20)$$

The solar energy absorbed by the absorber

$$S_w = \sigma_w \tau_g I_{\text{incident radiation}}. \quad (21)$$

substituted the above equations into Equations 7, 8, and 9 and the three equations solved simultaneously to give the $T_{\text{g(glass cover)}}, T_{\text{f(Air)}}, T_{\text{w(absorber)}}$.

The mean air temperature $T_{(\text{Air})}$ in the solar chimney channel is used to determine the average air density ρ_B which is then substituted in Equation 5; then, the flow rate in the solar chimney is a function of the chimney temperature.

Temperature predictions in an evaporative cooling wind tower

An evaporative cooling wind tower (ECWT) is used to let the hot dry air to enter from the outside and is cooled when it passes over a wet medium. When the air passes through the evaporative cooler, the component calls the TRNSYS psychrometrics to fully determine the state of the inlet air. The psychrometrics returns the inlet air wet bulb temperature and enthalpy backs the inlet temperature, pressure, relative humidity, and humidity ratio. The performance of the evaporative cooler is based on the concept of adiabatic process. The wet bulb temperature of the air entering and exiting the device is assumed to be constant, where the minimum outlet temperature equals the inlet wet bulb temperature. Figure 6 shows evaporative cooler schematic.

The energy balance equation expressing the heat removed from the air to evaporate the water is expressed as

$$m_{1a}h_{1\ \text{enthalpy.AirIn}} = m_{2a}h_{2\ \text{enthalpy.Airout}} + m_{2\text{vapour}}h_{2\text{vapour}}. \quad (22)$$

The rate of heat transfer from air to water in the wet panel surface, Q_{sensible} (the heat exchange by temperature difference), is given by Δy

$$Q_s = h\,A\,\Delta T_{\text{LM}} = mc_p(T_r - T_a), \quad (23)$$

and the heat gain of latent heat of air (heat exchange by humidity ratio) can be expressed as

$$Q_L = m_L h_L(\Delta w), \quad (24)$$

where w is the humidity ratio.

ΔT_{LM} is the log mean temperature difference which is given by [12,21]

$$\Delta T_{\text{LM}} = \frac{(T_r - T_a)}{\ln(T_r - T_{\text{wb}})/(T_a - T_{\text{wb}})}. \quad (25)$$

The performance of the systems with different flow type can be assessed on the saturation efficiency as

$$\eta\,\text{Saturation} = \frac{T_a - Tr}{T_a - T_{\text{wetbulb}}}. \quad (26)$$

The air temperature leaving the cooling device can be calculated from (26)

$$T_r = T_a - \eta\,\text{Saturation}(T_a - T_{\text{wb}}). \quad (27)$$

The wet bulb temperature of the air entering and exiting the device is assumed to be constant, where the minimum outlet temperature equals the inlet wet bulb temperature.

Since the performance in the evaporative cooler is an adiabatic process,

$$Q_{\text{Latent}} = Q_{\text{Sensible}}. \quad (28)$$

The humidity ratio of the air outlet from the device [22]

$$\text{RH}_{\text{out}} = \frac{X_{\text{inlet}} p_{\text{outside}}}{(0.622 + X_{\text{inlet}}) p_{\text{saturated.vapour.pressure}}}. \quad (29)$$

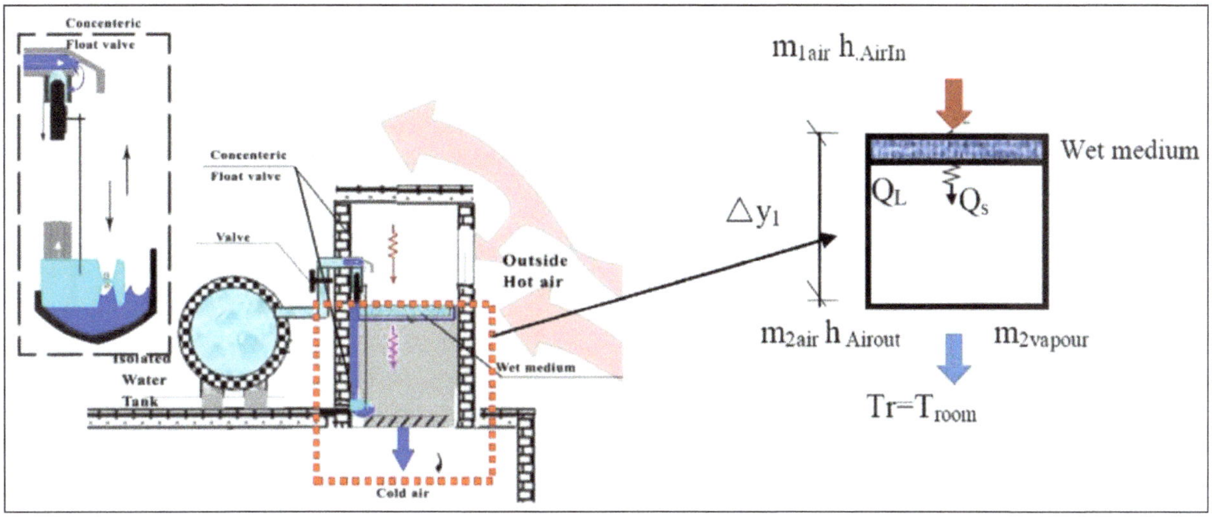

Figure 6 Evaporative cooler schematic.

The saturate vapor pressure (atm). Bouchahm, 2011 [23] is

$$P_{\text{saturated.vapour.pressure}} = 10^{\left(17.443 - \frac{2795}{T-273} - 3.868 \log_{10}(T-273)\right)}. \tag{30}$$

The corresponding energy removed from the air stream to evaporate water under these circumstances is given by Equation 31:

$$Q_{air} = \dot{m}_{cooler}(h_{.AirIn} - h_{.AirOut}), \tag{31}$$

where $h_{enthalpy} = C_p\ T + x(r + C_{P.vapor}T)$, expressed in kJ/kg (dry) [24]:

$$1.005\ T + x(2501.1 + 1.846T), \tag{32}$$

where $C_{p.air}$ = the specific heat of air (kj/kg.K), x = humidity ratio (kg/kg (dry)), r = evaporative latent heat,

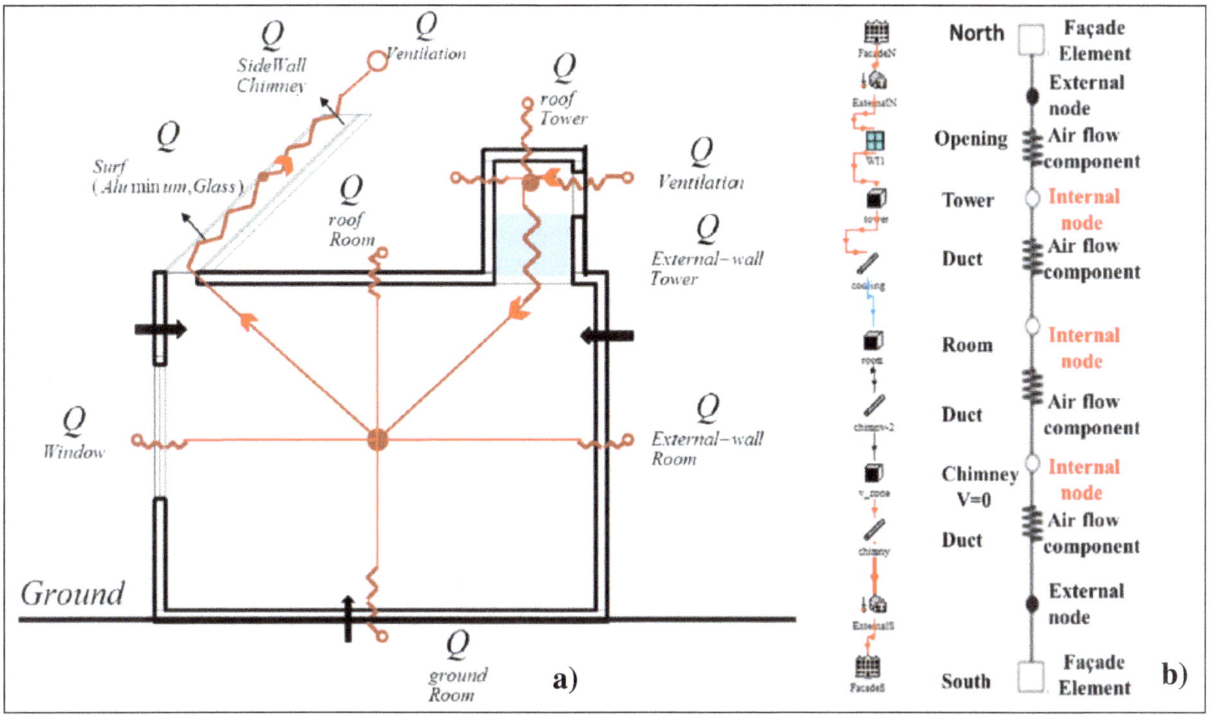

Figure 7 The building thermal model and the air flow network of the integrated model. (a) Heat flows in the ECWT, room, and solar chimney with combination of ventilation rate in TRNSYS model; **(b)** air flow network in COMIS with a description of the internal node, external node, and airflow components.

$C_{p.vapor}$ = the specific heat of vapor, and T = temperature (°C).

Air flow network and building thermal model

The ventilation model is integrated to the conventional building thermal model in order to have strong influence on the system performance. The internal heat load, solar load, and the heat transfer through walls, windows, roof, ground, and opening are considered and shown in Figure 7a, where Q is the heat transfer. The temperature of the ground at depth of 0.5 m is used in calculation for heat transfer from the ground and derived from Egyptian Typical Meteorological Year (ETMY) which is developed for standards development and energy simulation by US National Climatic Data Center (CA, USA) for a period of 12 years from 1991 until 2003 [25]. Also, the internal load is used in the calculation, where there are four occupants inside the single zone doing light work and seating with light load. Each occupant is assumed to have a sensible gain of 75 W and latent gain of 75 W according to ISO 7730 [26]. The occupants are assumed to be staying in the room for the whole day to apply the maximum load inside. The room light has 13 W/m^2. It has an energy saving lamp with convective part (visible) of 20%, and one device of 140 W is used.

The zones (ECWT, room, and SC) are linked to their ambiance via external nodes and to the north and south facades. Wind pressure data, wind speed, wind velocity profile for the building and for meteo, and orientation are defined. The COMIS program uses wind speed at meteo site in reference height to calculate the speed above the boundary layer using the wind velocity profile of the meteo station. The boundary layer is 60 m high for smooth terrain. Wind pressure coefficients (Cp values) belonging to a certain façade element can be given for several wind directions derived from 'CP-Generator' [13,27] according to the room with the integrated system with maximum height of 4.5 m and around the building. Figure 7b shows the air flow network in COMIS. Table 2 shows the building construction material description based on the ideal thermophysical properties [28] which was used in TRNBuild. The building thermal model is calculated based on the mathematical Equations 33 to 38.

The heat transfer through external walls (and roof) for room zone and tower zone is predicted as

$$\sum_{\text{external}} Q = \sum^{ne} \frac{A_z(T_{\text{star}} - T_z)}{R}, \tag{33}$$

where ne is the number of external walls attached to the zone, T_{star} is star temperature which can be used to calculate a net radiative and convection heat flux from the wall surface, R is the overall thermal resistance value including convective heat transfer resistance on both sides of the surface, and z is the zone [26].

The total gain to chimney zone from all surfaces is the sum of the combined heat transfers:

$$\sum_{\substack{\text{Surface}\\ \text{Chimney}}} Q = \sum A_s \dot{Q}_{\text{Comb\_z}} + S_{\substack{\text{outside}\\ \text{surface}}}, \tag{34}$$

where A_s is the inside area of the surface, and $S_{\text{outside surface}}$ is the solar radiation for outside surface.

The heat transfer across the windows is

$$\sum_{\text{Window}} Q = \sum A_z U_W (T_a - T_{\text{star}}) + I(\text{SHGF}), \tag{35}$$

where U_w is the overall heat transfer coefficient across the window (the rate of heat transfer through the window from outside to inside when it is hot per unit area and per unit temperature difference (W/m^{2}C), SHGF is

Table 2 Building material description used in calculation

Building part	Material part	Conductivity (kJ/h m^{-1} K^{-1})	U-value (W/m^{2}K^{-1})	Thickness (m)
Glass windows	Single glass	-	5.68	0.004
External walls	Common plaster (coating)	1.26	2.60	0.02
	Brick	3.60		0.10
	Common plaster (coating)	1.26		0.02
Roof	Insulation	0.2	0.443	0.05
	Concrete slab	4.2		0.12
	Cement plaster (coating)	4.50		0.01
Ground	Floor	-	0.797	0.10
	Insulation	0.2		0.02
	Concrete	4.2		0.10

Table 3 The results of the numerical simulation after using the new proposed system

Input			Outputs after cooling			
Ambient temperature (°C)	Solar radiation (W/m²)	Ambient RH (%)	Indoor temperature (°C)	Indoor RH (%)	Indoor ACH	Indoor (m³/h)
34	1,016	17	34	52.23	6.93	346.84
		25	25.64	58.65	6.95	347.75
38	1,016	17	27.45	51.39	6.80	340.36
		25	28.91	57.97	6.82	341.29
42	1,016	17	30.52	50.63	6.68	*334.05*
		25	32.13	57.53	6.70	335.09
34	425	17	24.33	52.36	5.45	272.95
		25	25.59	58.82	5.48	274.33
38	425	17	27.41	51.51	5.34	267.05
		25	28.83	58.22	5.37	268.53
42	425	17	30.48	50.74	5.22	261.39
		25	32.08	57.69	5.25	262.97
34	0	17	24.20	52.78	4.28	214.15
		25	25.45	59.31	4.31	215.87
38	0	17	27.28	51.91	4.17	208.75
		25	28.70	58.69	4.21	210.54
42	0	17	30.36	51.11	4.07	*203.57*
		25	31.95	58.14	4.10	205.45

the window's solar heat gain factor and its calculation is presented in details [29].

For external surfaces, the long wave radiation is considered using a fictive sky temperature T_{sky} [23]:

$$\dot{Q}_{Comb} = \dot{Q}_{convective} + \dot{Q}_{radiative} \tag{36a}$$

$$\dot{Q}_{convective} = h_{ConvOutside}\left(T_{AdjSurface} - T_{outsideSurface}\right) \tag{36b}$$

$$\dot{Q}_{Radiative} = \beta\varepsilon_{outside}\left(T^4_{OutsideSurface} - T^4_{fsky}\right) \tag{36c}$$

$$T_{fsky} = \left(1 - f_{sky}\right)T_{AdjSurface} + f_{sky}T_{sky}, \tag{36d}$$

where f_{sky} is the view factor to the sky, and ε is the long wave emissivity of outside surface = 0.9 for walls.

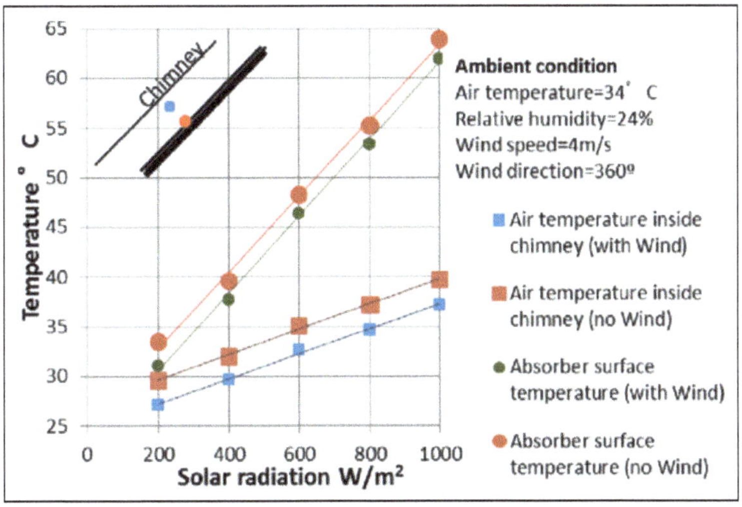

Figure 8 The variation of the absorber and chimney temperature with solar radiation.

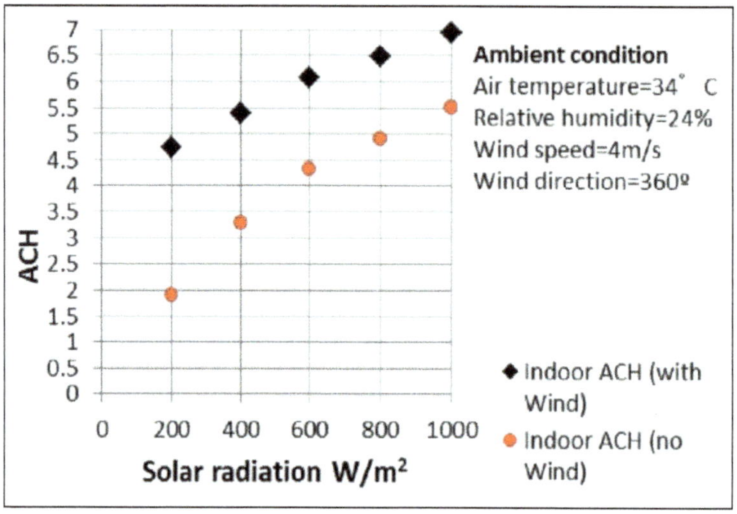

Figure 9 The variation of the ACH with solar radiation with different combinations.

The sky model used to calculate the radiation on inclined surface for the solar chimney by Perez model, which is considered the best available model.

Heat transfer due to ventilation in a zone is

$$\sum_{\text{Ventilation}} Q = \sum_{\text{Infiltration}} \dot{m}\, c_p T_{\text{outside}} - \sum_{\text{FromZone}} \dot{m}\, c_p T_{\text{zone}} + \sum_{\text{ToZone}} \dot{m}\, c_p T_{\text{zoneAdjacment}}. \qquad (37)$$

Then, the total heat load to maintain the zone at specified temperature can be calculated as

$$Q_{\text{load}} = \sum_{\text{Ventilation}} Q_{\text{Out}} + \sum_{\substack{\text{External\_wall}\\ \text{Tower}}} Q + \sum_{\substack{\text{External\_wall}\\ \text{Room}}} Q + \sum_{\substack{\text{ground}\\ \text{Room}}} Q + \sum_{\substack{\text{Window}\\ \text{Room}}} Q + \sum_{\substack{\text{Roof}\\ \text{Tower}}} Q + \sum_{\substack{\text{Roof}\\ \text{Room}}} Q + \sum_{\substack{\text{Side\_wall}\\ \text{Chimney}}} Q + \sum_{\substack{\text{Surface}\\ \text{Chimney}}} Q + \sum_{\substack{\text{People}\\ \text{Light}}} Q. \qquad (38)$$

Then, the zone temperature is iteratively predicted in the calculation.

Results and discussions

Steady state condition

The performance of the integrated system has been numerically studied, and the effect of the pressure coefficient on ACH and system performance was studied during the steady state condition. The investigation was at constant wind speed at 4 m/s (at Meteo), wind direction of 360°, and pressure coefficient (C_p) 0.081 for the north façade, and −0.040 for the south façade as shown in Table 3. Also, the effect of the different combinations of the three values of solar radiation, temperature, and relative humidity on indoor environment was studied. The main results show the increase of ACH with an increase of solar radiation. The system is capable of generating 130.5 m^3/h ventilation rates for a collector area of

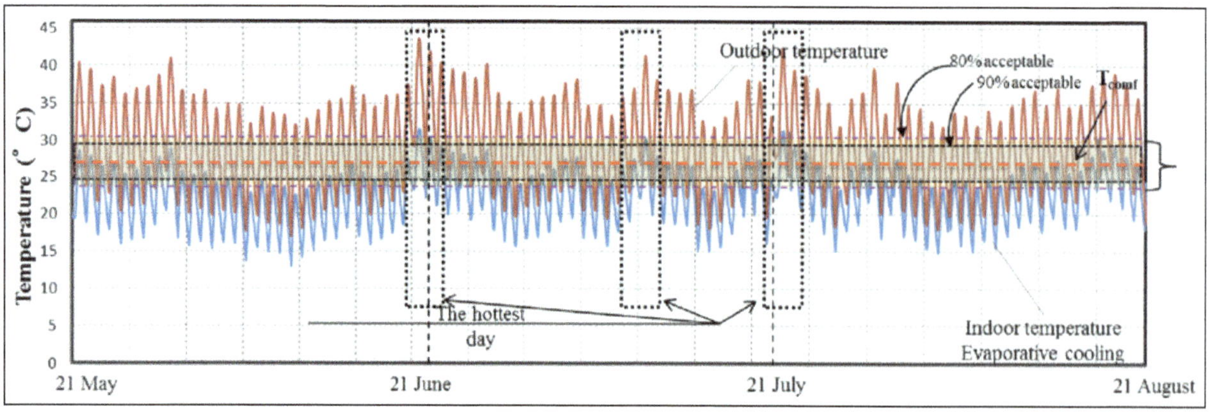

Figure 10 The performance of the system during 3 months of summer using standard weather data.

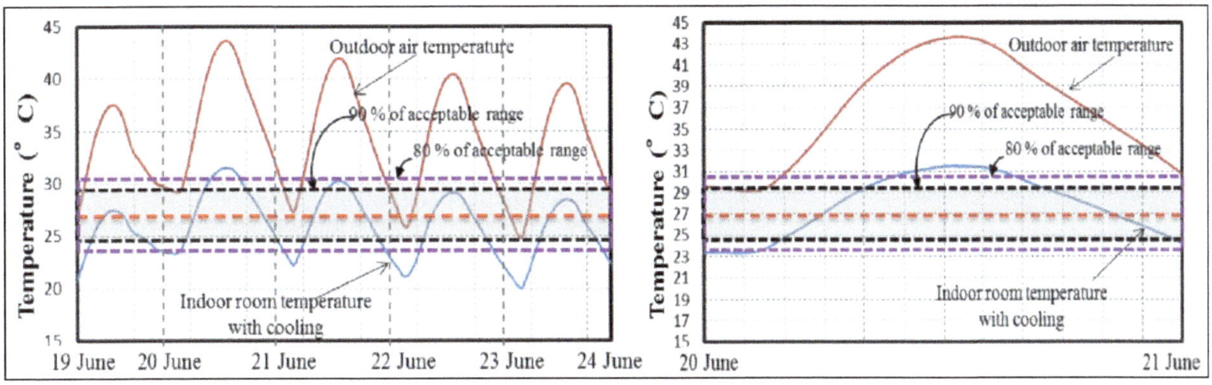

Figure 11 The performance of the system on 19 to 23 June with focus on the hottest day of June.

2.4 m^2 under the effect of solar radiation only. The values of these induced air flows depend on the geometry of the air collector and the performance parameters of the solar collector.

It can be seen from Figure 8 that the chimney cavity temperature is always higher than the ambient temperature with a very high temperature for an aluminum surface (black absorber) that reaches to 64°C at the radiation level of 1,000 W/m^2, then the surface temperature decreases to 40°C at the radiation level of 400 W/m^2. It has been observed that the black absorber possesses a high temperature value compared to chimney air temperature; this increase in temperature is due to more captured radiation and storage of high thermal energy.

This absorbed energy accelerates the flow of air through the chimney during the day and the night. As a result, this exit air flow will help more fresh air to enter to the space from the opening of the wind tower. Figure 9 shows the increase of ACH with an increase of solar radiation and pressure coefficient with maximum 7 ACH inside the room. Also, the minimum indoor air change rate is 2 ACH with 200 W/m^2 of solar radiation and no wind effect. Thus, the indoor ventilation rate reached zero with no radiation and no wind pressure coefficient. The temperature of the air is higher at the inlet and decreases at the outlet in the evaporative cooler (EC). This is because, initially, the dry air has a large potential to absorb the moisture and gets saturated as the airflow reaches the outlet of the ECWT.

According to ASHRAE 62–2001 standard (Atlanta, GA, USA), ventilation rates dependent on the floor area, whereas the minimum ACH is 0.35, but no less than 15 CFM/person. Also, the Egyptian Energy Code for residential buildings implies a minimum natural ventilation rate of 3 L/s-person (10.8 m^3/h) for the living room [30]. Therefore, the developed system is able to provide the desired ventilation and thermal needs of the occupants with/without solar radiation as shown in Table 3.

Real weather data

The real weather data of the New Assiut City which is taken from ideal weather data of 20 years [26] was applied to study the system performance during the summer season and the hottest days. In terms of thermal comfort, ASHRAE recognizes that the conditions

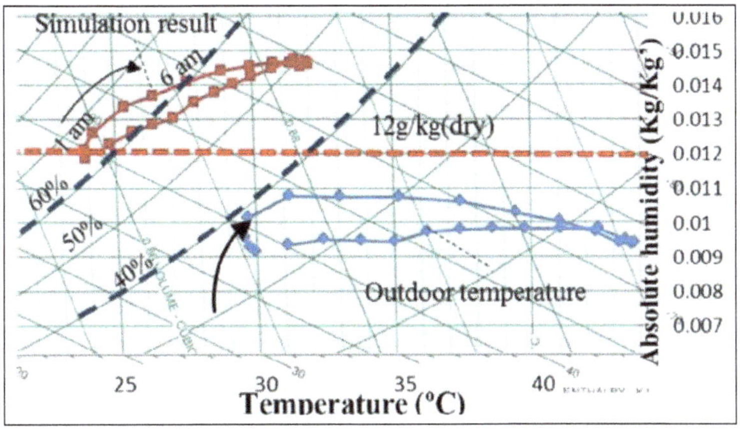

Figure 12 The humidity environment of the hottest day (20 June) using psychrometric chart of ASHRAE. Blue: outdoor temperature. Red: indoor temperature.

required for thermal comfort in spaces that are naturally conditioned are not necessarily the same as those required for other indoor spaces [31]. So, a new ACS for a naturally ventilated building has been proposed to be integrated within ASHRAE Standard 55, 2004. In the ACS, the mean monthly outdoor air temperature determines the acceptable indoor operative temperature. This relationship is expressed by the following formula: $To_{(comf)} = 0.31 Ta_{(out)} + 17.8$, where $To_{(comf)}$ is the optimum comfort operative temperature in °C and $Ta_{(out)}$ is the mean monthly outdoor air temperature in °C.

Therefore, most of the outlet temperature after passing through the wetted medium in ECWT is below the acceptable limits of Adaptive Comfort Standard (ACS) as shown in Figure 10. Only 5% of the results data exceed the upper limit of 80% acceptable range of ACS especially for the hottest days. An investigation of the system performance during the 5 days of June was done to focus on the hottest day in the summer season from the weather data file so that a clear pattern is seen in relation to adaptive comfort standard (ACS) as shown in Figure 11.

Arundel et al. proposed that the low or high humidity strongly affects the health and the optimum humidity level which is between 40% and 60% [32]. Therefore, the humidity environment in the living room and the bedroom is very important. Figure 12 shows most of the simulation results for the hottest day, which are located between 40% and 60% except before sunrise between 1 am and 6 am. Thus, this shows the effectiveness of a natural ventilation strategy with a direct evaporative cooling for an indoor environment. Finally, this proposed compact system could be an economical and practical passive alternative to the conventional air conditioning systems in hot and dry climates with no pollution released or energy consumed.

Conclusions

Important conclusions are drawn from the numerical operations of a new integrated system. Pressure coefficient and solar radiation in relation to the area of the collector are the main parameters that affect the air flow rate inside a single zone. When the temperature of the black absorber increases with/without pressure coefficient effects, the ACH increases with a strong relation to solar radiation as shown in Figure 8. The temperature of the indoor environment decreases by 10°C to 11.5°C compared to the outdoor temperature. This helps to achieve comfort during the hottest days of the summer season, with 95% of indoor temperature below the upper range of 80% acceptable comfort range during the period from 21 May to 21 August. The proposed system can be applied during daytime and nighttime, but it can be controlled and its use can be limited during nighttime especially when the inlet outdoor temperature is ≤30.5°C.

Overall, the results show how passive systems interact with the building and weather conditions. This paper studied the effect of a new integrated system for indoor comfort for the New Assiut City as a case study. The authors' future work will be a parametric investigation and optimization of the important operating parameter to achieve compact and high performance design so that guidelines for an efficient optimization process will be done to optimize cases with different locations, dimensions, and number of occupants.

Nomenclature

The list of nomenclature is presented in Table 4.

Table 4 Nomenclature

Symbol	Description	Unit
v	Air velocity	m/s
Q	Volume flow	m^3/s
T_{Wb}	Wet bulb depression	°C
Q_{air}	Energy transferred to or removed from the air stream	kJ/h
$\dot{m}_{coolere}$	Mass flow rate of the air through the cooling device	kJ/h
$\dot{m}$	Mass flow rate of the air	kg/s
h_{AirIn}	Enthalpy of air entering the device	J/kg
h_{AirOut}	Enthalpy of air exiting the device	J/kg
T	Temperature	K
t	Temperature	°C
ΔT	Wet bulb depression DRT-WBT	°C
ΔT_{LM}	Log mean temperature difference	°C
C_p	Specific heat at constant pressure	kJ/kg °C
$C_{psouth,north}$	Wind pressure coefficient	-
C_f	Specific heat of air	J/kg K
Δw	Humidity ratio	-
S_g	Solar radiation absorbed by glass	W/m^2
S_w	Solar radiation absorbed by absorber	W/m^2
h_{rwg}	Radiative heat transfer between absorber and the glass	W/m^2 K
h_{rs}	Radiative heat transfer between the glass and the sky	W/m^2 K
h_g	Convective heat transfer between glass and sky	W/m^2 K
h_{wind}	Convective heat transfer outside glass due to wind	W/m^2 K
h_w	Convective heat transfer between the absorber and the air	W/m^2 K
T_a	Temperature of ambient air	K
T_{sky}	Sky temperature	K
T_g	Mean glass temperature	K
$T_{f,air}$	Temperature of air in the chimney	K

Table 4 Nomenclature *(Continued)*

T_w	Temperature of the absorber	K
T_{room}	Room temperature	K
q''	Heat transfer to the air stream in the chimney	w/m^2
U	Overall convective heat transfer coefficient	W/m^2 K
V	Wind speed at reference height	m/s
I	Incident solar radiation	W/m^2
P	Atmospheric pressure	pa
$\dot{Q}$	Total heat transfer	W/m^2 K
G	Gravitational constant = 9.8	m/s^2
k	Thermal conductivity	W/m · K
Dimensionless terms		
Nu	Nusselt number ($h_f d/k_{air}$)	
Ra	Rayleigh number	
Greek symbols		
ε	Emissivity of glass = 0.9, emissivity of absorber = 0.95	
γ	Temperature weighting factor = 0.75	
$\eta_{Saturation}$	Saturation efficiency of the evaporative cooling	
ρ	Air density (kg/m^3)	
τ	Transmissivity	
Subscripts		
a	Ambient	
g	Glass	
w	Absorber	
Cav	Chimney cavity	
r	Room	

Competing interests
The authors declare that they have no competing interests.

Authors' contribution
All authors read and approved the final manuscript.

Authors' information
ASHA graduated in 2004 and received his M.S. in 2009 from the Department of Architecture, Assiut University, Egypt. He spent 2 years (2011 to 2013) as a researcher in Department of Architecture and Building Science, Tohoku University, Japan, for his of Ph.D. degree. YH is an Emeritus Professor, Department of Architecture and Building Science, Tohoku University. TG is an Associate Professor in the Department of Architecture and Building Science, Tohoku University. NE is a researcher, Department of Architecture and Building Science, Tohoku University. MMR is a Professor in the Department of Architecture, Assiut University. MAE is a professor in Department of Architecture, Assiut University.

Acknowledgements
The research leading to these results has received funding from the Egyptian Ministry of Higher Education. The author gives special thanks to Mr. Hans C. Phaff (TNO, The Netherlands), one of the COMIS developers for his scientific help with the simulation software, and to Mr. David Bradley, one of the developers of TRNSYS software.

References
1. IEA: Promoting energy efficiency investments. http://www.iea.org/publications/freepublications/publication/25recom_2011.pdf. (2012). Accessed June 2012
2. Oyedepo, SO, Adaramola, MS, Paul, SS: Analysis of wind speed data and wind energy potential in three selected locations in south-east Nigeria. Int. J. Energy Env. Eng. **3**(7), 4–10 (2012). doi:10.1186/2251-6832-3-7
3. Santamouris, M: Advances in Passive Cooling. Earth Scan, London (2007)
4. Bahadori, MN: An improved design of wind towers for natural ventilation and passive cooling. Sol. Ene. **35**(2), 119–129 (1985)
5. Watt, JR: Evaporative Air Conditioning Handbook. Chapman and Hall, USA (1986)
6. Khoukhi, M, Fezzioui, N: Thermal comfort design of traditional houses in hot dry region of Algeria. Int. J. Energy Env. Eng. **3**(5), 1–9 (2012). doi:10.1186/2251-6832-3-5
7. Hassan, AM, Goto, YH, Enteria, T, Abdelsamei, N, Eid, M, Radwan, MM: Analysis of thermal comfort for indoor environment of the new Assiut housing in Egypt. World Acad. Sci. Eng. Tech. **77**, 845–851 (2013). http://waset.org/journals/waset/v77/v77-200.pdf
8. Maerefat, M, Haghighi, AP: Natural cooling of stand-alone houses using solar chimney and evaporative cooling cavity. Renew. Ener. **35**, 2040–2052 (2010)
9. Alemu, TA, Saman, W, Belusko, M: A model for integrating passive and low energy airflow components into low rise buildings. Ener. Build. **49**, 148–157 (2012)
10. Idowu, OS, Olarenwaju, OM, Ifedayo, OI: Determination of optimum tilt angles for solar collectors in low-latitude tropical region. Int. J. Energy Env. Eng. **4**, 29 (2013). doi:10.1186/2251-6832-4-29
11. Camargo, JR, Ebinuma, CD, Silveira, JL: Experimental performance of a direct evaporative cooler operating during summer in a Brazilian city. Int. J. Ref. **28**, 1124–1132 (2005)
12. Elmetenania, S, Yousfia, ML, Merabetia, L, Belgrouna, Z, Chikouche, A: Investigation of an evaporative air cooler using solar energy under Algerian climate. Ener. Proc. **6**, 573–582 (2011)
13. Energy Performance of Buildings Group: COMIS Software Version 3.2: Conjunction of Multizone Infiltration Specialists. Applied Science Division, Lawrence Berkeley Laboratory, Berkeley (2005)
14. Feustel, HE, Raynor, AH: Fundamentals of the Multizone Airflow Model-COMIS. Annex V, Air Infiltration and Ventilation Center, Berkeley (1990)
15. Feustel, HE, Dieries, J: A survey of airflow models for multizone structures. Energy. Buildings. **18**, 79–100 (1992)
16. ASHRAE: Handbook of Fundamentals, p. 22. American Society of Heating, Refrigerating and Air-Conditioning Engineers, Atlanta (1985)
17. Ong, KS: A mathematical model of a solar chimney. Renew. Ener. **28**, 1047–1060 (2003)
18. Bassiouny, R, Koura, NSA: An analytical and numerical study of solar chimney use for room natural ventilation. Ener. Build. **40**, 865–873 (2008)
19. Hirunlabh, J, Kongduang, W, Nampraki, P, Khedari, J: Study of natural ventilation of houses by a metallic solar wall under tropical climate. Renew. Ener. **18**, 109–119 (1999)
20. Pica, RG, Volpes, R: An experimental investigation on natural convection of air in a vertical channel. Int. J. Heat. Mass. Transfer. **36**, 193–208 (2004)
21. Bellorio, MB, Pimenta, JM: Theoretical analysis of air conditioning by evaporative cooling influence on gas turbine cycles performance. Paper presented at the 18th international conference of Mechanical Engineers, Ouro Preto (2005)
22. Kulkarni, RK, Rajput, SP: Theoretical performance analysis of indirect-direct evaporative cooler in hot and dry climates. Int. J. Eng. Sci. Tech. **3**(2), 1239–1251 (2011)
23. Bouchahm, Y, Bourbia, F, Belhamri, A: Performance analysis and improvement of the use of wind tower in hot dry climate. Renew. Ener. **36**, 898–906 (2011)
24. Tanaka, IT, Terao, TH: Kenchiku Kankyo Kougaku, 3rd Edition, pp. 55–60, Japan (2006) [in Japanese]. ISBN 978-4-7530-1742-3C. http://www.inoueshoin.co.jp/books/ISBN4-7530-1742-7.html
25. US Department of Energy: Building Energy Software Tools Directory. http://apps1.eere.energy.gov/buildings/tools_directory/subjects_sub.cfm (2012). Accessed Sept 2012
26. Klein, SA, Beckman, WA, Mitchell, JW, Duffie, EN, Duffie, NA, Freeman, TL, Mitchell, JC, Braun, JE, Evans, BL, Kummer, JP, Urban, RE, Fiksel, A, Thornton, JW, Blair, NJ, Williams, PM, Bradley, DE, McDowell, TP, Kummert, M, Arias DA:

TRNSYS Version 16: A Transient System Simulation Program + Mathematical Equation. Solar Energy Laboratory, University of Wisconsin, Madison (2006)
27. Phaff, JC, Knoll, B, Gids, WF: Cp-Generator: pressure simulation program. http://cpgen.bouw.tno.nl/cp (2012). Accessed June 2012
28. Asan, H, Sancakter, YS: Effects of Wall's thermophysical properties on time lag and decrement factor. Ener. Build. **28**, 159–166 (1998)
29. Danny, HW, Li Joseph, CL: Solar heat gain factors and the implications to building designs in subtropical regions. Ener. Build. **32**, 47–55 (2000)
30. HBRC: Egyptian code for energy efficiency improvement in buildings. ECP306, M.O. Housing, HBRC Cairo (2005)
31. ASHRAE: Thermal Environmental Conditions for Human Occupancy. American Society of Heating Refrigerating and Air-Conditioning Engineers, Inc, Atlanta (2004)
32. Arundel, AV, Sterling, EM, Biggin, JH, Sterling, TD: Indirect health effects of relative humidity indoor environments. Env. Health Per. **65**, 351–361 (1986)

Accuracy analysis of software for the estimation and planning of photovoltaic installations

Petros J Axaopoulos[1], Emmanouil D Fylladitakis[2*] and Konstantinos Gkarakis[1]

Abstract

As the use of photovoltaics expands, with more and more commercial and residential users investing on solar energy systems around the globe, there is substantial demand for relatively simple, easy-to-use software packages for the planning and performance estimation of photovoltaic installations by installers and architects. In this paper, the calculative accuracy of TRNSYS, Archelios, Polysun, PVSyst, PV*SOL and PVGIS is being examined in comparison to the real electrical energy generated by a grid-connected 19.8kW$_p$ photovoltaic installation. The assessment has been performed by using the climatic data which have been recorded at the site of the real photovoltaics (PV) park over the same calendar year. Our results displayed that the software packages tend to overestimate the global irradiation received by the PV modules but still significantly underestimate the electrical energy generated by the installation.

Keywords: Photovoltaics; Solar irradiation; PVGIS; Energy Economics; PVSyst; PV*SOL; TRNSYS; Archelios; Polysun

Background

Albeit they are plagued by problems such as intermittency and relatively low efficiency [1], photovoltaics (PV) are the most popular form of renewable energy for electricity generation and their market penetration is expected to continually rise globally over the next few years [2]. The high demand has led to extensive research during the past few decades, improving the efficiency and other performance characteristics of the PV panels while significantly reducing their cost [3]. Due to their inert nature and motionless operation, PV systems are ideal for use in or near populated areas [4], especially on commercial and residential buildings [5,6], making them a vital addition to smart grids [7]. Small, distributed PV installations are considered to be a vital component to sustainable architecture [8].

The largest percentage of the PV market is held by private owners and small businesses, investing on small to medium grid-connected installations or building integrated systems [9,10], with the interest on small-scale residential applications continually growing across the globe [11-14], especially in the EU after the 2010/31/EU Energy Performance Building Directive [15], which suggests that buildings should require 'nearly zero energy' by 2020. As such, demand for software capable of performing energetic and economic analysis quickly and accurately was evident.

This commercial demand has led into the development of many PV analysis and planning software packages, mainly designed to be used for the composing of technical and economical essays by the PV installers and architects, especially during the initial design phases of a project [16]. Despite the vast growth of PV installations and the ever-growing number of available commercial software for their simulation and economic assessment, there has been little to no research regarding the calculative accuracy of commercial PV simulation software packages. Research studies are focused on developing methods which can be used to create new software packages, such as new simulation techniques [17,18], models which can be used to compute the effects of shading [19], methodologies which may be used for the estimation of solar irradiance on inclined surfaces [20], the modelling of PV panels [21] and the simulation of the power electronics involved in PV installations [22,23]. Undoubtedly, commercial PV simulation software packages are using combinations of these models and methodologies, each of which is scientifically sound; however, their accuracy as a complete software package against real-world data has not been assessed. Still, commercial software packages are being used as the basis of many scientific studies and papers [24-28]; therefore,

* Correspondence: Emmanouil.Fylladitakis@brunel.ac.uk
[2]Electronic and Computer Engineering Department, School of Engineering and Design, Brunel University, Uxbridge, London, Middlesex UB8 3PH, UK
Full list of author information is available at the end of the article

their accuracy should be thoroughly examined against real-world data and for various conditions.

In this article, we will be comparing the actual irradiance and energy data which have been extracted from the monitoring station of a 19.8-kW_p grid-connected PV installation to the results obtained from commercial PV simulation software packages which had the climatic data of the installation imported.

Photovoltaic installation

For the means of this study, long-term performance data have been extracted from an operational PV park in order to be compared to the artificial data created by the software. This free-standing PV park is situated at an open field at the foothills of Rhodope, near the settlement Dionis, in Thrace, Greece. The land plot measures nearly 2,000 m^2 large, without any obstructions which could cause shading near the installation. The exact geographical coordinates of the location are 40.98 N, 25.56 E.

Table 1 displays the specifications of each of the 90 SILCIO SE220 photovoltaic panels used by the PV park, which have a total surface of 150 m^2. Steel mounts are being used, placing the panels at a tilt angle of 30° and facing the true south. The installed power of the PV park is 19.8 kW, directly connected to the low voltage (230 V AC) distribution grid. For the means of connecting to the grid, three SMA Sunny Mini Central 7000TL have been used, each with a maximum power output of 7 kW and an efficiency of 97.7% at maximum output. Two strings of 15 panels each have been connected per inverter, leading to an inverter configuration oversized by about 6%.

The park has a Sunny Sensorbox weather station installed, which features Si module sensors for the global irradiation, PT100 sensors for the panel and ambient temperature, as well as a multi-directional anemometer for the wind speed. The solar irradiation sensors have been calibrated with the help of Kipp & Zonen CM 11 pyranometers [29].

Table 1 SILCIO SE220 PV panel specifications

Performance figure	STC rating
P_{nom}	220 W
V_{mpp}	28.8 V
I_{mpp}	7.65 A
V_{oc}	36.4 V
I_{sc}	8.3 A
NOCT	39.95 °C
μ_{Voc}	−0.30%/°C
μ_{Isc}	+0.047%/°C
μ_{Pmpp}	−0.39%/°C

Climatic data

The climatic data of the area were extracted from the PV park log files, all in hourly values and for a full calendar year (Additional file 1). These files contain the global irradiation received on horizontal surface (H), the irradiation on the inclined surface at 30° (H_{30}), the ambient temperature (T_{av}), the module temperature (T_c) and the wind speed (U_w). These data which have been converted to average monthly values are shown in Table 2.

The software

The climatic data extracted from the operational park have been entered into five commercial software packages which are available to us and are being alphabetically summarized in Table 3. PVGIS is not a commercial software and has been included in this study only in order to assess its accuracy as a quick assessment online tool.

TRNSYS is capable of receiving hourly climatic data, and thus, the radiation on the inclined plane H_{30}, the ambient temperature T and the wind speed U have all been imported as is, in their hourly values.

Archelios is the only commercial software which requires the input of the global, diffuse and direct irradiation on the horizontal plane, as well as the ambient temperature and wind speed, in monthly values, if the user seeks to generate a custom climatic data file. As such, the global irradiation on the horizontal should be split to diffuse and direct using either a mathematical model or the conversion ratio available for the area of the installation. As this study explores the accuracy of the simulation software assuming that it will be used as a daily tool and from users without high scientific background, the simplest option has been selected, and that is using the diffuse-to-global ratio available for the area, which has been extracted from climatic maps. The results are shown in Table 4.

Polysun requires the entry of the monthly global irradiation on the horizontal plane, ambient temperature and wind speed. PVSyst has similar requirements, but the wind speed is treated as optional. The monthly values of these three data sets have been entered into both of these software packages.

PV*SOL requires the entry of the monthly global irradiation on the horizontal plane and ambient temperature. Even though the algorithm of the software does take wind speed into account, there is no option to manually enter it into the software while creating a custom climatic data set and the software appears to be randomly generating wind profiles based on the location of the installation.

For the case of the PVGIS, which does not support the manual input of any climatic data, the integrated Climate-SAF database data is being used. The estimated total system losses have been entered to be 7%.

Table 2 Climatic data

Month	Global irradiation (H) (kWh/m^2)	Global irradiation on 30° (H_{30}) (kWh/m^2)	Ambient temperature (T) (°C)	Module temperature (T_C) (°C)	Wind speed (U) (m/s)
Jan	56.4	83.7	2.56	10.16	1.08
Feb	60.2	80.64	2.83	10.54	0.90
Mar	117.6	140.6	8.52	17.95	1.40
Apr	154.4	163.9	14.27	24.07	2.26
May	179.8	174.2	19.02	29.56	1.79
Jun	219.6	206	25.16	36.83	1.49
Jul	224.6	213	28.78	40.84	1.57
Aug	190.1	196	27.13	38.44	1.32
Sep	156.9	182.8	22.60	33.58	1.27
Oct	101.0	134.3	18.74	28.11	1.72
Nov	62.3	95	12.54	20.85	1.67
Dec	37.8	59.3	4.96	12.09	1.35

Methods

In all of the software used for the means of this study, three identical types of losses are taken into account: the cable losses which have been calculated to be 1%, the module mismatch losses which have been estimated to be 2% and the system quality losses which have been found to be 1.7%. The losses of the inverter are also being calculated by the model included in each software. For the case of PVGIS, as the software does not calculate the losses of an inverter, additional losses of 2.3% have also been added, bringing the total estimated losses up to 7%. All other losses that each commercial software may have the ability to calculate have been disregarded. The PV park is being treated as a free-standing installation with good ventilation.

We should note that the aforementioned software packages are being assessed only for their calculative accuracy. Other criteria, such as the user interface, features and support, are beyond the scope of this study. The final selection of a commercial software package should be a process based on multiple criteria.

In order to compare the goodness of fit between the data calculated from the simulation software and the data acquired by the logs of the installation, we used the following statistical parameters: root mean square error (RMSE), mean absolute deviation (MAD), absolute percentage error (MAPE) and model efficiency (EF).

Table 3 Photovoltaics simulation software examined

Software name	Version	Type
TRNSYS	16.1	Simulation
Archelios	13.01	Analysis and planning
Polysun	6.013	Analysis and planning
PVSyst	6.0.6	Analysis and planning
PV*SOL	6.0	Analysis and planning
PVGIS	4/6/2013	Online quick assessment

The root mean square error is given by Equation 1:

$$\mathrm{RMSE} = \sqrt{\frac{1}{n}\sum_{t=1}^{n}(H_t - F_t)^2} \tag{1}$$

Where H_t is the recorded value, F_t is the value derived from the simulation software and n is the number of periods (months). RMSE gives the deviation between the experimental and the calculated values, and it should be as close to zero as possible.

Table 4 Diffuse-to-global ratio, calculated beam and diffuse irradiation on the horizontal surface

Month	H_d/H	Beam irradiation H_b (kWh/m^2)	Diffuse irradiation H_d (kWh/m^2)
Jan	0.51	27.6	28.8
Feb	0.5	30.1	30.1
Mar	0.44	65.9	51.7
Apr	0.41	91.1	63.3
May	0.33	120.5	59.3
Jun	0.31	151.5	68.1
Jul	0.27	164.0	60.6
Aug	0.27	138.8	51.3
Sep	0.33	105.1	51.8
Oct	0.42	58.6	42.4
Nov	0.47	33.0	29.3
Dec	0.62	14.4	23.4

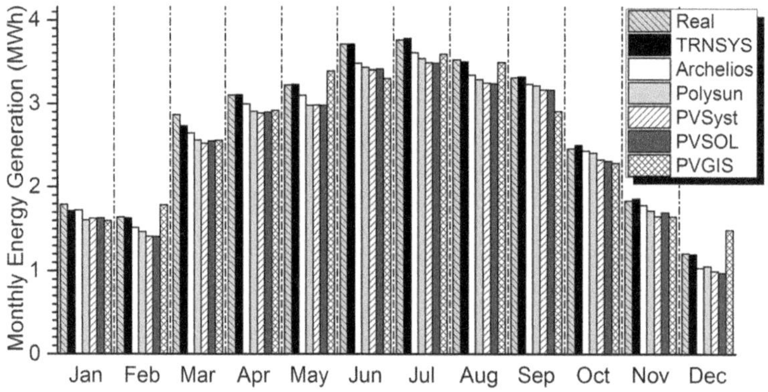

Figure 1 Monthly energy generation of the PV park and as calculated by the simulation applications.

The MAD parameter is used to measure statistical dispersion and is given by Equation 2:

$$\text{MAD} = \frac{1}{n}\sum_{t=1}^{n}|H_t - F_t| \qquad (2)$$

The MAPE parameter, also known as mean absolute percentage deviation or MAPD, is used to assess the accuracy of the model and is given by Equation 3:

$$\text{MAPE} = \frac{1}{n}\sum_{t=1}^{n}\left|\frac{H_t - F_t}{H_t}\right| \times 100\% \qquad (3)$$

The model efficiency test displays the goodness of fit between the calculated and experimental values, while the highest value it can get is 1. EF is given by Equation 4:

$$\text{EF} = \frac{\Sigma_{t=1}^{n}(H_t - z)^2 - \Sigma_{t=1}^{n}(F_t - H_t)^2}{\Sigma_{t=1}^{n}(H_t - z)^2} \qquad (4)$$

Where z is the mean value of the experimental data.

Results and discussion

Energy and solar irradiation results

Figure 1 displays the energy generation of the real PV park and the figures which each software calculated for a full calendar year. TRNSYS, which is using the irradiation data which have been recorded on the inclined plane, displays nearly perfect accuracy.

As it can be seen from Figure 2, which displays the deviation of the calculated results from the real production of the PV park, the four commercial PV analysis and planning applications underestimate the electrical energy generation of the installation every single month of the year, with the sole exception of PVGIS, which is making calculations based on its own climatic database and thus displays great fluctuations each month.

Figure 3 displays the real global irradiation received by the PV modules per month in comparison to that calculated by each software. Even though the commercial software packages underestimate the energy generation of the PV installation, they actually calculate the global irradiation received by the panels of the PV park to be higher than the real global irradiation on the inclined plane, as it can be seen in Figure 4. This is a paradox, indicating that the PV planning and analysis applications

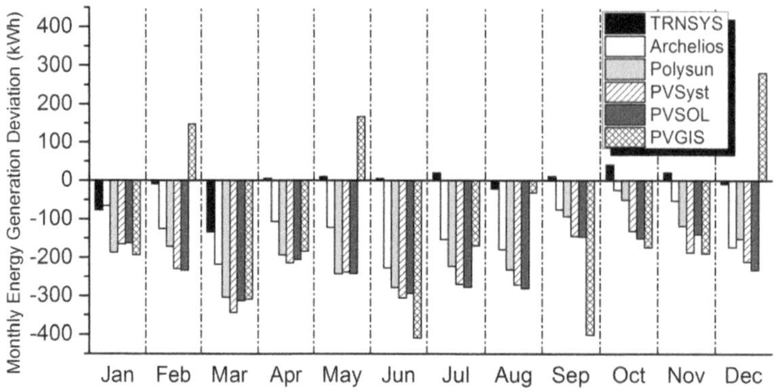

Figure 2 Monthly deviation of the calculated energy generation from the actual of the PV park.

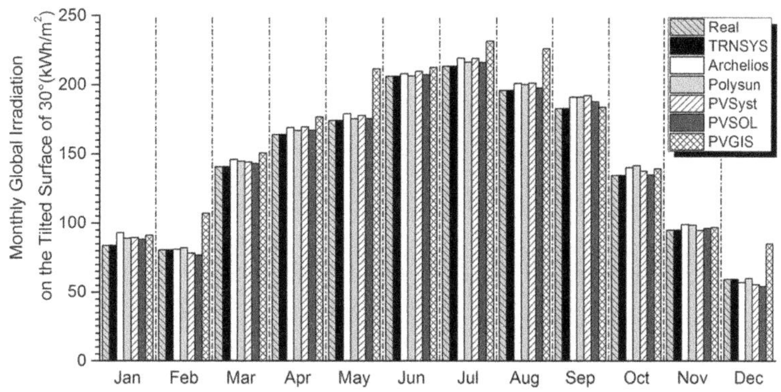

Figure 3 Monthly global irradiation on the 30° plane as calculated by the simulation applications.

overestimate the solar irradiation received by the PV panels of the park even though they underestimate the electricity generation. It also is interesting to note that even though PVGIS greatly overestimates the global irradiation received by the PV panels, the electricity generation results are significantly lower than the real output of the PV park.

As it can also be seen from Figure 4, while the global irradiation received by the PV modules that the software packages calculate coincides, for certain months, with the global irradiation recorded by the sensors of the park, the calculated electricity generation is significantly underestimated. As such, while the calculated global irradiation of Archelios for February, Polysun for June, PVSyst for November and PV*SOL for October coincides with the actual global irradiation received by the modules, the calculated electricity generation during those months is respectively 7.6%, 7.5%, 10.2% and 6.1% lower than that of the real PV park. Therefore, it becomes clear that regardless of the method each application is using to convert the provided global irradiation on the horizontal plane into the global irradiation on the plane of the PV modules, the main source of the electricity generation calculation error comes from the PV cell model which each software is using.

Table 5 displays the annual electrical energy generation and annual global irradiation received on the inclined plane.

Error analysis

According to the statistical parameters used, the data calculated from the simulation software better approximates the experimental data when the values of RMSE, MAD and MAPE are close to zero and the values of EF approach unity.

As TRNSYS is utilizing the recorded global irradiation on the inclined plane directly and suffers no irradiation conversion errors, it was expected that the accuracy of the energy generation simulation would be very high. As it can been seen from Table 6, the energetic results from TRNSYS display a model efficiency of 99.7%. Leaving TRNSYS aside and moving to the PV planning and analysis software, Archelios appears to be delivering the most accurate energy generation results over the course of a

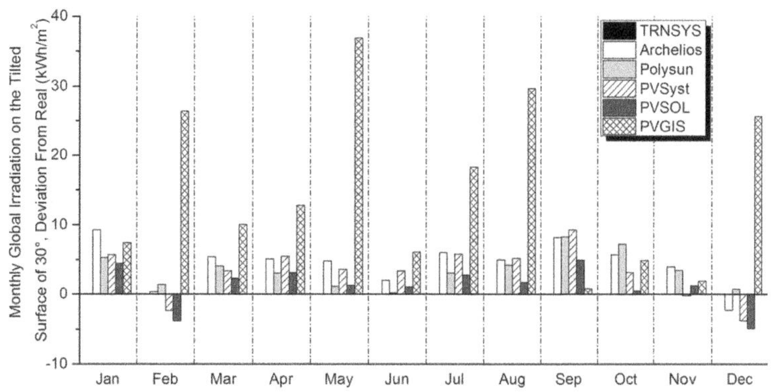

Figure 4 Monthly deviation of the calculated global irradiation on the 30° plane from the actual.

Table 5 Generated electricity and global irradiation received on the inclined plane per annum

	Annual electricity generation (kWh)	H_{30} (kWh/m^2)
Real data	32,443	1,729
TRNSYS	32,299	1,729
Archelios	30,923	1,783
Polysun	30,200	1,772
PVSyst	29,732	1,768
PV*SOL	29,765	1,744
PVGIS	30,980	1,910

year, followed by Polysun. However, the error becomes significant, with the MAPE of the results from Archelios surpassing 5.1%. PVSyst and PV*SOL display similar error results in all four tests, a MAPE of about 9.1% and a model efficiency of about 92.5%.

PVSyst and PV*SOL display similar error results when their energy generation figures are being tested; however, as it can be seen from Table 7, the latter appears significantly more accurate than the former when testing the calculated solar irradiation on the inclined plane. PV*SOL also displays the best received global irradiation calculation accuracy and the lowest irradiation overestimation out of all the tested applications. Despite it displaying the most accurate energetic results, the error analysis indicates that Archelios significantly overestimated the irradiation received by the PV installation.

Combining both the accuracy of the calculated energy generation and of the calculated received global irradiation, Polysun appears to be the most accurate software amongst the PV estimation and planning applications. Polysun displays the second most accurate energy generation results and, at the same time, the second most accurate received global irradiation results as well. On the other hand, Archelios displayed the most accurate energy generation results but had the least accurate received global irradiation calculation, significantly overestimating the irradiation received by the system, while PV*SOL displayed the most accurate calculation of the global irradiation received by the system but the accuracy of the energy generation results was the second worst, excluding PVGIS.

Table 6 Energy generation error analysis

Software	RMSE	MAD	MAPE	EF
TRNSYS	47.46	29.72	1.24%	99.69%
Archelios	141.23	126.67	5.14%	97.16%
Polysun	200.34	186.90	7.41%	94.30%
PVSyst	234.23	225.92	9.16%	92.47%
PV*SOL	230.92	223.18	9.08%	92.66%
PVGIS	244.88	221.42	9.24%	89.93%

Table 7 Solar irradiation at 30° error analysis

Software	RMSE	MAD	MAPE	EF
Archelios	5.41	4.85	3.70%	98.94%
Polysun	4.27	3.52	2.67%	99.31%
PVSyst	4.80	4.29	3.22%	99.21%
PV*SOL	3.07	2.67	2.47%	99.66%
PVGIS	18.96	15.07	12.81%	88.00%

Conclusions

The calculative accuracy of five commercial applications has been tested and evaluated through a set of four statistical parameters, namely RMSE, MAD, MAPE and EF. The statistical parameters have been used as indicators of the goodness of the estimation of the electrical energy and the global solar irradiation on the inclined surface.

The PV planning and analysis applications generally overestimate the irradiation received by the PV panels but, at the same time, underestimate the energy generation.

The software which is based and or depends on the PVGIS irradiation database may be significantly inaccurate, especially if the study is being performed for specific months or short time periods.

Even though the conversion of the global irradiation on the horizontal plane to the global irradiation on the plane of the modules is a source of calculation error, the main error source comes from the model of the PV cell. Finally, regardless of the deviations which the commercial software packages displayed, it can be considered that the results are generally useful and that their features and ease of use makes them a vital tool for planning and quickly assessing the performance of a PV installation.

Future studies may be performed to assess which parts of the complete simulation model used by the software packages are significant sources of error. Furthermore, by evaluating how each simulation parameter affects the accuracy of each model, the simulation models may be improved accordingly to provide more accurate results. As displayed in this paper, the main source of error comes from the model of the PV cell; future research may be performed to evaluate the accuracy of the PV simulation model used by each software package against real-world performance data and improve these models to better reflect real-world performance results. Furthermore, research on the error induced by the solar irradiation conversion model of each software package from the horizontal to the inclined plane should be performed, in order for the models of each software package to be improved so as to accurately convert the irradiation data from the horizontal to the inclined plane.

Additional file

Additional file 1: Solar Plant Data Log.

Competing interests

The authors declare that they have no competing interests.

Authors' contributions

AP performed the TRNSYS simulations and the analysis of the data. FE performed the literature review, the simulations with the commercial software packages, the error analysis and the English editing. GK performed the on-site PV park inspections and supplied the experimental data.

Author details

[1]Department of Energy Technology, Technological Educational Institute (TEI) of Athens, Ag. Spyridonos, Aegaleo 12210, Greece. [2]Electronic and Computer Engineering Department, School of Engineering and Design, Brunel University, Uxbridge, London, Middlesex UB8 3PH, UK.

References

1. Denholm, P, Margolis, RM: Evaluating the limits of solar photovoltaics (PV) in traditional electric power systems. Energy Policy **35**(5), 2852–2861 (2007)
2. Masson, G, Latour, M, Biancardi, D: Global market outlook for photovoltaics until 2016. European Photovoltaic Industry Association, Brussels (2012)
3. Tyagi, VV, Rahim, NAA, Rahim, NA, Selvaraj, JAL: Progress in solar PV technology: research and achievement. Renew. Sustain. Energy Rev. **20**(0), 443–461 (2013). doi:http://dx.doi.org/10.1016/j.rser.2012.09.028
4. Strzalka, A, Alam, N, Duminil, E, Coors, V, Eicker, U: Large scale integration of photovoltaics in cities. Appl. Energy **93**, 413–421 (2012)
5. Park, KE, Kang, GH, Kim, HI, Yu, GJ, Kim, JT: Analysis of thermal and electrical performance of semi-transparent photovoltaic (PV) module. Energy **35**(6), 2681–2687 (2010). doi:10.1016/j.energy.2009.07.019
6. Agrawal, B, Tiwari, GN: Life cycle cost assessment of building integrated photovoltaic thermal (BIPVT) systems. Energ. Buildings **42**(9), 1472–1481 (2010). doi:http://dx.doi.org/10.1016/j.enbuild.2010.03.017
7. Kanchev, H, Lu, D, Colas, F, Lazarov, V, Francois, B: Energy management and operational planning of a microgrid with a PV-based active generator for smart grid applications. Ind. Electron. IEEE Trans. **58**(10), 4583–4592 (2011)
8. Hegger, M, Fuchs, M, Stark, T, Zeumer, M: Energy manual. Birkhäuser (2008)
9. Bazilian, M, Onyeji, I, Liebreich, M, MacGill, I, Chase, J, Shah, J, Gielen, D, Arent, D, Landfear, D, Zhengrong, S: Re-considering the economics of photovoltaic power. Renew. Energy **53**, 329–338 (2013)
10. Lucas, R, Makhyoun, M: Evaluation of incentives for renewable energy in today's market. Proceedings of the Solar Conference, American Solar Energy Society, pp. 2887–2894. , Colo (2012)
11. Hammond, GP, Harajli, HA, Jones, CI, Winnett, AB: Whole systems appraisal of a UK Building Integrated Photovoltaic (BIPV) system: energy, environmental, and economic evaluations. Energy Policy **40**(0), 219–230 (2012). doi:http://dx.doi.org/10.1016/j.enpol.2011.09.048
12. Santos, IPD, Rüther, R: The potential of building-integrated (BIPV) and building-applied photovoltaics (BAPV) in single-family, urban residences at low latitudes in Brazil. Energ. Buildings **50**(0), 290–297 (2012). doi:http://dx.doi.org/10.1016/j.enbuild.2012.03.052
13. Defaix, PR, van Sark, WGJHM, Worrell, E, de Visser, E: Technical potential for photovoltaics on buildings in the EU-27. Sol. Energ. **86**(9), 2644–2653 (2012). doi:http://dx.doi.org/10.1016/j.solener.2012.06.007
14. Loulas, NM, Karteris, MM, Pilavachi, PA, Papadopoulos, AM: Photovoltaics in urban environment: a case study for typical apartment buildings in Greece. Renew. Energy **48**(0), 453–463 (2012). doi:http://dx.doi.org/10.1016/j.renene.2012.06.009
15. Commission, E: Directive 2010/31/EU of the European Parliament and of the Council of 19 May 2010 on the energy performance of buildings (recast). L153 Official J. Eur. Union **18**, 13–35 (2010)
16. Kanters, J, Horvat, M, Dubois, MC: Tools and methods used by architects for solar design. Energ. Buildings (2012). doi:http://dx.doi.org/10.1016/j.enbuild.2012.05.031
17. Minwon, P, In-Keun, Y: A novel real-time simulation technique of photovoltaic generation systems using RTDS. Energy Convers. IEEE Trans. **19**(1), 164–169 (2004). doi:10.1109/TEC.2003.821837
18. Gow, JA, Manning, CD: Development of a photovoltaic array model for use in power-electronics simulation studies. Elect. Pow. Appl. IEE Proc. **146**(2), 193–200 (1999). doi:10.1049/ip-epa:19990116
19. Alonso-García, MC, Ruiz, JM, Herrmann, W: Computer simulation of shading effects in photovoltaic arrays. Renew. Energy **31**(12), 1986–1993 (2006). doi:http://dx.doi.org/10.1016/j.renene.2005.09.030
20. Hay, JE, McKay, DC: Estimating solar irradiance on inclined surfaces: a review and assessment of methodologies. Int J Sol. Energ. **3**(4–5), 203–240 (1985). doi:10.1080/01425918508914395
21. Sera, D, Teodorescu, R, Rodriguez, P: PV panel model based on datasheet values, pp. 2392–2396. IEEE International Symposium on Industrial Electronics, Vigo (2007)
22. Ciobotaru, M, Kerekes, T, Teodorescu, R, Bouscayrol, A: PV inverter simulation using MATLAB/Simulink graphical environment and PLECS blockset, pp. 5313–5318. 32nd Annual Conference on IEEE Industrial Electronics, Paris (2006)
23. Maris, TI, Kourtesi, S, Ekonomou, L, Fotis, GP: Modeling of a single-phase photovoltaic inverter. Sol. Energ. Mat. Sol. C. **91**(18), 1713–1725 (2007). doi:http://dx.doi.org/10.1016/j.solmat.2007.05.027
24. Hasimah, AR, Khalid, MN, Mohammad, YH: Assessment of PV cell performance under actual Malaysia operating condition, Australasian Universities Power Engineering Conference. Adelaide **27–30**, 1–5 (2009)
25. Cheng, CL, Sanchez Jimenez, CS, Lee, MC: Research of BIPV optimal tilted angle, use of latitude concept for south orientated plans. Renew. Energy **34**(6), 1644–1650 (2009). doi:http://dx.doi.org/10.1016/j.renene.2008.10.025
26. Jahn, U, Nasse, W: Operational performance of grid-connected PV systems on buildings in Germany. Prog. Photovoltaics: Res. Appl. **12**(6), 441–448 (2004). doi:10.1002/pip.550
27. Bornatico, R, Pfeiffer, M, Witzig, A, Guzzella, L: Optimal sizing of a solar thermal building installation using particle swarm optimization. Energy **41**(1), 31–37 (2012)
28. Lundh, M, Wäckelgård, E, Ellegård, K: Design of hot water user profiles for Swedish conditions, pp. 2074–2078. Proceedings of ISES World Congress (Vol. I–Vol. V), Springer, Heidelberg (2007)
29. Kipp & Zonen: CM 11 Pyranometer manual. **2013**, 2013 (2013). http://www.kippzonen.com/Download/48/CM-11-Pyranometer-CM-14-Albedometer-Manual

Relationship between extreme temperature and electricity demand in Pakistan

Muhammad Ali[1*], Muhammad Jawed Iqbal[2] and Muhammad Sharif[3]

Abstract

Nowadays, different sectors of the economy are being significantly affected by the weather vagaries. Electricity market is one of the most sensitive sectors, due to the fact that electricity demand is connected to the numerous climatic variables, especially the atmospheric temperature. In this paper we have deduced the link between electricity consumption and mean monthly maximum temperature index in Pakistan, as a case study. ARIMA time series forecast model is developed for the temperature index. The forecast values of mean monthly maximum temperature shows an increasing trend. Linear trend model for electricity consumption is also developed as a function of temperature. Electricity consumption reveals a significant trend due to increase in temperature and socio- economic factors. The monthly behavior of our forecast values depicts that the electricity consumption is more for summer season, and this demand will be highest (6785.6 GWh) in July 2020, due to rise in temperature. Forecast model reveals that the electricity consumption (EC) and mean monthly maximum temperature are increasing with the passage of time.

Keywords: Maximum temperature index; Electricity consumption; Time series modeling; Forecast model

Background

It is well-known that electricity plays a critical role in economic growth, technological development and planning of a country. A study report of World Bank [1], states that no country in the world has succeeded in financial system, without using contemporary technology to produce energy [2]. Smith and Tirpak [3], showed the possible effects of climate change on the USA, across a range of sectors including electricity and that for Colombo by Andrew [4]. Many researchers estimated the impact of global warming on the energy expenditures in a region by Rosenthal [5]. Numerous economical activities have presented climatic changes, so that the predictable revenues may be seriously affected by extreme weather events. Extreme weather events are the infrequent or rare conditions of weather intensity in a locality, like heat waves, cold waves, tropical cyclone, flood, thunderstorm etc. The power sector is one of the most vulnerable to extreme weather, predominantly electricity consumption. Since electricity cannot be stored so the produced electricity must immediately be consumed, this implies that an appropriate model is needed to forecast future electricity demand by Valor [6].

In developin'g countries, there is a powerful positive correlation among wealth and energy especially electricity utilization. Though, the method of electricity production and consumption may generate air pollution and greenhouse gas which results global warming (Lee and Chiu [2], Ferguson [7]). Earlier studies mostly apply time-series or cross-sectional datasets to examine the appropriate topic of energy (Wolde-Rufael, [8,9]). Investigators have also started to utilize panel data to investigate the issues on energy (Lee and Lee [10]). Economic derivatives, such as future and alternative agreement on electricity, are usually engaged with this objective [11]. As electricity has become the basic need for survival in Pakistan. Unfortunately our country has been in deficit regarding electricity. Only 16% of rural population have grid-connected electricity, compared with 85% of the urban population [12]. The government called 2nd National Energy Conference on 9th April, 2012 and discussed critical issues and compelled to decide two holidays a week to recover the energy crises. Now it has become a big challenge for new government in Pakistan. Prime minister calls energy

* Correspondence: m.alishigri@gmail.com
[1]Mathematical Sciences Research Centre, Federal Urdu University of Arts Science and Technology, Karachi, Pakistan
Full list of author information is available at the end of the article

conference immediately, after taking over his responsibility. To overcome these problems an appropriate analysis of the link between electricity consumption and climatic variables, particularly air temperature, must be undertaken. The current work is an attempt to study the above relationship in Pakistan.

This communication is planned as, "Data description and methodology" gives an account of the data used in the study, and "Test for stationarity of temperature series" gives the estimates of the constancy test for temperature data. The time series model of temperature is constructed in "Time series modeling", while "Results and discussion" shows the forecast values of EC using regression technique. Finally, "Conclusion" summarizes the conclusions drawn from this investigation.

Data description and methodology

Electricity consumption in Pakistan

Currently Pakistan is facing severe energy crisis and power failures. The trend of power shortages has increased about 5000 MW, load shedding has been increased from 8 to 14 hours daily. Industrial growth has been decelerated and ultimately the whole economy has crashed down [13], so investigation of electricity consumption is an important study. A series of monthly electricity consumption (EC) of Pakistan in GWh, across the period from January 1990 through December 2010, has been used in this study. The data was recorded by Department of Federal Bureau of Statistics Pakistan. This data consists of electricity consumption in all economic sectors such as industrial, housing, and commercial sectors of Pakistan, because regional or sector wise disaggregated data was not available. Figure 1 shows that electricity demand has a considerable, growing trend that can be associated to demographic, community, and economic aspect, while a gradual increment in maximum temperature has also been observed.

The series of electricity demand shows seasonal effect, that can be examined with the monthly seasonal variation index (MSVI) and can be defined as;

$$\mathbf{MSVI}_{ij} = \frac{\mathbf{MEC}_{ij}}{\mathbf{MAE}_j} \tag{1}$$

where $MSVI_{ij}$ is the index value for month i in year j, MEC_{ij} is the monthly electricity consumption for month i in year j, and MAE_j is the monthly average electricity load for year j [6]. Figure 2 illustrate the average, maximum, and minimum MSVI values for each month of the year. Here the average values confirm the relative behavior of electricity consumption between different months while the variation between the highest and lowest amount shows the actual deviation from this mean behavior.

The monthly seasonality graph demonstrates that the electricity demand starts to rise up from March to August owing to increased use of air conditioning systems and is maximum in months of July and August while minimum in January and February. This gives graphical support for our assumption that electricity consumption increases with maximum temperature in our country

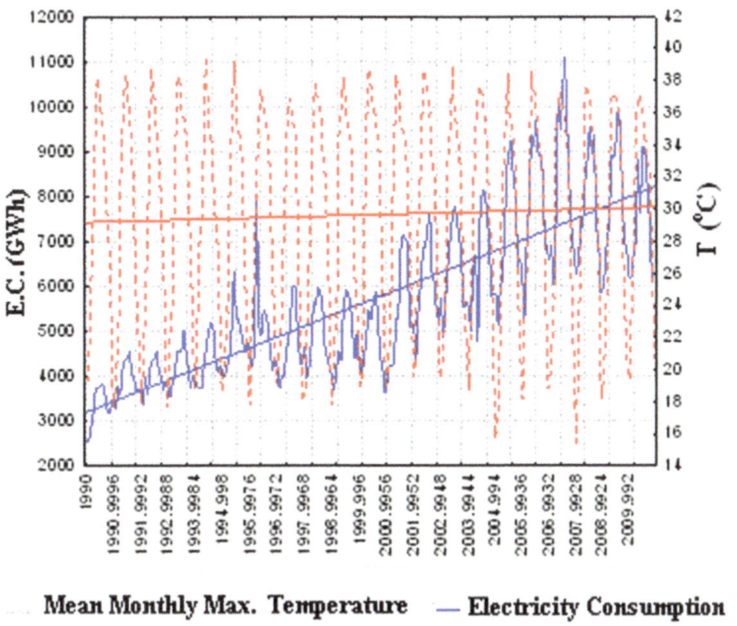

Figure 1 Relationship b/w electricity demand and maximum temperature from 1990 to 2010.

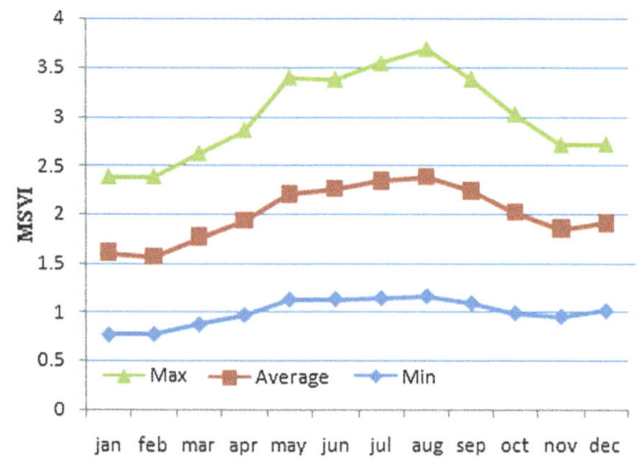

Figure 2 Monthly seasonal behavior of electricity consumption from 1990 through 2010.

while it has less correlation with minimum temperature also. Unfortunately the daily data for electricity consumption was not available due to which we are unable to say anything about the daily profile of electricity consumption e.g. in working days and holidays.

The weather variable

The results of past case studies reveal that the air temperature is the most important weather variable having an effect on electricity demand, while other variables i.e. humidity, wind, speed etc may also have an effect on temperature [6,14-17]. Since our country lies between latitudes 24°N to 36°N and longitudes 62°E to 78°E, so the maximum temperature affects electricity consumption a lot as compare to min temperature. Due to which we have estimated only the impact of maximum temperature on electricity consumption in this study. To show the average maximum temperature of Pakistan, we have taken the records of mean monthly maximum temperature of 18 meteorological stations (Chitral, Peshawar, Parachinar, Murree, Mianwali, Faisalabad, Lahore, Multan, Khanpur, Zhob, Queta, Khuzdar, Nokundi, Punjgur, Jiwani, Rohri, Nawabshah, Karachi and Badin) for the period from 1990 to 2010 which are spatially distributed throughout the country.

Since the accessible electricity consumption data are not present regionally, so a population weighted temperature index TI (°C) has been estimated from the mean monthly maximum temperatures of eighteen weather stations spatially spread across the country. The reason of selecting populations as weighting aspect is that weather affects the electricity consumption practices through the response of people to weather. i.e. depending on the heat, or coldness, people will increase or decrease the use of electric heating tools or air conditioners. Thus the higher the population has the higher the influence of climatic situation in electricity demand [6].

Test for stationarity of temperature series

Since we are using time series data so for the purpose of time series modeling first we have to check the stationarity of the data. A stochastic (or time-series) is called stationary if its mean and variance are constant over time. In addition, the value of the co-variance between the two time periods depends only on the lag between the two time periods and not the actual time at which the covariance is estimated. Although there are several tests of constancy, but we discuss only those that are prominently used in the literature i.e. Unit Root Test.

Unit root stochastic process

Consider a random walk model

$$Y_t = \rho Y_{t-1} + u_t \quad -1 \leq \rho \leq 1 \tag{2}$$

This model resembles the Markov first-order autoregressive model and depends upon the value of "ρ". if $\rho = 1$ then the above model shows nonstationarity or has unit root, where u_t is white noise error term.

Unit root test

This test starts from above equation and we simply regress Y_t on its (single period) lagged value Y_{t-1} and estimated if the predictable value of ρ is statistically identical to 1 then the series has unit root and we reject the null hypothesis.

For theoretical reasons, we manipulate eq. (2) as follows: subtract Yt -1 from both sides of eq. (2) we have

$$\begin{aligned} Y_t - Y_{t-1} &= \rho Y_{t-1} - Y_{t-1} + u_t \\ Y_t - Y_{t-1} &= Y_{t-1}(\rho - 1) + u_t \\ \Delta Y_t &= \delta Y_{t-1} + u_t \end{aligned} \tag{3}$$

where $\delta = (\rho - 1)$ and Δ is the first difference operator. Further, we estimate eq. (3) instead of eq. (2) and test (null) hypothesis that $\delta = 0$. If $\delta = 0$, then $\rho = 1$, which is the condition of unit root, implies that the time series under study is not stationary.

The Augmented Deckey Fuller (ADF) test

To tackle the autocorrelation problem, a test developed called Augmented Deckey Fuller (ADF) test. Now the test is conducted by "augmenting" the following three equations by adding the lagged values of the dependent variable Δ Yt

Y_t is a random walk with;

Intercept only i.e. β_1 is $\Delta Y_t = \beta_1 + \delta Y_{t-1} + u_t$ (4)

Trend and intercept i.e $\beta_2 t$ & β_1. $\Delta Y_t = \beta_1 + \beta_2 t + \delta Y_{t-1} + u_t$ (5)

No trend and no intercept. $\Delta Y_t = \delta Y_{t-1} + u_t$ (6)

Hypotheses

Null Hypotheses Ho: Variable has unit root (data is non stationary)

Here we can reject the null hypotheses in two ways:

i) If p-value is less than 5%
ii) If absolute value of Augmented Deckey Fuller test statistics is greater than the critical value at 1% and 5% significance level.

According to ADF test, the temperature series show no constancy in level forms because both the above conditions are not satisfied. Since temperature is not in stationary form so we will take first difference to make it stationary for time series modeling.

Now from Table 1,2 and 3 [18] the p-values are less than 5% and the absolute value of ADF test statistics is greater than the critical values at 1%, 5% and 10% level of significance also. So the first difference of temperature time series is stationary.

Time series modeling

Several researchers used AR(1) model to show daily maximum temperature fluctuations. Using a statistical package, E-VIEWS ver. 14, the parameter values for different orders of autoregressive (AR), moving average (MA) & auto regressive integrated moving average (ARIMA) models e.g. AR(1), AR(2), MA(1), MA(2), ARIMA(1,1,1), ARIMA (1,1,2), ARIMA(1,1,5), ARIMA(3,1,1), ARIMA(4,1,1) and ARIMA(4,1,4) were estimated. Now, in order to get optimized model, we use Akaike Information Criteria (AIC), which aims to optimize the exchange between goodness of fit and model parsimony. AIC can be represented mathematically as;

$$\mathbf{AIC} = \mathbf{ln}\left(\sigma^2\right) + \frac{\mathbf{2(p+q+1)}}{\mathbf{n}} \tag{7}$$

where n implies the sample size of the series and σ^2 is the mean squared error of the ARIMA model fit to the series. p is the order of AR parameter while q is the order of MA parameter [4]. Thus for ARIMA (4,1,4) model, p = 4, q = 4 and 1 shows the first difference. Here, our objective is to minimize AIC so that the required model can best explain fluctuating behavior of process under study.

For temperature data series the AIC shows that the ARIMA (4,1,4) is the best fitted model. The equation of this model is;

$$\begin{aligned} T_{\text{cal.}} = {} & \mathbf{0.00011} + \mathbf{2.571}T_{t\text{-}1} - \mathbf{3.16}T_{t\text{-}2} + \mathbf{2.063}T_{t\text{-}3} \\ & -\mathbf{0.707}T_{t\text{-}4} + 3.026U_{t\text{-}1} - \mathbf{3.55}U_{t\text{-}2} \\ & +\mathbf{1.835}U_{t\text{-}3} - \mathbf{0.313}U_{t\text{-}4} \end{aligned} \tag{8}$$

where T_{cal} Indicates the forecast values of temperature while T and U with subscript t show the time lagged values of autoregressive and moving average respectively. Table 4 gives the estimated future values of the monthly mean maximum temperature of Pakistan from Jan 2011 to December 2020 using above model (Eq. 8).

Table 1 First difference for Intercept form

Null hypothesis: D(Y) has a unit root			
Exogenous: constant			
Lag Length: 10 (Automatic based on SIC, MAXLAG=15)			
		t-Statistic	Prob.
Augmented Dickey-Fuller test statistic		−16.78416	0.0000
Test critical values:	1% level	−3.457515	
	5% level	−2.873390	
	10% level	−2.573160	

Table 2 First difference for Trend and intercept form

Null hypothesis: D(Y) has a unit root			
Exogenous: constant, linear trend			
Lag length: 10 (Automatic based on SIC, MAXLAG=15)			
		t-Statistic	Prob.
Augmented Dickey-Fuller test statistic		−16.74452	0.0000
Test critical values:	1% level	−3.996754	
	5% level	−3.428660	
	10% level	−3.137757	

Table 3 First difference for No Trend No Intercept form

Null hypothesis: D(Y) has a unit root			
Exogenous: none			
Lag length: 10 (Automatic based on SIC, MAXLAG=15)			
		t-Statistic	Prob.
Augmented Dickey-Fuller test statistic		−16.81727	0.0000
Test critical values:	1% level	−2.574633	
	5% level	−1.942153	
	10% level	−1.615818	

Table 4 Forecast values of mean monthly maximum temperature series for Pakistan from Jan, 2011 to Dec, 2020

	Mean monthly maximum temperature (°C)												
Year	**Jan**	**Feb**	**Mar**	**Apr**	**May**	**Jun**	**Jul**	**Aug**	**Sep**	**Oct**	**Nov**	**Dec**	**Average**
2011	20.81	22.15	25.32	29.85	34.72	38.39	39.38	37.24	32.90	28.01	24.00	21.60	21.60
2012	21.03	22.31	25.35	29.81	34.73	38.50	39.50	37.27	32.83	27.94	23.98	21.64	21.64
2013	21.10	22.39	25.43	29.90	34.81	38.56	39.52	37.25	32.79	27.91	23.98	21.67	21.67
2014	21.15	22.46	25.53	29.99	34.90	38.61	39.53	37.22	32.75	27.88	23.98	21.69	21.69
2015	21.20	22.54	25.62	30.10	34.99	38.67	39.54	37.19	32.72	27.86	23.98	21.72	21.72
2016	21.26	22.61	25.71	30.20	35.08	38.72	39.54	37.17	32.68	27.84	23.98	21.74	21.74
2017	21.31	22.68	25.81	30.30	35.17	38.77	39.55	37.14	32.65	27.82	23.98	21.77	21.77
2018	21.36	22.76	25.90	30.40	35.25	38.82	39.56	37.11	32.61	27.80	23.98	21.80	21.80
2019	21.41	22.83	25.99	30.50	35.34	38.87	39.56	37.08	32.58	27.78	23.99	21.82	21.82
2020	21.46	22.91	26.09	30.60	35.43	38.92	39.57	37.06	32.55	27.76	23.99	21.85	21.85
Average	21.21	22.56	25.67	30.16	35.04	38.68	39.53	37.17	32.71	27.86	23.98	21.73	

Results and discussion

Extreme temperature events affect electricity demands primarily by raising the number of air conditioner used, as people attempt to sustain a predictable level of comfort. Consequently, increase in extreme temperature events, results an increase in air conditioner use and power demand as well. Power utilities might have to approve new approach to fulfill the increased loads [4]. In addition the more demands can also sag the transmission lines with hot temperature.

Now, we will justify our hypothesis by using exploratory data analysis (EDA) for both electricity and temperature data series. EDA makes a general use of robust statistics and relies greatly on graphical techniques, which present an idea about the shape of the distribution of data under study. It also gives a widespread description of the underlying system in the form of models. To assess that how peak power demand strength to vary, it has to be correlated with maximum temperature. As, a correlation is made the probability of identical or exceeding certain electricity demand thresholds can be determined [4].

Figure 3 depicts a positive correlation between electricity consumption (EC) and temperature index (TI). The Pearson correlation of E.C. and TI is 0.412 with p-value (0.000) less than 5% and T statistics is 7.15. The correlation value is not very high because the electricity demand is always greater than its production in Pakistan. So the electricity production authorities are compelled to initiate load shedding to manage this problem, due to which exact correlation could not be found. Secondly, we could not collect the daily data and sector wise disaggregated data of EC in our country to show its exact relationship with temperature. Coefficient of determination is highest for the linear model as compared to quadratic and nonlinear models. This shows that a linear model

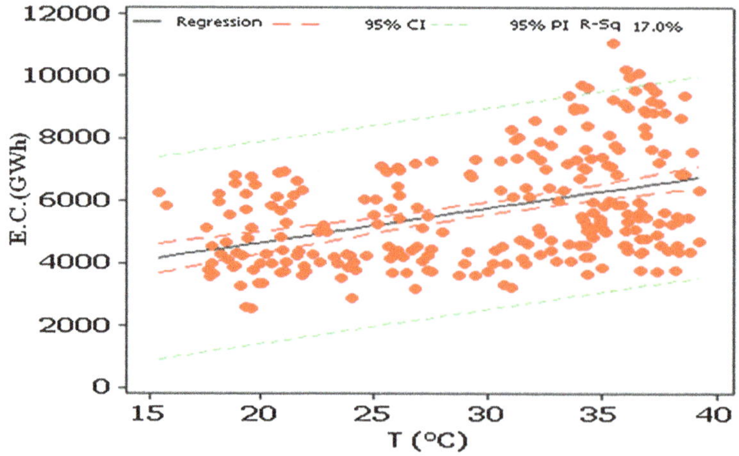

Figure 3 Correlation between EC and TI with best fitted regression line.

Table 5 Monthly forecast values of electricity consumption from Jan, 2011 to Dec, 2020 for Pakistan

Year	Electricity consumption (GWh)												Yearly average
	Jan	Feb	Mar	Apr	May	Jun	Jul	Aug	Sep	Oct	Nov	Dec	
2011	4735.9	4881.7	5228.6	5724.1	6256.3	6656.5	6765.1	6531.6	6056.5	5522.4	5084.5	4822.3	**5688.8**
2012	4759.7	4898.9	5231.8	5719.5	6257.4	6668.7	6778.4	6534.2	6049.4	5514.6	5082.3	4826.2	**5693.4**
2013	4767.4	4908.1	5241.0	5728.6	6266.1	6675.4	6780.8	6532.0	6045.1	5511.4	5082.0	4829.1	**5697.2**
2014	4773.2	4916.2	5251.0	5739.4	6275.7	6681.3	6781.8	6529.1	6041.1	5508.8	5082.0	4831.9	**5701.0**
2015	4778.7	4924.3	5261.2	5750.4	6285.4	6687.1	6782.5	6526.2	6037.2	5506.3	5082.0	4834.8	**5704.7**
2016	4784.2	4932.3	5271.3	5761.5	6295.0	6692.8	6783.3	6523.2	6033.4	5503.9	5082.1	4837.6	**5708.4**
2017	4789.8	4940.4	5281.5	5772.5	6304.6	6698.5	6783.9	6520.3	6029.5	5501.5	5082.3	4840.5	**5712.1**
2018	4795.4	4948.6	5291.7	5783.5	6314.1	6704.0	6784.5	6517.3	6025.7	5499.2	5082.5	4843.5	**5715.8**
2019	4801.0	4956.7	5302.0	5794.5	6323.7	6709.5	6785.1	6514.3	6022.0	5496.9	5082.7	4846.5	**5719.6**
2020	4806.7	4965.0	5312.2	5805.6	6333.1	6715.0	6785.6	6511.3	6018.2	5494.6	5083.0	4849.5	**5723.3**
Monthly average	**4779.2**	**4927.2**	**5267.2**	**5758.0**	**6291.1**	**6688.9**	**6781.1**	**6524.0**	**6035.8**	**5506.0**	**5082.5**	**4836.2**	

can better predict the values of E.C. in future. Figure 3 also reveals that when temperature increases above 30°C, the EC rapidly increases due to cooling appliances. The prediction equation for EC can be shown by following equation.

$$\mathrm{E.C.(T)} = \mathbf{2461 + 109.3T}(^\circ\mathrm{C}) \tag{9}$$

where T is the mean monthly maximum temperature and EC is the electricity consumption in Pakistan. Now being the best fitted model eq. (9), peak power demand corresponding to threshold temperatures has been computed and shown in Table 5. Since the correlation between electricity demand and mean monthly maximum temperature is not so good, implies that there are many other factors which are also accountable to increase the electricity consumption e.g. population, inflation rate of heating appliances, unit price of electricity, etc. In this article we have only attempted to calculate the expected change in EC, due to the increase of mean monthly maximum temperature for Pakistan by using the above forecast values of temperature from January 2011 to December 2020 (Table 4).

Now we use Figure 4 to compare the average monthly behavior of EC and maximum temperature for the predicted values in future. This figure also throws light on the correlation between these two parameters keeping influence of other parameters constant. This indicates that the EC is maximum for summer season and minimum for winter season. The cyclic trend in temperature is similar as that of EC, so one cannot neglect the influence of maximum temperature on electricity consumption.

Conclusion

This study demonstrates that the the electricity consumption (EC) increases with mean monthly maximum temperature having significant correlation coefficient 0.412. This correlation is not so good due to the insufficient electricity production in our country, the authorities are compelled to get load shedding and we could not get the monthly load shedding hour's data. The peak power demand is significantly increased during summer season as compared to winter season due to the uses of cooling appliances. The monthly behavior depicts that EC is more for summer season and highest demand will be 6785.6 (GWh) in July, 2020 due to increase in temperature. The average monthly behavior of forecast

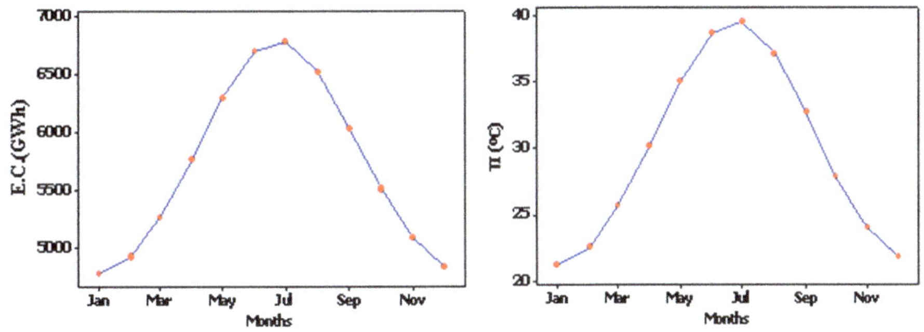

Figure 4 Monthly average forecast values of temperature (°C) & Electricity consumption (GWh) from 2011 to 2020.

values of temperature and electricity has been shown in Figure 4, which reveals that the maximum temperature has a great influence on EC keeping other parameters constant. The current study also suggests that ARIMA (4,1,4) model is the best fitted model for the future prediction of temperature series. Our calculation also shows that for an increase of 1°C temperature there will be an increase of 109.3 million-kWh electricity demand for Pakistan. So the power utilities have to adopt new strategies in order to meet future electricity demands.

Competing interests

The authors declare that they have no competing interests.

Authors' contributions

JI has given the techniques and ideas of the paper. He also removed technical as well as grammatical mistakes from the paper. MS helped me to analyze the data from statistical software. MA (corresponding author) has got data from Pak. Met dept. And Federal Bureau of statistics, analyzed the data, got forecast values of EC, handled all the comments from reviewer side and compiled the paper by getting suggestions from coauthors. All authors read and approved the final manuscript.

Acknowledgements

We are very thankful to all members of CDPC, Pakistan Meteorological Department and Federal Bureau of Statistics, Karachi for providing the required mean monthly maximum temperature and monthly electricity consumption data. The contents of this paper are the part of first author's PhD thesis.

Author details

[1]Mathematical Sciences Research Centre, Federal Urdu University of Arts Science and Technology, Karachi, Pakistan. [2]Institute of Space and Planetary Astrophysics (ISPA), University of Karachi, Karachi, Pakistan. [3]Department of Statistics, University of Karachi, Karachi, Pakistan.

References

1. Wolde-Rufael, Y: Electricity consumption and economic growth: a time series experience for 17 African countries. Energ. Policy **34**(10), 106–1114 (2006). Energy in Africa's Development
2. Lee, C-C, Chiu, Y-B: Electricity demand elasticities and temperature: evidence from panel smooth transition regression with instrumental variable approach. Energ. Econ. **33**, 896–902 (2011)
3. Smith, JB, Tirpak, D (eds.): The potential effects of global climate change on the United States. Report to congress, United States Environmental Protection Agency, Office of Policy, Planning and Evaluation, EPA-230-05-89-050. U.S. Government Printing Office; Office of Research and Development, Washington, DC (1989)
4. Colombo, AF, David, E, Karney, BW: Climate variability and the frequency of extreme temperature events for nine sites across Canada: implications for power usage. J. Clim. **12**, 2490–2502 (1999)
5. Rosenthal, DH, Gruenspecht, HK, Moran, EA: Effects of global warming on energy use for space heating and cooling in the United States. Energy J **16**(2), 77–96 (1995)
6. Valor, E, Meneu, V, Caselles, V: Daily air temperature and electricity load in Spain. J. Appl. Meteorol. **40**, 1413–1421 (2001)
7. Ferguson, R, Wilkinson, W, Hill, R: Electricity use and economic development. Energ. Policy **28**(13), 923–934 (2000)
8. Wolde-Rufael, Y: Energy consumption and economic growth: the experience of African countries revisited. Energ. Econ. **31**(2), 217–224 (2009)
9. Lee, C-C, Lee, J-D: A panel data analysis of the demand for total energy and electricity in OECD countries. Energ. J. **31**, 1–24 (2010)
10. Pilipovic, D: Energy risk: Valuing and managing energy derivatives. Vol. 300. McGraw-Hill, New York (1998)
11. Ghaffar, MA: The energy supply situation in the rural sector of Pakistan and the potential of renewable energy technologies. Renewable energy **6**(8), 941–976 (1995)
12. Alter, N, Syed, SH: An empirical analysis of electricity demand in Pakistan. Int. J. Energ. Econ. Pol. **1**(4), 116–139 (2011)
13. Hadley, SW, et al.: Future US Energy Use for 2000-2025 as Computed with Temperatures from a Global Climate Prediction Model and Energy Demand Model. In: 24th United States Association of Energy Economists/ International Association of Energy Economists North American Conference, Washington, DC. (2004)
14. Pardo, A, Meneu, V, Valor, E: Temperature and seasonality influences on Spanish electricity load. Energy Economics **24**(1), 55–70 (2002)
15. Li, X, Sailor, DJ: Electricity use sensitivity to climate and climate change. World Resource Review **7**(3) (1995)
16. Cancelo, JR, Espasa, A: Modelling and forecasting daily semes of electricity demand. Investigaciones economicas **20**(3), 359–376 (1996)
17. Hussain, MA, et al.: Forecast Models for Urban Extreme Temperatures: Karachi Region as a Case Study. Nucleus **47**(4), 301–311 (2010)
18. Gujarati, DN: Econometrics by example. Palgrave Macmillan, Hampshire, UK (2011)

21

Investigation on solar stills having floating plates

Hitesh Panchal[1,2*] and Pravin K Shah[3]

Abstract

Potable water is always a need of every human being in the world. Only one third of water is potable water, and two thirds of water is saline water, which is not used for drinking purposes. Solar still is a device, which converts the saline or brackish water into drinkable water. Here, in place of saline water, industrial wastewater is used to convert potable water. In this research work, three solar stills have developed by locally available materials and tested in climate conditions of Mehsana (latitude of 23°59′ and the longitude of 72°38′). Among three solar stills, two solar stills are consist of floating plates like aluminum and galvanized iron. Third solar still that is used as a reference solar still is called a conventional solar still. Three-month research showed that solar still, consists of aluminum floating plate, gave considerable daily distillate as well as cumulative output compared with galvanized iron floating plate solar still as well as conventional solar still. Hence, the performance of solar still could increase by the use of aluminum floating plate, which can be used as an energy absorbing plate.

Keywords: Solar still, Aluminum plate, Galvanized iron plate, Daily distillate output, Cumulative distillate output

Introduction

Solar desalination is a process of separation of pure water from saline water by the use of solar energy. The use of solar still is a simple and cheap method to obtain potable water. A conventional solar still has consisted of a black painted basin for holding the saline water and receiving solar radiation. Phase equilibrium has existed between the saline water and the air over it. Hence, the air over the water surface has been in saturated condition. Increasing of saline water temperature has caused the increase of partial pressure of water vapor at the water surface. The partial pressure of water vapor at the bottom of the glass cover has been very low as the temperature at the bottom of the glass cover has been lower than that of the water surface. The temperature difference between the water and glass cover is the partial pressure of water vapor transferring water vapor from the saline water surface to the glass cover. The still has been insulated with insulating material to reduce heat losses from the bottom as well as on the sides of the solar still. The rate of evaporation of water vapor from the water surface depends on the rate of condensation of water vapor in the glass cover and temperature difference between the saline water and glass cover.

Solar radiation can be used to distill and collect the drinking water, which is essential for the daily survival both at sea and in the deserts. It is light weight, compact, and easy to use. Also, this important and necessary piece of survival equipment has been used by the military for many years. The high increase of the earth's population by years has resulted in a large rise demand for potable water [1]. Added to that, problems of industrial and sewage pollutants to rivers, lakes, and groundwater exist. Solar distillation system will be one of the beneficial options for a short term solution for the needy arid communications facing shortage of potable water. Solar still is dependant on the level of solar radiation available. This level is in turn dependent on the location, season, and weather conditions [2].

Solar stills have two important features: technological simplicity and exploitation of a free heat source such as the sun. However, their productivity is relatively small, and throughout their operational history, various attempts even via relatively complex structures have been made to enhance their performance [3-7].

There are many different designs of solar stills that have been used by different researchers. They can be categorized in two systems: passive solar still and active solar still [7-12]. Many researchers concentrated on the

* Correspondence: Engineerhitesh2000@gmail.com
[1]KSV University, Gandhinagar Gujarat, 380025 India
[2]Department of Mechanical Engineering, Gujarat Power Engineering & Research Institute, Mehsana, Gujarat 380025 India
Full list of author information is available at the end of the article

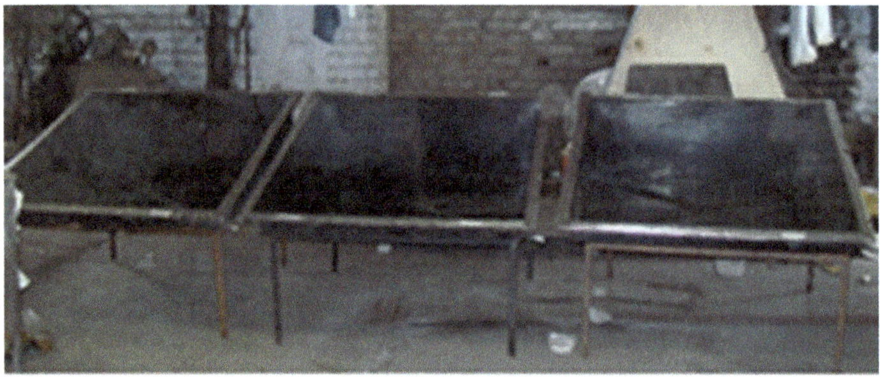

Figure 1 Experimental setup of solar stills.

factors affecting the still productivity, using the experimental techniques such as Al-Hayek and Bardan [5], Naim et.al. [8], Tripathi and Tiwari [9], Meukam et.al. [10], Nafey et.al. [11], Voropolus et.al. [12], and Madani and Zaki [13]. The effect of different designs of solar stills on yield has been conducted by Al-Hayek and Bardan [5]. They found that the productivity of asymmetric greenhouse-type solar still is more efficient. The output from the asymmetric-type solar still is 20 % more than the symmetric-type solar still. Also, they concluded that the water depth and solar radiations are major factors of the solar still to improve the productivity.

Naimet.al. [8] proved that the used of charcoal particles bed increases the productivity by 15 % over the wick-type solar stills. Nafeyet.al. [11] concluded that their experimental work using black rubber and black gravels within a single-sloped solar still as a storage medium, the black gravel, absorbs and releases the incident solar radiation faster than the black rubber; and also, they concluded that the black gravel of 20 to 30 mm increases the productivity by 19 % at conditions of 20 L of saline water and glass cover of 15°. Madaniand Zaki [13] used soot powder as absorber material instead of black painted concrete base to reduce the cost of productivity. They found that removing the basin's insulation reduced the productivity by 13 % to 17 %. Voropoluset.al. [12] experimentally investigated that the behavior of a solar still is coupled with hot water storage tank. They showed that the coupling system led to higher distilled output. Tripathi and Tiwari [9] studied that the effect of different water depths in the basin in the heat and mass transfer coefficient between water and condensing cover depends significantly on the water depth in the basin.

From the above literature reviews, it confirms that solar still is a very important device to convert brackish water into drinkable water. Main aim of this present study is to evaluate the effect of various energy absorbing plates in order to improve the distillate output of solar still through improvement in thermal conductivity. These plates are aluminum and copper plates in climate conditions of Mehsana Gujarat.

Experimental setup

In this research work, three solar stills were designed and fabricated to study and compare the performance of the solar stills, as shown in Figure 1. The first one is a solar still having aluminum plate, and the second is a solar still having galvanized iron plate, while the third is a conventional solar still. The conventional still (a single basin) has a basin area of 1 m^2 (50 × 200 cm). The still is made of iron sheets (2.5-mm thick). The whole basin surfaces are coated with black paint from inside to increase the absorptivity. Also, the still is insulated from the bottom to the sidewalls with sawdust 5-cm thick to reduce the heat loss from the still to the atmosphere. The insulation layer is supported by a wooden frame. The basin is covered with a glass sheet of 3-mm thick inclined at nearly 30° horizontally. Solar stills put on the terrace of 'GeetanjaliSociety', Mehsana. It has a latitude and longitude of Egypt (23°43′N,72°37′E) to maximize the amount of incident solar radiation and increases distillate output. The whole experimental setup is kept in the south direction to receive maximum solar radiation throughout the year.

Feed water tank of 50 × 50 × 50 cm^3 is used to feed water to all solar stills. The feed water tank is connected to the mainline which is divided into three feed waterlines. A

Table 1 Experimental instruments with accuracy, range, and percentage errors

Instrument	Accuracy	Range	%Error
Solarimeter	±1 W/m^2	0 to 4,000 W/m^2	0.5
Temperature indicator	±0.1 °C	0 to 200 °C	0.5
Measuring flask	±1 ml	0 to 5,000 ml	1
Anemometer	±0.1 m/s	0.2 to 40 m/s	1

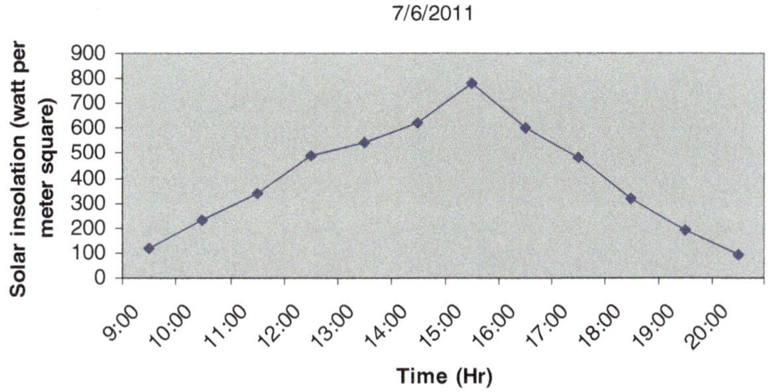

Figure 2 Comparison of solar insolation and time climate conditions of Mehsana (06 July 2011).

flow control valve is integrated at each line inlet in order to regulate the flow rate of water. The experimental setup is suitably instrumented to measure the temperatures at different points of the still (brine, absorber, and glass cover temperatures), total solar radiation, and the amount of distillate water. The temperatures have been measured using calibrated copper constantan-type thermocouples which were integrated with a modeler programmable logic control to measure all temperatures of the solar stills at the same time. The solar radiation intensity is measured instantaneously by a solarimeter. The digital airflow/volume meter is used to measure the wind velocity. Table 1 shows the different instruments as well as their accuracies used in experiment with solar stills.

Procedure

Experiments of solar stills were conducted at Geetanjali Society, Mehsana, Gujarat and carried out from 9 a.m. to 8 p. m. during July 2011. The solar radiation, atmospheric temperature, basin temperature, glass temperature, and distillate water were measured every 1 h for maintaining steady state conditions. However, the accumulated productivity during the 24 h is also measured in each experiment. All measurements were performed to evaluate the performance of the stills under the climate conditions of Mehsana. During the experiments, the ambient climatic conditions (solar radiation, ambient temperature, and wind velocity) were also measured.

Saline water in still is heated by solar radiation. The water vapor formed is condensed at the inner glass surface, and the water droplets are glided along the glass. The condensed water is collected in a calibrated flask. The depth of the saline water in the solar stills is maintained constant manually using the feed water tank and control valves. The present experimental study aims to study the effect of still surface area variation on fresh water productivity and the efficiency of single-slope basin solar stills.

Results & discussion

Depending upon the weather conditions of Mehsana, Gujarat, the wind speed is varied from 4.5 to 7 m/s during the month of March as well as April, and 1.5 to 4.5 m/s at different days of June. Variation of solar insolation versus time (hour) is shown in Figure 2. It remains the same for all solar stills because the solar insolation

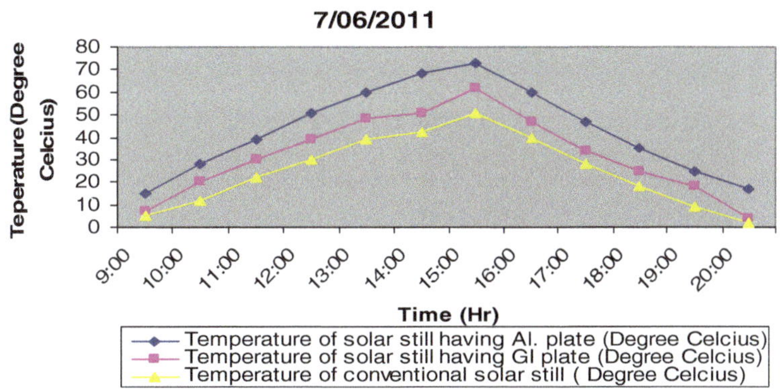

Figure 3 The comparison of hourly variation and time at constant water depth of 40 mm. Comparison of hourly variation of basin water temperature and time at constant water depth of 40 mm in climate conditions of Mehsana.

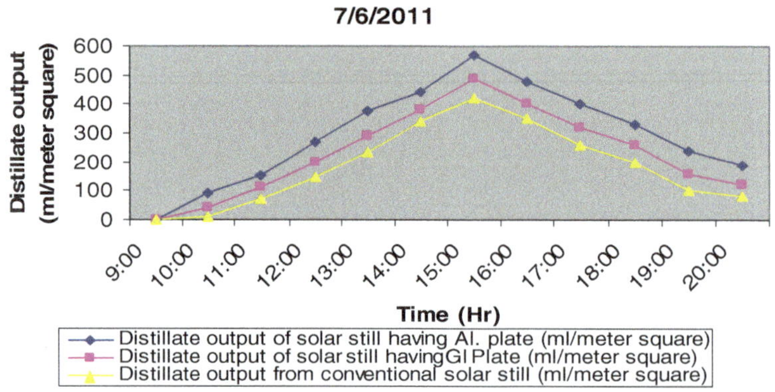

Figure 4 The comparison of hourly variation of distillxate output and time at constant water depth of 40 mm. Comparison of hourly variation of distillate output and time at constant water depth of 40 mm in climate conditions of Mehsana (06 July 2011).

incident on all solar stills is equal. Figure 3 shows the basin water temperatures of all solar stills versus time (hour). It is shown that temperatures at all points increased as the time increases until at maximum value during the afternoon (because maximum solar radiation falls on earth surface during the afternoon, hence, the maximum value is reached in the afternoon).It is also observed that the maximum value in basin temperature is reached during the period of 3:00 p.m. to 4 p.m. for all solar still due to great availability of solar radiation as well as warming of solar still from early morning to afternoon. It also shows that highest temperature is gained by solar still having aluminum plate compared with solar still having a GI plate as well as conventional solar still. It is increased due to its higher thermal conductivity.

Figure 4 shows comparison between hourly distillate outputs gained in all three solar stills like solar still having aluminum plate, GI plate as well as conventional solar still. It shows that maximum distillate output is gained in the afternoon. Also, it is observed that during the starting of readings, all the readings have an initial output of zero at 9:00 a.m., then they increased up to 3 p.m., and then decrease gradually. Here, the highest distillate output quantity is 29.50 L. From the quantity of brackish water inside the solar stills, it is shown that due to large amount of brackish water inside the conventional solar still, it requires highest time for evaporation as well as condensation; hence, the quantity of distillate output produced from conventional solar still is least and highest for solar still having aluminum plate. Figure 5 shows comparison of cumulative distillate water versus the time for all solar stills. Cumulative distillate output means total distillate output starting from 9:00 a.m. It is obtained by adding the distillate output from 9:00 a.m. to 10:00 a.m. and 10:00 a.m. to 11:00 a.m. and continues up to 8:00 p.m. It means that the cumulative output will be highest at 8:00 p.m. Also, it reaches its maximum of suspended aluminum plate solar still due to higher hourly distillate output. Table 2 shows the comparison of distillate output achieved by all solar stills

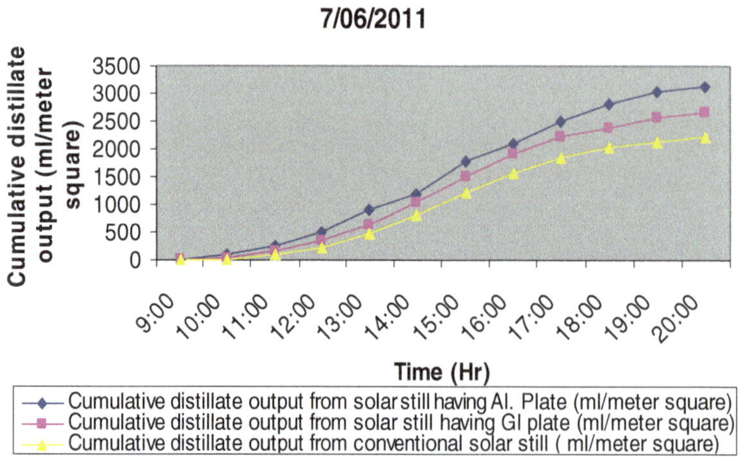

Figure 5 The comparison cumulative distillate output and time at constant water depth of 40 mm. Comparison cumulative distillate output and time at constant water depth of 40 mm in climate conditions of Mehsana (06 July 2011).

Table 2 The comparision of tested solar stills

Month	Conventional	GI	Aluminum
	solarstill (ml)	suspended plate(ml)	suspended plate (ml)
April 2011	2,300	2,600	3,212
May 2011	2,410	2,970	3,500
June 2011	2,540	3,100	3,703

Comparison of tested solar stills of 40 mm depth of water as well as 30 L storage of brackish water in solar stills during the months of April, May, and June 2011.

during the month of April, May, and June 2011. Here, clear sky conditions are considered. Results show that the highest average distillate output is gained during the month of June 2011 compared with the month of April as well as May 2011.

Conclusions

Solar still is a very simple device to obtain potable water. It can also somewhat reduce the demand of potable water, but due to its lower distillate output, it cannot be used in wider applications. Present work gave the following points in conclusion:

- Solar intensity is directly proportional to the distillate output of solar still.
- Suspended aluminum plate in solar still enhances distillate output compared with the suspended galvanized iron plate. It increased daily distillate output as well as cumulative distillate output.
- Highest distillate output is obtained in the month of June 2011. It was 3,703 by solar still having aluminum suspended plate and least obtained in the month of April; it was 2,300 by conventional solar still.
- Suspended aluminum plate is a good material to enhance the distillate output of solar still. It is cheap and easily available.

Competing interests

The authors declare that they have no competing interests.

Acknowledgments

I am very thankful to the management of Gujarat Power Corporation Limited (GPCL), Dr. AJ Mehta, Director of Gujarat Power Engineering & Research Institute for supporting me in the entire research work.

Author details

[1]KSV University, Gandhinagar Gujarat, 380025 India. [2]Department of Mechanical Engineering, Gujarat Power Engineering & Research Institute, Mehsana, Gujarat 380025 India. [3]Silver Oak College of Engineering & Technology, Ahmedabad, Gujarat, India.

Authors' contributions

HP made the experimental set up and readings of solar stills. PS checked the equations as well as updated the revised manuscript. All authors read and approved the final manuscript.

Authors' information

HP is a research scholar and PhD student of KSV University, Gandhinagar, Gujarat and assistant professor in the Department of Mechanical Engineering, Gujarat Power Engineering & Research Institute, Mehsana, Gujarat, India, and PKS is a principal at Silver Oak College of Engineering & Technology, Ahmedabad, Gujarat, India.

References

1. Tiwari, G.N., Singh, H.N.: Present status of solar distillation. Sol Energy. **75**, 367–373 (2003)
2. Rustam Mamrook, Omar.Barden.: Fuzzy sets implementation for the evaluation of factors affecting on solar still production. Desalination **203**, 394–402 (2007)
3. UNICEF: The state of world's children. Oxford UniversityPressm, UK (1995)
4. Malik, M.A.S., Kumar, A., Sodha, M.S.: Solar distillation. Pergamon Press, Oxford, UK (1982)
5. Al, Hayek, Badran, O.O.: The effect of using different designs of solar still on water distillation. Desalination **169**, 121–127 (2004)
6. Kalogirou, S.: Survey of desalination systems and system selection. Energy. **22**, 121–127 (1997)
7. Singh, A.K., Tiwari, G.N.: Thermal evaluation of regenerative active solar distillation under thermosyphon mode. Energy. Conv. Manag. **34**(8), 697–706 (1993)
8. Kawi, A.M.: Non conventional solar stills Part1. Non conventioanl solar still with charcoal as absorber medium. Desalination **153**, 55–64 (2002)
9. Tripathi, R., Tiwari, G.N.: Effect of water depth on internal heat and mass transfer for active solar distillation sytem. Desalination. **170**, 251–262 (2004)
10. Meukam, P., Nijomo, D., Gbane, A., Toure, S.: Experimental optimization of solar still: application to alchoholic distillation. Chem. Engg. Proc. **43**, 1569–1577 (2004)
11. Nafey, A.S., Abdelkadar, M., Andelmotalip, A.A.: MabroukAA: Solar still productivity enhancement. Energy Conv. Manag. **42**, 1401–1408 (2001)
12. Voropolus, K., Mathioulakis, E., Belessiotis, V.: Experimental investigation of a behaviour of a solar still coupled with hot water storage tank. Desalination. **156**, 315–322 (2003)
13. MadaniAA, Z.: GM: Experimental and theoretical study of single basin solar still with porous basin. Applied. Energy. **52**(2–3), 273–281 (1995)

Increasing wind energy penetration level using pumped hydro storage in island micro-grid system

Sheikh Mominul Islam*

Abstract

Ramea is a small island in southern Newfoundland. Since 2004, it has a wind-diesel hybrid power system to provide power for approximately 600 inhabitants. In this paper, wind speed data, load data, and sizing of pumped hydro system at Ramea, Newfoundland are presented. The dynamic model of wind turbine, pumped hydro system, and diesel generator are included in this paper. The dynamic model is simulated in SIMULINK/MATLAB to determine the system voltage and frequency variation and also to visualize different power outputs. Sizing of pumped hydro system indicates that a 150-kW pumped hydro storage system can be installed in Ramea to increase the renewable energy fraction to 37% which will reduce non-renewable fuel consumption on this island. Also, it is found that a pumped hydro energy storage system for Ramea is a much better choice than a hydrogen energy storage system. Such a system will have a higher overall efficiency and could be maintained using local technical expertise, therefore, a more appropriate technology for Ramea.

Keywords: HOMER, Hybrid, Electrolyzer, Pumped hydro

Background

In the early nineteenth century, settlers formed several small independent communities on the Ramea Islands. In the early 1940s, all inhabitants moved to Northwest Island and formed the Town of Ramea, which was incorporated in 1951. It has a population of 674, and the location provides an excellent harbor and was strategic for exploiting the fishery that was closed recently. Access to the community is through a paved road and by a ferry that runs from Burgeo to Ramea. The nearest community is Burgeo which is about 20 km. Residents of Ramea use furnace oil and wood to heat their homes in the winter. There are 335 customers of the diesel-generated electricity in Ramea. Community average load is 650 kW. The diesel plant uses about 18,000 L/week in the summer, but this could go as high as 22,000/week in the winter. Electricity is sold at $0.1495/unit. About 30 to 40 cars come and leave the island every day. Five 10,000-L oil tankers bring in diesel for the plant and furnace oil for the community every week. Until recently, a 2,775-kW diesel-generating plant and six 65-kW wind turbines were used to supply power to the town. Figure 1 shows location of the wind turbines on Ramea Island.

The distance between the six old 65-kW wind turbines and three new 100-kW wind turbines is about 2 km. Hydrogen energy storage system is next to the diesel plant. Figure 2 shows the single line diagram for Ramea distribution system.

Three Caterpillar D3512 diesel engines (Peoria, IL, USA) are the main power source of the isolated wind-diesel power plant [1]. Diesel generators are with the following specifications: 4.16 kV, 1,200 rpm, and 925 kW with a power factor of about 0.85. Each of the diesel units (CAT 3512, 1,400-hp engine with a 925-kW generator) is equipped with an automatic voltage regulator and a governor system. Two frequency control modes are used including (1) a speed-droop characteristic for fast load following capability and (2) an isochronous mode for load sharing and frequency regulation. One or two of the three diesel generators are normally required to supply the local community load. Parallel operation and cycling periods of the diesel generators are coordinated by the diesel generator's master controller. The

Correspondence: mominul511@gmail.com
Department of Engineering, Memorial University of Newfoundland, St. John's, Newfoundland and Labrador, A1B1V3, Canada

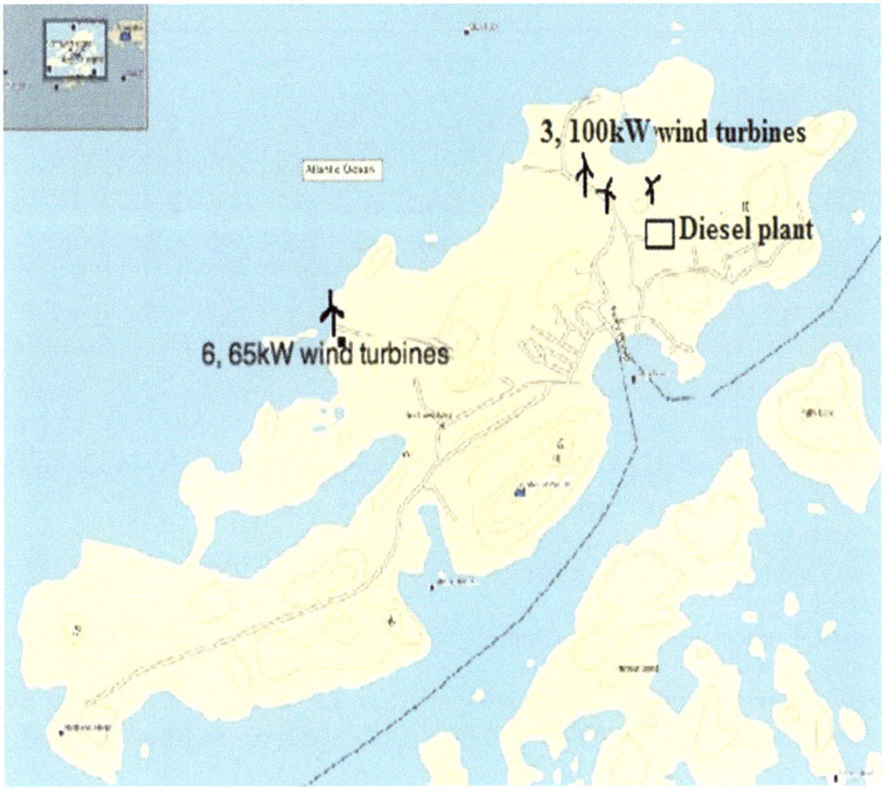

Figure 1 Location of wind turbines and diesel plant in Ramea.

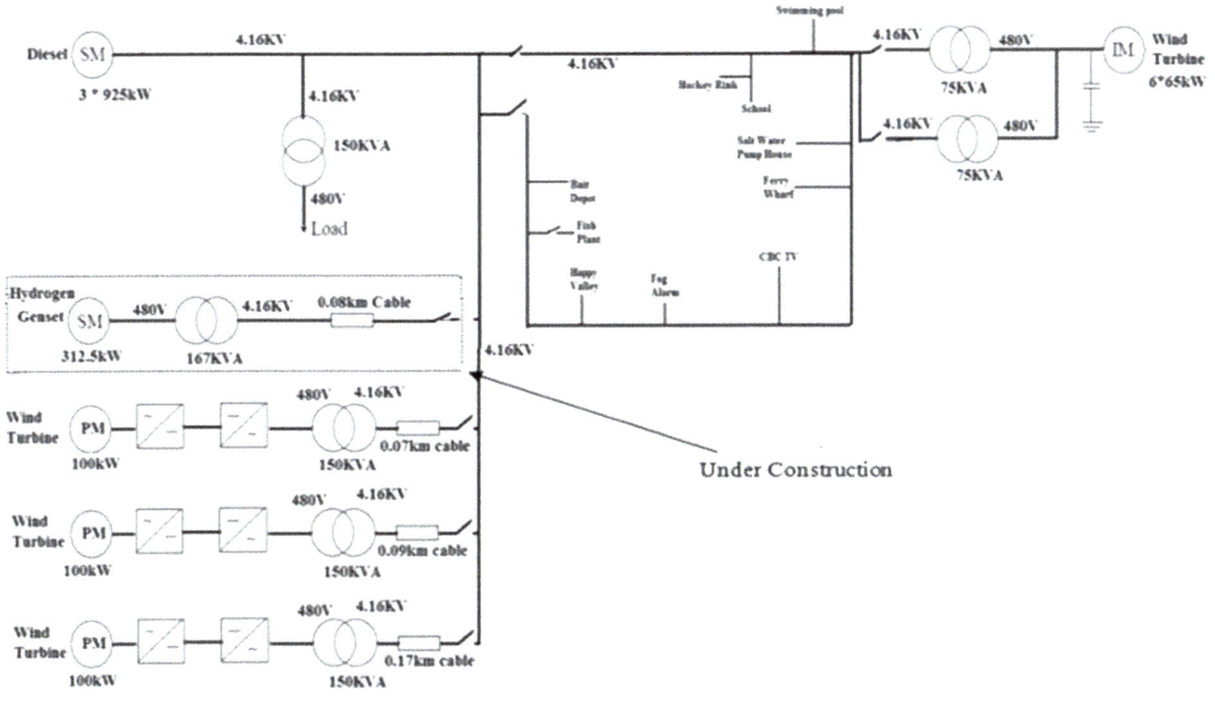

Figure 2 Single line diagram of the present Ramea distribution system.

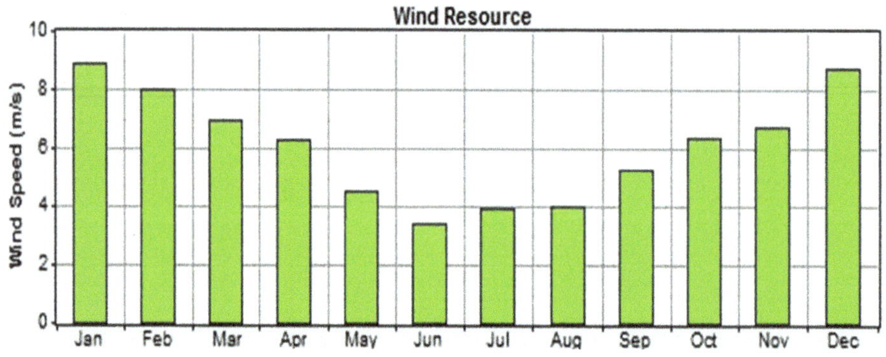

Figure 3 Annual wind speed data of Ramea.

wind energy system consists of six Windmatic wind turbines (OR, USA), a 200-kW controllable dump load, and six capacitor banks. The Windmatic WM15S is a horizontal axis, two-speed, upwind turbine which uses two induction generators, a 65-kW and a 13-kW unit for the energy conversion and direct connection to the distribution system. A 30-kvar fixed capacitor bank is connected in parallel with the output of each wind turbine to partially compensate for the reactive power needs of the induction generators. The start-up of the 65-kW wind turbines is currently assisted by the diesel plants; each wind turbine operates as a motor until it is accelerated beyond synchronous speed, at which point it begins to generate power. The diesel plant also provides the balance of the reactive power, while the capacitor banks are switched on/off to correct the wind plant's power factor to above 0.9.

Annual average wind speed at 10 m in height in Ramea is about 6.08 m/s. This report presents an overview of present pump hydro storage facilities in the world. Analysis of the present wind diesel hydrogen hybrid power system is included. A new pumped hydro storage system is proposed for the community. Based on the recent load and wind data, sizing of a pumped hydro storage (PHS) system is presented. For sizing and analysis, the National Renewable Energy Lab software called HOMER [2] is used. After determining the size of the PHS, the dynamic modeling and simulation of the PHS were carried out to determine the expected system voltage and frequency variations. System design and analysis is presented along with some future policy suggestions. Several studies and simulation have been carried out on pumped hydro storage system [3-7] in the past years.

Analysis of existing Ramea wind-diesel-hydrogen system

The new Ramea wind diesel hydrogen system is still under construction. In 2009, we visited the site and collected the

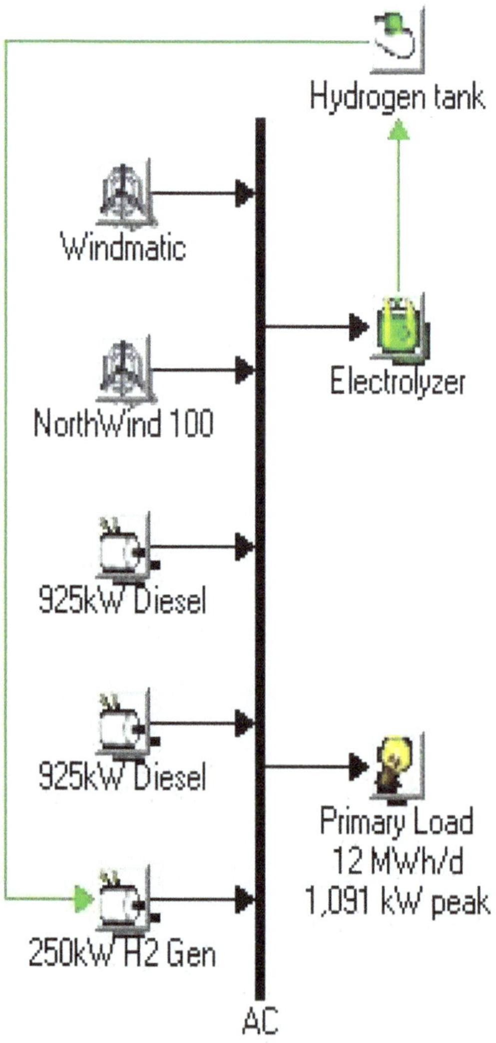

Figure 4 Ramea hybrid system.

site and system data. An analysis of the current configuration of Ramea hybrid system was done using HOMER.

Annual wind speed data for Ramea is shown in Figure 3. The purpose of this analysis is to determine the expected performance of the system. All system parameters and costs were obtained from NL Hydro (St. John's, Newfoundland and Labrador, Canada).

The inputs to HOMER are wind speed data, load data, wind turbine characteristic curve, electrolyzer and hydrogen storage details, hydrogen engine data, turbine costs, and system constraints. The data used in this research was collected from NL Hydro and Environment Canada [8]. From the wind speed data, it is found that the annual average wind speed is about 6.08 m/s at 10 m in height which indicates that this area has one of the highest wind resource regions in Canada. The maximum wind speed is 21.6 m/s. Winter months are windy as compared to summer months. The average wind speed in December and in January is 9 m/s. Community power requirement is also highest during the winter months. It means that using wind energy to reduce diesel consumption makes sense. Figure 4 shows the under construction wind-diesel-fuel cell hybrid power system in Ramea.

The electrical load profile of Ramea is shown in Figure 5. Daily profile indicates that load varies from 350 to 610 kW every day. Monthly profile indicates higher electrical load during the winter months than the summer months in a year.

Load profile also indicates that for a few days in winter, the load may be as high as 1,100 kW. The diesel plant has its own diesel storage tank, and it is also linked to a nearby larger storage tank maintained by a local oil supplier. Ever rising fuel cost is a major issue for Ramea. Community survival depends upon low cost energy. The simulated hybrid energy system is shown in Figure 6.

Figure 4 shows the current Ramea hybrid power system as simulated in HOMER. It has only an alternating current (AC) bus.

Excess electricity from the wind turbines is converted to hydrogen using an electrolyzer. Hydrogen is compressed and stored in three large tanks. Stored hydrogen is used to generate electricity when needed. All system parameters including costs were obtained from NL Hydro. System expected performance was analysed, and it is presented in Figure 5. It shows that the expected renewable fraction is 37% and the cost of electricity would be \$0.248/unit. One 925-kW diesel is always running. A second diesel would be needed only for 12 h in a year. A hydrogen generator will run only for 702 h in a year. Expected electrical performance is shown in Figure 5. It shows that the hydrogen generator will contribute only 1% of electricity to the system, while the electrolyzer will consume 10% of the system electricity. In other words, most of the electricity taken by the electrolyzer will be wasted in the system.

Sizing of a pumped hydro storage system

Sizing of the hybrid power system was done with the help of the optimization software HOMER. The hydrogen energy storage system shown in Figure 7 was removed, and it was replaced with a battery model in HOMER. Overall efficiency of the battery was reduced to 70% to represent a pumped hydro energy storage system. (Typical efficiency of a pumped hydro system is 70%). HOMER sized the system by making energy balance calculations for each of the 8,760 h in a year. Simulated hybrid energy system electrical performance is shown in Figure 8.

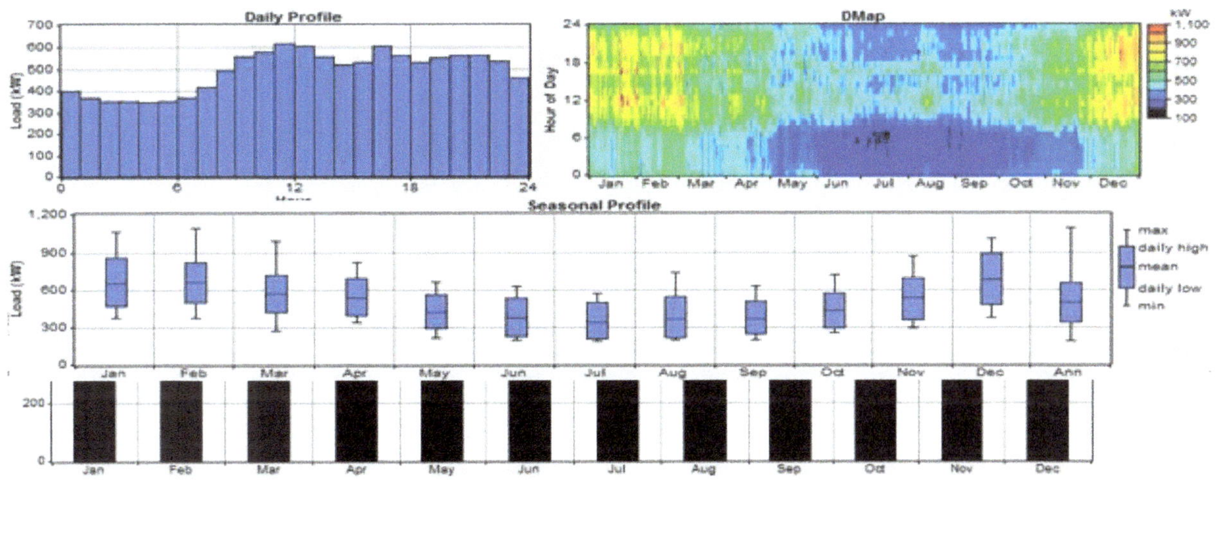

Figure 5 Expected electrical performance and load profile of wind-diesel-hydrogen hybrid power system in Ramea.

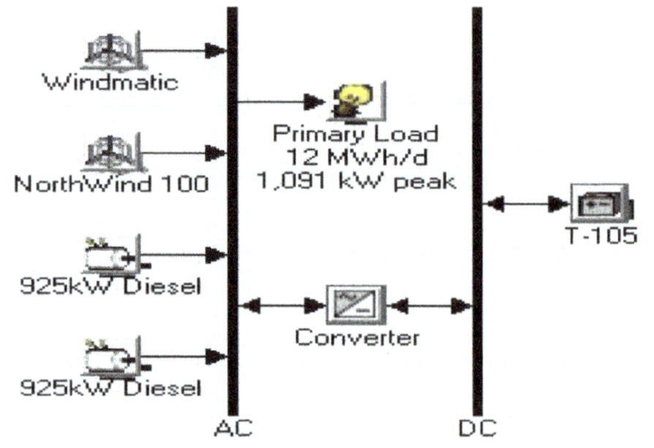

Figure 6 Ramea hybrid system with battery storage.

For each hour, HOMER compares the electric demand in the hour to the energy that the system can supply in that hour and calculates the flows of energy to and from each component of the system. The system cost calculations account for costs such as capital, replacement, operation and maintenance, fuel, and interest. In the HOMER simulation, it is assumed that the interest rate is 7% and the project lifetime is 25 years. With a condition of 40% renewable energy fraction, the optimized HOMER simulation results including sensitivity variables are shown in Figure 8. Renewable energy fraction for all cases is about 37%. From Figure 8, it is clear that yearly diesel consumption reduces with the battery size. The larger the battery sizes, the less diesel consumption. HOMER suggested that the best battery bank (consisting of Trojan T-105 batteries) should have 500 batteries. Based on the maximum power coming out of batteries, we selected a 250-kW converter between AC and DC buses. The nominal capacity of each T-105 battery is 1.35 kW h. Figure 9 describes the expected energy in and out of the battery storage system.

It shows that wind turbine will produce 37% of the total energy while the remaining 63% will come from the diesel plant. Expected diesel consumption is also shown in Figure 8, and it is less than what is shown in Figure 5. Figure 8 shows that most of the electricity will be produced by wind energy during the winter season, and less energy should be expected during the summer season. Therefore, during summer, most of the electricity demand will be met from the diesel. Figure 8 also shows that the total electrical energy required for that area is 4,281,096 kW h/year, and 37% of that will be met by the wind energy including the system peak demand. Figure 8 shows the monthly statistics and expected frequency histogram of battery state of charge. On average, the battery will be 70% charged (i.e., upper reservoir will be 70% full). The battery will be most used in July and August, while it will be least used in May. Effectively, this battery represents a pump hydro energy storage system with an overall efficiency of 70%. During our visit to Ramea in September 2009, we looked for small ponds or lakes which can be used for a pumped hydro installation. We found two small ponds whose elevation is a few meters above sea level. A head of few meters is not good for a pump hydro installation. The hills in Ramea can be used for such a purpose.

Sensitivity variables

Primary Load (kWh/d) 11,72! Wind Speed (m/s) 6.08 Diesel Price ($/L) 0.8

Double click on a system below for simulation results. Categori Overal Export... Details

W...	N...	D925 (kW)	D925 (kW)	T-105	Conv. (kW)	Initial Capital	Operating Cost ($/yr)	Total NPC	COE ($/kW...	Ren. Frac.	Diesel (L)	D925 (hrs)	D925 (hrs)	Batt. Lf. (yr)
6	3	925	925	500	250	$ 2,515,000	717,349	$ 10,874,687	0.218	0.37	783,529	7,406	4	6.3
6	3	925	925	400	250	$ 2,494,000	719,809	$ 10,882,348	0.218	0.37	790,927	7,497	3	6.9
6	3	925	925	600	250	$ 2,536,000	716,421	$ 10,884,867	0.218	0.37	779,219	7,353	5	6.3
6	3	925	925	700	250	$ 2,557,000	715,555	$ 10,895,774	0.218	0.37	776,015	7,314	4	6.5
6	3	925	925	300	250	$ 2,473,000	723,931	$ 10,909,385	0.219	0.36	800,916	7,619	4	8.8
6	3	925	925	800	250	$ 2,578,000	714,937	$ 10,909,582	0.219	0.37	773,429	7,283	2	6.8
6	3	925	925	900	250	$ 2,599,000	713,855	$ 10,917,965	0.219	0.37	770,929	7,252	3	7.2
6	3	925	925	1000	250	$ 2,620,000	713,220	$ 10,931,568	0.219	0.37	769,292	7,232	3	7.6
6	3	925	925	1100	250	$ 2,641,000	711,971	$ 10,938,014	0.219	0.37	767,470	7,210	2	8.1
6	3	925	925	1200	250	$ 2,662,000	711,365	$ 10,951,948	0.220	0.37	766,244	7,195	2	8.6
6	3	925	925	1300	250	$ 2,683,000	710,431	$ 10,962,066	0.220	0.37	764,807	7,177	3	9.1
6	3	925	925	1400	250	$ 2,704,000	709,815	$ 10,975,885	0.220	0.37	763,798	7,164	5	9.5
6	3	925	925	1500	250	$ 2,725,000	709,163	$ 10,989,287	0.220	0.38	762,766	7,152	3	10.0

Figure 7 HOMER-optimized results for the proposed hybrid system.

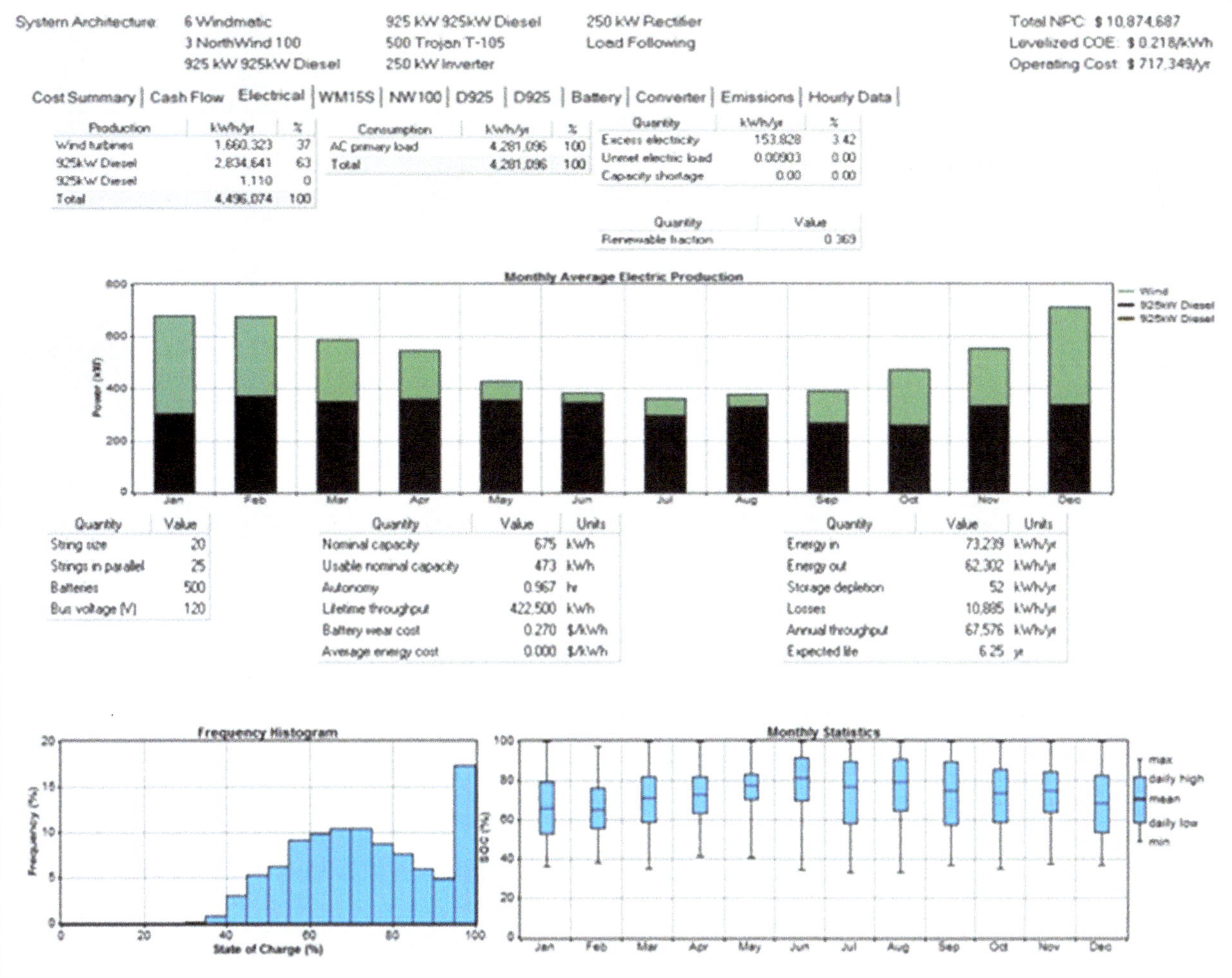

Figure 8 Expected electrical performance of the proposed hybrid power system.

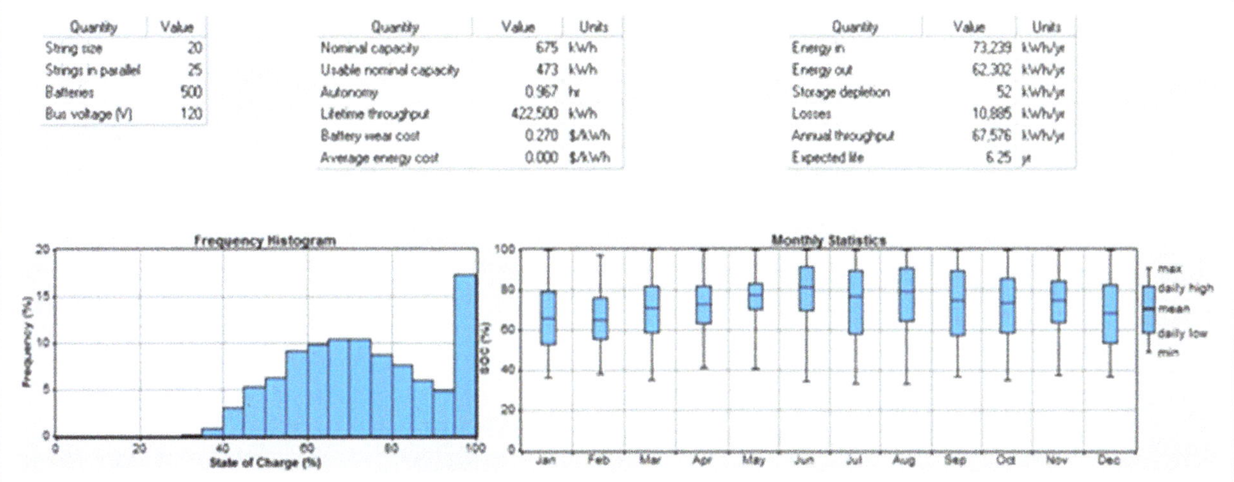

Figure 9 Expected energy in and out of the battery storage system.

It appears that a storage system needs to be built on a hill. In the east side of Ramea, we noted a hill named the Man of War (see Figure 10).

It is a potential place where a hydro reservoir can be built at 63 m in height as shown in Figure 11.

Access to this hill is not an issue as there is a telecommunication tower already installed for the community's communication on the next hill. The Man of War Hill is one of the highest hills in Ramea, and presently, nothing exists there. It has a significant area at the top and is very close to the water, i.e., required length of penstock would be minimal. In a hydro power plant, potential energy of the water is first converted to an equivalent amount of kinetic energy. The potential energy stored in the upper reservoir should be the same as the energy stored in the batteries suggested by HOMER. Figure 8 shows that we will need 500 Trojan T-105 batteries (each 1,350 W h). Hence, the potential energy stored in the upper reservoir should be as follows:

$$\begin{aligned}\text{Potentialenergy} &= \text{mgh} \rightarrow 1,350\times 3,600\times 500\\ &= \quad \text{mgh} \quad = (\text{volume}\times\text{density})\\ &\quad \times g\times h \rightarrow V = \frac{1,350\times 3,600\times 500}{9.81\times h\times 1,000}\end{aligned}$$

For a head, $h = 63$ m, required reservoir size will be $V = 3,932$ m^3 (equivalent to 500 batteries). Figure 10 shows the topographical location of Man of War Hill in Ramea. From Figure 10, it can be seen that about 2,000-m^2 area is available to build a hydro storage reservoir.

The height of the reservoir will be 2 m which is a reasonable height for a man-made concrete-based covered pool-type tank. The size of the hydro turbine is determined by observing the output of the converter shown in Figure 12.

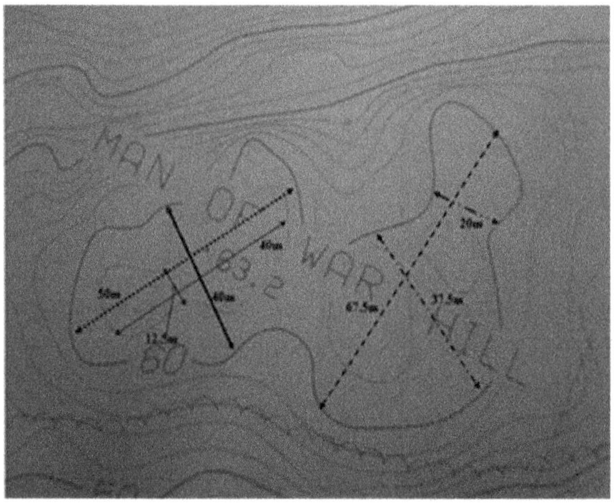

Figure 10 Topographical map of Man of War hill.

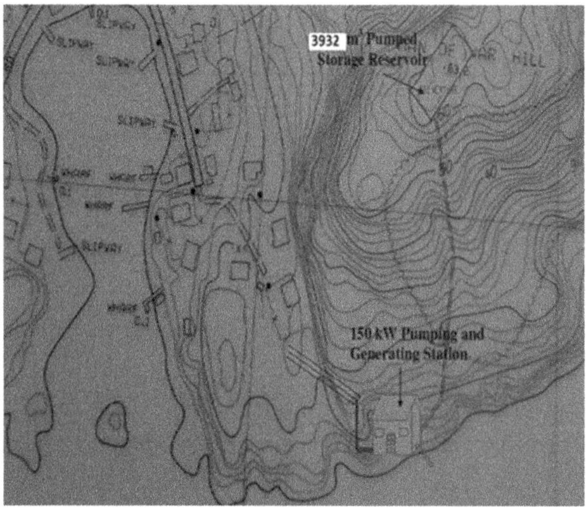

Figure 11 Possible location of the pump hydro energy storage system in Ramea.

From the HOMER time series data, it was found that during April, maximum power flow from the converter is 147 kW. Figure 13 shows expected output of the converter in June.

It shows that we might be taking power from the reservoir for more than 3 h.

Dynamic simulation of grid-connected PHS

The amount of power available from a micro hydro power system is directly related to the flow rate, head, and the acceleration due to gravity. From the above maximum daily average power of 147 kW to approximately 150 kW, the usable flow rate can be calculated using the equation $P = \eta \times Q \times H \times g$. Here, P = power output in kilowatts (150 kW), Q = usable flow rate in cubic meter per second, H = gross head in meters (63 m), g = 9.81 m/s^2, and η = hydro turbine efficiency which is equal to 70%. From the above equation, we determined the flow rate $Q = 0.347$ m^3/s. The minimum operating time for the hydro turbine will be equal to 3,932 m^3/0.347 m^3/s = 11331 s = 3.14 h. Figure 11 shows a possible site for a pump hydro facility. The proposed pumping and generating station could be about 100 m from the top reservoir and about 60 m from the existing 4.16-kV transmission line.

Hydro electricity generation is considered as an established renewable technology. A water flow from an upper to a lower level represents a hydraulic power potential. Pumped storage plants utilize a reversible pumping turbine to store hydro energy during off-peak electricity hours by pumping water from a lower reservoir to an upper reservoir. This stored energy is then used to generate electricity during peak hours, when electricity is costly to produce, by flowing water from

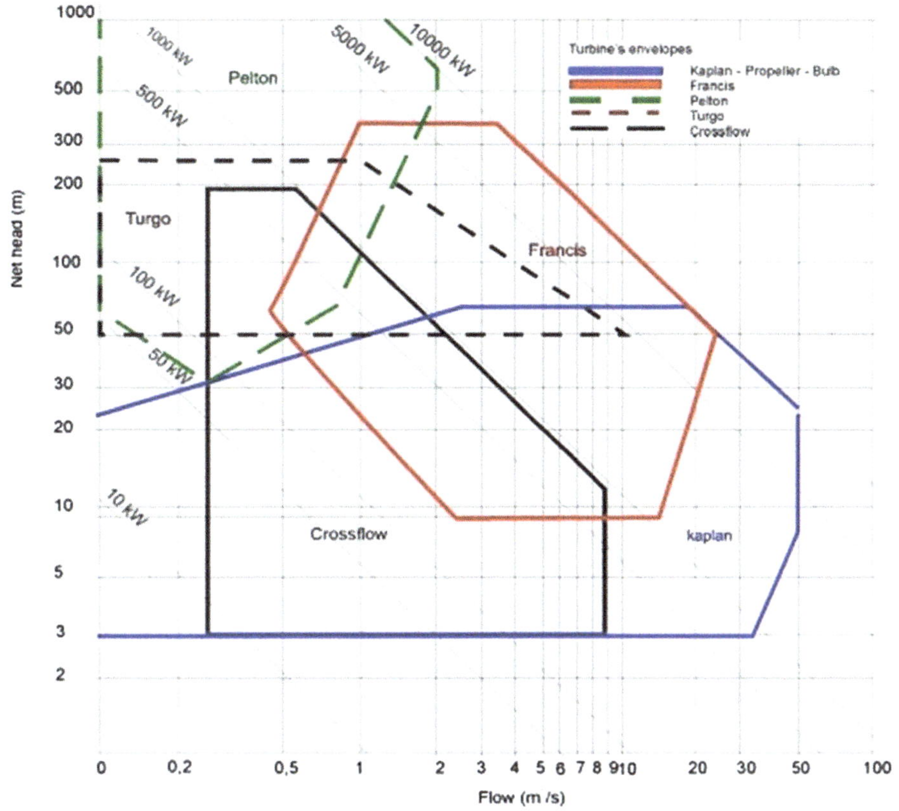

Figure 12 **Turbine selection based on head and discharge.**

the upper to the lower reservoir. The pumped hydro storage system will store excess energy during the off-peak time which will be produced by six 65-kW wind turbines, three 100-kW wind turbines, and a diesel plant. The stored energy will be used to produce electricity during peak times throughout the day. Hydro turbines can be broadly categorized into either impulse or reaction turbines. Figure 12 shows a guide for selection of

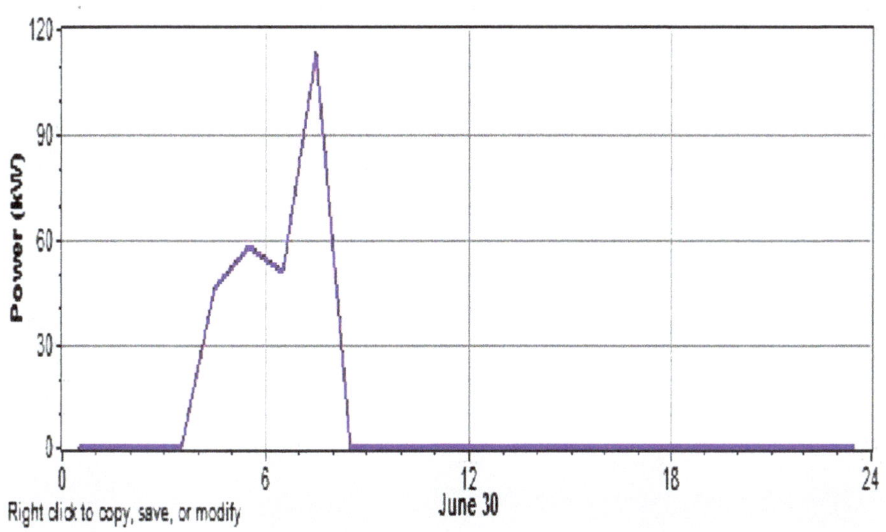

Figure 13 **Output of converter in June.**

hydro turbine. For the Ramea site, the expected flow rate is 0.347 m^3/s, and head is 63 m; therefore, the best selection from Figure 12 is a pelton- or turgo-type turbine.

The reference sites [9-11] can be used for a selection of turbine for the pumped hydro storage project in Ramea. Dynamic simulations of the proposed Ramea wind-diesel-pumped hydro system were done to determine the expected power quality. Simulation was also done to find out if an addition of a 150-kW pump hydro unit will greatly impact the stability of the system. Figure 14 shows diesel-wind-hydro hybrid power system in Ramea simulated in SIMULINK.

From Figure 8, it can be seen that only one diesel generator is running at a time; that is why dynamic simulations were done, considering only one diesel in the system. Figure 14 shows 390- and 300-kW blocks representing all 65-kW and all 100-kW wind turbines in Ramea, respectively. Here, the pump is considered as a 150-kW centrifugal pump with induction machine, and hydro turbine is considered as a 150-kW unit with a synchronous machine coupled to the system bus. Community load is considered as a constant load, and the system was simulated for 24 s. It took a quad core processor-based computer more than 3 h to complete one 24-s simulation. The system was studied for a varying wind speed.

In order to determine the system voltage and frequency variations, the following equations were used to generate a wind field [12].

$$V_w = V_t + V_a, \frac{dV_t}{dt} = -\frac{1}{T_v} V_t + m_w(t), T_v = \frac{10.5z}{V_o}$$

Here, V_w is the wind speed, which is composed of the average wind speed component (V_a) and turbulence component (V_t). The average wind speed value is chosen as 8.02 m/s; $m_w(t)$ is the random white noise which is used to generate the turbulence component within the wind field. Variable z is the turbine height, and V_o is the maiden wind speed. The wind speed characteristics are applied to the simulation, assuming a constant load of 500 kW at Ramea. During our simulation, it is considered that the pumping motor will operate for the first 5 s, and then, it will be switched off. At the 11th second, the hydro turbine will start, and it will operate for 5 s and then will switch off. It is understood that such fast switching will not happen in a real system. In a real system, such switching will happen in tens of minutes. It was not possible to do dynamic simulations for time in minutes. (a 1-min simulation took more than 3 days of computer time. The computer often hangs up, or

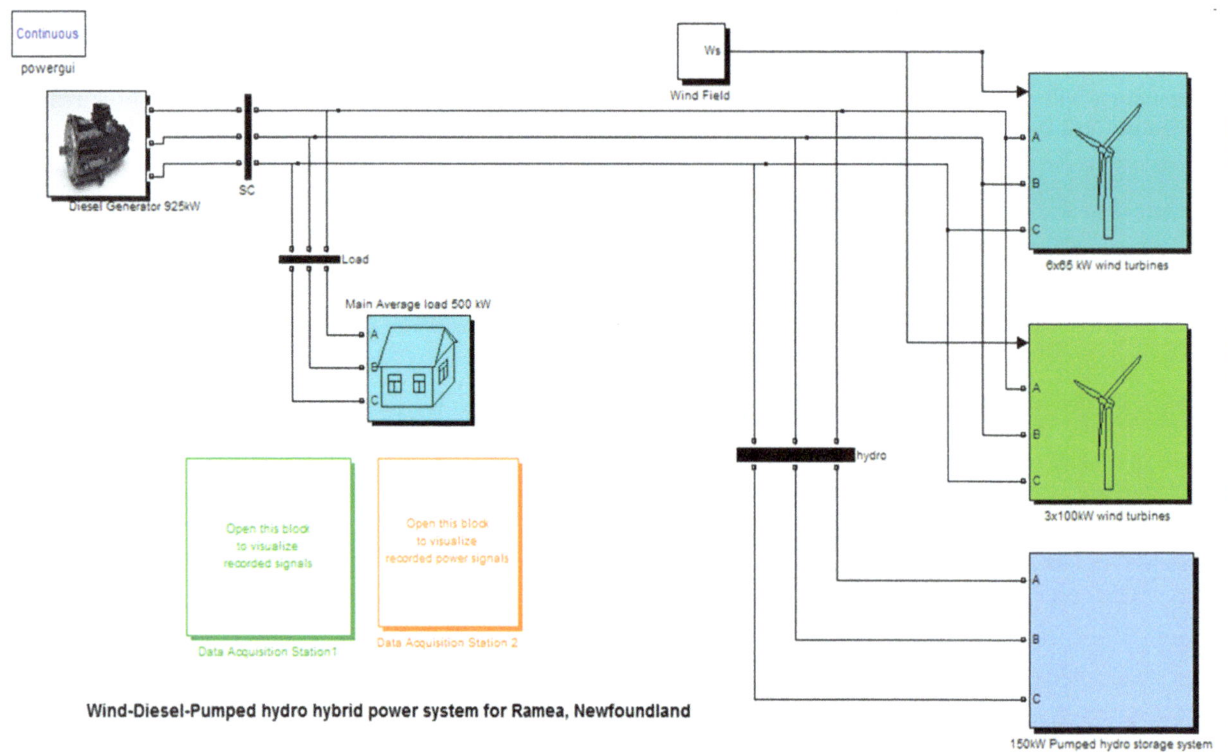

Figure 14 SIMULINK representation of the diesel-wind-hydro hybrid power system in Ramea.

simulation code crashes before simulation is done for 1 min.) Switching after 5 s well represents the transient in the system. Simulation for 24 s gave a good idea of expected transient in the system. Figure 15 shows power variations in the system.

From Figure 16, a voltage dip and a current surge can be seen for the first 5 s when induction motor is running the pump to store water in the upper reservoir.

The diesel engine tried to maintain the power balance by running faster, and the system frequency increased to 61 Hz. At $t = 5$ s, the pump was switched off, and the system came back to its original state within a few seconds. At the 11th second, the hydro turbine is switched on for 5 s. This leads to another transient in the system. Figure 16 also shows the expected variation in the system voltage, frequency, and current during this transient. It can also be noted in Figure 16 that a small variation in the wind speed is not leading to any significant transient in the system. Figure 15 shows the expected power variation in the system. It can be noted in Figures 15 and 16 that simulation was not perfectly converging between $t = 11$ s and $t = 16$ s. Significant time was spent to resolve this issue, but this remained unresolved. Various methods of integration were tried but all led to similar results.

Figure 15 shows the power output from all wind turbines. Wind plant operation results in fluctuating real and reactive power levels and will result in voltage and current transients. Changes in the mean power production and reactive power needs of a wind turbine can cause a steady state voltage and frequency change in the connected grid system. The voltage and frequency variation at load end is given in Figure 16. From Figure 16, it can be seen that the expected voltage variation and the frequency variation

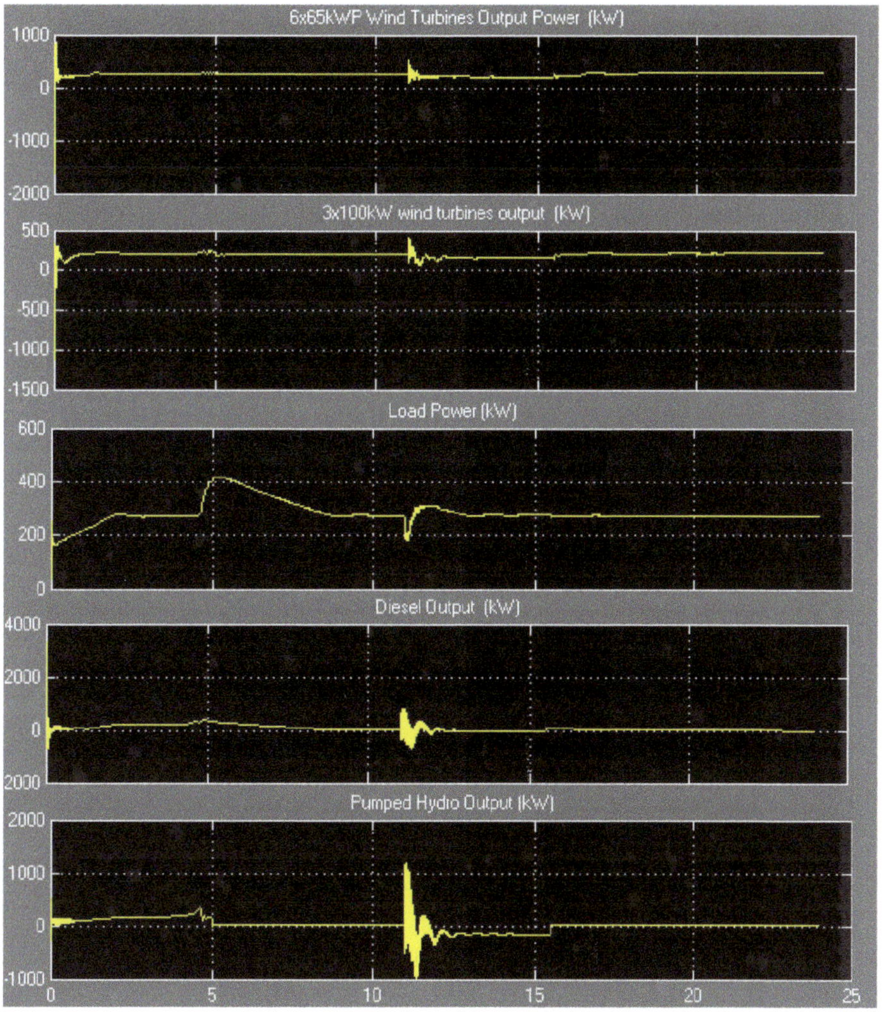

Figure 15 Expected power changes in a transient condition in the Ramea hybrid power system.

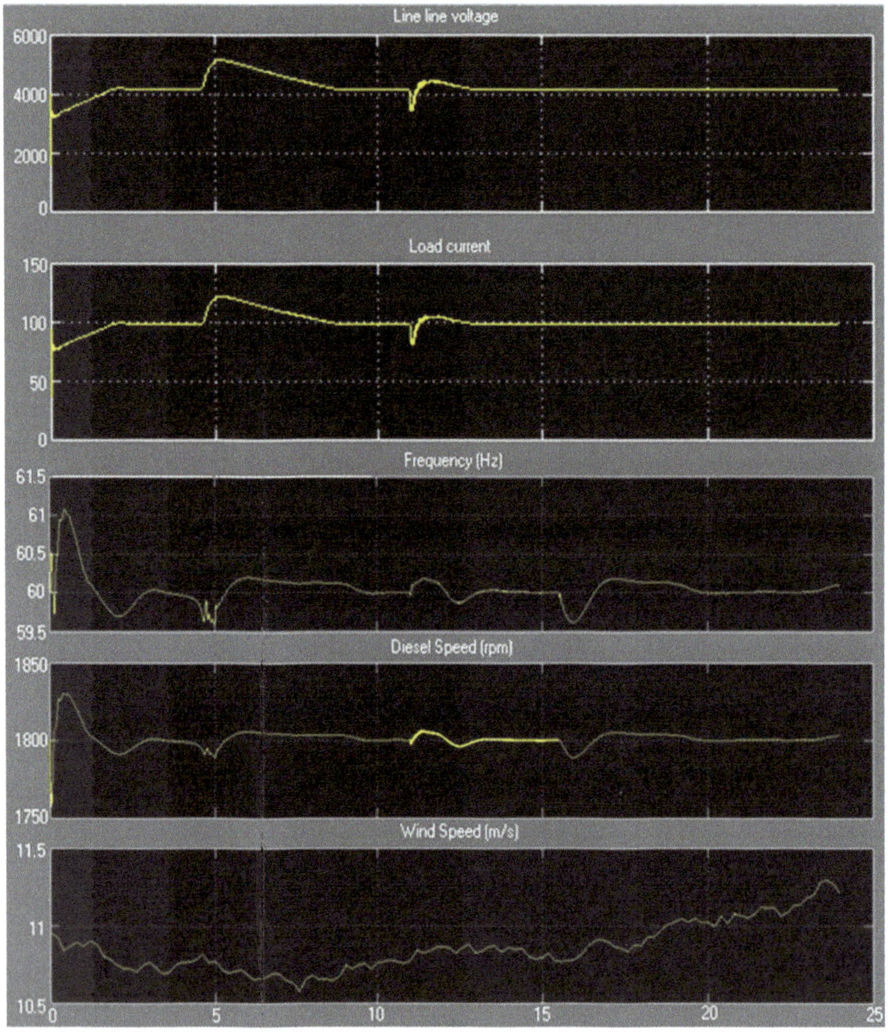

Figure 16 Expected transients in the Ramea hybrid power system.

due to wind speed variation is negligible although switching of large load or a hydro generation unit will cause large voltage and frequency variations in the system. According to power quality standard EN 50160, the voltage variation at the customer's end in remote locations is required to be within +10% [13]. Some voltage variation can be counteracted by adjusting the power factor of the wind turbines [14]. Therefore, the expected voltage and frequency variations in the proposed Ramea wind diesel pump hydro generation system are within current power quality standard.

Conclusions

The current hybrid power system in Ramea consists of a diesel plant, wind turbines, and a hydrogen-based energy storage system. Our analysis indicates that yearly contribution from the hydrogen storage to the system will be less than 1%. We propose a pumped hydro based energy storage system for Ramea. Our study shows that a 4000 m3 water reservoir could be built on the Man of War hill in Ramea for excess wind energy storage. This reservoir will have a head of 63 m and can use sea water. A 150 kW pumping and turbine station could be built near the base of Man of War hill next to the available water and already existing transmission line. Our analysis indicates that such a pumped hydro energy storage system will lead to a 37% renewable energy fraction and will result in acceptable electrical transients in the system. Using a synchronous machine-based turbine unit and an induction motor-based pumping unit, a pumped hydro energy system can be directly connected to the system. A variable speed system [15] is also a possibility. We believe such an energy storage system can be built and maintained using locally available technology and man-power. Therefore, we recommend a full feasibility study of such an energy storage system for Ramea, Newfoundland. We also believe that pumped hydro energy storage is the best energy storage option for many diesel communities in Newfoundland and Labrador.

Competing interests
The author declares that he has no competing interests.

Acknowledgements
The author thanks Newfoundland Labrador Hydro for providing the site load data and Harice Research Center MUN for providing financial support for this research.

Authors' information
SMI has completed his Master of Engineering degree from the Faculty of Engineering, Memorial University of Newfoundland. He has obtained his B.Sc. in Electrical and Electronic Engineering from Islamic University of Technology (IUT), Bangladesh. He has served as an instructor in Prime University and in American International University-Bangladesh. He worked as a Cell Planning Engineer for a mobile company in his country. His research interests are in HVac and HVdc power systems, renewable energy systems, hybrid power systems, control system, transmission line, distribution systems etc.

References
1. Katiraei, F, Natural Resources Canada CANMET Energy Technology Centre-Varennes, Abbey, C.: Diesel plant sizing and performance analysis of a remote wind-diesel microgrid. In: Power Engineering Society General Meeting, 2007, pp. 1–8. IEEE, Piscataway (2007)
2. HOMER energy: Hybrid renewable and distributed power design support. (2011). http://www.homerenergy.com/. Accessed 12 May 2011
3. Anagnostopoulos, JS, Papantonis, DE: Pumping station design for a pumped-storage wind-hydro power plan. In: Energy Conversion and Management, 19th International Conference on Efficiency, Cost, Optimization. Simul. Environ. Impact of Energy Systems **48**, 3009–3017 (2007)
4. Theodoropoulos, P, Zervos, A, Betzios, G: Hybrid systems using pump storage implementation in Ikaria island. In: Proceedings of OPET Island International Conference., (2001)
5. Katsaprakakis, AD, Christakis, GD, Zervos, A, Papantonis, D, Voutsinas, S: Pumped storage systems introduction in isolated power production systems. Renewable Energy. **33**, 467–490 (2008)
6. Bueno, C, Carta, JA: Wind powered pumped hydro storage systems, a means of increasing the penetration of renewable energy in the Canary Islands. Renewable and Sustainable Energy Rev. **10**, 312–340 (2006)
7. Katsaprakakis, DA, Dimitris, P, Christakis, G: A Wind Parks, Pumped Storage and Diesel Engines Power System for the electric power production in Astypalaia. In: Proceedings of EWEC 2006 European wind energy conference. (2006)
8. National Climate Data and Informative Archive. (2010). http://www.climate.weatheroffice.ec.gc.ca/climateData/canada_e.html. Accessed 08 May 2010
9. Toshiba International Corporation - Oceana: Product range. (2010). http://www.tic.toshiba.com.au/product_range/. Accessed 22 May 2010
10. WKV Wasserkraft Volk AG: Philosophy. (2010). http://www.wkvag.de/englisch/tech/tk_1_1.html. Accessed 20 July 2010
11. JLA & Co: (2010). http://www.jlahydro.be/turbines. Accessed 20 August 2010
12. Khan, MJ, Iqbal, MT: Analysis of a small wind-hydrogen stand-alone hybrid energy system. Appl. Energy J. **86**, 2429–2442 (2009)
13. Markiewicz, H, Klajn, Antoni, Voltage disturbances: standard EN 50160 -voltage characteristics in public distribution. Power Quality Application Guide, CDA (1995)
14. Tande, John Olav Giæver: Grid integration of wind farms. Wind Energy **6**, 281–295 (2003)
15. Suul, JA, Uhlen, K: Variable speed pumped storage hydropower for integration of wind energy in isolated grids case description and control strategies. NORPIE/2008, Nordic Workshop on Power and Industrial Electronics, (2008)

Efficient energy utilization as a tool for sustainable development in Nigeria

Sunday Olayinka Oyedepo

Abstract

This study takes a look at the national energy outlook of Nigeria. Energy utilization pattern of the country was investigated, and possible areas of energy conservation in the major economic sectors (industry, transportation, office and residential buildings) were considered. The study reveals that there is inefficient utilization of energy in the major economic sectors of the country. This study presented several energy conservation opportunities to cause energy savings and identified about six major areas through which energy conservation measures can effectively cause some savings in energy and allow for its stability. Such areas of focus for application of energy conservation measures include manufacturing/industrial setup, office and residential buildings, power generation and distribution, transportation, energy conservation through waste control etc. Various measures that need to be considered and appropriately addressed in moving towards energy sustainability in Nigeria have been recommended among which are energy use in ventilating equipment, lighting, electrically operated industrial machines and engines, design for energy-efficient buildings etc.

Keywords: Sustainable energy, Sustainable development, Energy, Efficient energy, Energy conservation

Background

Energy is the mainstay of Nigeria's economic growth and development. It plays a significant role in the nation's international diplomacy, and it serves as a tradable commodity for earning the national income, which is used to support government development programs. It also serves as an input into the production of goods and services in the nation's industry, transport, agriculture, health, and education sectors, as well as an instrument for politics, security, and diplomacy [1].

Some of the common energy carriers or sources are coal, petroleum, natural gas, nuclear fuels, biomass etc. Of all these, the most widely used energy sources are the hydrocarbon compounds or fossil fuels which account for more than 80% of global primary energy consumption [2].

Nigeria is richly blessed with primary energy resources. The country is endowed with the world's tenth largest reserves of crude oil currently estimated to be about 36 billion barrels (about 4.896 billion tonne of oil equivalent (toe)) in 2006. The country has also been described as more of a natural gas island than oil with an estimated endowment in 2006 put at about 166 trillion standard cubic feet (5,210 billion cubic meters). This includes associated and non-associated reserves, placing Nigeria among the top ten countries with the largest gas reserves in the world. Other significant primary energy resource endowment in Nigeria include tar sands - approximately 31 billion barrels oil equivalent (4.216 billion toe), coal and lignite - estimated to be 2.7 billion tonnes (1.882 billion toe), large hydropower potentials of approximately 10,000 MW, small hydropower potentials, provisionally estimated to be734 MW. Table 1 provides a brief summary of these endowments in Nigeria. The table contains recent estimates of other renewable potentials apart from hydropower [3].

Nigeria's coal reserves are large and estimated at 2 billion tonnes of which 650 million tonnes are proven reserves. About 95% of Nigeria's coal production has been consumed locally, mainly for railway transportation, electricity production, and industrial heating in cement production. Apart from the export potential of the Nigerian gas, local demand opportunities are power generation, cement industry, and iron and steel plants. The largest single consumer of natural gas in Nigeria is the Power Holding Company of Nigeria (PHCN), and it

Correspondence: Sunday.oyedepo@covenantuniversity.edu.ng
Mechanical Engineering Department, Covenant University, Ota 2023, Nigeria

Table 1 Nigeria's energy reserves and potentials (2005)

Resources type	Reserves	Reserves (BTOE)[c]
Crude oil	36.0 billion barrels	4.896
Natural gas	166 trillion SCF[a]	4.465
Coal and lignite	2.7 billion tonne	1.882
Tar sands	31 billion barrel of Oil equivalent	4.216
Subtotal fossil		15.459
Hydropower, large scale	10,000 MW	
Hydropower, small scale	734 MW	
Fuelwood	13,071,464 ha[b]	
Animal waste	61 million tonnes/year	
Crop residue	8.3 million tonnes/year	
Solar radiation	3.5 to 7.0 KWh/m^2/day	
Wind	2 to 4 m/s (annual average)	

[a]SCF, standard cubic feet; [b]Forest land estimate for 1981; [c]BTOE, billion tonnes of oil equivalent. Adapted from Dayo [3].

accounts for about 70% which is used to operate electricity-generating gas plants in the country. The consumption rate of petroleum product in Nigeria has increased tremendously from 1990 to 2004 with motor gasoline and diesel oil taking a significant lead. The consumption of petroleum products stood between 80% and 90% of the total commercial energy consumption for over 35 years. The growth rate over the period averaged at about 22%, with gasoline at 28%, kerosene at 19%, and diesel at 18%. Gasoline and diesel are mainly used for transportation which accounts for 87% [4].

Although Nigeria is relatively endowed with abundant fossil fuels, the energy situation in the country is yet to be structured and managed in such a way as to ensure sustainable energy development. As a nation that has limited technological capacity but sees industrialization as constituting a crucial leverage and precondition for meaningful development, Nigeria should be wise enough to manage her scarce energy resources judiciously. As a matter of utmost importance, industrialists, civil servants, researchers, government officers, and students inclusive in Nigeria should take advantage of opportunities in low level, low risk but high worth energy-efficient measures that reduce the bottom line of any business enterprise [5].

In so doing, a lead time will be created to pursue high-tech-driven production processes that will find support at maturity in an already established energy-efficient culture. Researchers have shown that energy supply and end-use efficiency in the developing countries are still only two-thirds to one-half of what would be considered 'best practice' in the industrialized world [6].

Despite the ample coal, oil, and natural gas reserves, at the present rate of extraction, it has been estimated that these reserves, by the next 40 years, will be depleted to the point where it would be uneconomical to continue exploration. It becomes imperative, therefore, that we start implementing energy conservation and efficiency measures in conversion systems while looking for alternative sources of energy [7].

The increasing role of energy efficiency as a catalyst for sustainable development is realism in the industrialized countries of the world. In Nigeria, the story is different at the moment as the huge benefits derivable from adoption of energy efficiency and conservation measures by various economic sectors remain largely untapped due largely to lack of awareness of the economic and social benefits of energy efficiency measures [8].

A sustainable energy system may be regarded as a cost-efficient, reliable, and environment-friendly energy system that effectively utilizes local resources and networks. It is not 'slow and inert' like a conventional energy system, but it is flexible in terms of new techno-economic and political solutions. The introduction of new solutions is also actively promoted [9].

This study will thus take a look at the various energy requirements of the various sectors *vis-a-vis* the energy conservation opportunities therein and also the measures to maximize these opportunities and thus conserve energy. The study identified some major areas through which energy conservation measures can effectively cause some savings in energy and allow for energy stability. The main area of focus for application of energy conservation measures considered in this study include the manufacturing/industrial setup, office and residential buildings, transportation, and power generation and distribution. Each of these areas is being examined for possible energy savings and building a sustainable, long-term, energy future in the country.

Energy crises in Nigeria

Throughout the world, electricity is the most widely used and desirable form of energy. It is a basic requirement for economic development and for adequate standard of living. As a country's population grows and economy expands, its demand for electrical energy multiplies. If this demand is not met adequately, then a shortage in supply occurs. This shortage can assume crisis proportions. According to Chigbue [10], electric power as a major component in the requirements for effective industrialization and development is grossly inadequate in Nigeria.

For many years now, Nigeria has been facing an extreme electricity shortage. This deficiency is multifaceted, with causes that are financial, structural, and socio-political, none of which are mutually exclusive. At present, the power industry in Nigeria is beset by major difficulties in the core areas of operation: generation, transmission, distribution, and marketing [11].

In spite of Nigeria's huge resource endowment in energy and enormous investment in the provision of energy infrastructure, the performance of the power sector has remained poor, in comparison with other developing economies. This assertion was confirmed by a World Bank [12] assessment study conducted on energy development in Nigeria, which compared the performance of Nigeria's power sector with those of 20 other developing countries. The study reveals that the sector had the highest percentage of system losses at 33% to 41%, with the lowest generating capacity factor at 20%, the lowest average revenue at US dollars of 1.56 kW h, the lowest rate of return at 8%, and the longest average accounts receivable period of 15 months.

There is no doubt that expensive and unreliable power remains a major concern to all sectors of the economy in Nigeria: the industrial, commercial, and domestic sectors especially. Multiple and unpredictable power cuts, which have become a daily occurrence in Nigeria, often result in equipment malfunctioning, which make it difficult to produce goods and provide service efficiently. As a result of this fundamental problem, industrial enterprises have been compelled to install their own electricity generation and transmission equipment, thereby adding considerably to their operating and capital costs [8].

Most businesses in Nigeria, large and small, end up relying on the generator for electricity to power their businesses. MTN - the South African mobile phone company and the largest mobile phone supplier in Nigeria - is estimated to have installed 6,000 generators to supply its base stations for up to 19 h/day. The company spends $5.5 million on diesel fuel to run the generators [8].

Energy consumption by sector in Nigeria

From the energy point of view, the Nigerian economy can be disaggregated into industry, transport, commercial, residential (household), and agricultural sectors. The household sector accounts for the largest share of energy use in the country - about 60%. This is largely due to the low level of development in all the other sectors [13]. Figure 1 shows the variation of energy consumption by sector in Nigeria from 1996 to 2005. It can be seen that the household sector has consistently accounted for over half of Nigeria's domestic energy consumption. Its shares varied from about 55% to about 61% [14].

The major energy-consuming activities in Nigerian households are cooking, lighting, and use of electrical appliances - in that order. Cooking accounts for a staggering 91% of household energy consumption; lighting uses up 6%, and the remaining 3% can be attributed to the use of basic electrical appliances such as televisions and pressing irons [7].

The predominant energy resources for domestic and commercial uses in Nigeria are fuel wood, charcoal, kerosene, cooking gas, and electricity [15]. Other sources, though less common, are sawdust and agricultural crop residues of corn stalk, cassava sticks, and in extreme cases, cow dung [16]. In Nigeria, kerosene and gas are the major cooking fuels especially for the urban dwellers. Majority of the people rely on kerosene stoves for domestic cooking while only a few use gas and electric cookers [17]. According to Enete and Alabi [18], distribution of household final energy consumption by type in Nigeria is electricity (4%), kerosene (13%), LPG (1%), and wood and others (82%).

Energy efficiency practice in Nigeria

Energy is an important production factor and therefore should be managed in parallel with land, labor, and capital. Energy-efficient production should be seen as a quick and cheaper source of new energy supply as the cost of providing energy can be several times the cost of saving it. Increasingly, energy efficiency is deemed to include not only the physical efficiency of the technical equipment and facilities, but also the overall economic efficiency of the energy system [5].

Energy efficiency means improvement in practices and products that reduce the energy necessary to provide services like lighting, cooling, heating, manufacturing, cooking, transport, entertainment etc. Energy efficiency products essentially help to do more work with less energy [19]. Energy efficiency is also defined as essentially using less energy to provide the same service [20]. In this sense, energy efficiency can also be thought of as a supply resource - often considered an important, cost effective near - to midterm supply option. Investment in energy efficiency can provide additional economic value by preserving the resource base (especially combined with pollution prevention technologies) mitigating environmental problems.

On the other hand, energy conservation defined as 'an attempt to reduce the amount of energy used for domestic and industrial purposes' is obviously synonymous with energy efficiency. It has been described as using energy more efficiently, whether through behavior, improved management, or the introduction of new technology [21]. Energy conservation is further defined as 'the strategy of adjusting and optimizing energy using systems and procedures so as to reduce energy requirements per unit of output (or well-being) while holding constant or reducing total costs of providing the output from these systems' [5]. It has sometimes been associated with efforts to curtail energy use at the cost of

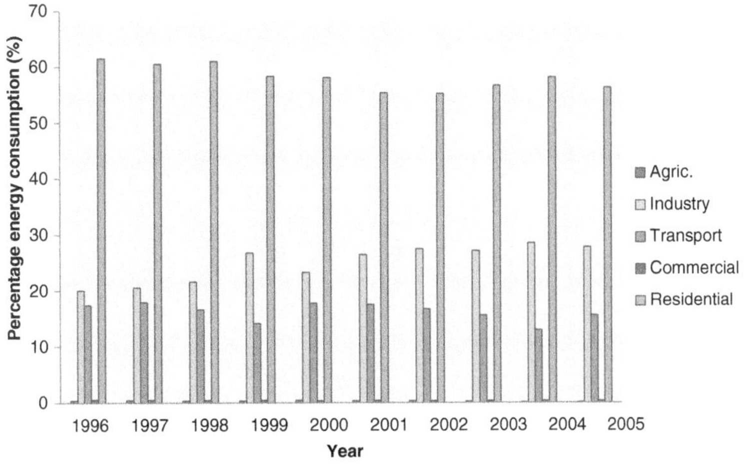

Figure 1 Sectoral energy consumption (PJ) from 1996 to 2005. Adapted from Dayo et al. [14].

economic activity and living standards, but it should be concerned exclusively with energy conservation as a means of increasing economic benefits. It can be seen from the various definitions and explanations that energy efficiency and energy conservation convey the same meaning and can be used interchangeably. The basic objective is the same - the reductions of energy costs or increase in energy supply.

Energy efficiency has become the key driver of sustainable development in many economies in the world [22]. If we use energy efficiently, it will help to reduce the building of more power stations. Thus, the money for building power stations will then be spent on other sectors of the economy. Moreover, more people will have access to energy; if we save energy in one part of the country, the energy saved can be made available in another part. In Nigeria, where the utility companies do not have enough energy to meet the needs of everybody at the same time, energy supply is alternated. With good energy management at the residential, public, and private sectors, there will be no need to alternate electricity supply.

Energy efficiency policy in Nigeria

The national energy policy and the draft energy master plan contain basic policies and strategies for energy efficiency and conservation in Nigeria. In specific terms the policy provides for the following:

- The promotion of energy efficiency and conservation in industrial, residential, and transport sectors,
- Designing a national program on industrial energy efficiency and conservation in collaboration with MAN and experts in higher institutions and research centers,
- Introduction of fuel efficiency labeling program in the transportation sector for various vehicle types,
- Establishing codes and standards for energy efficiency and conservation technologies, and
- Enforcing the codes and standards.

This policy is only on paper but never implemented so as to promote practice of energy efficiency and energy conservation principle in the country.

Inefficient utilization of energy in Nigeria

Energy efficiency does not mean that we should not use energy, but we should use energy in a manner that will minimize the amount of energy needed to provide services. This is possible if we improve in practices and products that we use. In Nigeria, a lot of energy is wasted because households, public and private offices, and industries use more energy than is actually necessary to fulfill their needs. One of the reasons is that they use old and inefficient equipment and production processes. The other reasons are unwholesome practices that lead to energy wastage. These are discussed in the following subsections.

Dominant use of incandescent light bulbs

The use of incandescent bulbs for lighting is energy intensive. Only about 5% of total energy used by an incandescent bulb is converted to light energy; the remaining 95% is converted to heat energy [23]. The energy rating of the incandescent bulbs found in the Nigerian market ranges from 40 to 200 W; thus, we have the ones for 40, 60, 100, and 200 W.

A major factor working against the shift from incandescent bulbs to energy-saving bulbs is the cost. Energy-saving bulbs are far more expensive than incandescent bulbs. The cost of energy-saving bulb in the Nigerian market ranges between N800 and N1,000. However, some substandard energy-saving bulbs could be purchased for about N200. On the other hand, the prices of incandescent bulbs range from N30 to N100. Energy consumed in Nigeria can be drastically reduced if Nigerians replace their incandescent bulbs with energy efficiency bulbs.

Putting on light to advertise goods

Many people who sell certain goods such as snack and electrical materials switch on light during the day to draw the attention of people to buy their goods. In the same way, operators of fast-food centers do the similar thing; they use incandescent bulbs to heat their food and at the same time draw the attention of people to their products. This practice is energy intensive and should be discouraged. In some of the fast-food centers, several incandescent bulbs are put on at the same time for aesthetic purposes and to create illumination during the day. These houses could have been built in a way that they use the natural light during the day.

Switching on outdoor lighting during the day

Many Nigerians do not put off their outdoor lighting during the day. This is particularly very common in commercial and residential areas in many major cities in Nigeria. Even in public institutions such as universities and government ministries were also found to have their outdoor lighting switched on during the day. A lot of energy can be saved if Nigerians cultivate the habit of putting off their outdoor lighting in the daytime. Energy saved from using the natural light instead of light bulbs during the day can be made available for use in offices and for industrial activities.

Proliferation of private water boreholes

In many major cities in Nigeria, many people now have boreholes in their houses. This arises because of the inability of government to provide water in many parts of the country. The use of privately owned boreholes is on the increase. In many cases, you find two or more boreholes in one street. The machine for pumping water from the aquifer is an energy intensive machine and can consume up to 2,000 W of electricity. Apart from consuming a lot of energy, these machines exert a lot of stress on PHCN facilities. In many parts of the world, water is conveyed from a central system through a network of pipes to residential, public, and private buildings. With this method, the energy used to take water from the ground and make it available to the people is greatly minimized.

Industrial activities in residential areas

Many cities in Nigeria are not properly planned. The practice of building industries in residential areas is unhealthy for power supply for residential use. With this kind of practice, utilities providing electricity are not able to plan on how to allocate energy to the various sectors. In addition, because of the high energy consumption of the equipment used in the industries, the equipment exert so much stress on the PHCN facilities which were initially installed to serve residential areas. In this kind of system, it is difficult to allocate energy for the two sectors in a way to maximally satisfy everybody. It is also difficult for utilities to do load shifting.

Setting appliances on standby mode

Many Nigerians do not know that leaving appliances on standby mode, the appliances still consume energy. Putting an electrical appliance on standby mode is not the same thing as putting it off. Electrical equipment consumes energy when on standby mode. Although the energy they consume is not the same as when they are switched on, putting them off when not in use can save some measure of energy. Consumers should be appropriately informed by the manufacturers of the energy electrical appliances at standby mode. A good way to do this is to inscribe it on a label and stick them to the appliances.

Simultaneous use of multiple appliances in public buildings

This is a very common practice among public officers in Nigeria, especially the senior staff. In one department or building, you will find refrigerators and air-conditioners at the same time in all offices, even those of junior staff. It is a common practice to find out that in government offices, you will find a refrigerator, air-conditioner, television set, photocopy machine, desktop computers, fans, electric kettle, and incandescent bulbs, and in many cases, these appliances are switched on at the same time. You go to another office in the same department or building, and you find similar things. The reason for this practice could be that public officers do not pay individually for electricity, and thus, they are not conscious of the way they use energy. Also, it has been revealed that many government buildings are not metered; thus, government officers are not accountable to the energy they use during office hours. In university dormitories, occupants use all kinds of electrical materials, and they do not have restrictions on the kind of equipment they use. It is a common practice for students to use all kinds of electrical heating equipment for cooking in student dormitories. The use of a particular heating equipment popularly called *hot plate* in student dormitories is very energy intensive and should be discouraged. Individual rooms in student dormitories are not metered; this encourages wastage as they are not held accountable for the energy they use.

Leaving appliance on when not in use

Many Nigerians do not put off their appliances when they are not in use. This practice can lead to significant wastage of energy in residential, private, and public buildings. The reason for this could be that many Nigerians do not really pay for the electricity they consume. In many houses, the meters installed by PHCN are no longer functioning. What PHCN officials do is to place these houses on estimated bill. This practice encourages the wastage of electricity since they do not really account for what they consume.

Multiple use of inefficient heating equipment

The use of heating equipment for cooking and heating water should be discouraged in the residential and private buildings. The government should encourage the use of solar heaters. Heating equipment consume about 60% of the energy used in houses. In places like hotels where several water heating equipment are installed in several rooms, sometimes numbering up to 100 rooms or more, the use of solar heaters in these buildings will help to save a lot of energy.

Purchase of secondhand appliances

The Nigerian market is flooded with all kinds of secondhand appliances. Over 90% of Nigerians use one secondhand product or the other. They are cheaper compared to the new ones. Many Nigerians are on the opinion that secondhand products are more durable than the new ones. This assertion could be based on the fact that there are a lot of substandard goods in the market and that the secondhand goods tend to last longer than them. Many of the secondhand products come from European and North American countries, and they may have been manufactured a long time ago. The efficiency of these products is quite doubtful, and the possibility exists that they may have been rejected by the former users to purchase more recent and efficient appliances. The secondhand market needs to be further studied to direct policies that will address the situation.

Barriers to energy efficiency development in Nigeria

The following are barriers to the development of energy efficiency in Nigeria:

- Lack of policy and legislation. Lack of policy and legislation to address the inefficient use of energy is such a key barrier to the development of energy efficiency. Policy and legislation will help to change behavior towards an energy-efficient economy. From the survey carried out by Etiosa [22], about 79% of respondents are not aware of any policy on energy efficiency made by the government. Private and public institutions should also be encouraged to make their own policy to promote the efficient use of energy. The government can make it mandatory for public and large-and small-scale private organizations to establish an energy management department or unit.
- Lack of awareness. Many Nigerians are not familiar with the term energy efficiency, while those who claim they are familiar with the term could not really define it properly. Awareness creation will go a long way to help people understand the concept and change their behavior.
- Lack of trained personnel and energy efficiency professionals. Inadequately trained personnel and professionals is another factor inhibiting the development of energy efficiency. Nigeria as a country lacks adequate energy efficiency experts who will drive the development of the concept and policy that will promote energy efficiency.
- Importation of used machines. The proliferation of imported secondhand appliances may hinder the use of efficient appliances. The reason is that these secondhand equipment are cheap and easily available; the new and efficient ones may be unable to compete with them in the market.
- Lack of research materials on energy efficiency. There is lack of research materials and data that will guide the development of policy that will strengthen the efficient use of energy. Also, there is lack of material to conduct training on energy efficiency.
- Inefficient metering system and low electricity pricing. The metering system in Nigeria is very inefficient and does not encourage consumers to pay the correct amount for the energy they consume. Many people who still use the old meters are now on estimation since these meters are faulty. The use of prepaid meters which was recently introduced by the PHCN will help change the behavior of consumers to use energy efficiently.
- Proliferation of inefficient equipment and desire to minimize initial cost. The desire to minimize initial cost forces many consumers to purchase cheap and inefficient appliances. For example, the cost of energy-saving bulbs in the Nigerian market is about N800 compared to an incandescent bulb which cost about N40. Many consumers will prefer to go for the cheaper ones not minding the long-term benefit of using efficiency bulbs.
- Low income. About 70% of Nigerians leave below the poverty line of $2/day. Many are not able to afford the cost of efficient appliances which are sometimes more expensive than the less efficient ones.

Energy conservation in various sectors in Nigeria

The need for energy is increasing and outstripping its supply in Nigeria. Therefore, and in view of these circumstances, primary energy conservation, rationalization, and efficient use are an immediate need. Getting all the possible energy from the fuel into the working fluid is the goal of efficient equipment operations. This not only saves money and produces higher productivity, but also influences the safety and life of the equipment and reduces pollution [24]. Steps taken to minimize energy consumption, or to use the energy more effectively, are steps in the right direction to preserve the global environment. Energy conservation measures or recommendations are often referred to more positively as opportunities. There are two primary criteria for energy conservation opportunities: that it should be easy to implement and that its payback be short. Ease of implementation and payback period have been used to classify energy conservation opportunities into three general categories: maintenance and operation measures, process improvement projects, and large capital projects [25].

Although energy conservation and efficiency are not resource *per se*, it is acknowledged that their adoption in the country can significantly mitigate the supply challenge. It is in recognition of this that the Federal Government of Nigeria recently approved the establishment of a National Centre for Energy Efficiency and Conservation [26]. The Center is charged with the responsibility for organizing and conducting research and development in energy efficiency and conservation. In this regard, the Center is to carry out the following functions:

1. Develop guidelines for energy-efficient end-use products and advise on their implementation;
2. Develop energy efficiency codes, standards, and specifications for domestic, industrial, and commercial facilities;
3. Gather, analyze, and manage energy supply and consumption data and information;
4. Serve as a Center for training of high-level manpower in energy efficiency and conservation;
5. Develop and execute pilot/demonstration project highlighting energy efficiency concepts;
6. Disseminate information on energy efficiency and conservation concepts through public awareness programs such as seminars, workshops, publications, etc.; and
7. Perform any other functions, as may be directed by the Federal Government in relation to energy efficiency and conservation in Nigeria.

Opportunities for energy conservation in buildings

If the building envelope and building materials were adequately taken care of to allow for a longer time period for daylighting, and maximum indoor space cooling, this would reduce the time of need of electrical energy for both lighting and cooling devices. Consequently, this will promote energy conservation. Apart from the building envelope, a lot of opportunities also exist for energy conservation in the lighting and cooling devices subsectors. For instance, due to the high first cost of fluorescent bulbs, incandescent bulbs (40- and 60-W rating) are still the predominant electric lighting device in the country. Hence, approximately 34.3% of total electricity use in urban households goes to lighting [27]. The tropical climate of Nigeria definitely makes space cooling an essential energy service. This is provided by electric fans and air-conditioners. Fans have a much higher market share than air-conditioners in Nigerian households because of their lower investment costs and lower electricity consumption. A study by Adegbulugbe and Akinbami [27] has revealed that electricity consumption by fan cooling ranges between 2% and 8% while the total average consumption is 7% of the total household electricity consumption in the various household income groups in the country. Similarly, the percentage share of electricity consumption by air-conditioners to the total household electricity consumption ranges between 0% and 2% with a total average of 1.5%. As a result of the downturn in economy in the past two decades, the Nigerian market has become a dumping ground for secondhand products from abroad. Definitely, due to overuse before shipment, the efficiencies of these products have dropped. These consequently promote energy inefficiency in our buildings. If better energy-efficient lighting devices such as compact fluorescent lamps (CFL) as well as better energy-efficient cooling devices such as fans and air-conditioners were promoted in the country, these would help in reducing both electric lighting and cooling energy. Putting altogether these energy-saving opportunities in the residential buildings alone, it is estimated that at least 10% of the total residential electrical energy use will be conserved. Similarly, about 10% of both the total industrial and commercial sectors' electricity demand could also be saved. Ultimately, these would lead to a reduction in greenhouse gas emissions in the country. Another means of energy conservation in buildings is to design the building with more consideration for its environment so that the need for active space heating or cooling by mechanical or electrical means is substantially reduced, if not, totally eliminated.

Opportunities for energy conservation in industries

In Nigeria, energy-saving opportunities in the industrial subsector of the economy have remained a matter for speculation over the years due to uncoordinated efforts at addressing issues relating to energy efficiency and management. It is in the bid to create necessary awareness on the huge potentials for energy savings in the sector that the Energy Commission of Nigeria in collaboration with UNIDO and other stakeholders have for some time now engaged themselves in industrial energy efficiency programs in Nigeria [8].

Strictly speaking, two forms of energy carriers are commonly used in the industry: electricity and heat. However, among all the energy forms, electricity is the most widely deployed in the industry for the transformation of raw materials into the desired end products. Electricity consumption in the industry is usually for lighting and motor power-drives of various kinds of equipment, such as pumps, fans, compressors, blowers, conveyors, air-conditioners, and various machine tools. It is also used in electric furnaces and electrolysis. Improvement in the efficiency of electric motors in particular can result to large energy and cost savings. On the other hand, thermal energy is mostly used in boilers for processed steam generation and in kilns such as in cement production [5].

Investigation carried out in some industries in Nigeria reveals areas of energy conservation (savings) in Nigerian industries.

The following are highlights of walk-through energy audits of some industries in Nigeria.

- A foundry industry. Walk-through energy audit carried out in this industry by Unachukwu [5] shows that there are opportunities for energy saving in the industry. For instance, the total current measured is above the rated value for the main breaker, leading to unacceptable overheating and frequent tripping of the breaker which is now superficially overcome by the installation of a big standing fan to dissipate the generated heat. This is a source of energy waste, and can be avoided by the replacement of the main breaker, especially due to the fact that the contacts may melt completely and results into shutdown and loss of production man-hours. Furthermore, based on the measured total dissolved solid (TDS) value of the high frequency induction furnace cooling water system, it is inferred that the ion exchange resin has expired and is therefore due for recharging or replacement to avoid scale formation and rust along the piping network.
- Plastic industry. In this industry, it was observed that energy-saving potentials in the plastic company include repair of badly damaged insulation; elimination of fuel, gas, oil, and water leaks; reduction in excessive heating and cooling; cleaning of dirty surfaces of heat exchangers, motors, and lamps; prompt replacement of worn-out belts; greater use of diffused light; and replacement of the large number of incandescent light bulbs with energy-efficient CFLs.
- In the bottling company. In this industry, it was observed that the electricity supply is 100% from three 800-kV A diesel generators, while thermal energy for the boiler is from low pour fuel oil (LPFO). Two out of the three diesel generators are run at a time (24 h/day), and the other stays on standby. To say the least, this scenario is a replicate of most industries in Nigeria which is indeed a sad commentary of the electricity supply situation in the country. In terms of energy efficiency, the compounding wastes along the energy supply line are better imagined. The scenario in the bottling company in relation to energy efficiency is that a 10-bar, 2-tonnes/h capacity, LPFO-fired steam boiler produces steam at a pressure of 4 to 5 bars (about 140°C to 150°C) used in bottle washing that requires hot water at a temperature of about 80°C to 90°C. It was observed that the steam produced at a high temperature of about 140°C has to be throttled to reduce the temperature to the required level for bottle washing. Ironically, the runoff water from the final washing stage comes out at a temperature of 60°C to 70°C and is emptied into the drain. While this practice is considered proper from the point of view of avoidance of contamination, it is suggested that a low pressure steam boiler operated at 2 bars can meet the steam requirement and thus save thermal energy. Furthermore, in the compressed air unit, the water-cooled single-stage compressor delivers at a temperature of about 80°C, while the cooling water comes off at 60°C and is again let off the drain. Opportunity for energy efficiency here is that the heat of the air compression can be recovered to heat the boiler feed water, and this may result to about 5% energy savings.
- In the beer manufacturing company. The investigation reveals the following areas of energy-saving opportunities: copious amounts of steam loss from the deaerator by deliberate action of operators; steam line leakages from loose joints and holes along the piping network; the Wort Kettle where there is loss of latent heat in the evaporation of water from the kettle; exposed steam lines where there is radiation loss from uninsulated parts; boiler fuel not sufficiently atomized for efficient combustion; oversized boiler that operates on part load most of the time; cooling tower where treated water is allowed to overflow, and there is thick ice formation along the NH_3 pipeline; brine motor pump where there is use of constant speed motor drive which runs continuously even at no load; large quantity of water waste at the bottle cleaning section; boiler TDS not monitored, feed water make up not measured, and condensate not recovered; low generator frequency at 47 Hz and low power factor at times.

Opportunities for energy conservation in transport

The transport sector is the third largest consumer of energy after the industrial sector. The share of energy consumption from transport sector varied from about 13% to about 18% from 1996 to 2005 [28]. In terms of consumption of petroleum products, it is by far the largest consumer sector and is therefore the sector with the highest impact in terms of foreign exchange costs. Thus, energy saving in this sector is of high priority particularly in financial and economic terms.

In Nigeria, many factors are responsible for high consumption of energy (fuel) in the transport sector. Some of these factors are as follows: owing of fleet of vehicles by the rich people and government officers (political leaders), purchasing of secondhand ('Tokunbo') vehicles, use of inefficient and uneconomic vehicles, use of vehicle with old engine, bad road network etc. To reduce energy consumption and greenhouse gas emissions, the government has to implement the following recommendations:

- Introduce mass transport services as operating in some cities (Lagos, Port Harcourt etc.) in the country: shift road freight to rail and from small vehicles to large vehicles.
- Promote the use of non-motorized transport (bicycles and pedestrian) for short distance.
- Promote availability of spare parts for maintaining efficient operation.
- Continuing improvement of road network. Continued efforts should be made to upgrade the road network particularly between the secondary towns, as this not only improves socioeconomic activity, but also reduces energy demand.
- Decentralize industrial development. Industrial investment should be encouraged by fiscal measures and should be following the pattern of secondary town development to avoid creating centralized industrial conurbation far from the location of the labor force, with resultant high transport requirements.
- Facilitate for railway rehabilitation. Given the generally favorable economics of rail transport when the capital outlay is no longer a significant part of cost, high priority should continue to be given to the restoration of the railway and the use of its terminal as an inland port for urban areas in Nigeria.
- Shift freight transport from road to rail and from using small vehicles to large ones.
- Standardize imported vehicles with respect to energy efficiency and environmental safety.

Energy conservation measures in Nigeria economic sectors

Possible energy conservation measures in six different Nigerian economic sectors considered in this study are given as follows.

Energy conservation measure in office and residential buildings

Possible energy conservation measures in office and residential buildings include the following:

1. Proper building orientation and symmetry. Building design should permit most of the spaces to be daylighted. Using daylighting reduces energy consumption by replacing electric lights with natural light. Buildings designed for daylighting typically use 40% to 60% less electricity for lighting needs than do conventional buildings.
2. Provision of enough windows for cross ventilation. In very hot climate, ventilation is very important. This will go a long way in reducing the use of air-conditioners at homes and offices. Although sunlight and daylight are free and readily accessible, their use without causing glare and overheating can be difficult. Glare can be avoided by using window sills, louvers, reflective blinds, and other devices to reflect light deep into the buildings. Thus, windows, with selective glazing that transmits the most visible light while reducing solar heat, should be favored. Considerable saving potentials for energy in Nigerian office and residential buildings are possible through cost-effective building design.

Lighting

Possible energy conservation measures through lighting in offices, homes, commercial centers, and industries in Nigeria include the following:

1. Relamping. It means substituting one lamp for another to save energy. New fixtures are available which produces superior energy savings, reliability, and longevity compared with incandescent lamps. CFLs are generally considered best for replacement of lower incandescent lamps at homes, offices, and commercial and industrial outfits. These lamps have efficacy ranging from 55 to 65 lm/W. The average rated lamp life is 10,000 h, which is ten times longer than that of a normal incandescent. They offer excellent color-rendering properties in addition to the very high luminous efficiency. Also, they offer energy-saving potential.
2. Installing lighting control systems in bathrooms, stores, and bedrooms. Lighting controls are devices for turning lights on and off or for dimming them. There is the need to install lighting control systems

such as photocells, timers, occupancy sensors, and dimmers in bathrooms, stores, bedrooms, and other areas not frequently used. This is to avoid wastage of energy in these areas.
3. Street light control. Street lighting accounts for more than 50% of all electricity consumed in Nigeria. Of this value, about 50% or more of the energy is wasted by obsolete equipment, inadequate maintenance, or inefficient use. Saving lighting energy requires either reducing electricity consumed by the light source or reducing the length of time.

Energy conservation measures in manufacturing and industrial processes

Three prominent broad areas had been identified for energy conservation measures in manufacturing and industrial processes in Nigeria. These include the following:

1. Improved housekeeping. Improved housekeeping with such factors as furnace maintenance, adjustment of lighting system operations, use of daylight, improving space conditions for lighting, and improving lamp and fixture efficiency are quantifiable measures of energy conservation.
2. Recovery of waste. This forms a significant savings in energy through recovery of waste heat - flue gas, exhaust steam, cogeneration of electricity etc. Heat reclamation is the recovery and utilization of energy that is otherwise wasted, which can be a substitute for a portion of the new energy that would normally be required for heating, cooling, and domestic hot water system. Heat recovery conserves fuels, reduces operating costs, and reduces peak loads.
3. Technological innovation. This border on major redesign of processes and products to yield greater efficiency of cycle operations.

Energy conservation measures in transportation

Possible energy conservation measures in transportation in Nigeria include the following:

1. Increasing the efficiency of the vehicle system through proper vehicle maintenance for better engine performance.
2. Rationing. Techniques used in rationing include restricting the uses of an item, for example, forbidding the use of gasoline to power pleasure boats, limiting the number of vehicles to be owned by every citizen to one, setting a maximum amount a person can spend for fuel (petrol or diesel) etc.
3. Improved technology through electric cars. These are automobiles propelled by one or more electric motors, drawing power from an onboard source of electricity. Electric cars are mechanically simpler and more durable than gasoline-powered cars, which store their energy on board typically in batteries but alternatively with capacitors or flywheel storage devices. It may also generate energy using a fuel cell or generator; they produce less pollution than do gasoline-powered cars. Energy conservation in electric cars, however, is so important that engineers found a way to recover the heat and use it for other heating purposes.
4. Encouraging people to use mass transport system and use of alternative energy source, e.g., fuel cells.

Energy conservation measures in household appliances

Possible energy conservation measures commonly used in household appliances in Nigeria include the following:

1. Air-conditioners. Check and clean the air-conditioner filter once a month, and make sure the air-conditioning unit is the proper size for the room it is cooling. Locate the air-conditioner on the north or east side of a house in a shady area, ventilate the house's attic to reduce heat buildup, and install ceiling fans to improve air circulation.
2. Refrigerators. Check door seals to make sure there are no air leaks, and clean condenser coils on the back of the refrigerator. Keep refrigerator away from the oven or dishwasher, and give the unit breathing room; turn the thermostat down to 2.8°C, and turn on the energy saver switch.
3. Water heater. Lower the heater setting in the range of 49°C and 54°C, insulate the water heater and any exposed hot water pipes, and use low-flow shower heads.
4. Computers. Turn off computers when not in use or set the computer to energy-saving mode.
5. Cloth washers and dryers. Use only with a full load; use warm or cold water, reserving hot water use only for heavily soiled clothes. Use only full loads for the dryer, and if a second load is necessary, dry that load immediately after the first to retain as much heat as possible. Clean lint filter before each load.
6. Ovens and stoves. Use microwave instead of oven where possible; food in glass dishes can be cooked at lower temperature. Preheating oven is usually unnecessary; on the stove top, cook with covered pans and match pan size to the size of the burner.

Energy conservation measures in power generation and distribution

This involves improvement in energy conversion technology for better efficiency, use of thermionic, thermoelectric in magnetohydrodynamic generators for better fuel saving.

Energy conservation measure through research and development

The Department of Energy (DOE) has responsibility for energy research. These programs, now distributed among a number of departments within the DOE, are concerned mainly with scientific and engineering research. They aim to develop better and cleaner methods for extracting and burning traditional fuels, such as coal and oil, and also to develop new sources of energy, such as solar power, liquid fuels from biomass, and nuclear fusion.

In Nigeria, demonstration plants have been funded for technologies such as coal liquefaction and coal gasification, fluidized bed combustion of coal [29], and improvement of methods for extracting oil from shale. Other concepts being explored are the conversion of solid waste into methane gas, the extraction of natural gas from coal seams, the use of fuel cells, magnetohydrodynamics, wind energy, and ocean thermal energy. Solar power research is being pursued in connection with efforts to design residential and commercial buildings that will use energy more efficiently. Some of the most promising devices are solar collectors (which employ water heated by the sun), solar mirrors, and photovoltaic cells capable of directly converting the energy of sunlight into electricity [29]. Improvement can be done through government assistance in releasing funds regularly for the energy research institutes in the country.

Conclusions

Energy is an important production factor and therefore should be managed in parallel with land, labor, and capital. Energy-efficient production process should be seen as a quick and cheaper source of new energy supply as the cost of providing energy can be several times the cost of saving it. Increasingly, energy efficiency is considered to include not only the physical efficiency of the technical equipment and facilities, but also the overall economic efficiency of the energy system. Hence, the adoption of energy efficiency measures in the major economic sectors (household, industrial, and transport) in Nigeria will enhance profitability, reduce greenhouse gas emissions, promote sustainable development, and improve corporate social responsibility. The time to begin aggressive campaigns for energy efficiency measures in the Nigerian economy chain is long overdue.

Presentation on the opportunities that are available in conserving energy in the various sectors - office building and residential areas, manufacturing industries, transportation, and electricity generation and distribution has been made in this work. The various areas where energy saving can be made have been identified in this study; these include energy use in the cooling system of office, residential, and industrial buildings, lighting, cooking, transport, household and office appliances, electrically operated industrial machines and heat engines such as pumps, motors, fans, boiler etc. Several guidelines and measures have been suggested to conserve energy in these areas, and if the guidelines and measures are strictly adhered to, substantive savings in energy will be made.

In order to ensure the sustainability of energy supply and subsequently of the country's sustainable economic development, the government has to intensify further the implementation of energy efficiency programs.

Competing interests

The author declares that there is no competing interest.

References

1. Sambo, AS: Strategic developments in renewable energy in Nigeria. International Association of Energy Economics. Third quarter, pp. 15–19 (2009)
2. Awwad, AA, Mohammed, AA: World energy roadmap - a perspective. WEC-2007 Energy Future in an interdependent world (2007)
3. Dayo, FB: Clean Energy Investment in Nigeria: The Domestic Context. International Institute for Sustainable Development (IISD), Manitoba. (2008)
4. Okoro, OI, Chikuni, E: Power sector reforms in Nigeria: opportunities and challenges. Journal of Energy in Southern Africa **18**(3), 52–57 (2007)
5. Unachukwu, GO: Energy efficiency measures investigation in cement company: BCC case study. Nigerian Journal of Renewable Energy **10**(1–2), 85–92 (2003)
6. Ramachandra, TV, Subramanian, OK: Energy requirements and conservation potential in industrial sector in Karnataka. International Journal of Ambient Energy **22**, 1 (2001)
7. Energy Commission of Nigeria and United Nations Development Programme [ECN-UNDP]: Renewable energy master plan: final draft report., (2005). http://www.iceednigeria.org/REMP%20Final%20Report.pdf. Accessed 17 June 2007
8. Uduma, K, Arciszewski, T: Sustainable energy development: the key to a stable Nigeria. Sustainability **2**, 1558–1570 (2010)
9. Alqnne, K, Saari, A: Distributed energy generation and sustainable development. Renew. Sustain. Energy Rev. **10**(6), 539–558 (2006)
10. Chigbue, NI: Reform of electric power sector: journey so far. A lecture delivered at the US Africa collaboration research sponsored by the National Science Foundation in Abuja, Nigeria (2006)
11. Idigbe, KI, Onohaebi, SO: Repositioning the power industry in Nigeria to guarantee reliability in operation and services. J of Engineering and Applied Sciences **4**(2), 119–125 (2009)
12. World Bank: World Development Indicators (CD-ROM) (2007)
13. Energy Commission of Nigeria: National Energy Policy. Federal Republic of Nigeria, Abuja (2003)
14. Dayo, FB, Adegbulugbe, AO, Ibitoye, F, Adenikinju, A, Voss, A: Estimating the Economic Benefits of Kyoto Protocol to the Nigerian Economy. UNIDO (Economy, Environment, Employment), Vienna (2004)
15. Famuyide, OO, Anamayi, SE, Usman, JM: Energy resources' pricing policy and its implications on forestry and environmental policy implementation in Nigeria. Continental J Sustainable Development **2**, 1–7 (2011)
16. Famuyide, OO, Adu, AO, Ojo, MO: Socio-economic impacts of deforestation in the Sudano-Sahelian Belt of Nigeria. Journal of Forestry Research and Management **1**(1–2), 94–106 (2004)
17. Abiodun, R: Fuel price hike spells doom for Nigeria's forest (2003). http: www.islamonline.net/English/index.shtml. Accessed 18 June 2011
18. Enete, CI, Alabi, MO: Potential impacts of global climate change on power and energy generation. Journal of Knowledge Management, Economics and Information, Technology **6**, 1–14 (2011)
19. Schutze, E, Worthington, R: Getting to Grips with Sustainable Energy. Publication of the Sustainable Energy and Climate Change Partnership (2002)

20. Rosen, MA: Towards energy sustainability: a quest of global proportion. Forum of Public Policy online: A Journal of the Oxford Round Table, Summer (2008)
21. UNIDO [United Nations Industrial Development Organization]: Sustainable Energy Regulation and Policymaking Training Manual (2006)
22. Etiosa, U: A Guide for Developing Policy and Legislation. International Rivers. Benin City, Community Research and Development Centre (2009)
23. Lebot, B: Energy efficiency and market transformation: a short overview of best practices. Abuja, A paper presented during the Inception Workshop of the UNDP-GEF Project to Promote Energy Efficiency in Residential and Public Building in Nigeria (14 July 2009)
24. Habib, MA, Said, SAM, Igbal, MO, El-Mahallawy, FM, Mahdi, EA: Energy conservation and early failure prediction in boilers and industrial furnaces. Symposium on Management of Energy Consumption in Industry, Chamber of Commerce, Damman, Kingdom of Saudi Arabia (2 October 1999)
25. Adeyemo, SB, Odukwe, AO: Energy conservation as a viable pathway towards energy stability. J. Eng. Appl. Sci. **3**(3), 233–238 (2008)
26. Sambo, AS: Matching electricity supply with demand in Nigeria. International Association for Energy Economics. Fourth quarter, pp. 32–38, (2008)
27. Adegbulugbe, AO, Akinbami, J-FK: Urban household energy use patterns in Nigeria. Natural Resources Forum **19**(2), 125–133 (1995)
28. Dayo, FB: Nigerian energy balances: 1990–2005. Technical Paper, Triple -E Systems Inc, Canada (2007)
29. Onyegegbu, SO: Renewable energy potentials and rural energy scenario in Nigeria. Report of National Stakeholders forum on Formulation of Strategy for Rural Industrialization and Development Through Renewable Energy Technology, pp. 14–15. Nicon Hilton Hotel, Abuja (November 2002)

Low-energy office buildings using existing technology: simulations with low internal heat gains

Kajsa Flodberg[1*], Åke Blomsterberg[1] and Marie-Claude Dubois[1]

Abstract

Although low-energy and nearly zero-energy residential houses have been built in Sweden in the past decade, there are very few examples of low-energy office buildings. This paper investigates the design features affecting energy use in office buildings and suggests the optimal low-energy design from a Swedish perspective. Dynamic simulations have been carried out with IDA ICE 4 on a typical narrow office building with perimeter cell rooms. The results from the parametric study reveal that the most important design features for energy saving are demand-controlled ventilation as well as limited glazing on the façade. Further energy-saving features are efficient lighting and office equipment which strongly reduce user-related electricity and cooling energy. Together, the simulation results suggest that about 48% energy can be saved compared to a new office building built according to the Swedish building code. Thus, it is possible, using a combination of simple and well-known building technologies and configurations, to have very low energy use in new office buildings. If renewable energy sources, such as solar energy and wind power, are added, there is a potential for the annual energy production to exceed the annual energy consumption and a net zero-energy building can be reached. One aspect of the results concerns user-related electricity, which becomes a major energy post in very low-energy offices and which is rarely regulated in building codes today. This results not only in high electricity use, but also in large internal heat gains and unnecessary high cooling loads given the high latitude and cold climate.

Keywords: office building, low energy, dynamic simulations, cooling, electricity, lighting, building envelope, HVAC, internal heat gains

Background

According to the European Union's Directive on Energy Performance of Buildings, all new buildings within the union must be nearly zero energy by the end of 2020 [1]. The most recent statistics for Sweden show that the total delivered energy to existing office buildings was around 210 kWh/m^2/year in 2005 [2], whereof the first half was electricity and the other half was district heating and cooling. Regarding new office buildings, the energy performance has been improved, in terms of reduced heating loads, but the electricity for lighting and equipment is still high. The high electricity use results in large primary energy demands in general and also in internal heat gains and heat surplus, leading to unnecessary cooling loads given the high latitude and cold climate.

Good examples of low-energy and nearly zero-energy residential houses have been built in Sweden during the past decade [3,4]. However, at the time of writing, there are very few examples of low-energy office buildings. In Germany, on the other hand, a number of passive and low-energy office buildings have been constructed and evaluated (see examples of demonstration projects in [5,6]). Also, research on energy efficiency potential for a passive office building in Germany has been carried out with dynamic simulations by Knissel [7]. Knissel shows that the primary energy requirement of an example building in Frankfurt can be reduced by 70% with high insulation levels in the building envelope, low electricity consumption for equipment and lighting, no active cooling, heat and humidity recovery and earth-to-air heat

* Correspondence: kajsa.flodberg@ncc.se
Department of Architecture and the Built Environment, Div. of Energy and Building Design, Lund University, Box 118, Lund 221 00, Sweden

exchangers. These German experiences are clearly important for the development of future zero-energy office buildings. However, they must be adapted to a Swedish context, as building techniques, climate conditions and indoor comfort criteria differ between the countries. For example, the sum of heating and cooling degree days is larger and the amount of useful daylight is smaller in Sweden compared to Germany, which affects the heating, cooling and lighting strategies. In order to bridge this gap, the project 'Energy-efficient office buildings with low internal heat gains: simulations and design guidelines' was initiated. The overall aim of the project is to provide knowledge to the Swedish building industry, supporting the development of cost-effective office buildings with good indoor climate and very low energy use. The main goal is to reduce the annual energy use by 50%, compared to the requirements in the Swedish building code.

This paper describes the second phase of the project, a parametric study carried out with dynamic simulations on a typical office building. The objectives of the study are to:

- Reveal important design features when designing a low-energy office building
- Present an office building with a good indoor climate, which uses less than half the energy compared to a new office building, including user-related electricity.

The obvious advantage with such reduction in energy is that the remaining demand can be met by renewable sources like wind, solar, geothermal or biomass, enabling a net zero-energy office building. The paper discusses recommendations for architects and engineers, regarding the design of future low-energy office buildings. Cost-effectiveness is considered, in a way, by using proven techniques that are available on the market today.

Methods

This section describes the overall method used for the dynamic simulations carried out with the software IDA ICE 4 on a model of a typical narrow office building with perimeter cell rooms. First, a reference building was modelled as a base case, designed to correspond to the energy regulations in the Swedish building code. Then, different design features were studied in a parametric study, and the results were analysed and compared to the base case. The parameters which were analysed were airtightness, insulation levels and thermal mass of the building envelope, glazing and solar control, cooling and ventilation strategies as well as control and installed power of lighting and electric equipment. Finally, the most effective design features were combined as the best case solution and simulated in order to obtain the maximum energy saving potential with a proven technique.

The parametric study did not include the study of different heating and cooling systems. Only the building's actual heating and cooling demand was investigated and district heating and cooling with a coefficient of performance (COP) of 1.0 was assumed.

The simulation software

IDA ICE 4 is a dynamic multi-zone simulation program for the study of indoor climate of individual zones within a building as well as whole-year energy consumption for the entire building. It uses the neutral model format language and hence enables the user to change and write new models. IDA ICE was developed in the mid-1980s at the Royal Institute of Technology (KTH) in Stockholm, Sweden, and now serves a global market. The simulation tool is provided by EQUA Solutions AB (Solna, Sweden), and it has been validated according to CEN 13791, ASHRAE 140–2004, CEN 15255, CEN 15265, CIBSE TM33, RADTEST and Envelope BESTEST [8].

The base case

The virtual reference building is a typical large office building with single office rooms along the façades and a central core with stairways, elevators and other facilities. The building is a six-storey building with a narrow shape (approximately 66 m × 16 m) with the short sides orientated to the east and the west (see Figures 1 and 2) [9]. The room height is 3.2 m and the floor height is 3.5 m with 0.3-m concrete intermediate floors and a thin ceiling. Each floor is 1,030 m^2 with a total heated net floor area of 6,180 m^2. More building data can be seen in Table 1. The base case input was chosen to correspond to the Swedish building code BBR18 [10]. In addition, the input is, to a great extent, in line with the standardized input parameters for energy calculations in office buildings (the SVEBY standard), provided by the Swedish Property Federation [11].

Figure 1 The reference building.

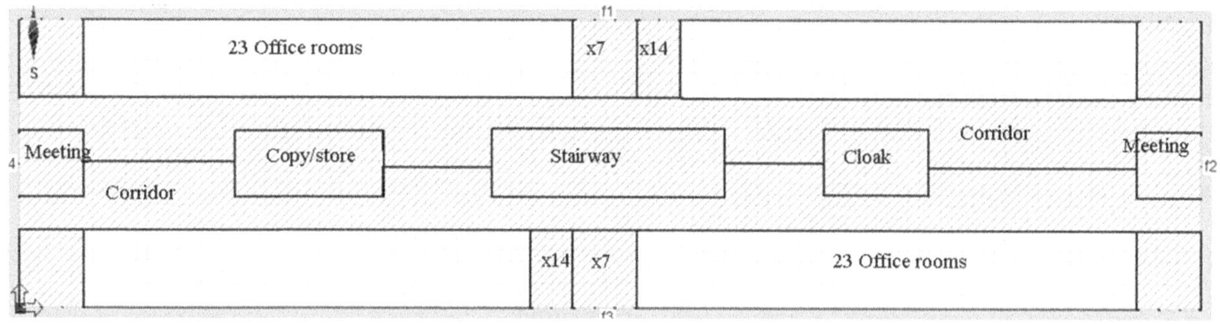

Figure 2 A typical floor plan with thermal zones modelled in IDA ICE.

Results and discussion

This section presents the results and analysis from the parametric study performed on the reference building. Results are displayed as the annual delivered energy for heating, domestic hot water, cooling, fan electricity and other facility electricity as well as user-related electricity for lighting and office equipment. Distribution losses for heating, hot water and cooling are included in the presented results.

The base case

The total delivered energy for the base case is 139 kWh/m^2/year including the user-related electricity for lighting and equipment (see Figure 3). Excluding these, the delivered energy is 92 kWh/m^2/year which is below the requirement in BBR 18 of 100 kWh/m^2/year plus the addition of large airflows. Hence, the base case achieves the regulation with a small margin. The most dominating posts are heating energy (48 kWh/m^2/year) and electricity for lighting and equipment (48 kWh/m^2/year). Even though internal heat gains from lights and equipment are quite large, and the cooling set-point is strict (23°C), the heating load dominates at this high latitude.

Building envelope design

In the first parametric setup, the building envelope was studied. Thermal mass, insulation levels, airtightness, window-to-wall ratio (WWR), orientation and solar shades were varied. The results are presented as total increase or decrease in heating, cooling and electricity, compared to the base case. The results in Figure 4 show that thermal mass has a rather small impact on the heating and cooling demand and that the saving potential for a heavy construction can save 2.5 kWh/m^2/year at most compared to the base case. This result indicates that the cooling load, due to solar gains and internal heat gains, is not large enough in countries at high latitudes to take advantage of thermal mass. Note that the reference building has a quite modest WWR (35%) compared to many modern office buildings, which implies that the solar heat gains are rather small. The result also indicates that there might already be enough thermal mass in the concrete floors alone, despite the ceilings and carpets. The reference building is large with many floors and maybe it cannot benefit from any more thermal mass. Regarding insulation levels and U-values, it is obviously more effective to choose passive house windows ($U = 0.9$ W/m^{2}°C) than passive house walls ($U = 0.1$ W/m^{2}°C), despite the rather modest WWR. However, this result depends on the base case starting points, which provided an improvement for the windows from 1.4 to 0.9 W/m^2°C and for the wall elements only from 0.2 to 0.1 W/m^2°C. The negative aspect with improved U-values is the increased cooling demand, but this is compensated by the even larger decrease in heating demand. With a combination of passive house windows and passive house walls, the total energy saving potential is 11 kWh/m^2/year compared to the base case. Finally, an improved airtightness turns out to have a large impact on the building's heating demand. The result is not surprising since the base case has a particularly leaky building envelope (1.6 l/s/m^2 envelope area at 50 Pa). According to the simulation results, the Swedish passive house criterion for airtightness is sharper than the international criterion, at least for the shape of the reference building [12,13].

Figure 5 shows the impact of a larger window area and different solar shading systems. These results indicate that a larger WWR has a significantly negative effect on energy savings, both for the heating and cooling demand. In total, an extra 25 kWh/m^2/year is needed for the case with WWR of 60% compared to the base case with WWR of 35%. Another negative effect with large glazing amounts is the risk of glare. According to a daylight study performed by Dubois and Flodberg [14] on a similar building, the optimal glazing-to-wall ratio (GWR) in Sweden is 20% to 40%, with the lower value preferable on the south façade where the risk of glare is higher. Furthermore, the study shows that increasing the GWR to more than 40% has a negligible effect on available daylight inside the building, and no electric lighting will therefore be saved. Hence, the results presented in

Table 1 Building data and base case input

Parameter		Simulation input	Comment
Climate conditions	Location	Stockholm 59.35°N, 17.95°E	
	Dry-bulb temperature, minimum/mean/maximum	−18.3°C:6.5°C:26.1°C	
	Horizon angle	15°	
Dimensions	Heated floor area (A_{temp})	6,180 m^2	
	Air volume	19,776 m^3	
	Envelope surface	5,193 m^2	
	Surface-to-volume ratio	0.26/m	
	Façade surface	3,133 m^2	
	WWR	35%	GWR 31%
	WFR	18%	
Building elements	External wall *U*-value	0.20 W/m^2°C	170 + 50-mm mineral wool
	External roof *U*-value	0.11 W/m^2°C	300-mm mineral wool
	External floor *U*-value (excluding ground resistance)	0.17 W/m^2°C	200-mm EPS
	Windows *U*-value (including frames)	1.4 W/m^2°C	Pilkington Suncool 2 glass
	Glazing	LT 72%, SHGC 43%	Pilkington Suncool 70/40
	Internal blinds	0.83 × SHGC	SHGC multiplier
	Total UA transmission	2,119 W/°C	
	Thermal bridges	445 W/°C	Calculated with HEAT2
	Air leakage rate	1.6 l/s/m^2 at 50 Pa	Envelope surface
Heating/cooling efficiency	Boiler COP	0.9	Total heat supply
	Heating coil COP	0.9	In air handling unit
	Domestic hot water COP	0.9	
	Domestic hot water use	2.0 kWh/m^2/year	SVEBY standard
	Chiller COP	0.9	Total cooling supply
	Cooling coil COP	0.9	In air handling unit
Thermal climate	Set-points for mean air temperature	22°C to 23°C	Normal target values
Ventilation	Ventilation operating hours	Weekdays, 0700 to 1900 hours	
	Constant air volume	1.5 l/s/m^2	SVEBY standard
	Heat exchanger efficiency	70%	Yearly average
	Total SFP	2.0 kW/(m^3/s)	BBR 18
	Supply air temperature	17°C	
Office operation	Office hours	Weekdays, 0800 to 1800 hours	1-h lunch break
	Occupant space	20 m^2/person	SVEBY standard
	Occupant heat	108 W/person	Sensible and latent heat
	Occupancy factor	0.7	SVEBY standard
Lighting	Installed power in office rooms and other spaces	10 and 6 W/m^2	
	Control	Manual switch on/off	
Computers and office equipment	Power on/standby	140/10 W/person	

WWR, window-to-wall ratio; WFR, window-to-floor ratio; SHGC, solar heat gain coefficient; COP, coefficient of performance; SFP, specific fan power.

this article are supported by Dubois and Flodberg's study, and it can be recommended to keep the glazing amount as small as possible in order to save energy and to avoid glare, but not less than 20% in order to secure enough daylight and view out. The building orientation, on the other hand, does not affect the whole building energy use because of the symmetry and compactness of the building. The study of solar shading devices indicates that the further out in the façade the blinds are placed, the more cooling energy can be saved, but in return, more heating energy will be needed. However, the overall effect is modest. Therefore, climatic conditions and

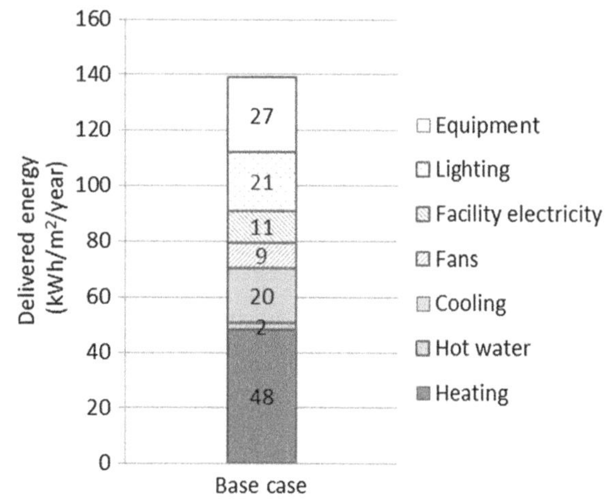

Figure 3 Total delivered energy for the base case.

the number of heating and cooling hours must be considered when selecting a solar shading strategy. It may not be profitable with external blinds if the daytime heating hours exceed the cooling hours or if external blinds are much more expensive due to high wind exposure. One possible, but rather expensive, solution is to have both internal and external solar shades and alternate these in order to optimize the solar heat gains in different seasons. It could also be an alternative to improve the glazing performance and select a glass with a low solar heat gain coefficient. However, there is a risk that the solar heat gains are reduced more than needed, creating an unnecessary heating load, and that the visual transmittance and window view are degraded.

HVAC strategies

Figure 6 shows the results from the study of temperature set-points, heat exchanger efficiencies, specific fan power (SFP), ventilation strategies and user-related electricity. The improvement in heat exchanger efficiency (eta) has a rather small impact on the reduction of heating. Improving the efficiency from 70% to 80% yields a saving of 3 kWh/m^2/year in heating energy, and improving the efficiency from 80% to 85% saves no heat at all. The explanation can be that the largest heating demand in an office occurs during the night when the air handling system is off. During office hours, the building is partly heated by internal gains and solar gains and the heat exchanger is even bypassed at times. The SFP was also improved in the parametric study. The base case fan efficiency, with a SFP of 2.0 kW/(m^3/s), was improved to 1.5 kW/(m^3/s). However, this only decreased the electric energy with 2 kWh/m^2/year. The greatest saving potential occurs when changing the CAV system into a variable air volume flow (VAV) system with airflows depending on indoor temperatures and CO_2 levels. For the reference building, a total of 21 kWh/m^2/year can be saved which is 15% of the total energy use. This result is in line with recommendations from the Passive House Institute, which states that comfort and a good indoor air quality should be ensured and provided by using just the necessary air quantities [15]. Modern Swedish office buildings often have strict indoor temperature targets at about 22°C to 23°C during working hours. The energy saving potential when allowing a larger mean air temperature range, for example 21°C to 24°C, is far from negligible. According to this study, 7 kWh/m^2/year (15%) of heating energy and 5 kWh/m^2/year (24%) of cooling energy can be saved by accepting a larger range in indoor temperatures. To avoid thermal dissatisfaction, it is important to keep the operative temperature close to the mean air temperature by avoiding, for example, solar radiation impinging on the occupants. The figure

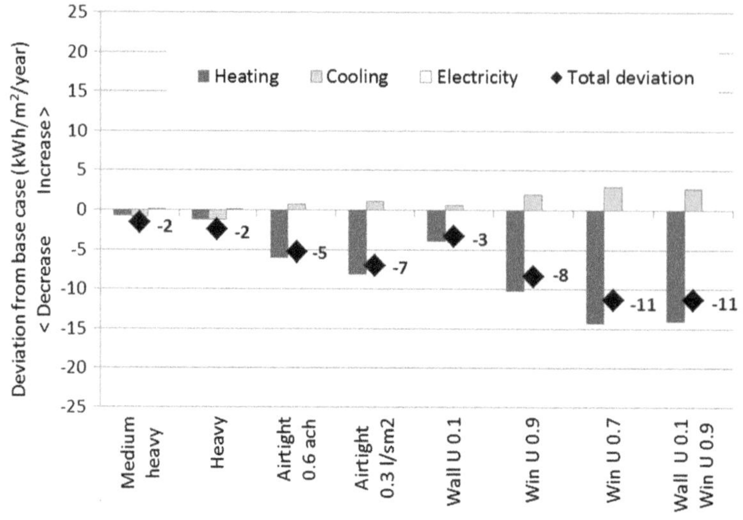

Figure 4 Impact of insulation, thermal inertia and airtightness on total delivered energy.

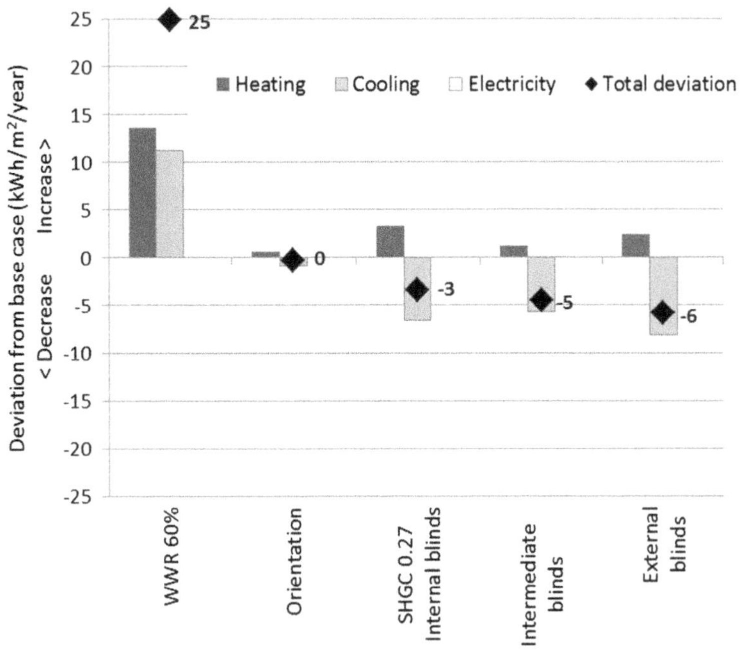

Figure 5 Impact of window area and solar shading systems on total delivered energy.

also shows the potential cooling effect from mechanical night ventilation, with variable airflows at night and heavy constructions. The energy saving potential is negligible (<2 kWh/m^2/year) compared to a similar model without night ventilation. This indicates that cooling with night ventilation might not be profitable in a Nordic country. Cooling with mechanical night ventilation actually saves some fan electricity in the simulation even though the fans operate more hours. This is because during a heat wave, night ventilation reduces the morning temperature; thus, the fans can operate with reduced airflows during the first working hours.

Lighting and electric equipment

The improvement in office equipment and lighting has a large impact on electricity, heating energy and cooling

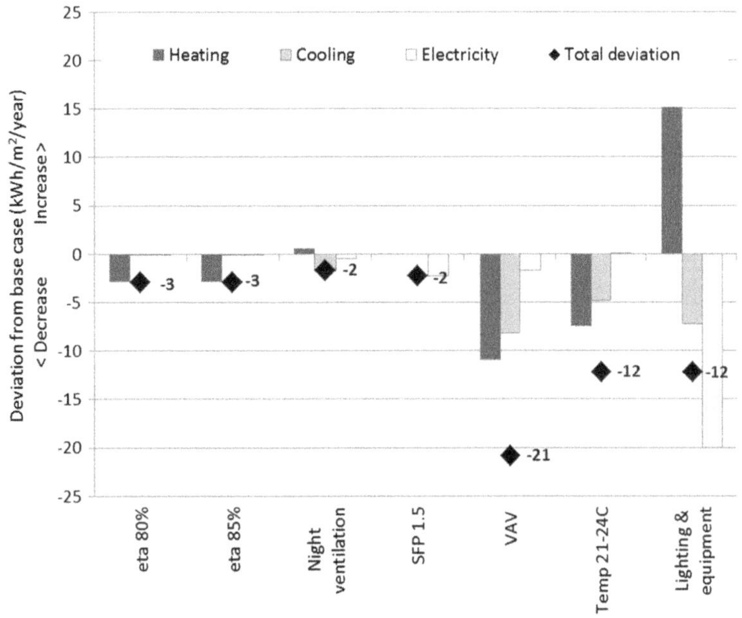

Figure 6 Impact of HVAC strategies and user-related electricity on total delivered energy.

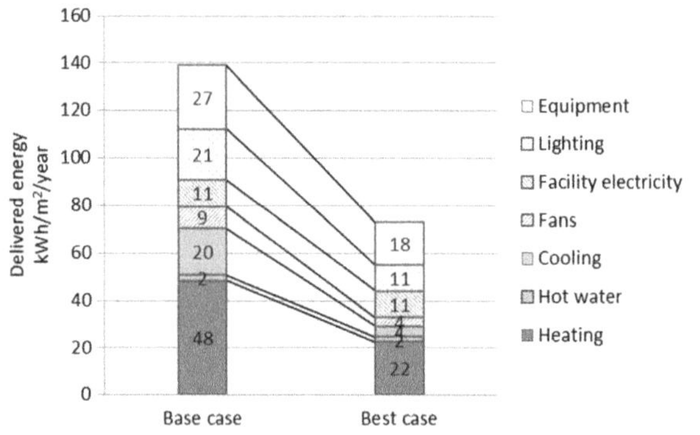

Figure 7 Energy saving potential with the best case solution.

energy, as was shown in Figure 6. When using more efficient office equipment and lighting, with reduced installed powers (EPD 55 W/room, LPD 8 and 4 W/m^2) and improved control (no standby losses at night and lighting with daylight dimming control). Compared to the base case, approximately 10 kWh/m^2/year of electric energy is saved when improving the office equipment and another 10 kWh/m^2/year is saved when improving the lighting system. Meanwhile, the cooling energy decreases and the heating energy increases due to reduced internal heat gains. The total energy saving potential, compared to the base case, is 12 kWh/m^2/year when both office equipment and lighting are improved. This result shows that it is desirable to reduce the internal gains even though the heating load increases. Furthermore, the results show that the available amount of daylight in Sweden is sufficient to reduce the electricity use for lighting to the same level as in Germany [7].

Best case scenario

Figure 7 presents the most efficient design features from the parametric study, combined as the 'best case scenario' with the intention to reach a low-energy solution (see design features in Table 2). By improving walls and windows, reducing window-to-wall ratios, introducing demand-controlled ventilation and lighting, allowing a larger range in temperature and installing more efficient equipment which are completely turned off outside office hours, the heating energy, cooling energy and electricity use can decrease significantly. The best case solution shows a great improvement in especially heating and electricity use. The space heating energy is reduced by 26 kWh/m^2/year (54% heating energy saved and 19% total energy saved). The total electricity use is reduced by 25 kWh/m^2/year (36% electricity saved and 18% total energy saved). The reduction in cooling energy is 15 kWh/m^2/year (77% cooling energy saved and 11% total energy saved). The total saving potential is 66 kWh/m^2/year (48%). This total energy use can probably be further reduced if an effort is made to reduce the remaining facility electricity, in particular energy for pumps which in this study was set to 9 kWh/m^2/year and not analysed further.

Table 2 Input for the best case scenario

Parameter	Simulation input
Airtightness	0.3 l/s/m^2
Wall *U*-value	0.1 W/m^2°C
Window *U*-value	0.9 W/m^2°C
Solar control	External blinds
Heat exchanger efficiency	80%
SFP	1.5 kW/(m^3/s)
Air flow	Variable (VAV)
Tenant electricity	50 W/person; off during the night
Lighting	8 W/m^2; daylight dimming and occupancy switch-off

Thermal conditions for the warmest and coldest days in the best case simulation are displayed in Figures 8 and 9. The mean air temperature is allowed to swing between 21°C and 24°C, and the operative temperature stays close to the mean air temperature, between 20.9°C and 24.3°C, during office hours.

Conclusions

Dynamic simulations were carried out with IDA ICE 4 on a typical narrow office building with peripheral individual rooms in Sweden. The simulations resulted in a very low-energy office building with a total end-use energy of 73 kWh/m^2/year for heating, cooling, facility electricity and user-related electricity. The result shows that 48% energy can be saved compared to a traditional modern office building, which means that the initial aim of this project was reached. If renewable energy sources, such as solar energy, geothermal energy and wind power, are added, there is a potential for the annual energy production to exceed the annual energy consumption and a net zero-energy building can be reached. The study showed that the following design features are essential for achieving this low-energy office building in Sweden:

- Reasonable WWR
- Demand-controlled ventilation
- Demand-controlled lighting and low-power equipment
- Wider temperature set-points
- Well-insulated and airtight building envelope.

These design features are not very expensive solutions, and nowadays, they are rather well mastered. The investment cost is slightly higher compared to the

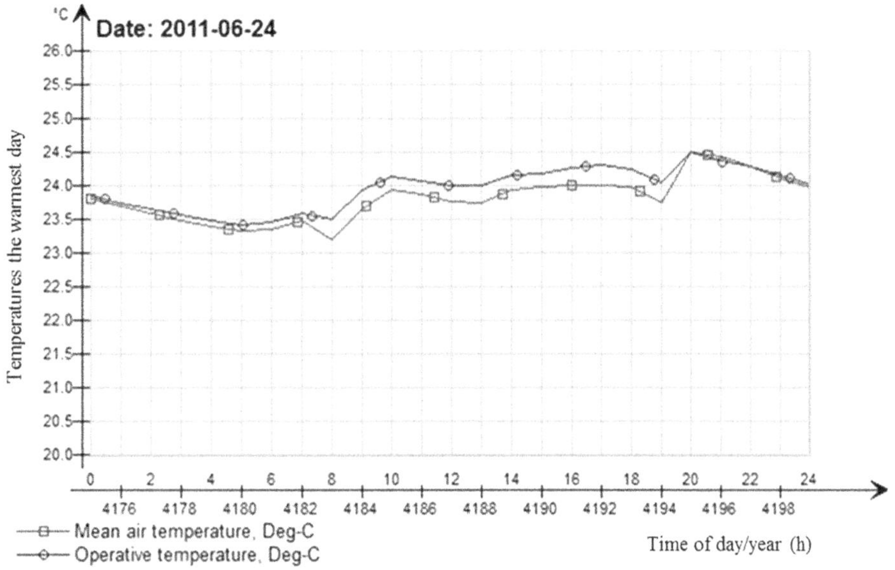

Figure 8 Indoor air temperature and operative temperature for the best case in the warmest day.

base case due to more expensive walls, windows and lighting control systems. The least established of the studied features is photoelectric dimming. Out of all studied design features, reducing user-related electricity is probably the greatest challenge since there has been limited focus on this issue earlier and it is difficult to control the tenant's use of office equipment over time. Power strips and multiple sockets may facilitate the reduction of 'off-mode' losses, but in addition, some kind of incentive is required in order to influence user behaviour. Displaying real-time electricity use is one possible aspect for raising awareness among users.

For low-energy offices, it is crucial to decrease user-related electricity and internal heat gains. A common perception in the building industry is that low-energy buildings would require additional energy when the internal gains are lowered, but this does not apply on office buildings which often include cooling systems. Not only is the user-related electricity diminished, but the cooling energy is also reduced and it will be easier to maintain the desired indoor climate.

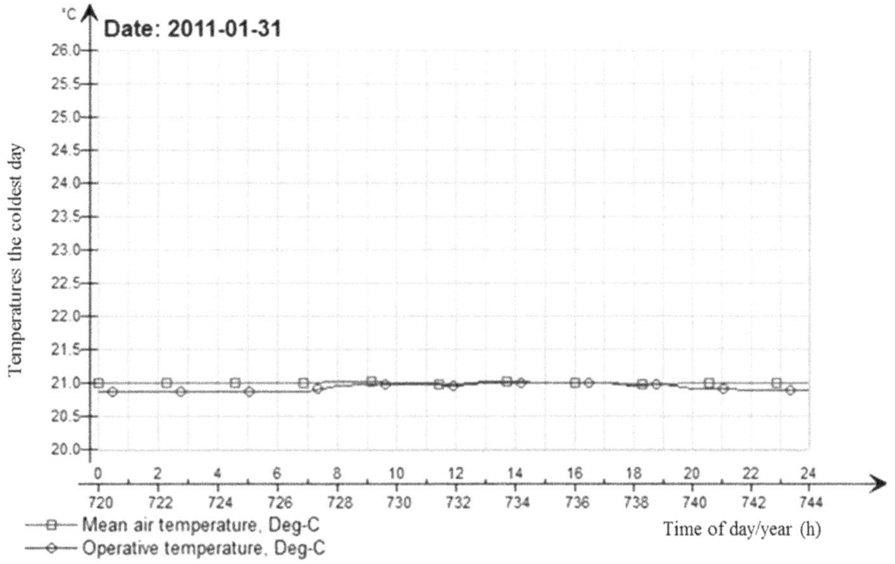

Figure 9 Indoor air temperature and operative temperature for the best case in the coldest day.

Abbreviations

COP: coefficient of performance; GWR: glazing-to-wall ratio; HVAC: heating, ventilation, air conditioning; SFP: specific fan power; SHGC: solar heat gain coefficient; VAV: variable air volume flow; WFR: window-to-floor ratio; WWR: window-to-wall ratio.

Competing interests

This work was supported by CERBOF (Centre for Energy and Resource Efficient Construction and Management of Buildings), SBUF (The Development Fund of the Swedish Construction Industry) and NCC Construction Sweden AB.

Authors' contributions

KF carried out the dynamic energy simulations in IDA ICE. ÅB put together the statistics for electricity use in Swedish office buildings. MCD carried out the daylight simulations with DAYSIM. All authors read and approved the final manuscript.

Acknowledgements

The authors would like to thank all participants of the reference group for their useful input and a special thanks to Katarína Heikkilä for proofreading this article. The authors also thank Equa Solutions AB for the support and for providing the IDA ICE licence.

References

1. European Parliament, Council of the European Union: Directive 2010/31/EU of the European Parliament and of the Council of 19 May 2010 on the energy performance of buildings. Off. J. Eur. Union **53**, 13–35 (2010)
2. Swedish Energy Agency: Improved energy statistics for non-residential buildings: "Stegvis STIL" Report for year 1, ER 2007:34. Swedish Energy Agency, Eskilstuna (2007)
3. Jansson, U: Passive houses in Sweden: experiences from design and construction phase. Lund University, (2008). Dissertation
4. Adalberth, K, Kamstedt, M: Passive house design as an important path to plus energy house. Paper presented at the 3rd Nordic passive house conference, Aalborg, 7–8 Oct 2010.
5. Voss, K, Löhnert, G, Herkel, S, Wagner, A, Wambsganß, M: Future office building. TÜV, Cologne (2005)
6. Pfafferott, JÜ, Herkel, S, Kalz, D, Zeuschner, A: Comparison of low-energy office buildings in summer using different thermal comfort criteria. Energy and Buildings **39**, 750–757 (2007)
7. Knissel, J: Energy efficient office building with reduced internal heat gains and thermal insulation of passive house standard. Institut Wohnen und Umwelt D83, Berlin (2002)
8. EQUA Solutions AB. http://www.equa-solutions.co.uk/ (2011). Accessed 29 Dec 2011
9. Poirazis, H: Single and double skin glazed buildings: analysis of energy use and indoor climate. Dissertation, Lund University (2008)
10. Boverket: Boverket's Building Regulations BBR 18 (BFS 2011:6). http://www.boverket.se/ (2011). Accessed 17 Jan 2012
11. Levin, P: Input for energy calculations in office buildings. Sveby (Standardize and verify the energy performance of buildings). http://www.sveby.org/ (2010). Accessed 17 Jan 2012
12. FEBY: Specification for passive houses, version 2009. http://www.energieffektivabyggnader.se/ (2009). Accessed 17 Jan 2012
13. Passive House Institute: Quality Approved Passive House Certification: criteria for non-residential passive house buildings. http://www.passiv.de/ (2011). Accessed 17 Jan 2012
14. Dubois, M-C, Flodberg, K: Daylight utilisation in perimeter office rooms at high latitudes: Investigation by computer simulations. Light Res. Technol (2012). doi:10.1177/1477153511428918
15. Passive House Institute: What is a passive house. http://www.passiv.de/ (2007). Accessed 17 Jan 2012

25

Economic analysis of wind energy conversion systems using levelized cost of electricity and present value cost methods in Nigeria

Olayinka S Ohunakin[1*], Olanrewaju M Oyewola[2] and Muyiwa S Adaramola[3]

Abstract

Technical electricity generation assessment and economic analysis of six wind energy conversion systems in the categories small, medium, and large (with power ratings of 20, 35, 275, 500, 1,000, and 2,000 kW) were examined in this study. Electricity cost values were estimated based on the levelized cost of electricity (LCOE) and present value cost (PVC) methods for six locations selected across all the geopolitical zones of Nigeria. This was done using wind speed data that span between 25 and 37 years, measured at the height of 10 m. The result showed that the annual average energy output ranges from 2.242 MW h in Uyo with P10-20 turbine to 12,521.55 MW h in Kano using Vestas V80-2 MW wind turbine. Furthermore, of all the selected sites, Kano gave the least costs of electricity production per kilowatt hour with Vestas V80-2 MW model at 67-m hub heights, while the highest is obtained in Uyo with GEV-HP (1 MW) model at 70-m hub heights for the LCOE and PVC height for both the LCOE and PVC methods. In addition, sensitivity of the selected parameters to the levelized cost of electricity was also carried out.

Keyword: Economic analysis, Wind energy, Wind energy conversion system, Capacity factor, Nigeria

Background

Wind, a form of clean energy among other renewable energy sources, has been experiencing rapid growth during the last two decades. The growth may be attributed to the promotion of models of renewable energy sources, energy supply security, fuel diversity concerns, ecological awareness, and economic reasons [1]. It has a major advantage of generating power near load centers, thus eliminating transmission losses especially in rural areas. Wind power capacity in recent years has increased in some countries [2] (Table 1), but the utilization rate in Nigeria is still limited to very few sites that are constructed for water pumping and other low-energy demanding activities, with none yet developed for grid electricity. This may be explained by the country's overdependence on oil and gas as sources of energy for export and other socio-economic activities. The decreasing reserves of these fossil fuels, together with environmental impact through pollution and ozone layer depletion, have called for the search of alternative sources of energy that are clean and renewable such as wind, solar, biomass, hydro, etc. In addition to these energy resources, solid waste can be classified as renewable energy source and used to generate energy in various forms, such as heat and electricity [3]. This energy source is also in large quantity across the country.

According to Adaramola et al. [4], annual mean wind speeds in Nigeria was found to vary between about 2 and 9.5 m/s with an overall annual mean wind speed of about 4.62 m/s. However, several research works have indicated promising and viable site for wind harvesting in the country (e.g., [5–9]). Some locations are adjudged as good sites for wind turbine installations for electricity generation and other types of wind applications. In spite of all the numerous work carried out so far on wind analysis in Nigeria, only few research studies have been focused on the economic analysis of wind applications in few sites in Nigeria. For instance, Ohunakin and Akinnawonu [10] conducted the economic feasibility of using wind energy conversion systems in Jos, Nigeria, using the present value cost (PVC) method, Ngala et al. [11] carried out an economic assessment on the viability of wind energy as a power source in

* Correspondence: olayinka.ohunakin@covenantuniversity.edu.ng
[1]Mechanical Engineering Department, Covenant University, P.M.B 1023, Ota, Ogun State 11001, Nigeria
Full list of author information is available at the end of the article

Table 1 Top ten installed wind power capacities (MW)

	Country	2000	2001	2002	2003	2004	2005	2006	2007	2008	2009	2010	2011
1	USA	2,554	4,275	4,685	6,374	6,740	9,149	11,603	16,818.8	25,170	35,159	40,200	46,919
2	Germany	6,113	8,754	12,001	14,609	16,628.8	18,427.5	20,622	22,247.4	23,903	25,777	27,214	29,060
3	Spain	2,235	3,337	4,830	6,202	8,263	10,027.9	11,630	15,145.1	16,754	19,149	20,676	21,674
4	China	340	401	468	566	764	1,266	2,599	5,912	12,210	25,104	44,733	62,733
5	India	1,167	1,407	1,702	2,110	2,985	4,430	6,270	7,850	9,645	10,925	13,064	16,084
6	Italy	427	682	788	904	1,265	1,718.3	2,123.4	2,726.1	3,736	4,850	5,797	6,747
7	France	79	85	147	198	390	757.2	1,567	2,455	3,404	4,410	5,660	6,800
8	UK	408	464	552	647.6	900.8	1,353	1,962.9	2,389	3,241	4,070	5,203	6,540
9	Denmark	2,300	2,417	2,880	3,110	3,117	3,128	3,136	3,125	3,180	3,465	3,752	3,871
10	Portugal	100	127	193.8	288.6	562	1,022	1,716	2,130	2,862	3,535	3,702	4,083

From [2].

Maiduguri, while Adaramola et al. [4] examined the economic viability of wind applications in selected sites in the north-central region using the levelized cost of electricity (LCOE) technique.

In this present study, a brief review of wind energy resource assessment in selected locations across Nigeria is presented. Investigation of wind energy generation and the economic analysis of wind turbine applications in Nigeria are also carried out by comparing LCOE and PVC methods of analysis for the various wind energy conversion system (WECS) with power capacities ranging from small to medium and large size. The levelized cost of an energy resource (per kilowatt hour) represents a constant cost per unit of generation computed to compare one unit's generation costs with other resources over similar periods. LCOE model takes into consideration the net present value of the current and future annual costs, whereas the PVC methods takes into consideration the current value of the total cost of energy investment during the entire lifetime of the energy conversion system. Six locations spreading across the country are selected for this analysis, and the locations are as follows: Kano (NW), Bauchi (NE), Uyo (SS), Owerri (SE), Minna (NC), and Iseyin (SW). The selected sites (Kano, Bauchi, Minna, Iseyin, Owerri, and Uyo) were found in previous works to be the best wind representative for each of the six geopolitical zones in Nigeria [5–9]. This information provided in this paper will be useful to wind energy developers, governments at various levels, and individuals, as well as private organizations that are interested in wind energy development.

Methods

The wind data used in this study were obtained from the Nigerian Meteorological Agency, Oshodi, Lagos. The data for the selected locations were captured at the height of 10 m by a cup-generator anemometer. The recorded wind speeds were obtained on hourly basis and thereafter computed as the mean of the speed for each month.

Wind energy studies in the selected locations

Wind distribution function adopted in this work is the Weibull function because it has been found to give a better fit with experimental data [7,12]. Its probability density function is expressed as:

$$f(u) = \left(\frac{k}{c}\right)\left(\frac{u}{c}\right)^{k-1} \exp\left[-\left(\frac{u}{k}\right)^{k}\right] \tag{1}$$

where k and c (m/s) are the Weibull shape and scale factors, respectively and u (m/s) is the wind speed.

Table 2 depicts the wind characteristics and geographical coordinates of the selected locations. It can be observed from this table that the minimum wind speed among the selected locations is 3.17 m/s in Uyo, while the maximum value is 7.77 m/s in Kano. In addition, the power densities (P_D) are 368.92, 120.52, 94.11, 54.96, 26.42, and 19.91 W/m^2 for Kano, Bauchi, Minna, Iseyin, Owerri, and Uyo, respectively. It can be seen that the maximum P_D is in Kano ($P_D = 368.92$ W/m$^2 > 300$ W/m^2) and falls under class 6 of the wind resource category based on the Battelle-Pacific Northwest Labs (PNL) wind power classification scheme, while the least exists in Uyo ($P_D = 19.91 < 100$ W/m^2) and found within class 1 range [13]. Furthermore, the most probable wind speed (U_{mp}) and wind speed carrying maximum energy (U_{max}) ranged from 3.31 to 7.92 m/s and 3.58 to 9.57 m/s in Uyo and Kano, respectively. The Weibull shape and scale parameters (k and c) showed that k varies from 2.32 (Minna) to 6.36 (Uyo), while c is between 3.42 and 8.57 m/s in Uyo and Kano, respectively. Detailed wind speed characteristics in other selected sites are shown in Table 2, and comprehensive work on these analyses can be found in, e.g., [5–9,11].

Kano has been found to be very suitable for wind turbine applications, while Bauchi may be considered marginally for wind power development though a higher hub height will be better preferred for economical wind power generation; Minna, Iseyin, Owerri, and Uyo, each

Table 2 Wind characteristics and geographical coordinates of the selected sites

Location	Mean wind speed (m/s)	Average power density (W/m²)	PNL classifications	k	c (m/s)	U_{mp} (m/s)	U_{max} (m/s)	Latitude (°N)	Longitude (°E)	Altitude (m)	Duration	Geopolitical location
Kano	7.77	368.92	6	4.04	8.57	7.92	9.57	12.03	08.12	472.5	1971 to 2007	NW
Bauchi	4.83	120.50	2	2.45	5.44	4.32	7.04	10.17	09.49	609.7	1971 to 2007	NE
Minna	4.291	94.11	1	2.32	4.84	3.73	6.42	09.37	06.32	256.4	1971 to 2007	NC
Iseyin	4.074	54.96	1	3.77	4.51	4.08	5.15	07.58	03.36	330.0	1983 to 2007	SW
Owerri	3.355	26.42	1	6.36	3.62	3.47	3.84	05.29	07.00	91.0	1977 to 2006	SE
Uyo	3.17	19.91	1	6.37	3.42	3.31	3.58	05.30	07.55	38.0	1981 to 2007	SS

location placed under class 1, may only be adequate for non-connected electrical and mechanical applications like battery charging and water pumping. From this assertion, all the regions are suitable for one form of wind energy application or the other; many sites in the North-West regions are suitable for grid-connected applications [7], and most locations in NC, SW, SE, and SS are more applicable to non-connected electrical applications.

Wind speed extrapolation and electrical power output

The selected wind turbines are designed to operate at different hub heights when compared to the available measured wind data; hence, the captured wind speed height (10 m) can be extrapolated to the turbine hub height through the power law expression given as:

$$\frac{U}{U_o} = \left(\frac{h}{h_o}\right)^{a} \tag{2}$$

where U is the wind speed at the wind turbine hub height h, U_o is the wind speed at original height h_o, and α is the surface roughness coefficient. In most cases, the accurate value of the surface roughness coefficient is not readily available or ascertained. Therefore, another approach is to use the Weibull probability function parameter values determined at the measured height and extrapolate them to the hub heights using the following expressions [14] as:

$$c(h) = c_o\left(\frac{h}{h_o}\right)^{n} \tag{3}$$

$$k(h) = k_o\left[1 - 0.088\ln\left(\frac{h_o}{10}\right)\right] / \left[1 - 0.088\ln\left(\frac{h}{10}\right)\right] \tag{4}$$

where c_o and k_o are the scale and shape factors, respectively, at the measurement height h_o, while h is the hub height. This approach is employed in this study to determine the capacity factor and mean power output from selected commercial wind turbines. This because it is easier to implement once the shape and scale factors of the Weibull function at the measured height has been determined. The exponent n is defined as [14]:

$$n = [0.37 - 0.088\ln(c_o)] / \left[1 - 0.088\ln\left(\frac{h}{10}\right)\right] \tag{5}$$

The performance of any installed wind turbine at any location can be evaluated by the mean power output ($P_{e,ave}$) over a period of time (usually, monthly and annually) and the capacity factor C_f (representing the fraction of the mean power output over a period of time to the rated electrical power P_{eR} of the turbine). The mean power output $P_{e,ave}$ and C_f can be calculated using the following expressions based on Weibull distribution function [15]:

$$P_{e,ave} = P_{eR}\left(\frac{e^{-\left(\frac{U_c}{c}\right)^k} - e^{-\left(\frac{U_r}{c}\right)^k}}{\left(\frac{U_r}{c}\right)^k - \left(\frac{U_c}{c}\right)^k} - e^{-\left(\frac{U_f}{c}\right)^k}\right) \tag{6}$$

$$C_f = \frac{P_{e,ave}}{P_{eR}} \tag{7}$$

U_c, U_r, and U_f are the respective cut-in wind speed, rated wind speed, and cutoff wind speed of the WECS. The capacity factor can be used to identify sites that are suitable for wind energy development and for the selection of wind turbine among available turbine to be installed in a site with known wind speed characteristics. The accumulated annual energy output (E_o) is given by:

$$E_o = P_{e,ave} \times 8760(kW\ h) \tag{8}$$

Economic analysis of selected turbines

Accurate estimate of all the costs involved in generating electricity over the life span of a WECS is essential for its economic viability in order to produce energy at a low operating cost [16]. The cost of a wind turbine is mostly set by the manufacturer, and it is shown to be dependent on the rated power (and control systems and other accessories) and varies from one manufacturer to another, as depicted in Table 3 [1,16,17]. As expected, new WECS always have low expenditures on operation and maintenance; however, the operation and, especially, the maintenance costs increase as the useful life of the power plant decreases [1]. Other factors which can influence the cost of electricity produced by WECS include cost of construction and other infrastructures, wind speed regime in selected location, turbine lifetime, and discount rate [1,17,18].

Out of the three different ways of quantifying the cost of wind turbines (cost per unit kilowatt, cost per unit rotor area, and cost per unit kilowatt hour of electricity produced) as expressed in [17], cost of electricity per unit kilowatt hour is adopted in this work and computed by comparing the LCOE with PVC to examine the economic analysis of the selected WECS. LCOE and PVC are computed using the following expressions given by [16,19] as:

$$\mathrm{LCOE} = \frac{\mathrm{CRF}}{E_{\mathrm{WT}}}\left(1 + C_{\mathrm{om(esc)}}\right) \quad Cost/kW\ h \tag{9}$$

$$\mathrm{PVC} = \frac{1}{E_{\mathrm{WT}}}\left[I + C_{\mathrm{om}}\left(\frac{1+i}{r_o - i}\right) \times \left[1 - \left(\frac{1+i}{1+r_o}\right)^n\right] - S\left(\frac{1+i}{1+r_o}\right)^n\right] \tag{10}$$

where I is the total capital/initial cost, $E_{\mathrm{WT}} = 8{,}760 P_{\mathrm{eR}} C_{\mathrm{f}}$ is the annual energy output of wind turbine in kilowatt hour, and CRF and $C_{\mathrm{om(esc)}}$ are the capital recovery factor and present worth of the annual cost throughout the lifetime of the WECS, respectively, and expressed in Equations 11 and 12 as given by[1]:

$$\mathrm{CRF} = \frac{(1+r)^n r}{(1+r)^n - 1} \tag{11}$$

$$C_{\mathrm{om(esc)}} = \frac{C_{\mathrm{om}}}{r - e_{\mathrm{om}}}\left[1 - \left(\frac{1 + e_{\mathrm{om}}}{1+r}\right)^n\right] \tag{12}$$

where C_{om}, e_{om}, n, r, r_o, and i are the operation and maintenance costs for the first year, escalation of operation and maintenance, useful lifetime of the turbine, discount rate, interest rate, and inflation rate, respectively. The discount rate can be corrected for inflation rate (i) and inflation escalation rate (e) using the following expressions [4,17]:

$$e_a = (1+e)(1+i) - 1 \tag{13}$$

e_α is the apparent escalation rate, and the real rate of discount (r) adjusted for both inflation and escalation can be obtained from the expression [4,17] as:

$$r = \frac{(1+r_o)}{(1+e_a)} - 1 \tag{14}$$

In using the LCOE and PVC methods to evaluate the costs of kilowatt hour of energy produced by the WECS at the selected sites, the following assumptions were taken into consideration:

1. The lifetime (n) of each turbine is considered to be 20 years.
2. Interest rate (r) and inflation rate (i) were taken as 15% and 12%, respectively.
3. The actual total of other initial costs depends on, among other factors, the available infrastructure in the selected location. For instance, it was reported that the cost of wind turbines and towers accounts for about 65% to 75% of the total initial costs in developed countries, while the cost of infrastructure, installation, and grid connection represents about 30% to 50% of the total capital cost in developing countries [20]. Considering the level of infrastructure development in most of the rural areas of Nigeria in locations included in this study, other initial costs such as that for land, infrastructure, installation, and grid integration are assumed to be 40% of the total initial cost (or about 67% of the wind turbine and tower cost).
4. Next to the wind turbine and other civil work initial capital costs, operating and maintenance cost (C_{om}) is next significant source of expenditure. The annual operation and maintenance costs have been reported to vary from about 1% to 7% of the initial system cost [20,21] and 2% to 16% of the wind turbine cost [20], depending on the rated power and age of the wind turbine. In this study, operating and maintenance cost is assumed to be 7% of the initial capital cost of the wind turbine installation system (system price/lifetime).Considering the higher assumed initial operation and maintenance costs, the escalation rate of operation and maintenance ($C_{\mathrm{om(esc)}}$) is assumed to be 0% in this analysis.
5. Scrap value was assumed to be 0% of the turbine price and civil work. It is expected that the unit price of electricity will be slightly higher based on this

Table 3 Cost of wind turbines based on the rated power

Wind turbine size (kW)	Specific cost (US$/kW)	Average specific cost (US$/kW)
10 to 20	2,200 to 2,900	2,550
20 to 200	1,500 to 2, 300	1,900
200>	1,000 to 1,600	1,300

Table 4 Characteristics of the selected wind turbines

Characteristics	P10-20	G3120	GEV-MP	P50-500	GEV-HP	Vestas V80-2 MW
Hub height (m)	36.6	42.7	60	70	70	67
Rated power, P_r (kW)	20	35	275	500	1,000	2,000
Rotor diameter (m)	10	19.2	32	50	62	80
Cut-in wind speed, U_C (m/s)	2.5	3.5	3.5	2.5	3	4
Rated wind speed, U_r (m/s)	10	8	15	12	15	16
Cutoff wind speed, U_f (m/s)	25	25	20	25	25	25

From [22–27].

assumption than if non-zero percentage is assumed for the scrap cost.

6. In practice, the annual mean wind speed in the selected locations may vary from year to year. However, it will be difficult to estimate the level of variation in the wind speed and, hence, the energy produced over the entire useful life of the wind energy conversion systems. Therefore, it is assumed in this study that the wind turbine produces the same amount of energy output in each year during its useful lifetime.

Results and discussion

Performance of wind turbines

Six (small, medium, and large) wind turbines with power ranging from 20 kW to 2 MW [22–27] were selected for performance assessment and economic analysis. The characteristic properties of the selected wind turbines using their respective designed hub heights are shown in Table 4. In [16], the factors influencing energy produced by any WECS at sites taken into consideration during the related time period are the power response of the WECS to different wind velocities, wind regime, and wind speed distribution. Hence, the annual average power and energy outputs, together with the capacity factors of the turbines for the selected locations, are computed using Equations 6, 7, and 8, and the computed results are listed in Table 5. It can be observed that the annual average energy output ranges from 2.242 MW h in Uyo with P10-20 turbine at 36.6-m hub heights to 12,521.55 MW h in Kano using Vestas V80-2 MW wind turbine at a 67-m hub height. Even though P50-500 and GEV-HP wind turbines have higher hub heights than Vestas V80-2 MW turbine, higher energy generation is possible with Vestas V80-2 MW because its higher rated power when compared with P50-500 and GEV-HP, respectively. Since the energy yield depends on the availability and strength of the wind speed at a given site and the particular WECS adopted [1], it can be readily seen that irrespective of the location, P10-20 turbine produces the least energy, with the most being given by V80-2 MW wind turbine. Hence, V80-2 MW model at 67-m hub height may be considered as the best type for energy generation at all the locations. Kano, among the considered sites in this work, exhibited the best situation from the energy production point using any of the selected WECS.

Figure 1 depicts the capacity factors for the selected wind turbines. The capacity factor reflects how effectively the turbine could harness the energy available in the wind spectra; it is expressed in an annual basis and as a function of the turbine as well as the wind regime characteristics [16]. The capacity factor for a reasonably efficient turbine at a potential site may range from 0.15 to 0.4 [7,28,29]. The least value of capacity factor, 0.22%, is obtained with GEV-HP turbine at 70-m hub heights, while the highest, 89.8%, is obtained in Kano with G3120 model at 42.7-m hub heights. Generally, from the figure, it can be observed that the lowest values of capacity factors across the WECS are obtained in Uyo,

Table 5 Mean annual power (kW/year) and accumulated annual energy outputs (MW h/year) for selected WECS

WECS (hub height)	Kano		Bauchi		Minna		Iseyin		Owerri		Uyo	
	Mean annual power output	Energy output	Mean annual power output	Energy output	Mean annual power output	Energy output	Mean annual power output	Energy output	Mean annual power output	Energy output	Mean annual power output	Energy output
P10-20 (36.6 m)	14.58	127.689	7.80	68.345	6.33	55.451	3.32	29.101	0.48	4.170	0.26	2.243
G3120 (42.7)	31.43	275.327	19.90	174.302	16.48	144.317	13.73	120.249	2.99	26.184	2.21	19.377
GEV-MP (60 m)	106.84	935.897	54.86	480.596	45.46	398.208	15.32	134.181	2.063	18.068	0.69	6.023
P50-500 (70 m)	213.20	1,867.632	110.75	970.170	92.50	810.300	32.20	282.072	4.45	38.982	1.55	13.578
GEV-HP (70 m)	356.90	3,126.444	184.30	1,614.468	153.90	1,348.164	49.90	437.124	6.70	58.692	2.20	19.272
Vestas V80-2 MW (67 m)	1,429.40	12,521.544	786.40	6,888.864	646.00	5,658.960	335.40	2,938.104	49.60	434.496	26.60	233.016

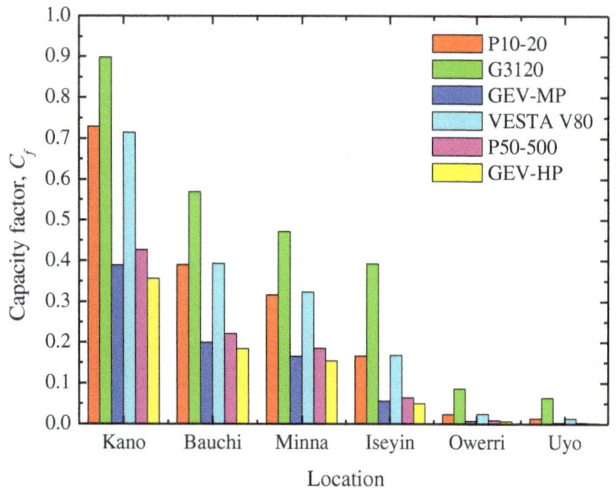

Figure 1 Capacity factors of the respective wind turbines in the selected locations.

whereas the highest values are recorded in Kano for all the models selected. Hence, considering the capacity factor of all the WECS for all the selected sites, Kano has a strong wind regime and hence a suitable potential for energy generation.

Economic analysis of selected wind turbines

Table 6 shows the economic result of comparing the LCOE and PVC method of cost analysis for the selected WECS in the respective locations. The computed values of LCOE and PVC are based on previously listed assumptions (see 'Economic analysis of selected wind turbines' in the 'Methods' section) and the average cost values of wind turbine bands shown in Table 3 (wind turbine cost band). The calculated cost values based on the two methods follow similar trends, with cost values increasing from Kano, Bauchi, Minna, Iseyin, Owerri, to Uyo, in that order, for all the turbine models adopted, as shown in the table. Whereas the least costs are estimated as US$0.0238/kW h and US$0.0182/kW h with Vestas V80-2 MW model at 67-m hub heights for the LCOE and PVC values, respectively, in Kano, the highest costs are determined as US$7.7275/kW h and US $5.9229/kW h in Uyo with GEV-HP (1 MW) model at 70-m hub heights for the LCOE and PVC values in that order. In addition, the lowest predicted cost values in all the locations with the selected turbine models is in Kano due to the high wind potential in this location. Hence, viable power generation is obtainable in high wind potential sites. Nigerian power tariff witnessed an upward review from 8.50 NGN (previously being implemented since July 1, 2008) to 10.00 NGN/kW h in April 1, 2011 [30], equivalent to US$0.0654 (US$1 ≈ 157.60 NGN [31]). Through this, the market became viable for renewable energy sources and independent power producers through a tariff system that ensures fair return on investment. Comparison of the new power tariff with the cost per kilowatt hour of electricity generated based on the LCOE and PVC methods (determined in the study) showed that Kano (models P10-20, G3120, GEV-MP (PVC), P50-500, GEV-HP (PVC), Vestas V80-2 MW), Bauchi (G3120, P50-500 (on PVC), Vestas V80-2 MW), Minna (G3120, Vestas V80-2 MW), and Iseyin with G3120 will be economically viable for wind power generation. With government supports in the form of hidden financial favors, subsidies, and any other incentives given to assist renewable energy exploitation in Nigeria, Kano, Bauchi, and Minna will be viable for power generation with all the wind turbine models.

It should be noted that the costs computed using LCOE methods are higher than those computed using the PVC method in all the location irrespective of the turbine model. This may be attributed to the following: (1) the sensitivity of LCOE to small changes in input variable and assumptions and (2) because the LCOE technique has discounted and annualized the PVC over the lifetime of the wind turbine. This is not available with PVC, thus leading to an underestimation of values as reflected in the computed values.

In order to investigate the effects of uncertainty or changes in some of the input parameters (such as wind turbine cost, interest rate, operation and maintenance cost, useful life of the wind turbine, and other initial capital cost) on the unit cost of electricity, sensitivity analysis was also carried out on GEV-MP wind turbine model in Kano (as a case study). The results of this

Table 6 Cost analysis per kilowatt hour (in US dollar) for WECS based on LCOE and PVC

	P10-20		G3120		GEV-MP		P50-500		GEV-HP		Vestas V80-2 MW	
	LCOE ($/kW h)	PVC ($/kW h)	LCOE ($/kW h)	PVC ($/kW h)	LCOE ($/kW h)	PVC ($/kW h)	LCOE ($/kW h)	PVC ($/kW h)	LCOE ($/kW h)	PVC ($/kW h)	LCOE ($/kW h)	PVC ($/kW h)
Kano	0.0458	0.0351	0.0277	0.0212	0.1467	0.0323	0.0399	0.0306	0.0662	0.0507	0.0238	0.0182
Bauchi	0.0855	0.0655	0.0437	0.0335	0.0852	0.0653	0.0768	0.0588	0.0922	0.0707	0.0432	0.0331
Minna	0.1054	0.0808	0.0528	0.0405	0.1028	0.0788	0.0919	0.0704	0.1105	0.0847	0.0526	0.0403
Iseyin	0.2008	0.1539	0.0634	0.0486	0.3052	0.2339	0.2640	0.2023	0.3407	0.2638	0.1014	0.0777
Owerri	1.4011	1.0739	0.2909	0.2230	2.2662	1.7370	1.9102	1.4641	2.5374	1.9448	0.6855	0.5254
Uyo	2.6052	1.9968	0.3931	0.3013	6.8002	5.2121	5.4840	4.2033	7.7275	5.9229	1.2782	0.9797

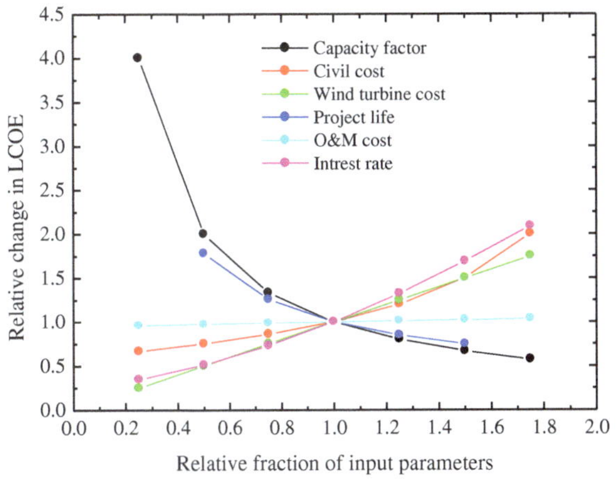

Figure 2 Sensitivity analysis of selected input parameters of LCOE for GEV-MP wind turbine model in Kano.

analysis are presented in Figure 2. It can be observed from this figure that the LCOE is sensitive to all the input parameters. The degree of sensitivity can be classified into two different groups. The first group comprises the capacity factor and useful life of the wind turbine. These parameters affect the LCOE very positively, that is, LCOE decreases as their values increase. This explains why a site with wind resource that provides high wind turbine capacity factor is desirable (economically). In addition, by increasing the project from 20 to 25 years, the LCOE is observed to decrease by about 15% (from US$0.0438/kW h to US$0.0370/kW h). The second group is made up of the wind turbine cost, the civil work and infrastructure costs, the interest rate, and the operation and maintenance costs. This figure indicates that with increasing value of these parameters, the LCOE increases, and hence, they have negative impact on the economic viability of wind energy system development in this location (and any other location in general). For example, there is an increase of about 20% in the unit cost of electricity if the percentage of civil work cost is increased from 40% to 50% of the total initial cost. It should be noted that the degree of influence of these parameters is not the same. It can be further observed that the cost of operation and maintenance has insignificant effect on the LCOE.

Conclusions

In this study, wind turbine performance assessment and economic analysis of selected commercial WECS were examined across all geopolitical zones in Nigeria. The findings are summarized as follows:

1. The annual average energy output ranges from 2.242 to 12,521.55 MW h in Uyo (P10-20 model) and Kano using Vestas V80-2 MW wind turbine. The highest energy generation is also possible with Vestas V80-2 MW because of its high rated wind speed among all the considered WECS.
2. In all the locations, P10-20 turbine model produces the least energy, with the largest being given by V80-2 MW wind turbine, thereby making V80-2 MW model at 65-m hub height considerably the best type for energy generation for all the locations.
3. The capacity factor has the least value in Uyo, 0.22%, with GEV-HP turbine at 70-m hub heights, whereas the highest is obtained in Kano, 89.8%, with G3120 model at 42.7-m hub heights.
4. The least costs of electricity production per kilowatt hour are estimated as US$0.0238/kW h and US $0.0182/kW h with Vestas V80-2 MW model at 67-m hub heights in Kano, while the highest is found to be US$7.7275/kW h and US$5.9229/kW h in Uyo with GEV-HP (1 MW) model at 70-m hub heights for the LCOE and PVC values, in that order.
5. The sensitivity analysis shows that increasing the capacity factor and useful life of the wind turbine can have a positive impact on the cost of electricity produced by the WECS, while other input parameters lead to increase in LCOE as their values increase.

Abbreviations

α: Surface roughness coefficient; *c*: The Weibull scale factor; C_{om}: Operation and maintenance cost for the first year; $C_{om(esc)}$: Present worth of the annual cost throughout the lifetime of the WECS; C_f: Capacity factor; c_o: The Weibull scale factor at the measurement height; *e*: Inflation escalation rate; e_a: Apparent escalation rate and the real rate of discount adjusted for both inflation and escalation; E_o: Accumulated annual energy output; e_{om}: Escalation of operation and maintenance; E_{WT}: Annual energy output of wind turbine; *h*: Turbine hub height; h_o: Original (measurement) height; *i*: Inflation rate; *I*: Total capital/initial cost; *k*: Dimensionless Weibull shape parameter; k_o: The Weibull shape factor at the measurement height; LCOE: Levelized cost of electricity; *n*: Useful lifetime of the turbine; NC: North-Central; NE: North-East; NW: North-West; P_D: Mean power density; $P_{e,ave}$: Mean power output; P_{eR}: Rated electrical power of the turbine; PNL: Battelle-Pacific Northwest Labs; PVC: Present value cost; *r*: Discount rate; r_o: Interest rate; SE: South-East; SS: South-South; SW: South-West; *U*: Wind speed at the wind turbine hub height; U_c: Cut-in wind speed of the WECS; U_f: Cutoff wind speed of the WECS; U_{max}: Wind speed carrying maximum energy; U_{mp}: Most probable wind speed; U_o: Wind speed at original height; U_r: Rated wind speed of the WECS.

Competing interests

The authors declare that they have no competing interests.

Authors' contributions

OSO collected the wind data and carried out the initial analysis and wind turbine calculations. Comprehensive work on turbine performance and economic analysis were performed by MSA. OMO checked the equations. The draft of 'Results and discussion' section was majorly drafted by MSA and supported by OSO and OMO. All authors read, edited, and eventually approved the final version manuscript.

Acknowledgments

The authors are grateful to the Nigerian Meteorological Agency (NIMET), Oshodi, Lagos, Nigeria, for providing the data for this study.

Author details
[1]Mechanical Engineering Department, Covenant University, P.M.B 1023, Ota, Ogun State 11001, Nigeria. [2]Department of Mechanical Engineering, University of Ibadan, Ibadan, Oyo State 900001, Nigeria. [3]Department of Energy and Process Engineering, Norwegian University of Science and Technology, Trondheim 7491, Norway.

References

1. Gökçek, M, Genç, MS: Evaluation of electricity generation and energy cost of wind energy conversion systems (WECS) in central Turkey. Appl Energ **86**, 2731–2739 (2009)
2. Wikipedia: Wind power by country. http://en.wikipedia.org/wiki/Wind_power_by_country#Installed_windpower_capacity (2012). Accessed 08 March 2012
3. Boukelia, TE, Mecibah, MS: Solid waste as renewable source of energy: current and future possibility in Algeria. IJEEE **3**, 17 (2012). doi:10.1186/2251-6832-3-17
4. Adaramola, MS, Paul, SS, Oyedepo, SO: Assessment of electricity generation and energy cost of wind energy conversion systems in North-Central Nigeria. Energ. Convers. Manage **52**, 3363–3368 (2011)
5. Ohunakin, OS: Wind resource evaluation in six selected high altitude locations in Nigeria. Renew Energ **36**, 3273–3281 (2011)
6. Oyedepo, SO, Adaramola, MS, Paul, SS: Analysis of wind speed data and wind energy potential in three selected locations in South East Nigeria. IJEEE **3**, 7 (2012). doi:10.1186/2251-6832-3-7
7. Ohunakin, OS, Adaramola, MS, Oyewola, OM: Wind energy evaluation for electricity generation using WECS in seven selected locations in Nigeria. Applied Energ **88**, 3197–3206 (2011)
8. Ohunakin, OS: Assessment of wind energy resources for electricity generation using WECS in North-Central region, Nigeria. Renew. Sust. Energ. Rev. **15**, 1968–1976 (2011)
9. Ohunakin, OS: Wind resources in North-East geopolitical zone Nigeria: an assessment of the monthly and seasonal characteristics. Renew. Sust. Energ. Rev. **15**, 1977–1987 (2011)
10. Ohunakin, OS, Akinnawonu, OO: Assessment of wind energy potential and the economics of wind power generation in Jos, Plateau State, Nigeria. Energ. Sustain. Dev (2011). doi:10.1016/j.esd.2011.10.004
11. Ngala, GM, Alkali, B, Aji, MA: Viability of wind energy as a power generation source in Maiduguri Borno State, Nigeria. Renew Energ **32**, 2242–2246 (2007)
12. Kollu, R, Rayapudi, N, Narasimham, SVL, Pakkurthi, KM: Mixture probability distribution functions to model wind speed distributions. IJEEE **3**, 27 (2012). doi:10.1186/2251-6832-3-27
13. Ilinca, A, McCarthy, E, Chaumel, JL, Retiveau, JL: Wind potential assessment of Quebec Province. Renew Energ **28**, 1881–1897 (2003)
14. Justus, CG, Hargraves, WR, Mikhail, A, Graber, D: Methods for estimating wind speed frequency distributions. J. Appl. Meteorol **17**, 350–353 (1978)
15. Akpinar, EK, Akpinar, S: An assessment on seasonal analysis of wind energy characteristics and wind turbine characteristics. Energ. Convers. Manage **46**, 1848–1867 (2005)
16. Gökçek, M, Erdem, HH, Bayülken, A: A techno-economical evaluation for installation of suitable wind energy plants in Western Marmara Turkey. Energ. Explor. Exploit. **25**, 407–428 (2007)
17. Mathew, S: Wind Energy: Fundamentals. Resource Analysis And Economics, Springer, Heidelberg (2006)
18. Morthorst, PE: Cost And Price, Wind Energy—The Facts, vol. 2. The European Wind Energy Association, Brussels (2004)
19. Hanitsch, R, Ahmed Shata, AS: Evaluation of wind energy potential and electricity generation on the coast of Mediterranean Sea in Egypt. Renew Energ **31**, 1183–1202 (2006)
20. Manwell, JF, McGowan, JG, Rogers, AL: Wind Energy Explained: Theory, Design And Application, 2nd edn. Wiley, Chichester (2009)
21. Nouni, MR, Mullick, SC, Kandpal, TC: Techno-economics of small wind electric generator projects for decentralized power supply in India. Energ Policy **35**, 2491–2506 (2007)
22. Polaris America: http://www.powerenergypr.com/documents/ModelP10-20kWSpecSheet.pdf. (2012). Accessed 19 March 2012
23. Endurance Wind Power: http://www.endurancewindpower.com/g3120.html (2012). Accessed 19 March 2012
24. Vergnet Wind Turbine: http://www.thewindpower.net/turbine_en_88_vergnet_gev-mp-275-32.php. (2012). Accessed 19 March 2012
25. Vergnet Wind Turbine: http://www.thewindpower.net/turbine_en_289_vergnet_gev-hp.php (2012). Accessed 19 March 2012
26. Polaris America: http://www.hasmak.com.tr/ruzgar_turbinleri/500kw_ruzgar_turbinleri.pdf. (2012). Accessed 19 March 2012
27. Vestas V80-2MW: http://www.vestas.com/en/wind-power-plants/procurement/turbine-overview/v80-2.0-mw.aspx. (2012). Accessed 19 March 2012
28. Wind power and the UK wind resource: Environmental Change Institute. Oxford University, (2005). http://www.eci.ox.ac.uk/publications/downloads/sinden05-dtiwindreport.pdf. (2012). Accessed 14 March 2012
29. Renewable Energy Research Laboratory, University of Massachusetts at Amherst: Wind: capacitor factor, intermittency and what happens when the wind doesn't blow? http://www.umass.edu/windenergy/publications/published/communityWindFactSheets/RERL_Fact_Sheet_2a_Capacity_Factor.pdf. (2012). Accessed 14 March 2012
30. Nigerian Electricity Regulatory Commission (NERC): Consultation paper for the 2011 major review of the Multi Year Tariff Order (MYTO). http://www.nercng.org/index.php?option=com_content&task=view&id=69&Itemid=84. (2011). Accessed 29 September 2011
31. XE universal currency converter: http://www.xe.com/. (2012). Accessed 19 March 2012

Permissions

All chapters in this book were first published in IJEEE, by Springer; hereby published with permission under the Creative Commons Attribution License or equivalent. Every chapter published in this book has been scrutinized by our experts. Their significance has been extensively debated. The topics covered herein carry significant findings which will fuel the growth of the discipline. They may even be implemented as practical applications or may be referred to as a beginning point for another development.

The contributors of this book come from diverse backgrounds, making this book a truly international effort. This book will bring forth new frontiers with its revolutionizing research information and detailed analysis of the nascent developments around the world.

We would like to thank all the contributing authors for lending their expertise to make the book truly unique. They have played a crucial role in the development of this book. Without their invaluable contributions this book wouldn't have been possible. They have made vital efforts to compile up to date information on the varied aspects of this subject to make this book a valuable addition to the collection of many professionals and students.

This book was conceptualized with the vision of imparting up-to-date information and advanced data in this field. To ensure the same, a matchless editorial board was set up. Every individual on the board went through rigorous rounds of assessment to prove their worth. After which they invested a large part of their time researching and compiling the most relevant data for our readers.

The editorial board has been involved in producing this book since its inception. They have spent rigorous hours researching and exploring the diverse topics which have resulted in the successful publishing of this book. They have passed on their knowledge of decades through this book. To expedite this challenging task, the publisher supported the team at every step. A small team of assistant editors was also appointed to further simplify the editing procedure and attain best results for the readers.

Apart from the editorial board, the designing team has also invested a significant amount of their time in understanding the subject and creating the most relevant covers. They scrutinized every image to scout for the most suitable representation of the subject and create an appropriate cover for the book.

The publishing team has been an ardent support to the editorial, designing and production team. Their endless efforts to recruit the best for this project, has resulted in the accomplishment of this book. They are a veteran in the field of academics and their pool of knowledge is as vast as their experience in printing. Their expertise and guidance has proved useful at every step. Their uncompromising quality standards have made this book an exceptional effort. Their encouragement from time to time has been an inspiration for everyone.

The publisher and the editorial board hope that this book will prove to be a valuable piece of knowledge for researchers, students, practitioners and scholars across the globe.

List of Contributors

Sreekrishnaperumal Thanga Ramesh
Department of Civil Engineering, National Institute of Technology, Tiruchirappalli, Tamilnadu Pin: 620015, India

Rajan Gandhimathi
Department of Civil Engineering, National Institute of Technology, Tiruchirappalli, Tamilnadu Pin: 620015, India

Puthiya Veetil Nidheesh
Department of Civil Engineering, National Institute of Technology, Tiruchirappalli, Tamilnadu Pin: 620015, India

Shanmugam Rajakumar
Department of Civil Engineering, National Institute of Technology, Tiruchirappalli, Tamilnadu Pin: 620015, India

Subramani Prateepkumar
Department of Civil Engineering, National Institute of Technology, Tiruchirappalli, Tamilnadu Pin: 620015, India

Chenicheri Chandroth Sreejith
Department of Mechanical Engineering, National Institute of Technology Calicut, NIT Campus (P.O.), Kozhikode 673601, India

Chandrasekharan Muraleedharan
Department of Mechanical Engineering, National Institute of Technology Calicut, NIT Campus (P.O.), Kozhikode 673601, India

Palatel Arun
Department of Mechanical Engineering, National Institute of Technology Calicut, NIT Campus (P.O.), Kozhikode 673601, India

Nora R Aimaretti
Instituto de Investigaciones en Catálisis y Petroquímica, Facultad de Ingeniería Química, Universidad Nacional del Litoral, Santa Fe 3000, Argentina
Laboratorio de Investigaciones Aplicadas, Facultad de Química, Universidad del Centro Educativo Latinoamericano, Av Pellegrini 1332, Rosario 2000, Argentina

Adriana L Clementz
Instituto de Investigaciones en Catálisis y Petroquímica, Facultad de Ingeniería Química, Universidad Nacional del Litoral, Santa Fe 3000, Argentina
Laboratorio de Investigaciones Aplicadas, Facultad de Química, Universidad del Centro Educativo Latinoamericano, Av Pellegrini 1332, Rosario 2000, Argentina

Agustín Codevilla
Laboratorio de Investigaciones Aplicadas, Facultad de Química, Universidad del Centro Educativo Latinoamericano, Av Pellegrini 1332, Rosario 2000, Argentina

María L Rojas
Departamento de Química Inorgánica y Química Técnica, Facultad de Ciencias, Universidad Nacional de Educación a Distancia, Madrid 28040, Spain

Juan C Yori
Instituto de Investigaciones en Catálisis y Petroquímica, Facultad de Ingeniería Química, Universidad Nacional del Litoral, Santa Fe 3000, Argentina

Charles H Heide
Department of Mathematics and Physics, University of Wisconsin-Parkside, Kenosha, WI 53141, USA

Pirooz Mohazzabi
Department of Mathematics and Physics, University of Wisconsin-Parkside, Kenosha, WI 53141, USA

Olugbenga Olufemi Awolu
Department of Food Science and Technology, Federal University of Technology, Akure, 23401, Nigeria

Stephen Kolawole Layokun
Department of Chemical Engineering, Obafemi Awolowo University, Ile Ife, Nigeria

Antonio Gagliano
Department of Industrial Engineering, University of Catania, Viale Andrea Doria, 6, 95125 Catania, Italy

Francesco Nocera
Department of Industrial Engineering, University of Catania, Viale Andrea Doria, 6, 95125 Catania, Italy

Francesco Patania
Department of Industrial Engineering, University of Catania, Viale Andrea Doria, 6, 95125 Catania, Italy

Alfonso Capizzi
Department of Industrial Engineering, University of Catania, Viale Andrea Doria, 6, 95125 Catania, Italy

Chandrakumar B Pardhi
Department of Mechanical Engineering, Sagar Institute of Science, Technology and Research (SISTec-R), Bhopal 462044, India

Jiwanlal L Bhagoria
Department of Mechanical Engineering, Maulana Azad National Institute of Technology (MANIT), Bhopal 462051, India

Rodolfo Araneo
D.I.A.E.E. - Electrical Engineering Division, "Sapienza" University of Rome, Via Eudossiana 18, Rome 00184, Italy

Umberto Grasselli
D.I.A.E.E. - Electrical Engineering Division, "Sapienza" University of Rome, Via Eudossiana 18, Rome 00184, Italy

Salvatore Celozzi
D.I.A.E.E. - Electrical Engineering Division, "Sapienza" University of Rome, Via Eudossiana 18, Rome 00184, Italy

Abdelfattah Bouatem
Department of Mechanical Engineering, Moulay Ismail University, ENSAM-MEKENS, PO. BOX. 4024, Meknes 50003, Morocco

Ahmed Almers
Department of Energetic, Moulay Ismail University, ENSAM-MEKENS, P.O. BOX. 4024, Meknes 50003, Morocco

Noureddine Boutammachte
Department of Energetic, Moulay Ismail University, ENSAM-MEKENS, P.O. BOX. 4024, Meknes 50003, Morocco

Shanmugasundaram Kandasamy
Department of Physics, Karpagam University, Coimbatore, Tamilnadu 641 021, India

Manikandan Vellingiri
Department of Physics, Karpagam University, Coimbatore, Tamilnadu 641 021, India

Shanmugan Sengottain
Department of Physics, Karpagam University, Coimbatore, Tamilnadu 641 021, India

Janarthanan Balasundaram
Department of Physics, Karpagam University, Coimbatore, Tamilnadu 641 021, India

Ismaila Badmus
Mechanical Engineering Department, Yaba College of Technology, Yaba, Lagos, 6457, Nigeria

Ajiboye Saheeb Osunleke
Chemical Engineering Department, Obafemi Awolowo University, Ile-Ife, 220005, Nigeria

Richard Olayiwola Fagbenle
Mechanical Engineering Department, University of Ibadan, Ibadan, 200284, Nigeria

Miracle Olanrewaju Oyewola
Mechanical Engineering Department, University of Ibadan, Ibadan, 200284, Nigeria

Rahnuma Siddiqui
Renewable Energy Research Laboratory, Department of Physics, University of Lucknow, Lucknow 226007, India

Usha Bajpai
Renewable Energy Research Laboratory, Department of Physics, University of Lucknow, Lucknow 226007, India

Ahmad Marasabessy
Agrotechnology and Food Sciences Group, Biobased Commodity Chemicals, Wageningen University, P.O. Box 17, Wageningen 6700 AA, The Netherlands
Department of Chemical Engineering, RijksuniversiteitGroningen, Nijenborgh 4, Groningen 9747 AG, The Netherlands
Agency forthe Assessment and Application of Technology (BPPT), Jl. MH Thamrin 8, Jakarta 10340, Indonesia

A Maarten J Kootstra
Agrotechnology and Food Sciences Group, Biobased Commodity Chemicals, Wageningen University, P.O. Box 17, Wageningen 6700 AA, The Netherlands
Bioprocess Engineering Group, Wageningen University, P.O. Box 17, Wageningen 6700 AA, The Netherlands

Johan PM Sanders
Agrotechnology and Food Sciences Group, Biobased Commodity Chemicals, Wageningen University, P.O. Box 17, Wageningen 6700 AA, The Netherlands

Ruud A Weusthuis
Agrotechnology and Food Sciences Group, Biobased Commodity Chemicals, Wageningen University, P.O. Box 17, Wageningen 6700 AA, The Netherlands

Samuel E Agarry
Biochemical Engineering and Biotechnology Research Laboratory, Department of Chemical Engineering, Ladoke Akintola University of Technology, Ogbomoso-Ilorin Road, Ogbomoso, Oyo State 260001, Nigeria

Oladipupo O Ogunleye
Biochemical Engineering and Biotechnology Research Laboratory, Department of Chemical Engineering, Ladoke Akintola University of Technology, Ogbomoso-Ilorin Road, Ogbomoso, Oyo State 260001, Nigeria

Partha Kayal
Department of Electrical Engineering, Bengal Engineering & Science University, Shibpur, Howrah 711103, India

Chandan Kumar Chanda
Department of Electrical Engineering, Bengal Engineering & Science University, Shibpur, Howrah 711103, India

Afshin Takdastan
Department of Environmental Health and Environmental Technology Research Centre, Ahvaz Jundishapur University of Medical Sciences, Ahvaz, Iran

Azadeh Eslami
Department of Water and Wastewater Engineering, Sazabpardazan Consultant Engineering, Ahvaz, Iran

Hind El Hassani
Department of Physics, Faculty of Sciences, Moulay Ismail University, Meknès 50050, Morocco

Nour-eddine Boutammachte
Department of Energy, Ecole Nationale Supérieure d'Arts et Métiers, Moulay Ismail University, Meknès 50050, Morocco

Juergen Knorr
Institute of Energy, Faculty of Mechanical Engineering, University of Technology, Dresden 01062, Germany

Mohamed El Hannaoui
Department of Physics, Faculty of Sciences, Moulay Ismail University, Meknès 50050, Morocco

Amr Sayed Hassan Abdallah
Graduate School of Engineering, Tohoku University, Aoba 6-6-11-1201, Sendai 980-8579, Japan

Hiroshi Yoshino
Graduate School of Engineering, Tohoku University, Aoba 6-6-11-1201, Sendai 980-8579, Japan

Tomonobu Goto
Graduate School of Engineering, Tohoku University, Aoba 6-6-11-1201, Sendai 980-8579, Japan

Napoleon Enteria
Graduate School of Engineering, Tohoku University, Aoba 6-6-11-1201, Sendai 980-8579, Japan

Magdy M Radwan
Department of Architecture, Assiut University, P.O. Box 71516, Assyut, Egypt

M Abdelsamei Eid
Department of Architecture, Assiut University, P.O. Box 71516, Assyut, Egypt

Petros J Axaopoulos
Department of Energy Technology, Technological Educational Institute (TEI) of Athens, Ag. Spyridonos, Aegaleo 12210, Greece

Emmanouil D Fylladitakis
Electronic and Computer Engineering Department, School of Engineering and Design, Brunel University, Uxbridge, London, Middlesex UB8 3PH, UK

Konstantinos Gkarakis
Department of Energy Technology, Technological Educational Institute (TEI) of Athens, Ag. Spyridonos, Aegaleo 12210, Greece

Muhammad Ali
Mathematical Sciences Research Centre, Federal Urdu University of Arts Science and Technology, Karachi, Pakistan

Muhammad Jawed Iqbal
Institute of Space and Planetary Astrophysics (ISPA), University of Karachi, Karachi, Pakistan

Muhammad Sharif
Department of Statistics, University of Karachi, Karachi, Pakistan

Hitesh Panchal
KSV University, Gandhinagar Gujarat, 380025 India
Department of Mechanical Engineering, Gujarat Power Engineering & Research Institute, Mehsana, Gujarat 380025 India

Pravin K Shah
Silver Oak College of Engineering & Technology, Ahmedabad, Gujarat, India

Sheikh Mominul Islam
Department of Engineering, Memorial University of Newfoundland, St. John's, Newfoundland and Labrador, A1B1V3, Canada

Sunday Olayinka Oyedepo
Mechanical Engineering Department, Covenant University, Ota 2023, Nigeria

Kajsa Flodberg
Department of Architecture and the Built Environment, Div. of Energy and Building Design, Lund University, Box 118, Lund 221 00, Sweden

Åke Blomsterberg
Department of Architecture and the Built Environment, Div. of Energy and Building Design, Lund University, Box 118, Lund 221 00, Sweden

Marie-Claude Dubois
Department of Architecture and the Built Environment, Div. of Energy and Building Design, Lund University, Box 118, Lund 221 00, Sweden

Olayinka S Ohunakin
Mechanical Engineering Department, Covenant University, P.M.B 1023, Ota, Ogun State 11001, Nigeria

Olanrewaju M Oyewola
Department of Mechanical Engineering, University of Ibadan, Ibadan, Oyo State 900001, Nigeria

Muyiwa S Adaramola
Department of Energy and Process Engineering, Norwegian University of Science and Technology, Trondheim 7491, Norway